Exploring Design, Technology, & Engineering

Third Edition

R. Thomas Wright
Professor Emeritus, Industry and Technology
Ball State University
Muncie, Indiana

Ryan A. Brown
Assistant Professor
Illinois State University
Normal, Illinois

Publisher
The Goodheart-Willcox Company, Inc.
Tinley Park, Illinois
www.g-w.com

Manufactured in the United States of America.

Library of Congress Catalog Card Number 2010033181

ISBN 978-1-60525-420-3

1 2 3 4 5 6 7 8 9 – 12 – 15 14 13 12 11 10

The Goodheart-Willcox Company, Inc. Safety Notice: The reader is expressly advised to carefully read, understand, and apply all safety precautions and warnings described in this book or that might also be indicated in undertaking the activities and exercises described herein to minimize risk of personal injury or injury to others. Common sense and good judgment should also be exercised and applied to help avoid all potential hazards. The reader should always refer to the appropriate manufacturer's technical information, directions, and recommendations; then proceed with care to follow specific equipment operating instructions. The reader should understand these notices and cautions are not exhaustive.

Library of Congress Cataloging-in-Publication Data

Wright, R. Thomas
Exploring design, technology & engineering / R. Thomas Wright, Ryan A. Brown. p. cm.
Rev. ed. of Technology, design and applications / R. Thomas Wright, Ryan A. Brown. 2008.
Includes index.
ISBN 978-1-60525-420-3
1. Technology--Textbooks. 2. Industrial design--Textbooks. I. Brown, Ryan A. II. Title.

T47.W738 2012
600--dc22 2010033181

FOREWORD

Exploring Design, Technology, & Engineering, by R. Thomas Wright and Ryan A. Brown, presents the core of technological knowledge and skills demanded to actively participate in our ever-improving society. Studying and applying the lessons included in this textbook provides you with the solid content and the hands-on and minds-on experiences that will enable you to understand contemporary technology, while preparing you to make good decisions regarding future technology opportunities and options. This book was written and illustrated to support the growing importance of technology in our democratic society. ***Exploring Design, Technology, & Engineering*** is entirely based on the *Standards for Technological Literacy: Content for the Study of Technology*. The organization and content of ***Exploring Design, Technology, & Engineering*** align with the objectives described in the *Standards for Technological Literacy*, regarding what the content of technology education should be.

Exploring Design, Technology, & Engineering presents curriculum content in five easily understood sections. Therefore, the reader masters the essential core of technological knowledge. The first section defines *technology* and explains technology as a system. The second section describes the input resources of tools, materials, energy, information, people, time, and capital. The third section covers creating technology, the invention and design processes, and problem solving in a technological world. The fourth section explains technology contexts, utilizing comprehensive coverage, while providing hands-on applications for the essential technology core areas of agriculture, construction, energy, information and communication, manufacturing, medicine, and transportation. The fifth section relates technology to our society and our future opportunities and challenges.

Additionally, ***Exploring Design, Technology, & Engineering*** provides exceptional hands-on learning opportunities supporting the Technology Student Association (TSA) competitive events. Active participation with TSA provides not only exciting student learning experiences, but also a framework for individual growth and leadership opportunities. You are encouraged to study ***Exploring Design, Technology, & Engineering*** to expand your understanding of the role technology plays in our contemporary society and the role it will play in our future. By applying the lessons, skills, and content of this book, you will be well equipped to make positive decisions to build our technological world for a better tomorrow.

The Publisher

INTRODUCTION

What is technology? Why is it important in our lives? We need only to look around to see how technology changes our lives. Tools and machines make our work easier. We have automobiles, cellular telephones, computers, and many other products to save us time and improve the quality of our lives.

Construction of all types provides shelter and convenience for all our activities. Houses and apartments keep us comfortable and protect us from the elements. Bridges allow us to cross rivers. We can choose many different kinds of vehicles for travel. Radios, telephones, computers, satellites, and television all help us keep in touch with each other and the rest of the world. All these advantages are the results of advances in technology.

Technology is the knowledge of doing. This knowledge is a means of extending human abilities. Technology allows us to make useful products better and more easily. This knowledge enables us to build structures on Earth and in space. Technology lets us move people and goods more easily.

This book, ***Exploring Design, Technology, & Engineering***, introduces you to the various technologies. ***Exploring Design, Technology, & Engineering*** explains the technologies as systems. These systems have inputs (such as people and materials), processes, outputs, goals, and constraints. You will be able to learn about the effects of technology and see that, while most effects of technology are good, some are not. As a result, you will be able to form opinions and make decisions about how to use technology wisely.

Exploring Design, Technology, & Engineering does more than tell you about technology. At the end of each section, you have a chance to apply what you have learned through carefully designed activities. In some activities, you will build and test products. You might even use the products in competition with other students. In other activities, you might be introduced firsthand to mass production or the use of tools. We hope this combination of information and doing activities is a worthwhile experience for you.

R. Thomas Wright

Ryan A. Brown

About the Authors

Dr. R. Thomas Wright is one of the leading figures in technology-education curriculum development in the United States. He is the author or coauthor of many Goodheart-Willcox technology textbooks. Dr. Wright is the author of *Manufacturing and Automation Technology*, *Processes of Manufacturing*, and *Technology & Engineering*. He has served the profession through many professional offices, including President of the International Technology and Engineering Educators Association (ITEEA) and President of the Council on Technology Teacher Education (CTTE). His work has been recognized through the ITEEA Academy of Fellows Award, the ITEEA Award of Distinction, the CTTE Technology Teacher Educator of the Year, the Epsilon Pi Tau Laureate Citation, the Epsilon Pi Tau Distinguished Service Citation, the Sagamore of the Wabash Award from the Governor of Indiana, the Bell Ringer Award from the Indiana Superintendent of Public Instruction, the Ball State University Faculty of the Year Award, the Ball State University George and Frances Ball Distinguished Professorship, and the Educational Exhibitors Association–SHIP (EEA-SHIP) Citation.

Dr. Wright's educational background includes a bachelor's degree from Stout State University, a master's of science degree from Ball State University, and a doctoral degree from the University of Maryland. His teaching experience consists of 3 years as a junior high instructor in California and 37 years as a university instructor at Ball State University. In addition, he has also been a visiting professor at Colorado State University; Oregon State University; and Edith Cowan University in Perth, Australia.

Dr. Ryan A. Brown is an assistant professor in the Department of Curriculum and Instruction and an associate director of the Center for Mathematics, Science, and Technology at Illinois State University. He currently teaches courses for preservice teachers on topics such as instructional methods and assessment. Previously, he taught a variety of courses at the secondary level, including design processes, transportation systems, and fundamentals of engineering. Dr. Brown coauthored *Energy, Power, and Transportation Technology* with Dr. Len S. Litowitz. He has also written titles in both the *Humans Innovating Technology Series (HITS)* and the *Kids Inventing Technology Series (KITS)* for ITEEA, as well as in the *Activity!* series for the Center for Implementing Technology in Education. Dr. Brown's educational background includes a bachelor's degree and master's degree from Ball State University and a doctorate degree from Indiana University. Dr. Brown; his wife, Heather; and his sons, Benjamin and Samuel, reside in Normal, Illinois.

TSA Modular Activities

The Technology Student Association (TSA) is a nonprofit, national student organization devoted to teaching technology education to young people. TSA's mission is to inspire the association's student members to prepare for careers in a technology-driven economy and culture. The demand for technological expertise is escalating in American industry. Therefore, TSA's teachers strive to promote technological literacy, leadership, and problem solving to their student membership.

The TSA Modular Activities are based on the TSA competitive events current at the time of writing. Please refer to *The Official TSA Competitive Events Guide* for actual regulations for current TSA competitive events. This guide is periodically updated. TSA publishes two *Official TSA Competitive Events Guides*: one for middle school events and one for high school events. To obtain additional information about starting a TSA chapter at your school, to order *The Official TSA Competitive Events Guide*, or to learn more about TSA and technology education, contact TSA:

Technology Student Association
1914 Association Drive
Reston, VA 20191-1540
www.tsaweb.org

The Career Clusters

The Career Clusters are 16 groups of different types of occupational and career specialties, which are further divided into pathways. Looking ahead at these pathways will help determine the course of study for your chosen career. The Career Cluster icons are being used with permission of the:

States' Career Clusters Initiative, 2010, www.careerclusters.org.

Acknowledgments

The authors and publisher wish to thank the following companies, organizations, and individuals for their generous contributions of photographic images, artwork, and resource material:

Agricultural Research Service, U.S. Department of Agriculture
Alaska Airlines
Alcoa
All-Glass Aquarium
AMEC Wind
American Association of Blacks in Energy
American Baseball Company
American Petroleum Institute
American Red Cross
Aminoil UAS, Inc.
Amoco Corp.
AMP, Inc.
AMS Phoenix/Jim Dore
Andersen Corp.
Army ROTC
Art Design International (ADI) of St. Hubert, Quebec
Artromick International
Ashlar Vellum 2000
Asphalt Roofing Manufacturers Association
AT&T Network Systems
Bethlehem Steel Corp.
Bobcat Company
Boeing
Bren Instruments Inc.
Buick
Carolina Power and Light
Case IH
Caterpillar, Inc.
Cessna Aircraft
The Chicago Tribune Co.
Chrysler Corp.
Cincinnati Milacron
Clorox Co.
Coachman Industries
The Coca-Cola Company
Colorado State University
Conoco, Inc.
Custom Modeling & Graphics Studio, Inc.
Daimler
Dana Corporation
Datastream
Deere and Company
Dell Inc.
Derek Jensen
Design Central
Design Edge
Dremel
Duracell
Eastman Chemical Company
EIS Div., Parker-Hannifin
Eli Lilly and Company
Elke Wetzig
En-Vision America
Faro Technologies, Inc.
Federal Emergency Management Agency
Fine Art Lamps
FMC Corporation

Ford Motor Company
Forrest M. Mimms III
Frist Center for the Visual Arts
Gannett Co.
GE Medical Systems
Goodyear Tire and Rubber Company
Graphic Arts Technical Foundation
Habitat for Humanity International
Harris Corp.
Hibbing Taconite Company
Hirdinge, Inc.
Honda
IDEO
IKEA Home Furnishings
Illinois State University
Ingenico
Inland Steel Company
Intel
Jack Klasey
Jeff Sokalski
Jerry E. Howell
Kalb
Keith Nelson
Ken Hammond
Lakeside Equipment Corporation
The LEGO Group
Leon Goldik
Lexington Homes
LTV Steel Company
Lucent Technologies, Inc./Bell Labs
Mark Chrapla
Medlife
Microsoft Corp.
Mila Zinkova
Monster Cable Products, Inc.
Motorola, Inc.
MTS Corporation
Nano-Tex
Napa Valley Balloons, Inc.
NASA
NASA's Marshall Space Flight Center
Natchez Trace Parkway
National Park System
Natural Gas Supply Assoc.
NOAA
Northern Telecom
Ohio Art Co.
Olivier Tétard
OSHA
Owens-Brockway
Owens Corning
Piaggio Aviation
Product Development Technologies, PDT
Rick English
Rollerblade
R. R. Donnelley & Sons Company
The San Diego Low Speed Wind Tunnel
Sauder Woodworking
Science@NASA
SportsArt
Standard Oil of California
Standard Oil of Ohio
StarTrac
Steven Moeder
Stratasys, Inc.
Tandy Co.
Technology Student Association (TSA)
Tillamook County Creamery Assoc.
Timothy Hursley
U.S. Department of Agriculture
U.S. Department of Labor
U.S. Park Service
U.S. Patent and Trademark Office
Underwriter's Laboratory
Vapo Oy
Velux-America
Visteon
VX Corporation
WB Automotive
Werner Co.
Westinghouse Electric Corp.
Weyerhaeuser Co.
Whirlpool Corporation
Wisconsin Department of Natural Resources
Xerox
Zureks

TECHNOLOGY HEADLINE:
HOLOGRAPHY

[illegible]ography, a technology invented in the [illegible]0s, continues to evolve today, offering [illegible]ossibilities in the world of communica[illegible] [illegible]raphy is a science that focuses on the [illegible]nd recreation of an image in such a [illegible]ppears to be in 3D. Lasers are used [illegible] object upon a special photographic [illegible]. When the plate is moved, the 3D image, or hologram, appears to move as if it were a 3D object.

Holograms have many common applications. They are used to validate certain types of currency, such as the Euro. Barcode scanners found in grocery and department stores worldwide employ holographic lens technology. Holograms also appear on government-issued IDs and credit cards to make it harder to create forgeries. They are even used as works of art! As you move any of these items back and forth, the design of the hologram seems to "jump out" at you.

Despite numerous modern-day applications, scientists continue to work with holograms, developing new uses for this interesting technology. Advancements in holographic technology will improve current uses. For example, the holograms used as the secure "signature" on driver's licenses are currently the same for every license issued by a particular state. In the future, these holograms will be unique to each individual's license. These holograms can be checked electronically to ensure they are legitimate forms of identification.

Another developing possibility in the field of holography is the use of holograms to make a better television and motion picture image. Liquid crystal displays (or LCDs) in televisions will become brighter and whiter. This improvement will produce a picture that boasts better quality overall.

Holography will also advance cell phone technology. After downloading a special application, smart phone users will be able to record video messages that will then play as 3D representations of the recorded item. Perhaps it will come in the form of a friend listing off items needed from the grocery store, or a self-recording reminding you of an upcoming doctor's appointment. When this holographic message is replayed, it will pop out from the screen as a miniature, 3D version of the recorded person. Rather than an audio recording or text message, you will receive a recording that is like a visual voicemail. This type of advanced communication could also be applied to video chatting—it will be as if the person you are talking to online is actually sitting right next to you!

Although the technology of holography was first introduced in 1947, it continues to grow and change today. The science of holography can be applied to many fields. This diversity presents many exciting uses in the future. Who knows what scientists will come up with next—the possibilities are endless!

Technology Headline features in each section highlight an emerging technology.

CHAPTER 2
Technolog[illegible]
Syst[illegible]

Objectives identify the topics covered and goals to be achieved by students.

Did You Know? features in every chapter offer bits of trivia about the content covered in the chapter.

DID YOU KNOW?

- The solar system is a very large natural system. This system consists of the Sun, eight official planets, over 165 moons, five dwarf planets, and many other small bodies.
- A computer's operating system is the first software seen when a computer is turned on. This system is the last software seen when the computer is turned off. The ope[illegible] system directs all the p[illegible] used on a com[illegible] manag[illegible] [illegible] microwave [illegible]ed operating [illegible]. The computer in a microwave oven runs a single program all the time.

Key Words list new vocabulary covered in the chapter, enhancing student recognition of important concepts.

OBJECTIVES

The information given in the chapter will help you do the following:

- Explain the concept of a system.
- Compare natural systems and technological systems.
- Summarize how a system works.
- Give examples of some major goals of technological systems.
- Summarize the seven major inputs to technological systems.
- Give examples of the major types of processes used in technological systems.
- Explain the major types of outputs of a technological system.
- Explain the two types of feedback systems.

KEY WORDS

These words are used in this chapter. Do you know what they mean?

closed-loop system
desired output
energy
feedback
feedback system
goal
input
intended output
knowledge
machine
material
natural system
open-loop system
output
pollution
process
system
technological system
time
undesirable output

PREPARING TO READ

As you read this chapter, outline the details of the five basic steps that make up a system.

You hear about systems all the time. The weather forecaster talks about a new weather system moving into your region. A doctor tells you your digestive system is upset. The auto mechanic explains that the ignition system of your family's car needs work. So what is a system?

Simply put, a ***system*** is a [illegible]oup of parts working together to [illegible] a task. See Figure 2-1. T[illegible] [illegible]l and

A system is ...

Preparing to Read features provide students with questions to think about while reading the chapter.

... a group of parts that work together to do a task.

[illegible] machine is a technological system people developed. (Cincinnati Milacron)

Technology Explained features briefly explain how technological devices and systems work.

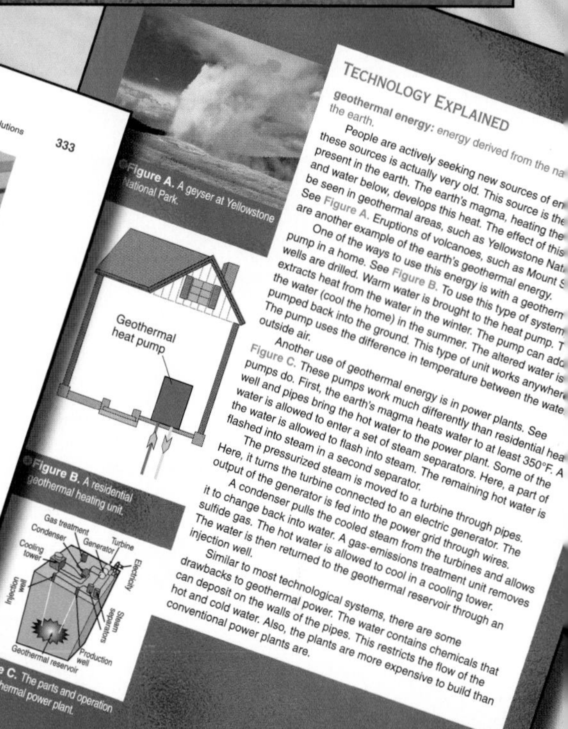

Safety notes identify activities that can result in personal injury, if proper procedures or safety measures are not followed.

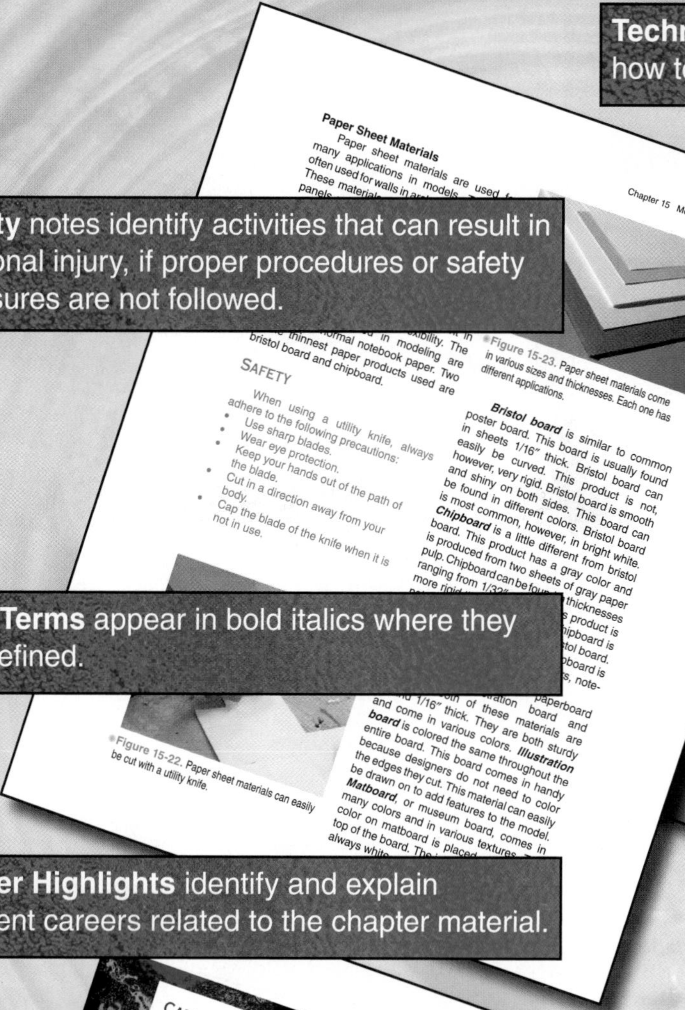

New Terms appear in bold italics where they are defined.

Career Highlights identify and explain different careers related to the chapter material.

Think Green features briefly explain environmental concepts related to technology.

STEM Connections at the end of each chapter encourage students to apply math and science concepts to real-life situations and develop skills related to chapter content.

STEM

STEM Connections

Science

Research science careers in private industries, not-for-profit agencies, universities, and governmental agencies. One source for information is the U.S. Department of Labor Internet site. (Use a search engine and enter "Department of Labor.")

Mathematics

Gather data from the U.S. Department of Labor Internet site. Prepare graphs comparing starting wages or salaries for 10 different jobs.

Curricular Connections

Social Studies

Investigate the types of jobs available in your school. Determine the levels of authority and responsibility each has.

Activities

1. Draw a chart listing your abilities, interests, and values. Include what you are good at doing, as well as what you dislike. Try to be honest and look at yourself as you think others might see you.
2. Find several careers you think you might like. Gather information about the requirements of these careers. Describe the duties, in terms of working with people, information and ideas, and machines.
3. Compare your abilities, likes, and strengths with the career information you have collected. Do not expect a perfect match. The effort will, however, start you on the way to career planning.
4. With two or three other students, create an imaginary business. Decide what product or service you will offer to the public. Write a one-page summary of the company's role in the free enterprise system. Agree on the image you want your company to have. Develop a marketing plan that complies with this image, paying particular attention to ensuring customer satisfaction.
5. Choose a career area that interests you. Research its academic and professional advancement requirements. Prepare a verbal presentation with PowerPoint slides for the class describing the requirements.
6. Develop a schedule for a major assignment in one of your classes. Set deadlines for achieving project milestones. Schedule time to work on the project.

Curricular Connections in every chapter offer suggestions of activities and assignments that connect the content to other subject areas.

Activities at the end of each chapter encourage students to apply concepts to real-life situations and develop skills related to chapter content.

Test Your Knowledge questions help students review the topics and the material covered in the chapter.

men's, women's, and children's hands. This information is valuable for glove makers. Likewise, the data on human-torso sizes is important to clothing-manufacturing companies. The data on the width of people is useful for aircraft and stadium-seating designers.

Experimental research is typical of the research scientists conduct. See Figure 7-13. This approach structures activities so changes or improvements can be measured. For example, a team might carefully study how people are doing a job. The job might be modified to see if the production improves. The quality of the product, the time required to do the job, or the amount of scrap and waste might be compared.

Historical research describes what was possible. Descriptive research describes what is possible. Experimental research describes what can be possible. See Figure 7-14.

Figure 7-13. Scientists often conduct experimental research.

Summary

Technological information and knowledge have helped people control and modify the environments around them. They have helped people build better ways of life. This information and knowledge have given people better food, clothing, and shelter. They have made movement from one place to another easier. Technological information and knowledge have increased the standard of living and life expectancy of the U.S. population.

Summary provides the student a review of major concepts covered in the chapter.

Figure 7-14. The focus of each type of research.

Test Your Knowledge

Do not write in this book. Place your answers to this test on a separate sheet of paper.

1. A need is a requirement to live. True or false?
2. List four examples of a want.
3. A(n) _______ is a situation that can be made better through an improvement.
4. The use of a new product or system or an existing product or system in a new way is a(n):
   A. problem.
   B. opportunity.
   C. need.
   D. want.
5. Name and describe the three types of problems.
6. Problems can be found only at home. True or false?
7. It is important to write a problem statement clearly. True or false?
8. What is the function of criteria?
9. List and explain the five types of constraints.
10. A design brief includes a problem statement, criteria, and constraints. True or false?

Reading Organizer

Make a chart on a separate sheet of paper. List at least three different problems, the type of those problems, and their solutions.

Problem	Type of Problem	Solution to Problem
Example: Illiteracy	Social problem	Education

Chapter 1 Activity A

What Is Technology?

The Challenge

Lift a 4-oz. weight 36″ vertically, using the energy that the air blowing from a standard box fan produces. See Figure 1A-1.

Figure 1A-1. Use wind energy to lift a 4-oz. weight.

Introduction

Technology is used to help people do work, live better, or in some other way, control and modify the environment. A common task facing all people is moving weight. We lift, pull, shove, or slide objects from one location to another. Early humans used their physical strength to complete these tasks. People have limited strength, however, and are limited to the size and weight of the object they can move. Modern life uses a number of technological devices to lift objects from one level to another. Each of these devices has an energy source and a unique technique for lifting the weight. These devices extend a person's ability to do work and are, therefore, technology.

Materials

Develop a technological device using any or all of the following materials:

- A 1/4″-diameter dowel.
- Bond paper.
- Copper wire.
- Index paper.
- Masking tape.
- Paper clips.
- Poster board.
- Soda straws.
- String.
- A thread spool.
- Tongue depressors.
- White glue.

Challenging Your Learning

How could you change solution to make it work better?

Chapter Activities correlate to the Tech Lab Workbook and provide suggested laboratory activities requiring creativity and critical thinking skills.

TSA Modular Activity

This activity develops the skills used in TSA's Inventions and Innovations event.

Inventions and Innovations

Activity Overview

In this activity, you will investigate and determine the need for an invention, develop an idea for the invention, and then present your idea using a stand-alone multimedia presentation; a documentation notebook; a model, or prototype; and an oral presentation.

Materials

- Presentation software.
- A three-ring binder with 8 1/2″ × 11″ pages.
- Materials for the model, or prototype (will vary greatly).

Background Information

Selection. Before selecting the theme for your project, consider past inventions and innovations and current needs in each of the major divisions of technology:

- Medicine.
- Agriculture and biotechnology.
- Energy and power.
- Information and communication.
- Transportation.
- Manufacturing.
- Construction.

Use brainstorming techniques to identify several possible inventions and innovations ... area. Select an idea for a final invention to meet an identified need. The ... completely new, or it can be an improvement to an existing device, ... information about the identified need. Work on ... and construct a model, or prototype,

TSA Modular Activities features are additional activities intended to develop skills used in TSA's competitive events.

... What is the name of materials having rigid structures?
... of the following are engineering materials. You can choose more

Reading Organizer in every chapter helps students to organize what they've learned about chapter concepts.

6. ... Resources ...
7. Match the processes on the ...
8. Match ... example of each type of processing.
   ____ Producing plastic. A. Mechanical processing.
   ____ Making lumber. B. Chemical processing.
   ____ Producing steel. C. Thermal processing.
9. What is an industrial material?
10. Name at least three products of primary processing activities.
11. Explain what the term property means, in the context of this chapter.
12. List the seven properties of materials and define each.
13. Name three things that can happen to materials that are not properly stored.

Reading Organizer

Draw a bubble diagram for each main idea in the chapter. Make each of the main ideas the central bubble, while using details in smaller bubbles to surround the main points. An example from this chapter is shown as an example.

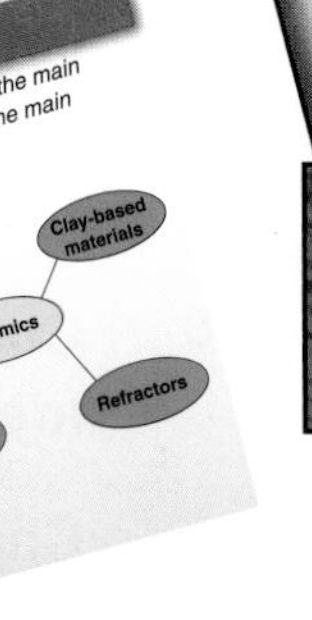

BRIEF CONTENTS

SECTION 4 TECHNOLOGICAL CONTEXTS 410

SECTION 5 TECHNOLOGY AND SOCIETY. 626

CONTENTS

SECTION 3 CREATING TECHNOLOGY . . 202

SECTION 4 TECHNOLOGICAL CONTEXTS 410

FEATURES

Technology Headlines

Career Highlights

Technology Explained

Think Green

TSA Modular Activities

STEM and Curricular Connections

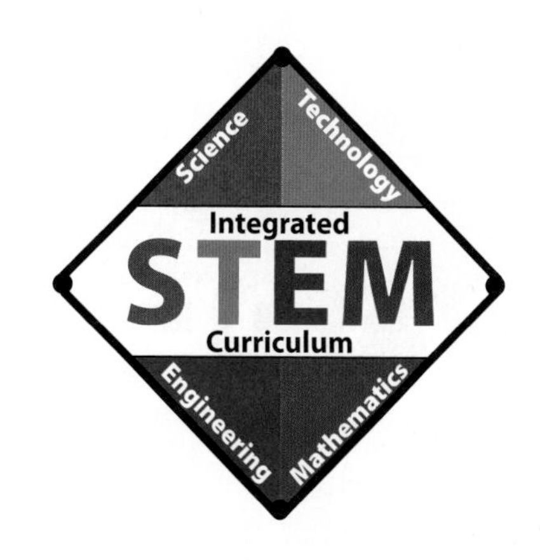

STEM Connections and Curricular Connections are found at the end of the chapters. These activities tie the technological topic to a related language arts, mathematics, science, or social studies skill. The following chart identifies the types of activities contained in each chapter:

Chapter	All Subjects	Language Arts	Mathematics	Science	Social Studies
1	✓			✓	✓
2	✓			✓	✓
3			✓	✓	✓
4			✓	✓	✓
5			✓	✓	✓
6		✓	✓	✓	✓
7			✓	✓	✓
8			✓	✓	✓
9		✓		✓	✓
10		✓		✓	✓
11		✓		✓	✓
12		✓	✓	✓	
13		✓	✓	✓	
14			✓	✓	
15			✓	✓	✓
16		✓		✓	✓
17		✓	✓	✓	
18		✓		✓	
19			✓	✓	✓
20			✓	✓	✓
21			✓	✓	✓
22	✓		✓	✓	✓
23			✓	✓	✓
24			✓	✓	✓
25			✓	✓	✓
26			✓	✓	✓
27		✓			✓

Section 1 Scope of Technology

This book will help you understand how we control our environment. If we are cold, we can use devices we invent to provide heat. If we wish to travel farther than we can walk, we can use fast and efficient methods of transportation systems that we have developed. This ability to create change in our environment is the result of applying knowledge to solve problems. As a result of these problem-solving abilities, we are always changing the way we live, work, and move. New inventions and discoveries are a part of our everyday lives.

What is new for us today soon becomes commonplace. One million years ago, fire was a new tool. Fire was used to fashion crude stone tools, cook food, and keep warm. A little more than 50 years ago, computers took up whole rooms. They were big, loud, and slow. Today, computers are operated by chips that can fit on your fingertip. These chips can be found in everything from our telephone systems to our automobiles.

As you study the following three chapters, you will begin to see how important technology is to you. You will begin to see technology as a series of carefully organized efforts. Also, you will begin to see there are right ways and wrong ways to use technology.

Technology Headline:

Hands-Free Video Games

Once enjoyed solely by kids and teenagers, video games have boomed into a phenomenon the whole family can enjoy! Grandpa bowls, Mom uses the workout games—you get the picture. Video game development companies are constantly one-upping their competition by releasing hot new games and must-have consoles. New possibilities in hands-free gaming are exciting gamers who embrace the newest advances in this technology. In this application, your avatar—an electronic character whose actions you control—mimics your every move without the use of traditional controllers.

Design of the hands-free gaming systems may vary, but the overall concept remains the same: the movement of a player's body and eyes, as well as their voice, can be tracked, recorded, and eventually processed as a means for controlling their on-screen character. In some models, an infrared sensor is mounted on a specially designed pair of glasses. The sensor records all eye movement, and a special webcam and computer translate eye movement into on-screen movement.

Don't feel like wearing a funny pair of glasses? Another prototype that tracks eye movement does so through electrodes the player attaches to his or her face. With either game system, the eyes move in various directions, indicating the direction a character or component should be moved. Take a simple game of air hockey; the paddle follows eye motion up or down and moves accordingly in an attempt to redirect the puck at the opponent. Because this application requires nothing more than eye movement, it opens the gaming world up to more participants. Physically impaired individuals would be able to take part in a way that might have previously been impossible. Success with this endeavor could be applied to other areas, perhaps improving motorized wheelchair controls.

In another version of the hands-free video game, 3D cameras capture players' movement. When playing a soccer game, players will kick an imaginary ball or throw their hands out to block an opponent's goal. Voice and face recognition also play a part. The console will recognize faces and log the player in automatically. The game's speech recognition capabilities will allow it to understand verbal commands. For example, in a game that simulates painting, a player could change the color of his or her paint by simply stating "blue" or later, "green" as he or she raises a hand to create brush strokes.

With these new developments, players could interact with their game in an exciting new way. In a sense, they could step into a new world or experience something they never have before, simply by popping in a different game.

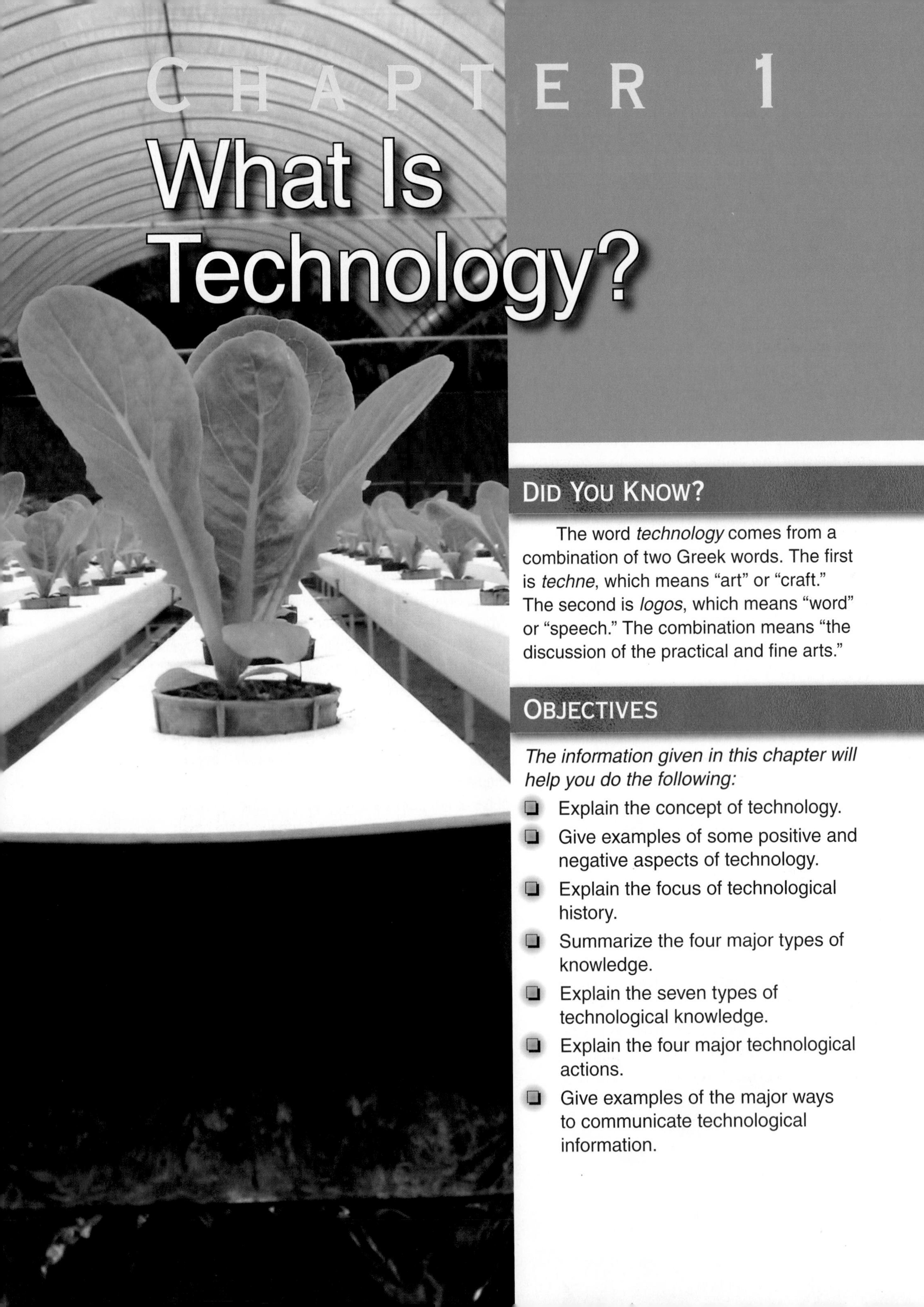

CHAPTER 1
What Is Technology?

DID YOU KNOW?

The word *technology* comes from a combination of two Greek words. The first is *techne*, which means "art" or "craft." The second is *logos*, which means "word" or "speech." The combination means "the discussion of the practical and fine arts."

OBJECTIVES

The information given in this chapter will help you do the following:

- Explain the concept of technology.
- Give examples of some positive and negative aspects of technology.
- Explain the focus of technological history.
- Summarize the four major types of knowledge.
- Explain the seven types of technological knowledge.
- Explain the four major technological actions.
- Give examples of the major ways to communicate technological information.

Key Words

These words are used in this chapter. Do you know what they mean?

agricultural knowledge
assess
construction knowledge
descriptive knowledge
design
energy knowledge
humanities knowledge
information and communication knowledge
language
manufacturing knowledge
medical knowledge
produce
scientific knowledge
technical drawing
technological knowledge
technology
transportation knowledge
use
vocabulary

Preparing to Read

As you read this chapter, think of the types of knowledge discussed. Think about everyday examples of each of the types, and use the Reading Organizer at the end of the chapter to organize what you have learned.

Almost everyone has heard and used the word ***technology***, yet the word means different things to different people. See Figure 1-1. Technology, to many people, means equipment and tools. People often think of things such as robots, spaceships, and computers, when they use the word *technology*. In this context, technology is the hardware that has changed life for better or worse.

Figure 1-1. Technology means many things to different people. What does it mean to you?

To other people, technology causes many of the world's problems. These people see technology as the origin of inventions that have polluted the air and water around us. These opponents of technology believe it has caused problems such as crime, unemployment, and global warming.

To still other people, technology is the hope for a better life. This group of people sees new technological products and systems as the solutions to many of society's immediate and future problems. These products and systems can help do things that could not be done without the help of technology. This group of people points out that technology can improve personal lives by providing efficient transportation, rapid communication, comfortable housing, and plentiful food. People can travel faster with the use of engines. Water pumps can move water to distant locations where it is needed. Cures for diseases can be provided to decrease health risks. Machines can be used to help with hard labor.

Tools and processes designed by people to help us live better

Figure 1-2. A simple definition of technology.

To some extent, all three of these views are correct. See Figure 1-2. Technology involves tools and processes that help us live better lives. These tools and processes can also cause changes that make life more complicated and stressful. Technology can provide more food and clothing, but it can be used to create a society in which some people live in great comfort and others live in poverty. Used well, technology can be an agent of good. Used poorly, it can cause great harm to people and the environment. To fully understand technology, you must understand that it involves knowledge and actions and that it has a history and vocabulary of its own.

TECHNOLOGY AND HISTORY

Often the study of history is one of wars and kings, as history is usually presented from a political viewpoint. The information in most history textbooks focuses on who was in power and the wars fought to solidify that power. Another way to view history, however, is through technological history. This type of history describes how people have lived and attempted to cope with the natural world throughout the course of time. See Figure 1-3. Technological history looks at the types of homes in which people have lived. This history also describes how humans have transported themselves and their belongings. This type of history explains how people have grown and preserved food. Technological history also presents the communication systems they have *used* and the ways in which they have made common utensils and tools.

Figure 1-3. Early humans used simple technologies to improve their lives.

This view of history accepts that people have always wanted to live better lives. Early humans lived in caves and crude shelters. They gathered fruits and berries or killed game for food. These people had harsh and unpredictable lives. They had to move whenever food became scarce. To help live more efficiently, they developed crude tools. To make weapons, these early humans tied sharp stones to sticks. These new tools allowed them to become better hunters. These people tanned animal hides and made crude clothing from them. They sewed these garments together using bones for needles and tendons for thread. These early humans also developed stone axes and hammers, allowing them to make better houses.

As civilization advanced, people started to grow and harvest plants. This allowed them to have more reliable sources of food, so they no longer had to move constantly. They selected tree limbs with certain shapes that could be used as digging sticks. See Figure 1-4. This early plow

Figure 1-4. The invention of the plow was the start of efficient farming. (Deere and Company)

Career Highlight

Technology Teachers

The Job: Technology teachers help students learn information and apply concepts about the development, production, and use of technology. They must have the ability to select, organize, and deliver appropriate material to students. Teachers must possess excellent communication skills and understand their subject area well. They need to know how to help students solve problems, work in groups, and document progress and solutions.

Working Conditions: These teachers generally work in laboratory-type classrooms equipped with a variety of tools and machines. They can experience rewards from students successfully completing work. Technology teachers can also, however, become frustrated from having to deal with unmotivated or disrespectful students. Most teachers work in individual classrooms, causing them to be somewhat isolated from their colleagues. Many teachers work more than 40 hours a week.

Education and Training: All public school teachers need to be licensed by the state in which they teach. Technology teachers must hold a bachelor's degree from an approved teacher-education program. They must have completed a set number of classes and a practice teaching assignment. In addition, most states require applicants to pass a basic competency test before the applicants receive a teaching license. Also, many states require teachers to participate in continuing education programs.

Career Cluster: Education & Training

Career Pathway: Teaching/Training

allowed them to till the soil. Agriculture was born. With the plow and other early farming tools, the foundation for modern civilization was set. People could gather in villages and cultivate nearby land. Without the ability to grow large amounts of food in one place, concentrated population centers would have been impossible. The development of tools to gather and harvest crops followed the invention of the plow. Later, irrigation was developed along large rivers in ancient Egypt and other areas. Dams were built. Ditches were dug. Simple devices to raise the water from one level to another were developed. Irrigation projects were some of the first engineered works people built. The knowledge used in these projects was applied to other building projects, such as the pyramids.

The skills and knowledge developed for one project were applied to other projects. Each new project allowed people to develop more skills and knowledge. The process of applying old knowledge and skills while developing new knowledge and skills is a characteristic of technology. This is true because technology is a product of the human mind. Technology involves the engineering spirit and is the result of human ambition, vision, and control. People have developed all forms of technology to meet the needs and wants of other people. These developments are the result of human needs and the ability to be creative. Creativity and innovation allow us to generate new products and systems to make life easier. See Figure 1-5.

Technology and Knowledge

People use knowledge every day. They use it to make decisions, complete work, and entertain themselves. There are several types of knowledge. First, there is knowledge about the world around us, and it describes the plant and animal kingdoms and the land, water, and air we depend on for life. This knowledge explains the laws and principles governing the universe. We call this type of knowledge ***scientific knowledge***. See Figure 1-6.

Figure 1-5. People develop technology for other people to use.

Figure 1-6. Scientific knowledge describes the natural world.

Figure 1-7. Technological knowledge is used to develop the human-built world. (Northern Telecom)

Second, people have knowledge of the society around them. They usually have religious ideals and beliefs and know what is considered right and wrong. Most people know about the histories of their families, communities, and world. They understand how their societies were formed and how they are organized and governed. This type of knowledge is called ***humanities knowledge***.

Also, people have a way to use signs and symbols to communicate with each other. They use an alphabet to form words that can be spoken or written. We call these symbols ***language***. People can also use numbers to show relationships and values. They can use lines to show shapes and forms in drawings. We call this type of knowledge ***descriptive knowledge***. This knowledge helps people describe objects and events.

Finally, people have knowledge of how to use tools and materials. They use this knowledge to make things and construct buildings. This knowledge helps them process information, grow and harvest crops, convert and use energy, and heal diseases. The knowledge of how to use tools and materials is used to design, produce, and use things other people create. This knowledge is the knowledge of the human-built world and is called ***technological knowledge***. See Figure 1-7.

There are many types of technological knowledge. They can be categorized in a number of ways. See Figure 1-8. One example divides technological knowledge into these categories:

Types of Technological Knowledge
Agricultural
Construction
Energy
Information and communication
Medical
Manufacturing
Transportation

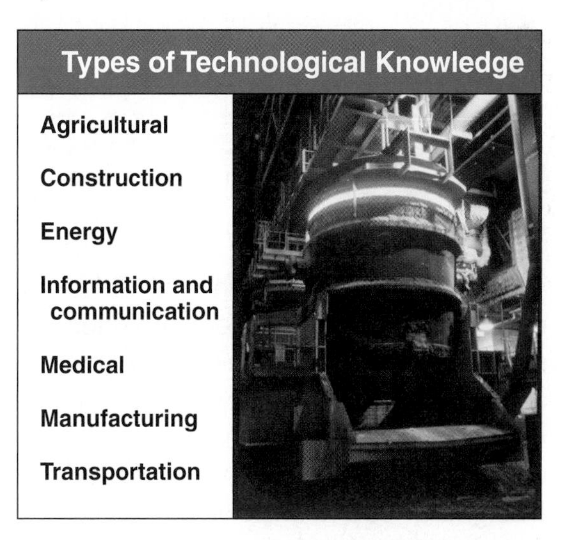

Figure 1-8. Some types of technological knowledge.

- ***Agricultural knowledge.*** This is the knowledge about using machines and systems to raise and process foods.
- ***Construction knowledge.*** This is the knowledge about using machines and systems to erect buildings and other structures.
- ***Energy knowledge.*** This is the knowledge about using machines and systems to convert, transmit, and apply energy.

- ***Information and communication knowledge.*** This is the knowledge about using machines and systems to collect, process, and exchange information and ideas.
- ***Manufacturing knowledge.*** This is the knowledge about using machines and systems to convert natural materials into products.
- ***Medical knowledge.*** This is the knowledge about using machines and systems to treat diseases and maintain the health of living beings.
- ***Transportation knowledge.*** This is the knowledge about using machines and systems to move people and cargo.

Technology and Action

People apply technological knowledge and skills in different activities and settings. Consumers use technological knowledge to select devices to fit their needs. They use technological skills to operate and maintain these products. Designers use technological skills and knowledge to create or improve products and systems. Workers use technological abilities and knowledge to make products to meet our needs. Some products, however, are developed before a need is identified. Corporations can use technological knowledge and abilities to create demand for these unnecessary products. They do this by bringing the products onto the market and advertising them. These and hundreds of other activities fall into four categories of technological actions. See Figure 1-9. These actions change resources into outputs we want and need:

- Designing and engineering technological products and systems.
- ***Producing*** products and systems with the use of tools and machines.
- Using the products and systems to meet human needs and wants.
- ***Assessing*** the impacts of products and systems on people, society, and the environment.

Designing Actions

Technology is created by the purposeful actions of people. All technology started in the minds of people. These people identified problems needing solutions or opportunities that could be met. Products and systems were then developed. They

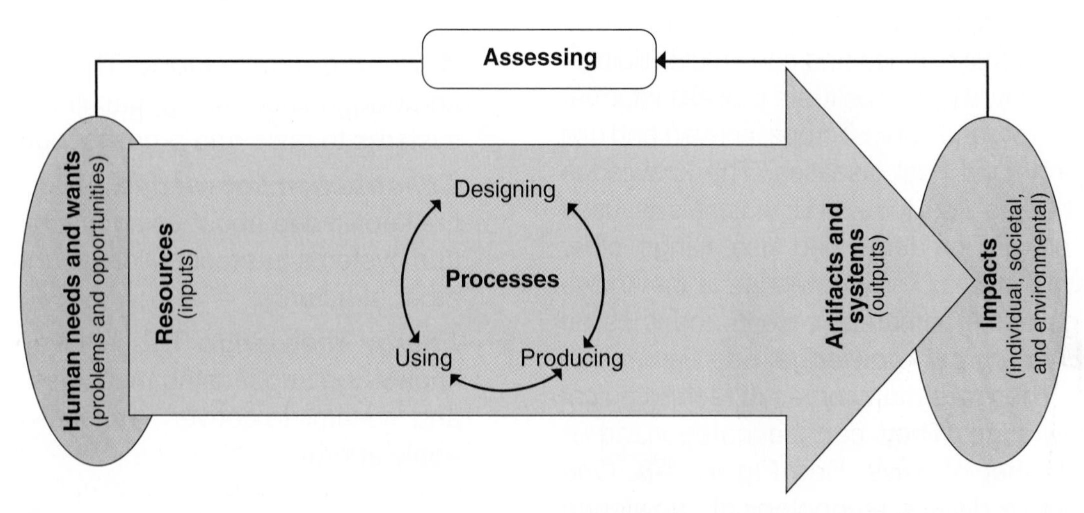

Figure 1-9. Technology involves designing, producing, using, and assessing actions.

were designed to meet the problems or address the opportunities. A common technique used to create products or systems is ***design***. A common design process is presented in Chapters 9 through 18.

Producing Actions

Designs, in themselves, are of little value to people. They become valuable when they are converted into products or systems. There is vast knowledge about how to use tools to make or grow things. This is the basis for a variety of actions called *producing*. Production actions are used to build products, erect buildings, grow and harvest crops, communicate information, and transport cargo. Information about these and other production actions is included in Chapters 19 through 25.

Using Actions

People use technological knowledge and products to meet many demands. See Figure 1-10. They use products in their roles as workers, consumers, family members, and citizens. For most people, using technology is a daily action. Using technology involves selecting an appropriate product and determining which technological service to use. Product service and repair decisions are also using actions.

Figure 1-10. People use technological devices to meet their needs. (Deere and Company)

Assessing Actions

Using technology is more than simply meeting human wants. Technology must fit within the social and political systems of a community or nation. Also, it must be in harmony with the environment. This means people must assess the impacts and importance of each form of technology they use. They must decide what is effective or appropriate technology. The description of what is appropriate, however, changes from country to country and over time.

Technology and Language

Each major type of knowledge has its own language. Scientists use words many of us find hard to understand. Likewise, historians have their own terms and ways of communication.

Technology has two communication tools unique to it. It has a ***vocabulary*** that has been developed to describe actions with tools and the products of these actions. Suppose you had a dictionary from the 1800s. You would not find words such as *microchip*, *airplane*, *television*, and *photocopy* in it. These are new words developed to describe new technological devices. Other new words were coined to describe tools and actions.

New technological words are developed every year. They are first part of the language of technology. Later, some of them become part of the common language. For example, *debug* was a technical term in computer technology. Early computers had exposed wires and hot vacuum tubes that attracted bugs. When a bug got close to the hot tube, the heat killed it. The dead bugs built up on the circuits and shorted them out. People had to open the computer cabinets and physically remove the dead bugs. This process was called *debugging*. Later, the term was used to describe correcting circuit and software problems in computers. Today, some people use the term to describe diagnosing problems in any system.

Second, technology uses ***technical drawings*** as a form of communication. These drawings are tools of engineers and architects, showing how products are to be made or buildings are to be built. They also can show how communities are to be developed or parks are to be landscaped.

Drawings can be as simple as a sketch of a designer's ideas. They can be so complex that computer systems are used to help in their development. See Figure 1-11. Chapter 17 discusses technical, or engineering, drawings more completely.

Figure 1-11. Drawings are part of the language of technology. (Ford Motor Company)

Summary

Technology plays a part is everyone's life. People use it to meet their needs and wants. Technology is a series of actions in which products and systems are designed, produced, used, and assessed. These actions are the actions that can be accomplished with the knowledge of tools and materials. Technology has its own history and ways of communication. These actions are unique and important to each of us.

STEM Connections

Science

Research a technological device that has helped scientists learn more about the universe. Describe the device, its use, and when it was developed.

Curricular Connections

All Subjects

Read a current-events magazine or newspaper article. Highlight examples of descriptive, humanities, scientific, and technological knowledge. Use a different colored pencil or marker for each type of knowledge.

Social Studies

Ask several older people what they think technology is. Record their answers. Later, arrange the comments under three headings: Computers and Hardware, Source of Society's Problems, and Hope for a Better Life. Note: A person's responses might have entries under more than one heading.

Activities

1. Science is the knowledge of the natural world. Technology is the knowledge and actions used to create the designed world. Divide a sheet of paper into two columns. Write *Science* at the top of the left-hand column and *Technology* at the top of the right-hand column. On your way to school, list things you see explained by each of these:
   A. Scientific laws or theories. In the left-hand column, for example, you might list an apple falling to the ground, a flower dying, or a building casting shade on the sidewalk.
   B. Technological principles or applications. In the right-hand column, for example, you might list a pothole being fixed, a car being towed away, a building being painted, or a billboard being installed.
2. Develop a simple information sheet that would help people understand the four major actions of technology.
3. Develop an assortment of photos and other pictures showing the various types of technological knowledge.

Test Your Knowledge

Do not write in this book. Place your answers to this test on a separate sheet of paper.

1. What are some negative effects technology can cause?
2. Why is the study of technological history important?
3. Select the statement best defining *technological knowledge*:
   A. Knowledge of the laws and principles of the natural world.
   B. Knowledge about developing and making products.
   C. Knowledge about social systems people have developed.
   D. Knowledge about numerical relationships.
4. Identify each of the following types of knowledge:
   A. Knowledge about the world around us.
   B. Knowledge about society.
   C. Knowledge about communicating with other people.
   D. Knowledge of how to use tools and materials.
5. Describe each of the following terms:
   A. *Agricultural knowledge.*
   B. *Construction knowledge.*
   C. *Energy knowledge.*
   D. *Information and communication knowledge.*
   E. *Medical knowledge.*
   F. *Manufacturing knowledge.*
   G. *Transportation knowledge.*
6. Actions used to develop new or improved products or systems are called _____.
7. Actions used to build products, erect buildings, grow and harvest crops, communicate information, and transport cargo are called _____.
8. Actions involved in selecting, operating, and servicing products are called _____.
9. Actions used to determine the value or impact of a technology are called _____.
10. The language of technology involves _____ and _____.

Reading Organizer

Copy the following chart on a sheet of paper. List examples for each of the types of knowledge shown.

Types of Knowledge	Examples
Scientific knowledge	*Example:* Understanding scientific laws or theories
Humanities knowledge	
Descriptive knowledge	
Technological knowledge	

Chapter 1 Activity A

What Is Technology?

The Challenge

Lift a 4-oz. weight 36″ vertically, using the energy that the air blowing from a standard box fan produces. See Figure 1A-1.

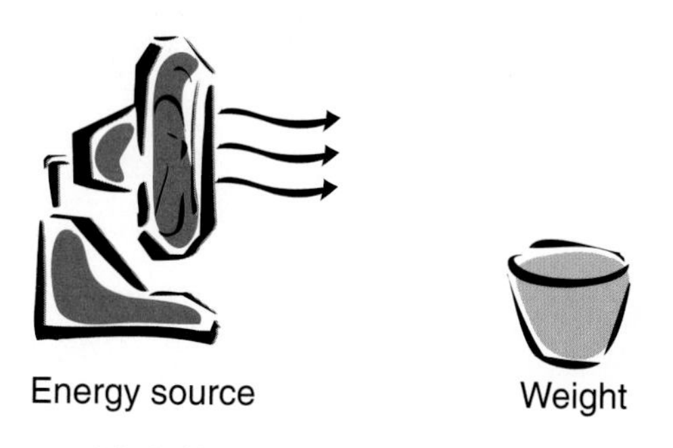

Figure 1A-1. Use wind energy to lift a 4-oz. weight.

Introduction

Technology is used to help people do work, live better, or in some other way, control and modify the environment. A common task facing all people is moving weight. We lift, pull, shove, or slide objects from one location to another. Early humans used their physical strength to complete these tasks. People have limited strength, however, and are limited to the size and weight of the object they can move. Modern life uses a number of technological devices to lift objects from one level to another. Each of these devices has an energy source and a unique technique for lifting the weight. These devices extend a person's ability to do work and are, therefore, technology.

Materials

Develop a technological device using any or all of the following materials:

- A 1/4″-diameter dowel.
- Bond paper.
- Copper wire.
- Index paper.
- Masking tape.
- Paper clips.
- Poster board.
- Soda straws.
- String.
- A thread spool.
- Tongue depressors.
- White glue.

Challenging Your Learning

How could you change your solution to make it work better?

CHAPTER 2
Technology as a System

Did You Know?

- The solar system is a very large natural system. This system consists of the Sun, eight official planets, over 165 moons, five dwarf planets, and many other small bodies.
- A computer's operating system is the first software seen when a computer is turned on. This system is the last software seen when the computer is turned off. The operating system directs all the programs used on a computer. This system manages the hardware and software elements of the computer system. The operating system provides a way for applications to deal with the hardware.
- Not all computers have operating systems. Computers controlling simple objects, such as microwave ovens, do not need operating systems. The computer in a microwave oven runs a single program all the time.

Objectives

The information given in the chapter will help you do the following:

- Explain the concept of a system.
- Compare natural systems and technological systems.
- Summarize how a system works.
- Give examples of some major goals of technological systems.
- Summarize the seven major inputs to technological systems.
- Give examples of the major types of processes used in technological systems.
- Explain the major types of outputs of a technological system.
- Explain the two types of feedback systems.

Key Words

These words are used in this chapter. Do you know what they mean?

closed-loop system
desired output
energy
feedback
feedback system
goal
input
intended output
knowledge
machine
material
natural system
open-loop system
output
pollution
process
system
technological system
time
undesirable output

Preparing to Read

As you read this chapter, outline the details of the five basic steps that make up a system.

You hear about systems all the time. The weather forecaster talks about a new weather system moving into your region. A doctor tells you your digestive system is upset. The auto mechanic explains that the ignition system of your family's car needs work. So what is a system?

Simply put, a ***system*** is a group of parts working together to complete a task. See Figure 2-1. There are both natural and

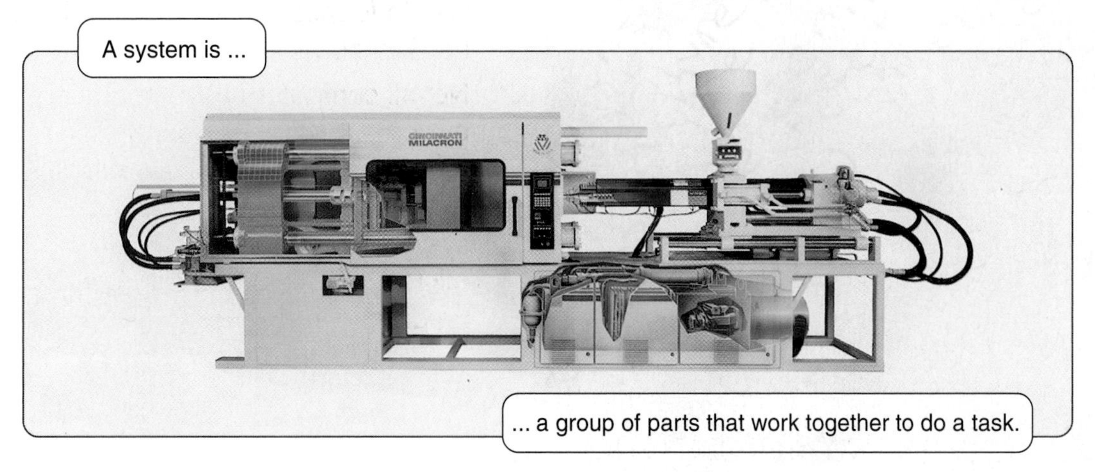

A system is ...
CINCINNATI MILACRON
... a group of parts that work together to do a task.

Figure 2-1. This machine is a technological system people developed. (Cincinnati Milacron)

human-made systems. ***Natural systems*** appear in nature without human interference. They are part of the world around us. These systems include the universe. Natural systems also include all the living things populating Earth. People study natural systems as they explore science. Typical natural systems are the transpiration systems of plants, the circulatory systems of animals, the solar system, and ecosystems found in lakes and ponds. See Figure 2-2.

Human-made, or designed, systems are used throughout society. People create these systems to meet their needs. There are many types of human-made systems. Political systems are designed to develop laws and regulations governing human actions. Judicial systems are created to enforce these laws and regulations. Economic systems are used to exchange items of value. Social systems are created to help people interact with one another. ***Technological systems*** are used to make the artifacts and services people want or need. See Figure 2-3.

Technological systems include agricultural, communication, transportation, manufacturing, construction, and energy conversion systems. Computers and information processing systems are technological systems. Irrigation systems found on farms and radiation treatment systems found in hospitals are technological systems as well. Technological systems can be linked to one another and often interact with each other. Sometimes the output of one system is the input to another system, and sometimes one system controls another system. Often, several systems work together to produce the technology we use.

Systems thinking requires considering how each part connects to the others. Analyzing a system can be done by looking at its individual parts. This can also be done by looking at how the system as a whole interacts with other systems.

All systems have a basic structure. See Figure 2-4. Generally, they are made up of five basic elements:

- ***Goals.*** These are the reasons for developing and operating the system.
- ***Inputs.*** These are the resources the system uses to meet the identified goals.
- ***Processes.*** These are the actions taken to use the inputs to meet the goals.
- ***Outputs.*** These are the results obtained by operating the system.

Figure 2-2. The universe is made up of natural systems.

Figure 2-3. People build technological systems, such as this product storage system, to meet their needs. (©iStockphoto.com/dlewis33)

Figure 2-4. This automatic manufacturing system has inputs, processes, outputs, and feedback. (©iStockphoto.com/DanDriedger)

- ***Feedback.*** This includes the adjustments made to the system to control the outputs.

All five parts of a system are extremely important to the proper operation of the system. Malfunctions of any one part can affect the function and quality of the entire system. When part of a system fails or works incorrectly, the results can range from a nuisance to a catastrophe. It is vital to understand how the elements of a system all work together because they are so interrelated. This can be done by examining a common system.

Let's consider a typical heating system found in a home. See Figure 2-5. The primary goal of the system is to keep the building comfortable during cold weather. The system is designed to maintain an even, livable temperature in the house. The main input to a heating system is an energy source. A fuel, such as natural gas, propane, coal, wood, or fuel oil, provides the ***energy***. The process involves converting the fuel's energy into heat. This conversion uses a natural process called *combustion*. The fuel is burned to convert its stored energy into heat energy.

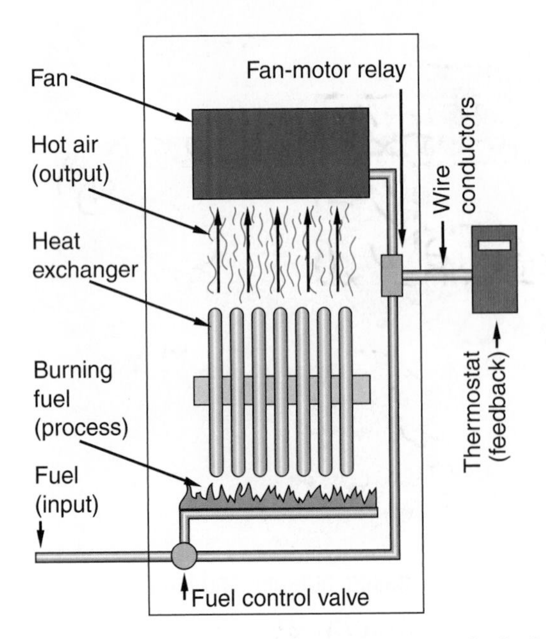

Figure 2-5. A heating system is a technological system developed to keep buildings warm.

This heat energy is the ***desired output*** of the system. Other outputs might include carbon dioxide gas, nitrogen oxide, water, or ash. These outputs are not needed. They are, however, unavoidable.

An important part of most heating systems is a temperature regulator. This regulator is called a *thermostat*. The thermostat keeps the temperature within a set range so the house is neither too hot nor too cold. The temperature regulator has a part that senses the room's temperature. When the temperature reaches the lower limit of the thermostat's setting, it turns the heating unit on. When the room has reached the upper limit of the thermostat's setting, it turns the heater off. The thermostat is part of the feedback loop of the system. This temperature regulator uses data (room temperature) to control the output of the system (heat). Feedback loops adjust and control the operations of systems.

Although it is not considered one of the main elements of a system, a successful system also must include some form of maintenance. Maintenance is the process of inspecting and servicing a product or system on a regular basis to ensure it continues to work correctly, to prolong its life, or to improve its capability. All technological systems ultimately break down. Maintenance, however, decreases the risk of early failure. If maintenance is not done, failure is guaranteed. The rate of failure depends on such issues as how complex the system is, what kinds of environment it must function in, and how well it was initially made.

System Goals

Technological systems have one primary goal. This is to meet the needs and wants of people. Without this goal, any technological activity is doomed to failure.

Many technological actions, however, are carried out as part of the economic system. Businesses are involved in performing the actions. Companies and individual proprietors are involved in the systems. They transport people, build buildings, deliver medical services, raise crops, generate electricity, and make products. These companies and individuals expect to be paid for their efforts. They expect to make a profit. Profits, the second goal of technology, are the rewards for creative ideas, personal labor, and financial risks.

A third goal of technological activity is to make a positive contribution to society. This means we expect that all technological activities help make life better. We use them to help people in their various life pursuits. The systems should contribute to a clean environment. They should also promote community, national, and international goals.

System Inputs

Technology consists of human-built systems and includes all the material things people have created over the years. Creating these products and systems takes resources. People use resources in creative ways to meet their needs and wants. The resources for creating technology are called *system inputs*. See Figure 2-6. These inputs can be grouped into seven major categories:

- People.
- Natural resources (***materials***).
- Machines.

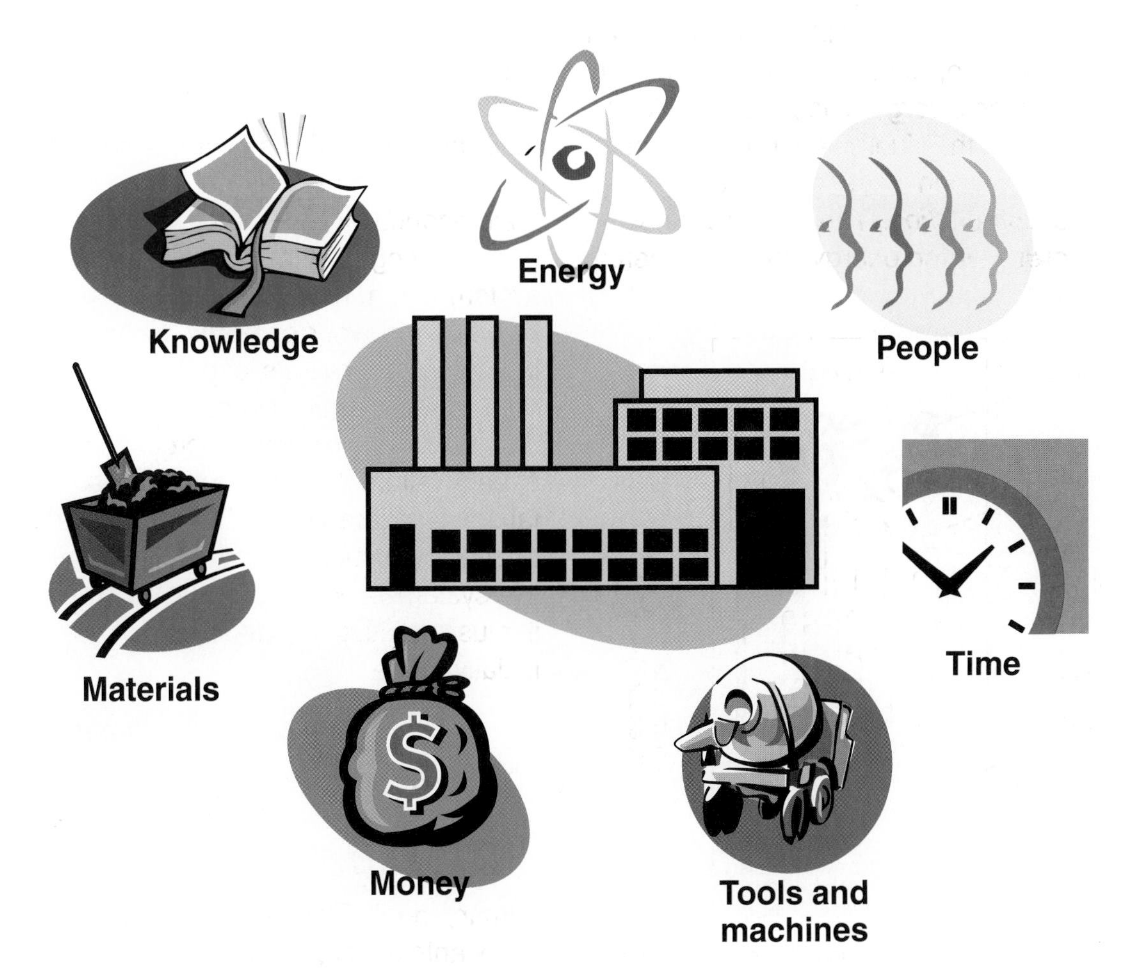

Figure 2-6. These are the inputs to technological systems.

- ***Knowledge*** (information).
- Energy.
- Financial resources (money).
- Time.

Each of these inputs is essential for developing new or improved technological devices. These inputs are used to create and produce the things people design and use.

Technology and People

People have developed and produced all forms of technology. Technological devices and systems are the results of people's ideas and work. Without people's minds and efforts, there would be no technology. Likewise, people use technology. If people did not want a device or technological system, there would be no need to develop it. In short, people develop and produce technology for other people to use.

They bring many skills and roles to technology. See Figure 2-7. Designers and engineers develop technological devices and systems. Machine operators and laborers make devices and systems. Advertisers develop commercials to promote the use of products. Managers organize and supervise development and production processes. Retailers operate stores and e-commerce sites. Customers buy and use products and systems. Politicians and public officials regulate the production and use of technology. Without people, there would be no technology.

Technology and Natural Resources

Technology includes the devices and systems people design, build, and use. These things are made out of materials found on Earth and in the air. See Figure 2-8. Materials can be classified as gases, liquids, or solids.

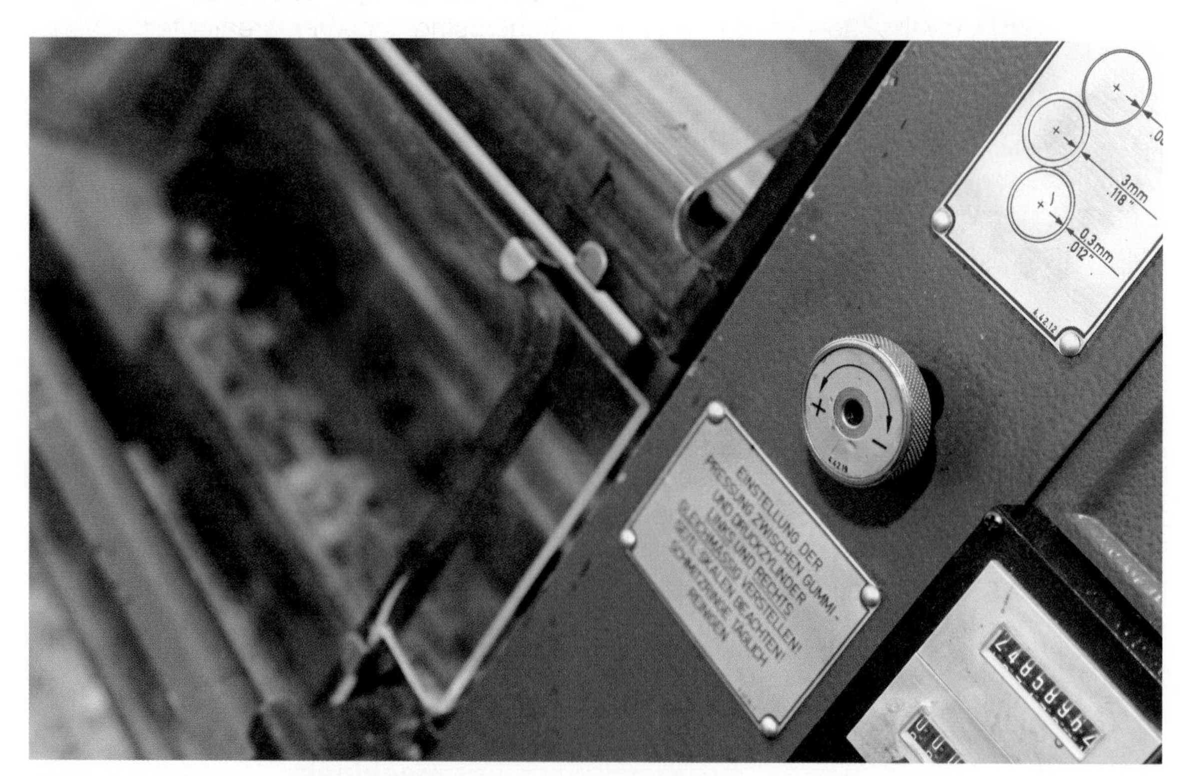

Figure 2-7. People develop, operate, control, and use technological products and systems. This is a control panel to control a printing press.

Figure 2-8. Material resources start with natural resources found on Earth and in the air. (American Petroleum Institute)

Gases can be the fuels used to make devices. Also, gases can become part of products. Typical gases used in technology are natural gas fuels and inert gases. Liquids are also used as fuels. They can be used as finished products as well. Liquids can be lubricants used during production. The products themselves might use them. Typical liquids used in technology are gasoline and water. Solid materials are shaped and formed to become structures and products. They might become frames for buildings. Solids might become structures of products. Typical solids used in technology are metal sheets and bars, glass sheets, lumber, and plastic pellets and sheets.

Technology and Machines

Technology requires ***machines*** to shape materials, process information, convert energy, and transport cargo. See Figure 2-9. Machines are used to construct roads, produce bread, and broadcast television programs. They are used to carry people and cargo, make furniture, and fabricate computers. Machines extend humans' abilities to do work. They help us grow and harvest crops, heal illnesses, and manufacture products. Machines help us transport goods, convey messages, and build buildings.

Technology and Knowledge

Without knowledge, we would not have technology. People apply their knowledge about materials and processes to design and make things. They use knowledge about using tools to create technology. People use this knowledge to use technology. They use design knowledge to create products and systems. People apply production knowledge to use machines to make these products from materials. See Figure 2-10. They use knowledge to select,

Figure 2-9. This ship is an example of a technological machine.

Figure 2-10. This model maker applies technological knowledge and skill to build a prototype of a new product.

use, service, and repair devices. People use knowledge to measure or predict the impacts technology has on other people and the environment.

Technology and Energy

All technological systems are designed to do a task. Most of these applications require energy. Energy is needed to make furniture from lumber, broadcast television programs, harvest grain, build a road, and haul gravel to a building site. Technological systems could not operate without energy. This energy might come from inexhaustible sources, such as wind and water. Energy is also available from renewable sources, such as wood and corn, and energy is found in exhaustible sources, such as coal, petroleum, and natural gas, as well.

Technology and Financial Resources

People design technological systems. They also build these systems. To do this, materials must be located and purchased. Machines and buildings must be bought or leased. Energy supplies must be obtained. People must be employed. Patents must be sought. All these activities require finances. It takes money before a technological system can be created, operated, and maintained. See Figure 2-11.

Technology and Time

It takes ***time*** to move a creative idea from inside a designer's mind to a product. Time is needed to design and engineer the product or system and to develop a production facility to make the device. This resource is needed to build and test the product or system, to ship products to stores, and to install new products. Time is a key resource in developing and operating technological systems. This resource is measured in machine time, work hours, product life, or service intervals.

System Processes

The understanding of how a system works is vital if one is to operate and maintain the system successfully. Technological systems are designed to be operated and maintained in order to achieve a given purpose and create a desired output. Producing these outputs involves processes. See Figure 2-12. Different technologies entail different sets of processes. Typically, technological systems use two types:

- Production processes.
- Management processes.

These two types of processes work together to meet the system's goals, such as transporting cargo, growing crops, communicating information, or manufacturing products.

Figure 2-11. Labor, materials, and machines, such as these on a construction site, cost money. They require financial resources to obtain.

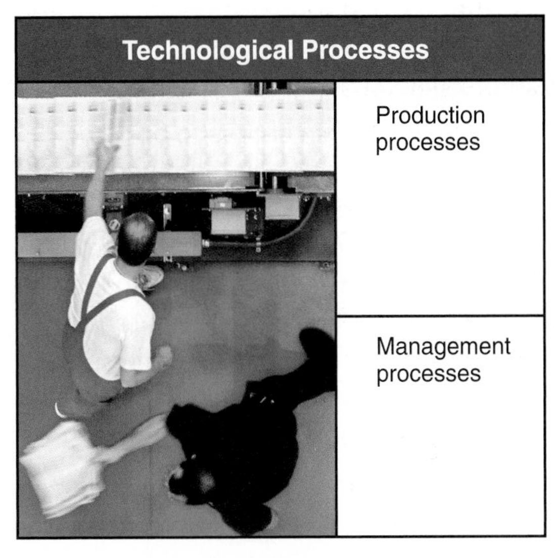

Figure 2-12. Technological systems involve production and management processes. This state-of-the-art printing press requires both types to function smoothly. (Gannett Co.)

Production Processes

Production processes are actions changing inputs into outputs. See Figure 2-13. These processes include a number of actions:

- Growing and harvesting crops (agricultural technology).
- Changing materials into products (manufacturing technology).
- Changing materials and manufactured products into constructed works (construction technology).
- Converting information and ideas into printed or broadcast messages (communication technology).
- Using devices to improve health and fight diseases (medical technology).

1. Loading the containers
2. Moving the containers to the port
3. The containers arriving at the port
4. Loading the containers onto a ship

Figure A. These are the steps used in container shipping by sea. Once the ship reaches its destination port, these steps are reversed. The containers are unloaded from the ship and then moved to their individual destinations.

Figure B. Wide-body aircraft allow for containers to be shipped by air. (©iStockphoto.com/mevans)

Technology Explained

container shipping: using sealed containers to group and contain items for bulk shipment.

Shipping cargo across oceans has become a commonplace activity. In earlier times, each crate or box was lifted individually onto a ship and stored in the hold. This technique was time-consuming and expensive. Loading each crate individually tied the ship up in port for a number of days at the end of each trip.

To increase the efficiency of ocean shipping, a new type of ship was needed. To meet this need, the containership was developed in the 1960s. This type of ship uses steel containers 20′ or 40′ (6 or 12 m) long, 8.5′ (2.6 m) high, and 8′ (2.4 m) wide. Special high cube containers are the same length and width. They are, however, 9.5′ tall. Each container can hold up to 30 tons (27 metric tons) of cargo.

Container shipping works by grouping a number of small shipments going to one place into one large load. The grouping of shipments by destination is called *consolidation*. Each item is handled only twice: once when it is placed in the container and again when it is removed from the container.

This type of shipping has several steps. See **Figure A.** First, the cargo is picked up from the shipper. A number of different shipments destined for the same location are put together. These shipments are loaded into a container. The container is transported to the dock. Containers can be placed on a set of wheels to form a semitrailer, which a truck then pulls. The truck can pull the container to the dock. Trucks sometimes haul containers to rail yards. There, the containers are lifted off the set of wheels and set onto railcars. They make the rest of their journey by rail.

Often, entire trains are made up of railcars loaded with containers. The train moves the containers to the port, where they are stored in large stacks. When the ship comes to port, the containers are moved to the dock. Large cranes lift them and place them on the ship. Using this process, dockworkers can load an entire ship in less than a day. Using the older loading techniques, it could take up to five days to unload a ship and another five days to load it.

In addition to being faster, container shipments do not require warehousing at dock sites. Each container is a small, weatherproof "warehouse." Container shipping is now being adapted for air transportation. See **Figure B.** The development of wide-body airplanes allows the use of large containers. Container shipping is popular with overnight parcel companies.

During this time, people learned to grind grain, spin yarn, and weave fabric. They started making clay storage and cooking vessels. Hammering shaped soft metals, such as copper, silver, and gold. Later, ores (metal-bearing rocks) were heated (smelted) to obtain the pure metals. Copper and tin or lead were combined to make bronze (an alloy). Metal tools replaced stone ones. This stage of human history is called the ***Bronze Age***. This stage started in an area near Thailand as early as 4500 BC. The Bronze Age was widely seen in the Western world by 3000 BC.

Irrigation systems were developed during this period. This advancement allowed farming to increase the food supply. Cities grew. Building techniques developed. See Figure 3-4. People moved goods on water, using simple boats with oars or sails. Pack animals were used on land. Carts with wheels appeared for the first time. Fired-clay bricks were developed to supplement building stones.

The Iron Age—1000 to 500 BC

Iron ore had been available to developing cultures for a long period of time. People could not, however, get the metal from the rocks. They could not develop the required temperature of about 2800°F. The development of better furnaces changed this. People learned to smelt iron around 1000 BC. By 500 BC, the metal was widely available in the Western world. This metal introduced a period called the ***Iron Age***. The actual time frame for this age varies by culture. The use of iron as a primary tool-making material did not impact all areas at the same time.

The new metal led to better agriculture, with the development of the iron-tipped plowshare. Cooking ware and weapons improved. During this period, new building techniques developed. See Figure 3-5. Fired-clay bricks and tiles were widely produced and came into common use.

Figure 3-4. Many construction techniques were developed during the Bronze Age. The Egyptian pyramids are examples of these techniques.

Figure 3-5. The Roman Colosseum was first used in 80 AD, after eight years of construction. This amphitheater rises 65′ above the streets. The Colosseum is 610′ long and 515′ wide.

The traditional building materials, such as limestone and marble, also became widely used. Large buildings, temples, and monuments appeared.

Also, ships were improved with new types of sails and a keel. Boats became oceangoing vessels. Carefully laid out road systems were developed to connect major cities. See Figure 3-6.

Figure 3-6. This is a view of a street in Pompeii. An eruption of the volcano Mount Vesuvius destroyed it in 79 AD. Notice the sidewalks, cobblestone streets, and remains of the building.

The Industrial Age—1750 to the Late 1900s

The use of iron tools led to rapid advancements in technology. Village crafts grew in number. Many villages had craftsmen who worked with leather and metal. Other trades included making rope, barrels, candles, and soap. Demand for the products of these workers grew.

New machines and sources of power were developed to support the industry of the day. The steam engine was invented and perfected. Metal cutting and shaping machines were developed. These and other developments led to a period called the *Industrial Revolution* or ***Industrial Age***. This was a period in which most Western countries changed from rural to urban. The primary focus moved from farming to manufacturing.

The Industrial Revolution impacted different cultures at different times. This revolution is generally believed to have started in Great Britain about 1750. The Industrial Age began in the United States around 1850. The revolution traveled to Japan and Russia during the first half of the twentieth century. Many developing (Third

Career Highlight

Construction Managers

The Job: Construction managers have a number of different titles, including constructor, construction superintendent, general superintendent, and project manager. They plan and coordinate the work done on construction projects. Construction managers manage the people, materials, equipment, budgets, and schedules for a specific project. Generally, they are involved in processes from the conceptual development through final construction.

Working Conditions: These managers work out of a main company office or a field office at the construction site. Many management decisions are made on the job site. Managers with large companies might travel from state to state as the company is awarded project contracts.

Education and Training: People are promoted into construction-manager positions after having considerable experience as construction craft workers, construction supervisors, or owners of independent contracting firms. Construction managers need a solid foundation in mathematics, science, computer use, building practices, and management.

Career Cluster: Architecture & Construction

Career Pathway: Construction

World) countries are just now having their industrial revolutions.

This revolution was primarily born in the textile industry of England. The first textile factories started to appear in 1740. Over the next 100 years, factory-made cotton garments replaced the woolen garments the British wore. The factory system in England aided this change. Cheap cotton from the American colonies also played a part in this shift. Eli Whitney's invention of the cotton gin in 1793 greatly aided this development.

The factory concept quickly spread to other activities. See Figure 3-7. Steel mills, clay-tile factories, and other plants soon appeared. Along with new factories came new transportation and communication technologies. The railroad and automobile were developed. Telegraph, telephone, and radio (wireless) communications were developed. Printing became common with the advent of moveable type and the printing press.

A key to the Industrial Revolution was the variety of new power sources. Reliance on waterpower disappeared with the inventions of the steam engine and electric motor. Later, the steam engine gave way to gasoline and diesel engines. Homes were wired for electricity. Natural gas and heating oil provided fuel for heating systems.

Cities grew. Farms became mechanized. See Figure 3-8. Tractors, combines, row cultivators, and other implements made farming more efficient. Skyscrapers were developed. Also, for the first time,

Figure 3-7. The factory system was developed to produce textiles. This system soon spread to many other industries. Here is a continuous production process producing glass containers. (Owens-Brockway)

Figure 3-8. Modern farm machinery allows fewer people to grow more food. People who were once needed to grow food can now move to the cities. There, they can work in factories or do other work. (Deere and Company)

large numbers of people worked for other people. The factory system developed division of labor. Owners were not laborers the way they were in the older village crafts. A system of managers and workers developed. With it, labor unions were born.

The Information Age—Late 1900s to Present Day

The twentieth century saw rapid changes in technology. Many countries became industrial powers with the development of assembly lines and mass production. Companies made conscious efforts to innovate. They worked hard to improve their production processes. A new material called *plastic* came into use. Synthetic fibers were developed. Products became plentiful. A large middle class with money to spend developed.

The computer was developed in 1944. The transistor was invented in 1947. Late in the century, however, the real impacts of these technologies were felt. The microchip and personal computer were then developed. With these devices came a revolution as important as the Industrial Revolution. This was the Information Revolution. With it, the ***Information Age*** was born. Again, the focus of societies changed. Economies built on manufacturing started to focus on information processing. Manufacturing did not disappear. The ways products were manufactured, however, changed. Automation entered the workplace in many areas. See Figure 3-9. Computers, rather than people, directly controlled many machines.

New and larger farm machines were developed, which allowed farmers to work greater acreage. The number of people involved in agriculture continued to drop. Travel increased with the jet aircraft and high-speed rail systems. The use of the automobile for personal transportation increased. Instant communication appeared with the communication satellite and cellular (cell) phone. People became more closely tied to technology. Therefore, some people call the period we are living in the *Technological Age*.

Figure 3-9. Automation and computer process control are parts of the Information Age. This man is working in an IT control room.

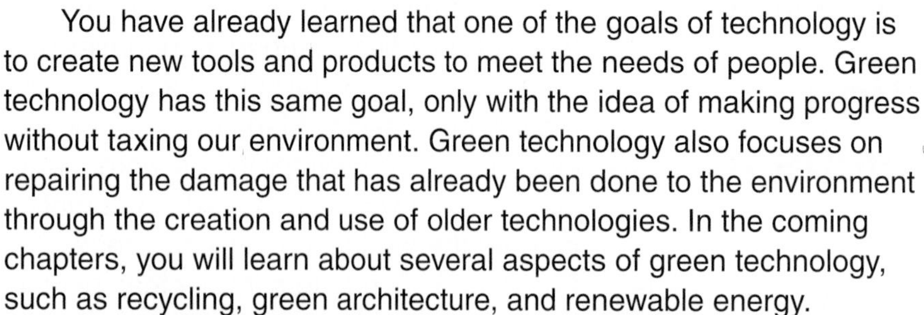

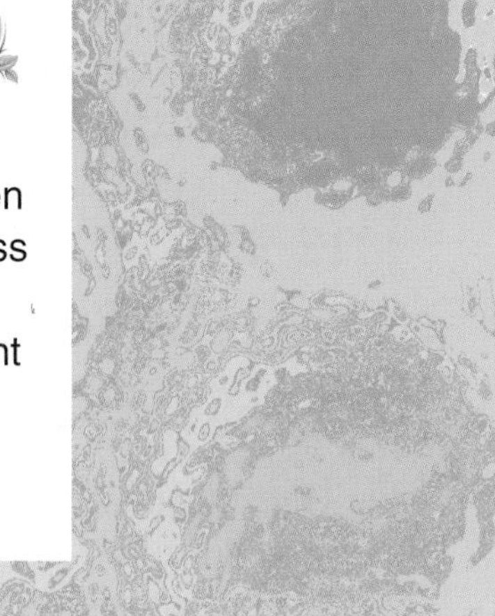

Think Green

Green Technology Overview

You have already learned that one of the goals of technology is to create new tools and products to meet the needs of people. Green technology has this same goal, only with the idea of making progress without taxing our environment. Green technology also focuses on repairing the damage that has already been done to the environment through the creation and use of older technologies. In the coming chapters, you will learn about several aspects of green technology, such as recycling, green architecture, and renewable energy.

Societal Context

Technology has always had an impact on society. Likewise, society and its expectations have impacted technology. People have different views because of the technology they use. Also, social and cultural priorities and values dictate which technological devices are developed and used.

Three major societal factors affect technology. First, there must be a societal need for the technology. See **Figure 3-10.** People develop technologies to serve their needs. All through history, the demands, values, and interests of individuals, businesses, industries, and societies have resulted in new technologies. Without the need, there would be no reason for developing the device or system.

Second, the proper resources must be available. The society must have capital (money), labor (workers), and materials available. The people must possess the knowledge needed to develop and use the technology. Proper energy resources must be available. Creative ideas for new technologies die without proper resources.

Third, the society must value the technology. Meeting societal expectations is the major influence behind the acceptance and use of products and systems. The people must feel the technology benefits them and agrees with their economic, political, cultural, and environmental concerns. They must be receptive to new ideas, devices, and ways of doing things.

Scope of Society

In ancient times, the society meant the tribe. People lived in close-knit groups and traveled little. They seldom received news from other groups. In more recent times, people have developed strong national identities. They consider themselves citizens of a nation. Present times require a broader view. Rapid communication and transportation systems have made us citizens of the world. The events in one nation or region impact all other regions. Pollution, hunger, and war ignore national boundaries. We now live in a global society, due in a large part, to technology. Understanding technology in terms of its impacts on people and societies requires several views. This understanding requires looking at technology as it relates to individuals, communities, nations, and the world.

Technology provides...
...better materials and products,
...rapid and safe transportation,
...and comfortable housing.

Figure 3-10. Technology meets societal needs, such as housing, transportation, and useful products.

Technology and the Individual

Technology impacts people in several ways. These devices and systems change the way people live and view their world. The use of technology influences people in many aspects, affecting things such as their well-being, comfort, choices, and opinions about technology's development and use. People's knowledge about and attitudes toward the uses of technology vary greatly. Their moral, social, and political beliefs often influence how they feel about technological developments.

In early history, people were satisfied with some simple technology. They were happy when they had a few tools to hunt with, a small shelter, and simple clothes. The use of inventions and innovations led to changes in society. Later generations needed more material things to be happy. They wanted more products of technology to help them. As these more recent generations acquired new technology, their needs and wants changed even more. The development of technology sometimes creates the demand for new technology.

Let's consider communication as a means of exchanging information and ideas. Once, simple writing was enough. Hieroglyphics (ancient Egyptian writing) on walls and buildings were considered adequate. Paper, ink, and pens were then invented (technological advancements). People could carry their written message with them. They started to expect more information. The few educated people started to exchange information freely. Knowledge passed from one generation to another more easily. People could exchange information with those in other areas.

In the mid-1400s, moveable type and the printing press were developed. Printed books and newspapers became available. More people could receive printed messages. Information could be the property of all people. This encouraged more people to learn to read. Technology changed their views of basic education. These devices and systems also changed the way people thought about their world. Knowledge started to become power.

Later, the telegraph, telephone, and radio were developed. The age of telecommunications was born. Television, communication satellites, the facsimile (fax) machine, and the cell phone followed. People today are not satisfied with only the written word. In fact, large numbers of people read very little. Technology has allowed them to avoid reading. Many people have become passive information gatherers. They depend on electronic media (radio and television) as sources of information. Many people feel they must be in constant contact with their friends and business associates. They use the cell phone, electronic mail (e-mail), and the Internet many hours a day.

Technology has changed our views about communication. See Figure 3-11. We no longer want occasional sources of information. People want immediate and constant contact with others. Likewise, technology has changed our views about transportation. We want quick, flexible, and inexpensive systems. Our views of products have also changed. We expect a wide selection of easy-to-use devices. People expect buildings to be heated in the winter and cooled in the summer. We want comfortable levels of light. These and other examples indicate that our expectations have changed. New technological products and systems are developed because of these expectations.

Technology and Society

Society can be viewed from several perspectives. Young children view their families as society. As they grow, their perspectives widen. Children start seeing their communities as society. Later, they might see the nation as society. The most developed view sees us as part of a global society. See Figure 3-12.

Simple Drawings

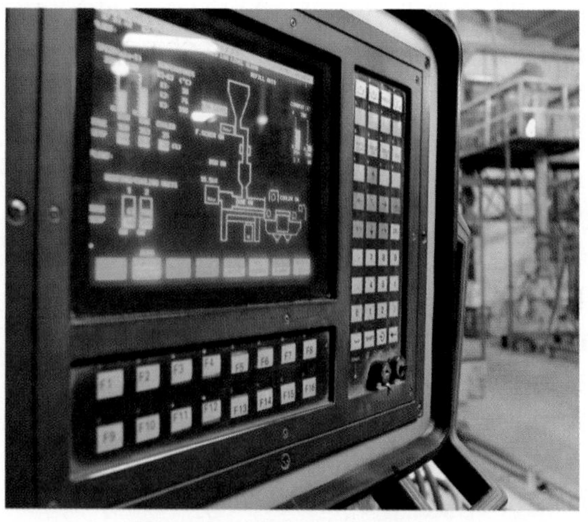

Computer-Based Systems

Figure 3-11. Communication has changed from simple writing and drawings to computer-based systems.

Figure 3-12. Over time, society's focus has changed from local to national to global.

Global society means everyone on Earth is part of one large society. Everyone's needs and wants are considered, as leaders shape policies. These leaders understand that the actions of one country affect all people. They guide policy and technological development with a global view.

The type of society people live in is partly established by technology. The regions having the resources (land, knowledge, people, materials, and machines) and desire (values and acceptance) for technology have developed more rapidly. These are called the *industrialized countries*, or ***First World countries***. Towns, states, or sections within a country having resources and desire have also prospered.

Where the resources are poor and human will is absent, technological advancement has been slower. Such countries are called ***Third World countries***. Lack of education and poor political leadership have limited their growth. Resistant personal or religious beliefs, low personal incomes, and overpopulation also slow growth. Likewise, towns and states having poor resources and leadership that has resisted change have failed to prosper.

Having all the new technology is not, however, always good. In fact, technology, by itself, is neither good nor bad. How people use it in society determines its worth. Choices about the use of products and systems can result in desirable or undesirable outcomes. Nuclear technology can be used for good in treating diseases. This technology can be used for destruction in war. Coal-fired generating plants provide electric power needed for homes and industry. They also, however, produce pollution, a source of acid rain. This has negative impacts on forests. The impacts of technology on individuals and people are discussed more fully in Chapter 26.

Functional Context

There are many types of technology developed for specific uses. The specialization of function has been at the core of many technological advancements. Each type of technology serves a different function. A product, a system, or an environment developed for one situation, however, might be useful to another situation as well. In this book, the functional uses of technology center on seven technologies. See Figure 3-13. Each of the common functional areas (or contexts) for technology is discussed in a chapter in Section 4 of this book:

- ***Agricultural technology.*** This area involves developing and using devices and systems to plant, grow, and harvest crops. Also, it includes raising livestock for food and other useful products. Broadly defined, this area can include farming, fishing, and forestry activities. Agriculture includes related biotechnology activities. These pursuits use living organisms to produce technological products.
- ***Communication and information technology.*** This context involves developing and using devices and systems to gather, process, and share information and ideas. These systems might involve communication using print or electronic (radio, television, and Internet) media. They might involve information-processing techniques used in computer applications and e-mail systems.
- ***Construction technology.*** This area involves using systems and processes to erect structures on the sites where they will be used. These structures might be buildings or civil works, such as dams, roadways, and power transmission lines. They might be residential, commercial, governmental, or industrial structures.
- ***Energy and power technology.*** This context involves developing and using devices and systems to convert, transmit, and use energy. These devices can be used to provide power, heat, light, or sound. Energy and power technology is used in residential, commercial, industrial, and governmental settings.

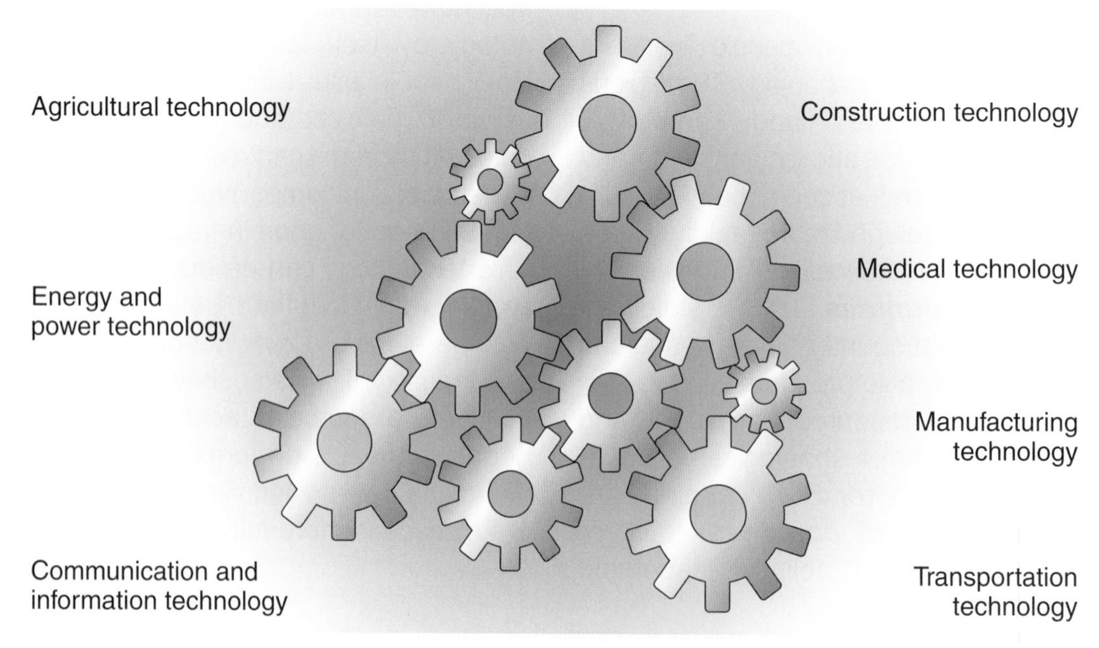

Figure 3-13. The major types of technologies.

- ***Manufacturing technology.*** This area involves developing and using systems and processes to convert materials into products in a factory. Manufacturing technology includes finding and extracting natural resources. This technology involves converting these resources into industrial materials. Finally, manufacturing technology includes converting these materials into products.
- ***Medical technology.*** This context involves developing and using devices and systems to promote health and curing illnesses. This area includes devices used to study the human body. Also, devices replacing parts of the body are part of this area. Finally, devices used to treat illnesses and medical conditions are part of this technology. The area of workplace health and well-being can be included in this area.
- ***Transportation technology.*** This area involves developing and using devices and systems to move people and cargo from an origin point to a destination. This technology includes the means of transportation (such as vehicles or pipelines) and the support systems (terminals or routes, for example). This area includes land, water, air, and space transportation systems.

Summary

The study of technology is valuable in many ways. Studying technology's history lets us see how and why things were invented and used. Studying the relationship between technology and society allows us to see how technology impacts people and cultures. The functions, or areas, of technology show us how people are using technology today. With this understanding, people can decide which technology is good and which is not appropriate. This knowledge can help them direct technological development and reduce negative impacts. Knowledge can give people the power to control the direction technology is going.

STEM Connections

Science

Select a major type of technology, such as agricultural, medical, or transportation. Identify and explain a major scientific discovery aiding in the development of the technology you selected.

Mathematics

Select a type of technology and explain how people using that technology use mathematics. For example, in construction technology, surveyors use trigonometry to measure the distances across rivers.

Curricular Connections

Social Studies

Make a display or poster presenting the major technologies shaping each age. Include the Stone Age, Bronze Age, Iron Age, Industrial Age, and Information Age.

Social Studies

Trace the development of a major type of technology over recorded history. You can select a technology such as housing, transportation, or food preservation.

Activities

1. Prepare a chart with the five major ages (Stone, Bronze, Iron, Industrial, and Information) across the top. List the major types of technologies (agricultural, communication and information, construction, energy and power, medical, manufacturing, and transportation) along the side. List the major technological advancements that have happened for each technology in each age. You might want to work in groups to complete this activity.
2. Build a model of a technological advancement contributing to the growth of one of the technology areas (agricultural, communication and information, construction, energy and power, medical, manufacturing, or transportation).
3. Build a diorama showing life for the people in one of the ages (Stone, Bronze, Iron, Industrial, or Information). Show aspects such as a typical home and transportation devices.

Test Your Knowledge

Do not write in this book. Place your answers to this test on a separate sheet of paper.

1. Name the three major contexts in which technology can be viewed.
2. How did the Stone Age get its name?
3. Copper and tin were major metals used in the _____ Age.
4. Select the statement best describing the Iron Age:
   A. The period of history when the focus moved from farming to manufacturing.
   B. The period of history when the focus moved from manufacturing to information processing.
   C. The period of history when villages and towns started to appear and metal tools replaced stone ones.
   D. The period of history when building techniques improved, road systems were developed, and new cooking ware and agricultural tools were created.
5. The steam engine was a major development in the _____ Age.
6. The computer is the driving technology in the _____ Age.
7. We live in a (local, national, global) society.
8. Summarize some ways in which technology has changed our expectations of things, such as communication and transportation.
9. Technology, in itself, is neither good nor bad. True or false?
10. Define each of the following:
    A. *Agricultural technology.*
    B. *Communication and information technology.*
    C. *Construction technology.*
    D. *Energy and power technology.*
    E. *Manufacturing technology.*
    F. *Medical technology.*
    G. *Transportation technology.*

Reading Organizer

Copy the following chart to a separate sheet of paper. In the Technological Object column, list at least 5 everyday technological objects you use. In the Technological Category column, write which of the seven areas of technology your examples fall into.

Technological Object	Technological Category
Example: Cell phone	Communication and Information Technology

SECTION 2 TECHNOLOGY AND SOCIETY

Resources are the many kinds of supplies, materials, and services used to get a task done. Whether you are writing a letter, cooking dinner, or building a house, you need resources. Consider the simple process of writing a letter. What resources do you need?

First, there is the matter of tools or machines. At the very least, you need a pencil. If you use a pencil, you might need a sharpener. You might even decide to use a machine to write your letter—a computer with word processing software.

Next, you need a chair for sitting and a surface on which to write or type. Also, you need paper and an envelope. Mailing your letter requires help from people at the post office. If you use your computer, you need energy for power.

Anything else? What about information? There is a good chance you will write about something you have heard, read, or experienced.

Technology is the system that puts the resources to use. No matter what we are doing, we use the same basic resources. You will see how resources are tapped and tied into one another as you read the next five chapters.

Technology Headline:

Holography

Holography, a technology invented in the late 1940s, continues to evolve today, offering exciting possibilities in the world of communication. Holography is a science that focuses on the recording and recreation of an image in such a way that it appears to be in 3D. Lasers are used to record the object upon a special photographic plate. When the plate is moved, the 3D image, or hologram, appears to move as if it were a 3D object.

Holograms have many common applications. They are used to validate certain types of currency, such as the Euro. Barcode scanners found in grocery and department stores worldwide employ holographic lens technology. Holograms also appear on government-issued IDs and credit cards to make it harder to create forgeries. They are even used as works of art! As you move any of these items back and forth, the design of the hologram seems to "jump out" at you.

Despite numerous modern-day applications, scientists continue to work with holograms, developing new uses for this interesting technology. Advancements in holographic technology will improve current uses. For example, the holograms used as the secure "signature" on driver's licenses are currently the same for every license issued by a particular state. In the future, these holograms will be unique to each individual's license. These holograms can be checked electronically to ensure they are legitimate forms of identification.

Another developing possibility in the field of holography is the use of holograms to make a better television and motion picture image. Liquid crystal displays (or LCDs) in televisions will become brighter and whiter. This improvement will produce a picture that boasts better quality overall.

Holography will also advance cell phone technology. After downloading a special application, smart phone users will be able to record video messages that will then play as 3D representations of the recorded item. Perhaps it will come in the form of a friend listing off items needed from the grocery store, or a self-recording reminding you of an upcoming doctor's appointment. When this holographic message is replayed, it will pop out from the screen as a miniature, 3D version of the recorded person. Rather than an audio recording or text message, you will receive a recording that is like a visual voicemail. This type of advanced communication could also be applied to video chatting—it will be as if the person you are talking to online is actually sitting right next to you!

Although the technology of holography was first introduced in 1947, it continues to grow and change today. The science of holography can be applied to many fields. This diversity presents many exciting uses in the future. Who knows what scientists will come up with next—the possibilities are endless!

CHAPTER 4

Tools and Technology

Did You Know?

The metric system was developed in France during the reign of Napoleon in the 1790s. This system missed being adopted by the United States by one vote in the Continental Congress in the late 1700s or early 1800s.

Objectives

The information given in the chapter will help you do the following:

- Summarize the unique abilities of humans.
- Explain how humans are toolmakers and tool users.
- Compare tools, mechanisms, and machines.
- Give examples of the six major types of primary tools.
- Summarize the use of mechanisms as force and distance multipliers.
- Explain the six mechanisms, or simple machines.
- Recall the major parts of a machine tool.
- Give examples of the major types of machine tools.
- Explain how to properly maintain and store tools.

Key Words

These words are used in this chapter. Do you know what they mean?

distance multiplier
drilling machine
force multiplier
grinding and sanding machines
inclined plane
lever
machine tool
mechanism
milling and sawing machines
planing and shaping machines
pulley
screw
shearing machine
simple machine
tool
turning machine
wedge
wheel and axle

Preparing to Read

In this chapter, you will learn about the different categories of tools and machines. As you read the chapter, outline the different types of tools and machines discussed.

Humans are unique in the world of living things. People have the ability to reason. Using this ability, we can predict events and plan actions. For example, you can say, "If this is true, that must also be true," "If I do this, I can expect that to happen," or "I must do this first and then do that next."

People can also think about the future. They can plan activities over a period of time and decide to do something at a later date. In addition, humans can think of different ways to do the same task. They can offer alternatives for action. People can adjust their actions to meet different situations and events. Their minds connect things, actions, and relationships. All these traits combine to produce an ability only humans have. This ability is called *rational thought*.

A second unique ability humans have is complex language. People use this language to describe their thoughts to other people. Early language was based on sounds that became the spoken language. Later, symbols were used to represent these sounds. These symbols evolved into an alphabet, the basis for written language. See **Figure 4-1.**

Humans also have value systems. They have a sense of what is right and what is wrong. People can think, "It is okay to do this. I shouldn't, however, do that." This reasoning is called *moral judgment*.

The abilities to reason, use complex language, and make moral judgments are positive differences between humans and other mammals. Humans, however, lack some basic survival traits and innate knowledge other animals have. People lack an inherited "blueprint" of action on how to lead our lives. Animals can survive with their built-in knowledge, called *instincts*. Beavers instinctively know how to build dams. Bees use their instincts to collect pollen. Salmon know how to return to their spawning ground after several years in the ocean. Birds know the path from their summer homes to winter habitats. Humans, however, have to learn most of

α β γ δ ϵ ζ η
ϑ ι κ λ μ ν ξ
ο π ρ σ τ υ φ
χ ψ ω

Α Β Γ Δ Ε Ζ Η
Θ Ι Κ Λ Μ Ν Ξ
Ο Π Ρ Σ Τ Υ Φ
Χ Ψ Ω

Figure 4-1. The Greek alphabet.

these things. We cannot instinctively build homes, grow food, or make clothing.

In fact, humans are poorly equipped physically to meet the challenges of the world. People cannot lift much weight, run very fast, withstand cold temperatures very well, or see very far. Other animals are better equipped to survive in the natural environment. Eagles can see movement of prey from hundreds of feet in the air. Lions have great speed to catch food. Geese have the strength to fly hundreds of miles toward warmer climates in the winter. Ants can lift several times their weight. Bears have heavy fur to fight the cold. In contrast, modern humans cannot survive with only their natural equipment. They would starve or freeze to death without help.

Toolmakers

To overcome physical weaknesses, humans have another special ability. They can design, make, and use tools. See Figure 4-2. ***Tools*** are devices people develop and use to do specific tasks.

Over the history of humankind, people have developed many types of tools. Early humans used things they could find to make their tools. They used sticks, rocks, bones, and other natural elements for tools. These humans used pointed sticks to hunt animals and spear fish, tree limbs for clubs, and animal bones as needles to make clothing.

As civilization advanced, tree limbs were used to plow the soil. Weapons were made from copper and other metals. Clay vessels were shaped and fired. Hammers were made from the newly developed steel.

Today, people have tools for every job. Dentists have a set of tools to work on their patients' teeth. Surgeons have special tools to perform operations. Auto mechanics use different tools from what plumbers use. Have you ever seen a carpenter at work? Carpenters use many different types of tools. They use a heavy

Figure 4-2. Humans have designed, built, and used many tools and machines. (Caterpillar, Inc.)

hammer to drive stakes into the ground. A smaller hammer or an air nailer is used to nail 2 × 4s together. Still another hammer or nailer is used to attach the trim around windows and doors. Each profession has its own set of tools. These tools can be understood better by looking at the major differences among a tool, a mechanism, and a machine. See Figure 4-3:

- A tool is a device people design and use to complete a task.
- A ***mechanism*** is a device people design and use to adjust or power a tool.
- A machine is a combination of tools and devices people design and use to complete complex tasks.

Figure 4-4. Over time, humans have developed many types of tools. This offshore drilling platform is a special tool of the petroleum industry. (American Petroleum Institute)

TOOLS

Throughout history, humans have survived by using their hands, arms, and brains. In earliest times, the hand held a rock to form a crude hammer. The brain directed the arm to move in a controlled manner. Later, a stick was attached to the rock to make the first hammer. Exchanging the rock with a sharp stone produced a hatchet. Each of these examples is a human-made device called a *tool*. See Figure 4-4. Over time, humans have developed many different types of tools:

- Tools used with language, such as pens, pencils, and printing presses.
- Tools used in education, such as projectors, models, and maps.
- Tools used in business and trade, such as calculators, scales, and money.
- Tools used in religion, such as vessels and special clothing.
- Tools used in art, such as easels, paintbrushes, modeling tools, and sculpture chisels.

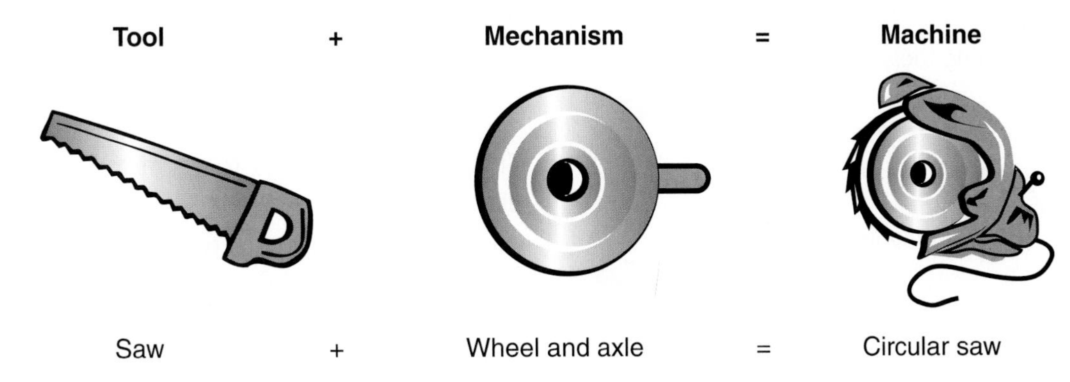

Figure 4-3. People use tools, mechanisms, and machines to do work. (Combining a tool and a mechanism produces a machine.)

- Tools used by the government, such as military weapons and police equipment.
- Tools used in games and sports, such as basketball goals, tennis rackets, soccer balls, and baseball bats.
- Tools used in the pure sciences, such as telescopes, microscopes, and chemical apparatus.
- Tools used in technological systems, such as agricultural, communication, construction, energy conversion, manufacturing, medical, and transportation equipment.
- Tools used in managing companies, such as computers and word processors.

Each of these groups of tools makes human action more efficient. These tools extend the human potential to do jobs.

All tools can be traced back to another type of tool. See Figure 4-5. We call these older tools *primary tools*. They are the tools people use to make new tools. Without them, humans could not develop new tools and machines to make lives easier.

Figure 4-5. Types of tools. We can use them to build other tools and mechanisms.

Measuring Tools

Early humans lived in small groups that moved from place to place, searching for food. These people needed only a few tools to hunt game and gather food. As the population grew, however, the nomadic life gave way to permanent settlements. With this growth of civilization, people needed new tools to measure things. They had to plot fields to farm. These people had to measure material as they made houses. Grain placed in the village storehouses had to be weighed.

Two other types of measurement are also important. People measure distances and relationships. We measure the size of objects (distance) and how one surface relates to another (for example, squareness).

Measuring Distances

Measuring distance tells us how far it is from one point to another. Humans have developed a number of different tools to measure distances. See Figure 4-6. Some of these tools are considered precision measurement devices. They provide very accurate measurements. Most precision measurement devices measure distances in thousandths (1/1000) of an inch or smaller. Other measurement devices are nonprecision devices. They give us measurements in fractions (such as 1/8, 1/16, and 1/32) of an inch. See Figure 4-7.

The ruler gives us standard measurements. Precision measurements are often made with a micrometer. To use a micrometer, the part to be measured is placed between the spindle and the anvil. The spindle is brought into contact with the part. The measurement is read on the micrometer barrel. See Figure 4-8.

Measuring Relationships

The relationship between two surfaces is often important. People want to know if a part is square. They need to know that a

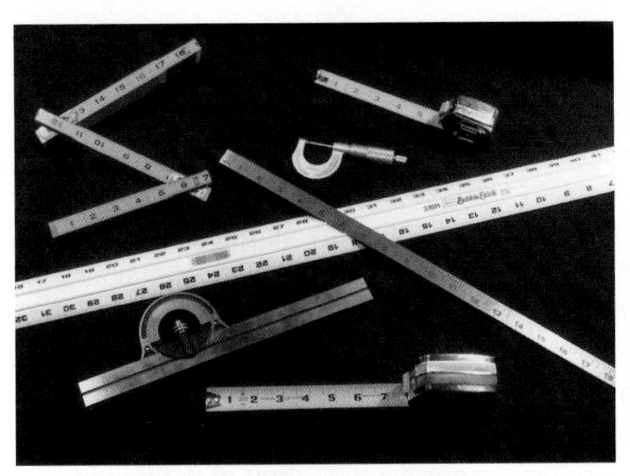

Common Measuring Tools

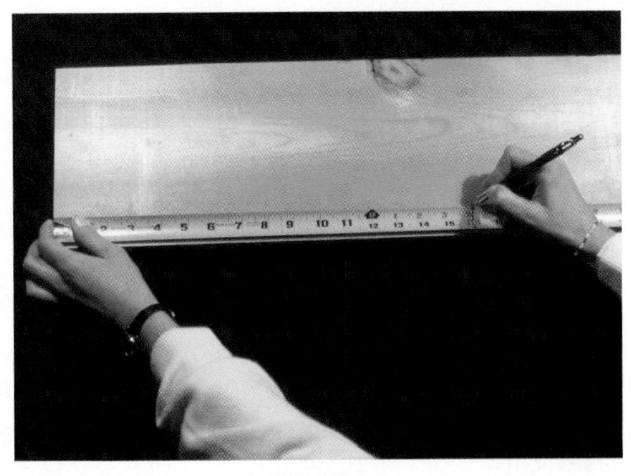

A Steel Tape

Figure 4-6. There are many kinds of common measuring tools. A steel tape measures a board to length.

Precision Measurement

Nonprecision Measurement

Figure 4-7. The dyes in this stamping press were developed using precision measurement. The package designer here is using standard measurements. (Dana Corporation)

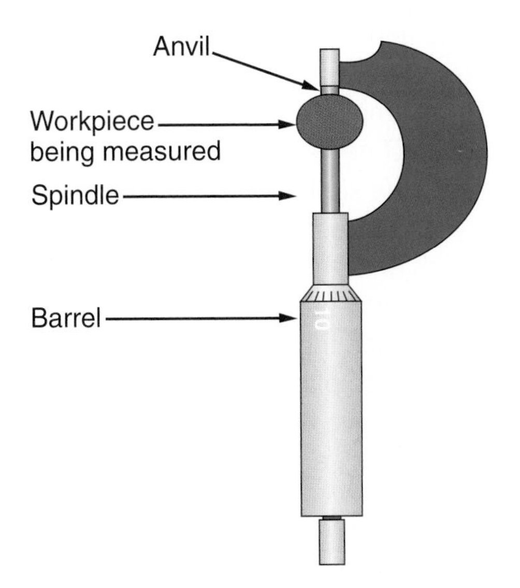

Figure 4-8. A micrometer is used for precision measurement.

side is 90° (at a right angle) to an end. In another case, it is very important that the ends of parts of a picture frame are at 45°. These measurements are made with tools called *squares*. See Figure 4-9.

Safety

Be careful when using cutting tools. Never rub fingers over cutting edges. Keep your hands behind any slicing or shearing tools.

Cutting Tools

Cutting tools remove material to size and shape parts. Each tool cuts away unwanted material until the part has a

TECHNOLOGY EXPLAINED

laser: a device that emits a beam of coherent, monochromatic light.

We are all probably familiar with lasers in one form or another. The word *laser* stands for "light amplification by stimulated emission of radiation." A laser is a device that amplifies (strengthens) light. This device is based on the findings of Albert Einstein and Niels Bohr. When atoms are exposed to an outside source of energy, such as electricity or light, the electrons become excited. This raises the electrons to a higher energy level within the atom. When the electrons fall back to their original energy level, they give off light. More light is emitted when this light hits another atom.

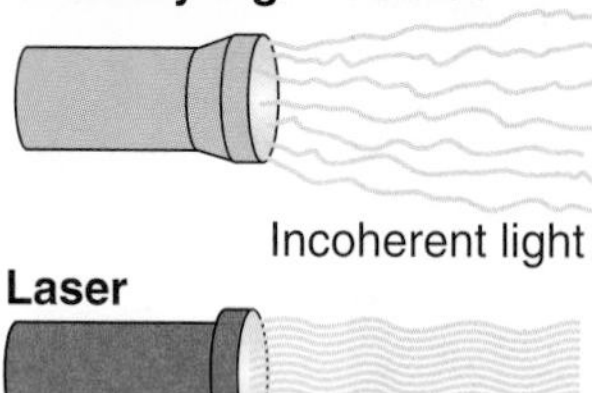

Figure A. Lasers produce coherent light, meaning the light travels in the same direction as the triggering light. This light, thus, amplifies the triggering light. In incoherent light systems, such as the Sun and the electric lightbulb, light travels in different directions, thus reducing its power.

Scientists discovered that all the atoms of a material give off light that is the same wavelength. The light has one color because it has one wavelength. Thus, laser light is monochromatic. Laser light is also coherent, meaning all the wave crests line up. See **Figure A.** Beams of laser light are very intense and do not spread out as they pass through the air. The heat from the light can become so strong, it can burn a hole in a rock.

The first practical laser used a ruby rod. See **Figure B.** Ruby is an aluminum-oxide crystal. Some chromium atoms that replace a few aluminum atoms in the crystal cause its deep red color. The ruby rod is polished at each end. The ends are coated with a reflective material. One of the ends has a thinner coating than the other. To start the laser, outside energy excites the chromium atoms. Some of the light waves the chromium atoms emit strike the reflective ends. The light waves hitting the ends bounce back through the ruby rod. The waves strike other atoms, which become excited. The light is amplified as it bounces back and forth between the reflective ends of the rod. Once the light becomes strong enough, some light passes through the end of the rod with the thin coating.

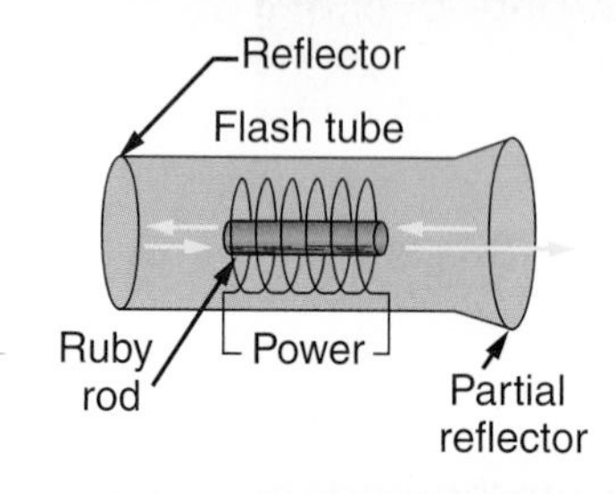

Figure B. The first laser used a ruby rod excited by a flash tube. Both ends of the rod are reflectors. One end, however, is a partial reflector. This allows the laser light to escape the rod when it has enough energy.

We use many types of lasers today. Lasers are grouped by the amplifying medium they use. For example, the ruby laser used a ruby rod as the amplifying medium. Other lasers use gases, dyes, or semiconductor materials to amplify the light. All lasers operate on the same basic principle as the ruby laser. Each type of laser gives off light that has a different wavelength and, therefore, a different color.

We use lasers in more applications every day. For example, we use them for measurement. We also use them to send messages through fiber-optic cables. Compact disc players and some computer disc drives use lasers to store and retrieve data. People also use lasers for surgery, for welding, for cutting, and in bar code readers.

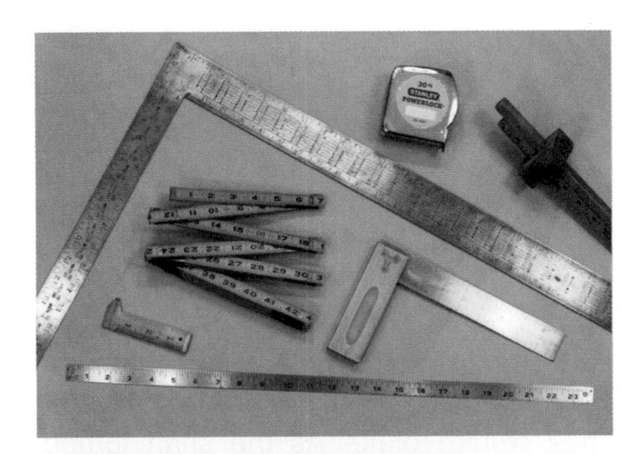

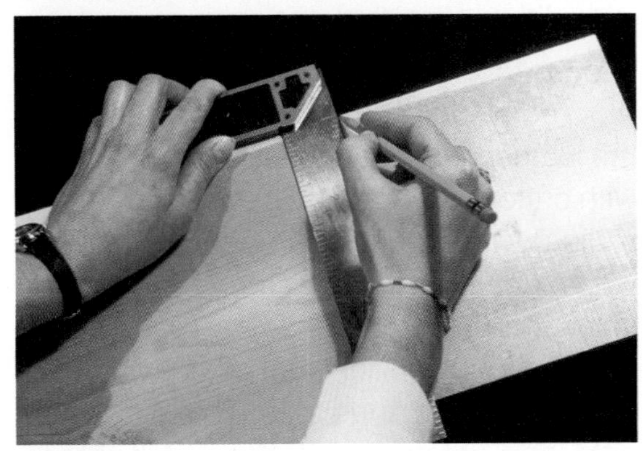

Figure 4-9. Squares are tools measuring angles. Note how a try square is used to mark a 90° angle on a board. Squares are often combined with rules.

desired shape. Cutting tools include three major types. These are sawing, slicing, and shearing tools. See Figure 4-10.

Sawing Tools

A saw uses a set of teeth to cut the material. The tooth is a sharp projection (point) on a body. The teeth can be arranged along a strip. Examples are hand, coping, scroll, or band saws. Other saws have the teeth on a disc.

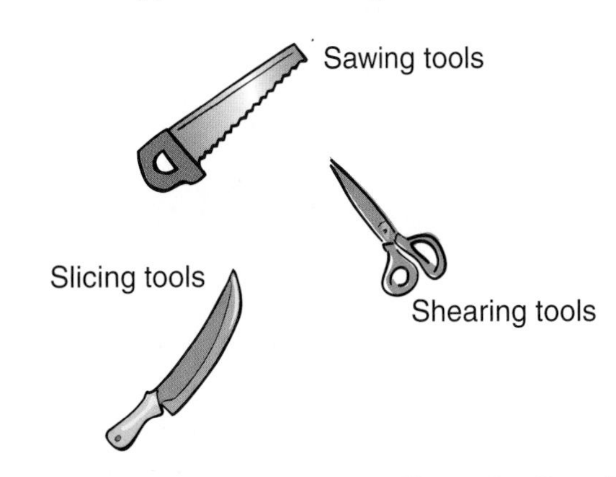

Figure 4-10. These are the different families of cutting tools.

Slicing Tools

Slicing tools use a sharp, wedge-shaped edge to separate the material. The wedge cuts away unwanted material in the form of shavings. Typical slicing tools are knives, chisels, carving tools, and woodworking planes. See Figure 4-11.

Shearing Tools

Shearing tools fracture material between two opposing edges. The workpiece is placed between the edges. The knives coming together cause the material to separate. Common shearing tools are tin snips and scissors. See Figure 4-12.

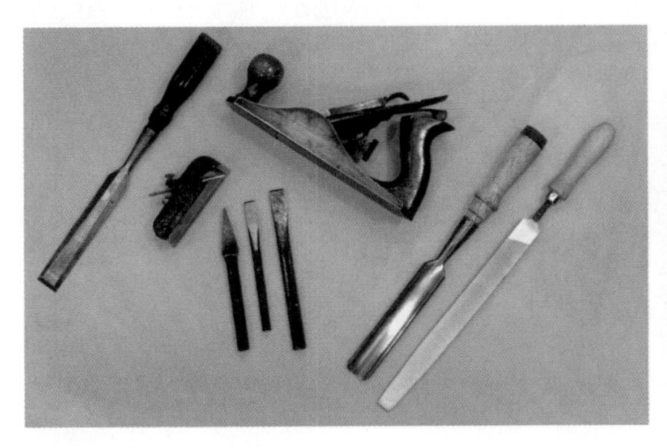

Figure 4-11. Typical cutting tools include planes, chisels, and files. Hand planes smooth board surfaces.

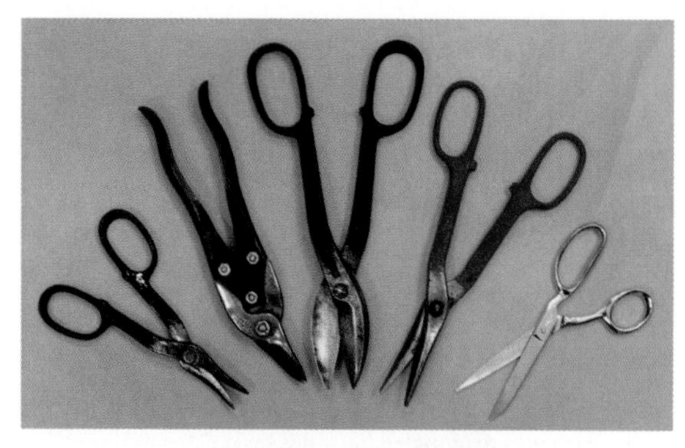

Varieties of Tin Snips

Cutting Sheet Metal

Figure 4-12. Shearing tools use two sharp opposing edges to cut the materials. There are several varieties of tin snips. Tin snips cut sheet metal.

Drilling Tools

One of the first tools ancient humans developed was the drill. The first one was no more than a sharp, pointed stone attached to a wooden shaft. The tool user rotated the shaft in the palm of the hands. The stone tip was pointed downward and rested on the piece to be drilled. When the hands were rubbed back and forth, the shaft would rotate. As the shaft turned, downward pressure forced the stone point into the work.

Today, drilling tools and machines use the same action. A steel shaft is formed with cutters on its end. The shaft is rotated, while downward pressure forces it into the workpiece. This action cuts a series of chips, producing a hole.

Common drilling tools are twist drills, spade bits, and auger bits. These are often held and rotated in a hand drill, brace, portable electric drill, or drill press. See **Figure 4-13.**

Gripping Tools

The human hand is a natural gripping device. People hold and twist things with their hands. The hand, however, has

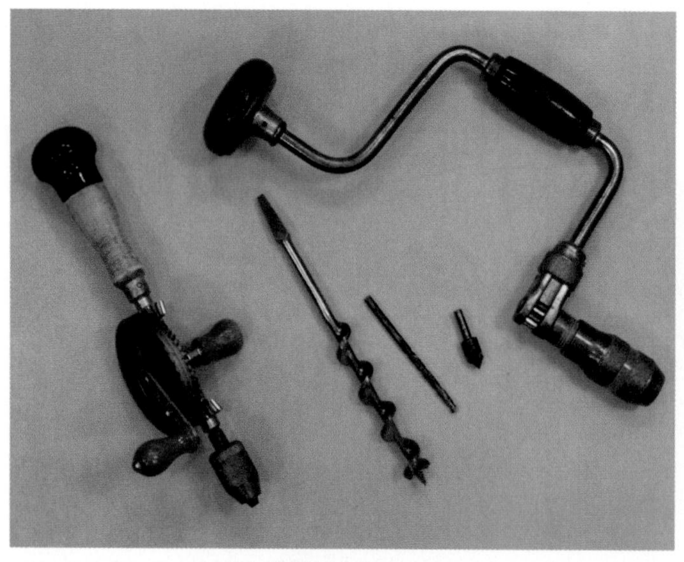

Drilling Tools

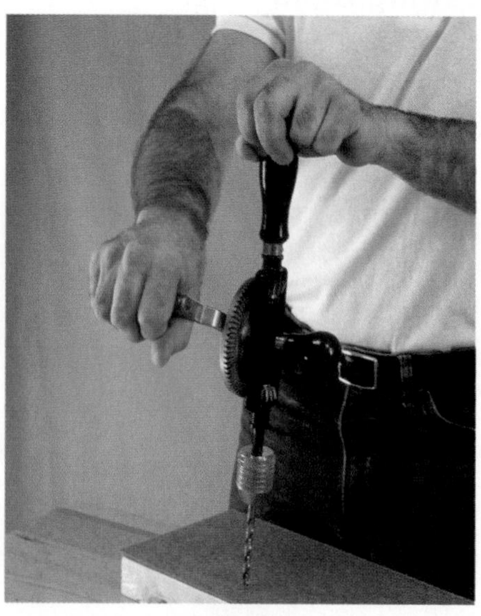

A Drill

Figure 4-13. Here is a sample of drilling tools. Drills must be rotated to make them cut.

Career Highlight

Fishing-Boat Captains

The Job: A fishing-boat captain manages the overall fishing operation. This includes deciding the fish to seek, the location of appropriate fishing grounds, the fishing method used, the duration of the trip, and the sale of the catch. The captain checks the fishing boat for seaworthiness, purchases needed supplies, hires the crew, and obtains necessary licenses.

Working Conditions: Fishing operations are conducted in many regions and under a variety of environmental conditions. Often, weather can cause hazardous conditions. Many times, help is not readily available when injuries occur.

Education and Training: Captains of large commercial fishing vessels and charter sportfishing boats must complete an approved training course. The U.S. Coast Guard must license them. Although it is not required, students can enroll in secondary school vocational or technical programs or community college and university programs in fishery technology.

Career Cluster: Agriculture, Food & Natural Resources

Career Pathway: Natural Resources Systems

serious limitations. The human hand can exert only a limited amount of force. Also, the human hand is easily damaged when it holds rough materials. Therefore, holding and gripping tools have been developed to extend the power of the hand. Pliers are familiar gripping tools.

Holding Tools

Holding tools replace the gripping power of the hand. They are used to hold an object in place. These tools squeeze the part and keep it from moving. Typical holding tools are vises and clamps. See Figure 4-14.

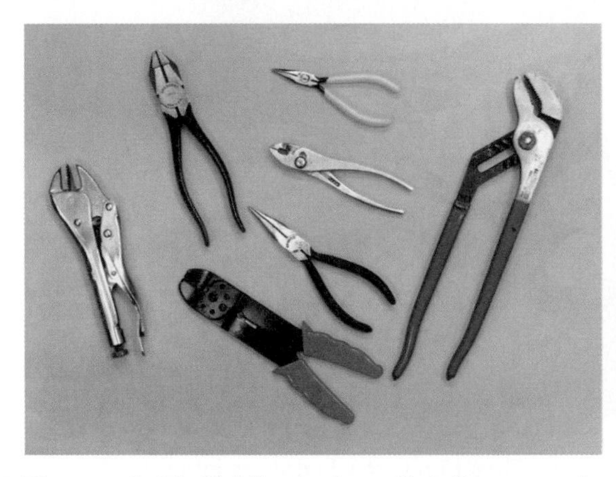

Figure 4-14. Holding tools multiply the squeezing force of the hand. There are many types.

Turning Tools

Turning tools are used to position objects or tighten fasteners. Typical turning tools are wrenches and screwdrivers. Wrenches are made adjustable or in fixed sizes. Fixed-size wrenches usually come in sets to fit common bolts and nuts. They are either open-end or box styles. Commonly used adjustable wrenches are open-end wrenches and pipe wrenches. See **Figure 4-15.**

Screwdrivers are commonly used to set or tighten wood, sheet metal, and machine screws. They are available in several lengths and blade sizes. Screwdrivers are also made to fit different slots in screw heads. There are screwdrivers fitting standard screws and Phillips-head screws. See **Figure 4-16.**

Pounding Tools

From early times, people needed to apply a striking force to objects. They needed to pound things into the ground, break them, or shape them. This need to pound objects gave rise to the development of pounding tools. Typically, these tools consist of a heavy head attached to a handle.

Today, we call these tools *hammers*. Hammers come in many types and styles:

- Claw hammers for driving all types of nails.
- Sledgehammers for driving stakes.
- Shingling hammers for attaching roof shingles.
- Mallets for driving woodworking chisels.
- Ball-peen hammers for striking cold chisels.

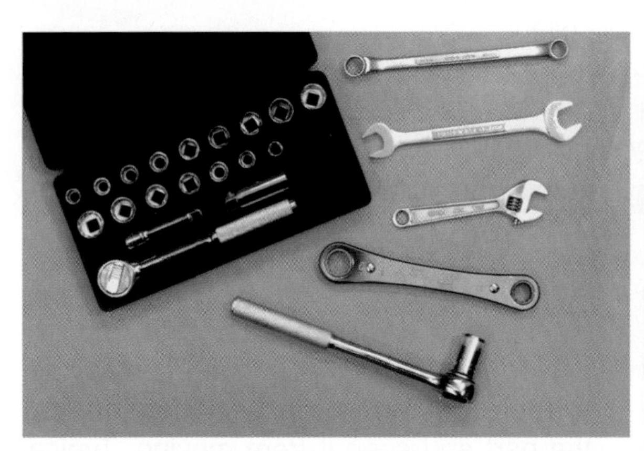

Figure 4-15. These tools are useful for assembling parts using bolts and nuts.

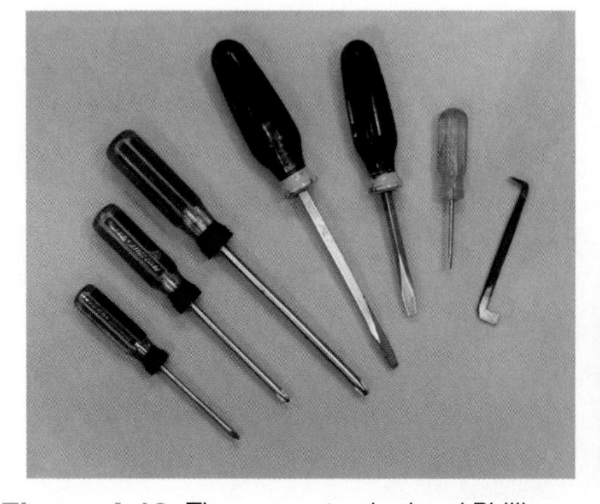

Figure 4-16. These are standard and Phillips screwdrivers. They turn screw fasteners into parts.

- Rubber or plastic mallets for striking parts to align them.

Several types of hammers are illustrated in Figure 4-17.

Polishing Tools

Many parts and products require smooth surfaces. This type of surface finish can be produced using scrapers and abrasive grits. These tools remove small amounts of material to improve the surface of the material. Hand and cabinet scrapers use a curled edge (burr) to scrape away the unwanted material.

Sanding and grinding tools use abrasives (mineral grit) to create uniform scratches on the material. The action replaces dents and large scratches with small, straight scratches. As these scratches become very small, they cannot be seen. There is a degree of roughness. The eye and the hand, however, tell us the material is smooth. Typical abrasive tools are loose and sheet abrasives, sharpening stones, and buffing compounds.

Mechanisms

Hand tools served early humans well. As civilization grew, however, demand for food and products also grew. People using hand tools could not meet the increasing demand. Machines to produce more goods were needed. These improvements required technological advancements.

There is not, however, a direct step from a hand tool to a machine. The hand tool must be combined with a mechanism to produce a machine. What is a mechanism? A mechanism is a basic device controlling or adding power to a tool. Science calls these mechanisms ***simple machines***. These simple machines, or basic mechanisms, multiply the force applied or distance traveled. An understanding of them helps us understand more complex machines. There are six simple machines that work off two basic principles. Levers, wheels and axles, and pulleys use the lever principle. Inclined planes, wedges, and screws use the inclined plane principle. See Figure 4-18.

Principle	Simple Machine
Lever	Lever Wheel and axle Pulley
Inclined plane	Inclined plane Wedge Screw

Figure 4-18. The six simple machines are based on two basic principles.

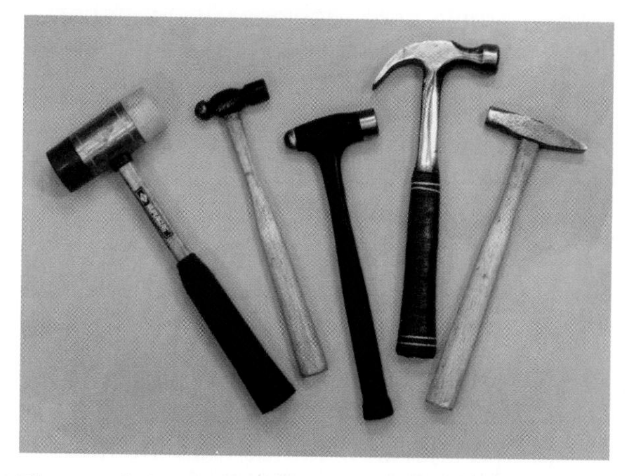

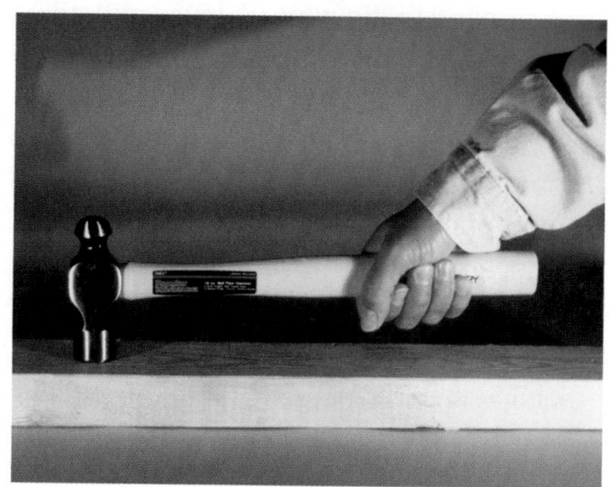

Figure 4-17. Here are some of the many hammers we use. They pound and drive certain types of fasteners.

Career Highlight

Mechanical Engineering Technicians

The Job: Mechanical engineering technicians use their knowledge of science, mathematics, technology, and engineering to solve the technical problems many manufacturing companies face. They work with engineers to design, develop, and test manufacturing equipment and consumer products. These technicians might work in product development, process development, production planning, plant layout, or other similar areas.

Working Conditions: Engineering technicians generally work a standard 40-hour workweek. They work in product-development laboratories and manufacturing plants.

Education and Training: Most mechanical engineering technicians have a two-year associate's degree or a four-year bachelor's degree in engineering technology. These programs include courses in science, engineering, manufacturing materials and processes, and mathematics.

Career Cluster: Manufacturing

Career Pathway: Manufacturing Production Process Development

Levers

Have you ever seen someone pry open a crate using a crowbar? If you have, you have seen a ***lever*** in action. See Figure 4-19. A lever, similar to all simple machines, is a force or distance multiplier. This simple machine can increase the force applied to the work. A lever makes us stronger than we really are. This is using it as a ***force multiplier***.

A lever can also let us change the amount of movement created. A small amount of movement at one end produces a large amount of movement at the other end. This is using a lever as a ***distance multiplier***.

This simple machine is a device consisting of a lever arm and a fulcrum. The fulcrum is a pivot point on which the lever arm rotates. A lever consists of the lever arm, the fulcrum, the load to be moved, and the force to be applied. There are three basic arrangements for these components of a lever. These arrangements are called *classes of levers*. See Figure 4-20. Scissors and pry bars use the principle of the first-class lever. Wheelbarrows and hand trucks are devices using second-class levers. When you use brooms or baseball bats, you are using third-class levers.

Levers can be force or distance multipliers:

- **Force multiplier.** See Figure 4-21 (top). The fulcrum of this first-class lever is located near the load to be moved. Thus, a small amount of force, moving a greater distance, moves a load a shorter distance than the force movement.

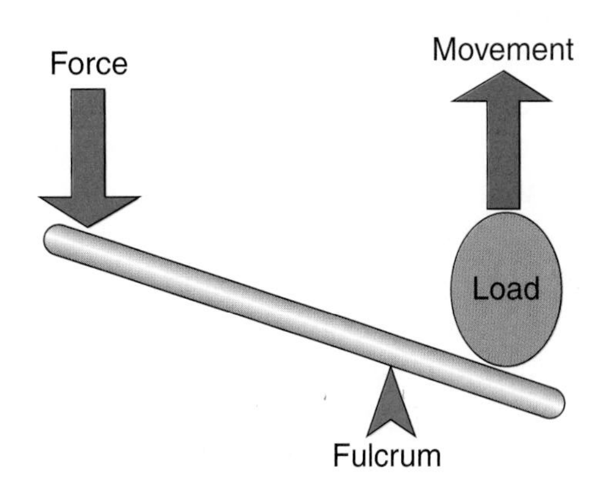

Figure 4-19. Levers multiply force. The bar on the left is an application of the lever illustrated on the right.

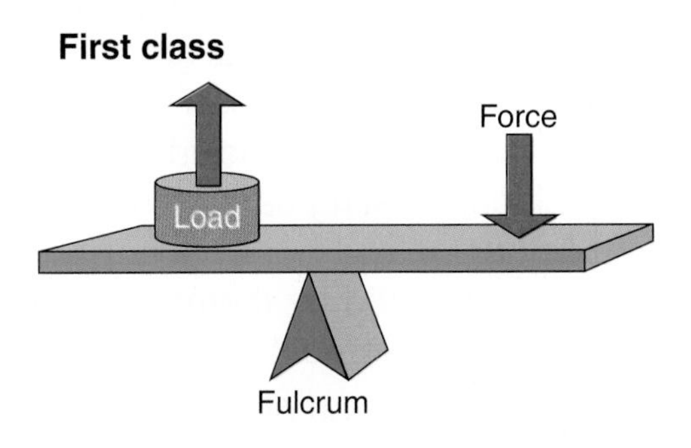

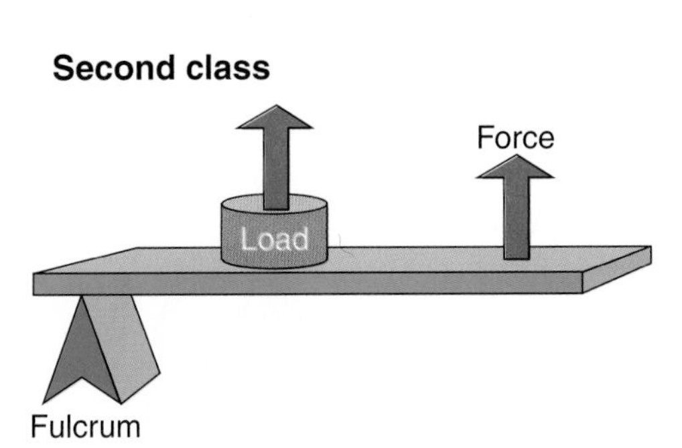

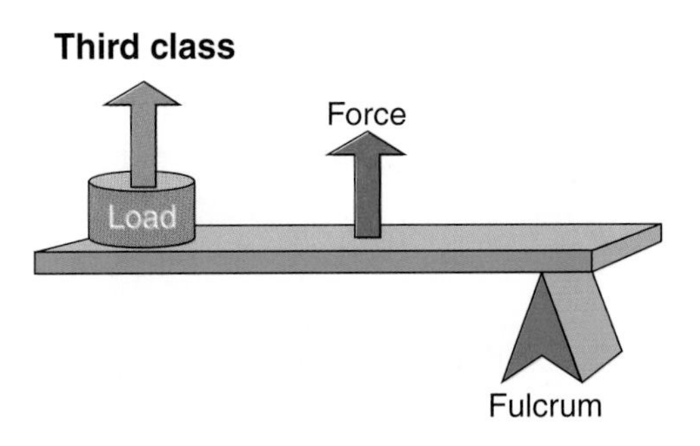

Figure 4-20. These are the three classes of levers. Try to think of tools you use fitting these classes.

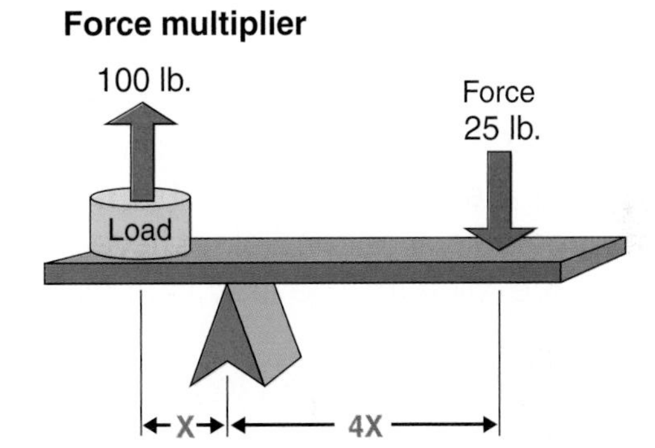

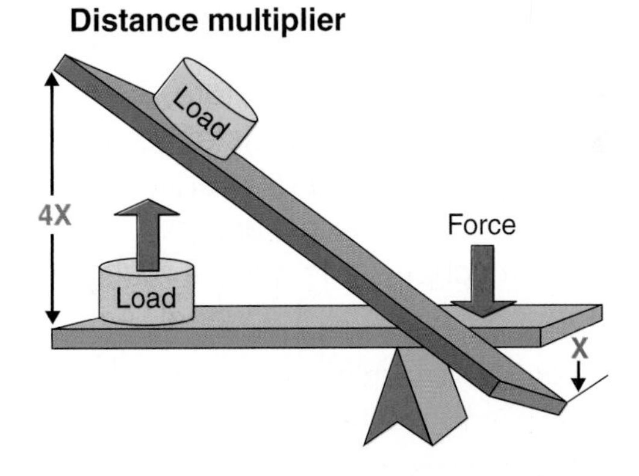

Figure 4-21. We use levers to multiply force and distance. On the top, a 25-pound force is multiplied four times. On the bottom, great force applied on the right moves a short distance to move loads four times further.

- **Distance multiplier.** See Figure 4-21 (bottom). The fulcrum is located near the force applied to the lever. Thus, a force moving a short distance can cause a load to be moved a much greater distance. Of course, a much larger force must be applied to multiply the movement of the load.

Look around the school laboratory for examples of levers in use. Did you identify pliers, tin snips, or claw hammers (when pulling a nail)?

Wheels and Axles

The ***wheel and axle*** is another mechanism, or simple machine. This mechanism consists of a shaft attached to the center of a disc. See Figure 4-22. This mechanism operates a second-class lever. The axle is the fulcrum. The load is applied to the wheel or axle.

If the force is applied to the axle, the mechanism becomes a distance multiplier. One revolution of the axle causes the wheel to rotate one time. The circumference of (distance around) the wheel, however, is greater than the axle. Therefore, the mechanism moves a greater distance. Bicycle drives and automotive differentials (gears turning the axle) use this action.

The mechanism is a force multiplier if the force is applied to the wheel. A screwdriver is a good example of this action. Try to drive a screw by gripping the shaft (axle) of a screwdriver. Repeat the task, gripping its handle (wheel). You will find it is much easier to turn the mechanism by applying the force to the wheel. This principle is used for automobile steering wheels, other control knobs and wheels, and woodworking braces. See Figure 4-23. Note how wheels and axles can be used as both force and distance multipliers. Can you think of other examples?

Pulleys

A ***pulley*** is a third type of simple machine, or mechanism. This simple machine is a wheel with a grooved rim attached to a loose axle. People use pulleys for many tasks, including raising sails on ships, hoisting cargo, and pulling parts together.

Pulleys are used by themselves or in sets. They can do three things:

- Change the direction of force.
- Multiply force.
- Multiply distance.

Figure 4-22. The wheel and axle is an application of the lever.

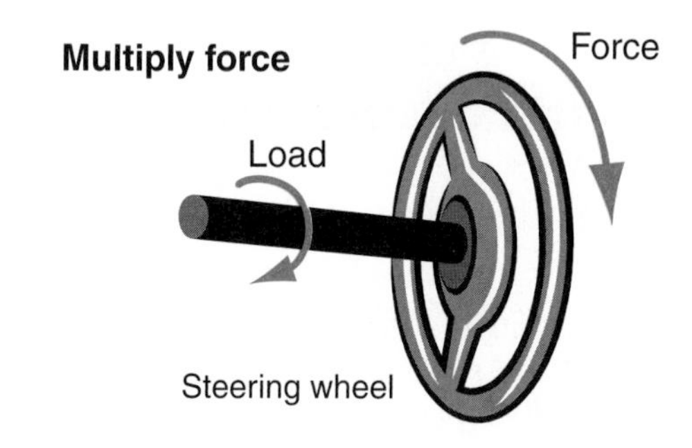

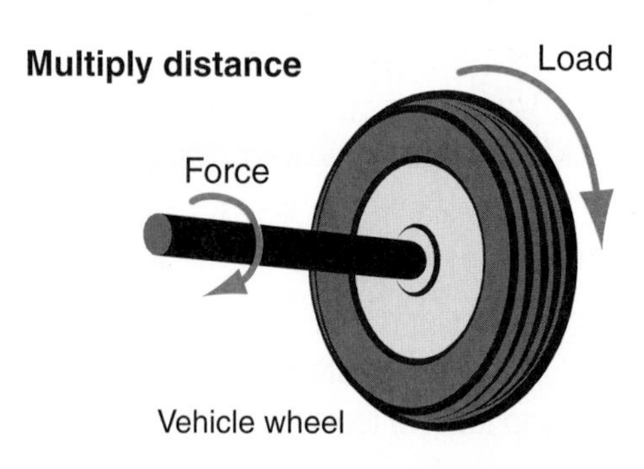

Figure 4-23. Similar to the lever, the wheel and axle can be either a force multiplier or a distance multiplier.

Career Highlight

Welders

The Job: Welders apply heat to metal pieces to melt and fuse them together. They use many types of welding, including arc, gas tungsten arc, and gas metal arc. Skilled welders often work from drawings and use their knowledge and skill to produce the specified weld.

Working Conditions: These workers normally work in well-ventilated areas. Welders can be exposed to a number of hazards, however, including the intense arc light, poisonous fumes, and hot materials. They must wear safety shoes, goggles, hoods with protective lenses, and other devices.

Education and Training: Skilled welders are developed using a number of training systems. These systems include on-the-job training, specialized welding schools, and apprenticeships. Apprenticeships combine on-the-job training with classroom instruction.

Career Cluster: Manufacturing

Career Pathway: Production

See Figure 4-24. Notice how pulleys are used to do each of these jobs.

Inclined Planes

An ***inclined plane*** is the fourth type of mechanism, or simple machine. This mechanism uses a sloped surface. See Figure 4-25. This mechanism operates on the principle that moving up a slope is easier than lifting straight up.

A simple experiment tests this principle. Pull a smooth weight up a slope. Use a scale to measure the force. Lift the weight. You will find that the longer the slope is, the easier it is to move the object.

Wedges

A ***wedge*** is the fifth type of simple machine, or mechanism. This simple machine is a set of two inclined planes. See Figure 4-26. This mechanism is used in many simple hand tools. The wood chisel, knife, ax, splitting wedge (for firewood), and cold chisel operate on the wedge principle. Also, nails are wedges.

Screws

A ***screw*** is the last type of mechanism, or simple machine. This mechanism is actually an inclined plane wrapped on a round shaft. The threads move slowly up the shaft as they go around it. The screw is a great force multiplier. It takes a great deal of rotating motion to move a nut a short distance onto a bolt.

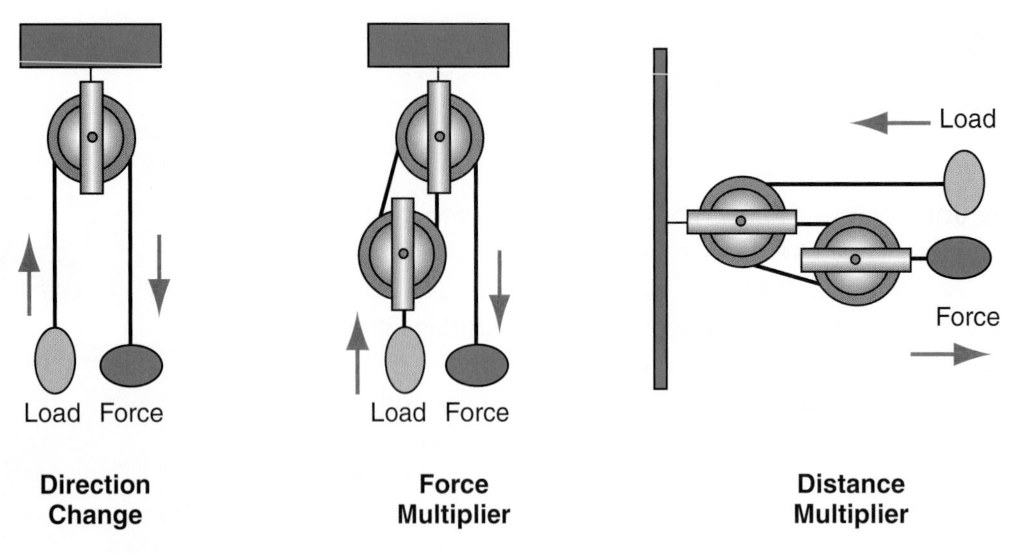

Figure 4-24. Pulleys change direction, multiply force, or multiply distance.

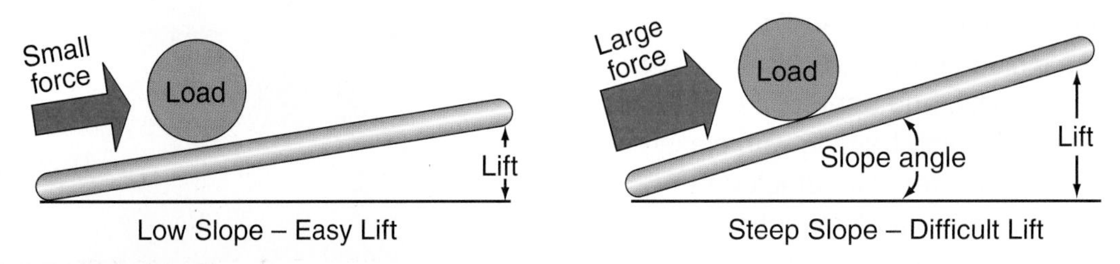

Figure 4-25. Here is an inclined plane. Loads can be lifted with less force. The lower the slope, the easier the load is to move.

Figure 4-26. People use wedges to split logs. Do you see that the wedge is two inclined planes put together?

Consider a 1/2″ × 12 bolt (1/2″ diameter with 12 threads per inch). To move the nut 1″, the bolt must be turned 12 times. A point on the circumference of the bolt would move almost 19″.

MACHINES

Every type of technology uses machines of one type or another. There are machines used in agriculture. People use machines to till soil, plant and cultivate crops, and harvest output. There are machines to irrigate plants, apply insecticides and herbicides, and store harvests.

Machines are used in communications. Printing presses produce newspapers and magazines. Transmitters send signals through the airways to radios and television sets. Switching gears interconnect our telephones.

There are energy conversion machines. They change energy from one form to another. These machines change the energy in coal into electricity. They change the energy in petroleum into mechanical energy to power vehicles. Energy conversion machines change electrical energy into heat and light.

Machines are used in construction. Bulldozers prepare construction sites. Cranes lift structural steel into place. Cement mixers prepare concrete for use. Power trowels smooth the concrete, while saws cut joints into the concrete.

There are machines used in manufacturing. They change the forms and shapes of materials. These machines make products. Presses stamp out parts. Welders fuse metal. Robots move parts from place to place.

Machines are used in medicine and health care. They restore or improve the health of people and animals. Machines in hospitals monitor patients' vital signs. X-ray machines help doctors diagnose illnesses. Pharmaceutical (drug-manufacturing) companies use machines to develop drugs to treat illnesses. Veterinarians use machines to treat diseases in pets and farm animals.

There are machines used in transportation systems. Trucks, trains, ships, and airplanes move people and cargo. Conveyors load and unload cargo from vehicles. Computers maintain reservation information.

Many of these machines are presented later in this book. All these machines, however, can be traced back to other types of machines. These other machines are a special type of manufacturing machine. They are called ***machine tools***. Machine tools are the machines that make other machines. Without them, there would be no agricultural, communication, construction, energy conversion, general manufacturing, medical, or transportation machines. Machine tools change raw materials into parts. These parts later become machines and other products. See Figure 4-27. Machine tools have four major elements:

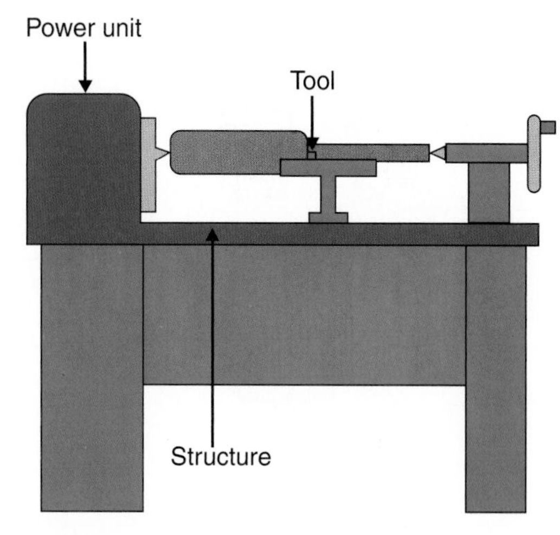

Figure 4-27. Every machine tool has these three basic parts.

- A basic structure (such as a frame, bed, or table).
- A power unit (such as an electric motor or a hydraulic drive).
- A control unit (such as a feed, speed, or depth-of-cut control).
- A tool (a device to produce a cut).

Machine tools can be grouped into six major classes:

- Turning machines.
- Drilling machines.
- Milling and sawing machines.
- Shaping and planing machines.
- Grinding and sanding machines.
- Shearing machines.

See Figure 4-28. Each machine tool has its own way of operating. The machine cuts materials into shapes using different motions and tools.

Turning Machines

Turning machines were some of the first machines to be developed. They are almost as old as civilization itself. The potter's wheel is an example of an early turning machine. Later turning machines are lathes.

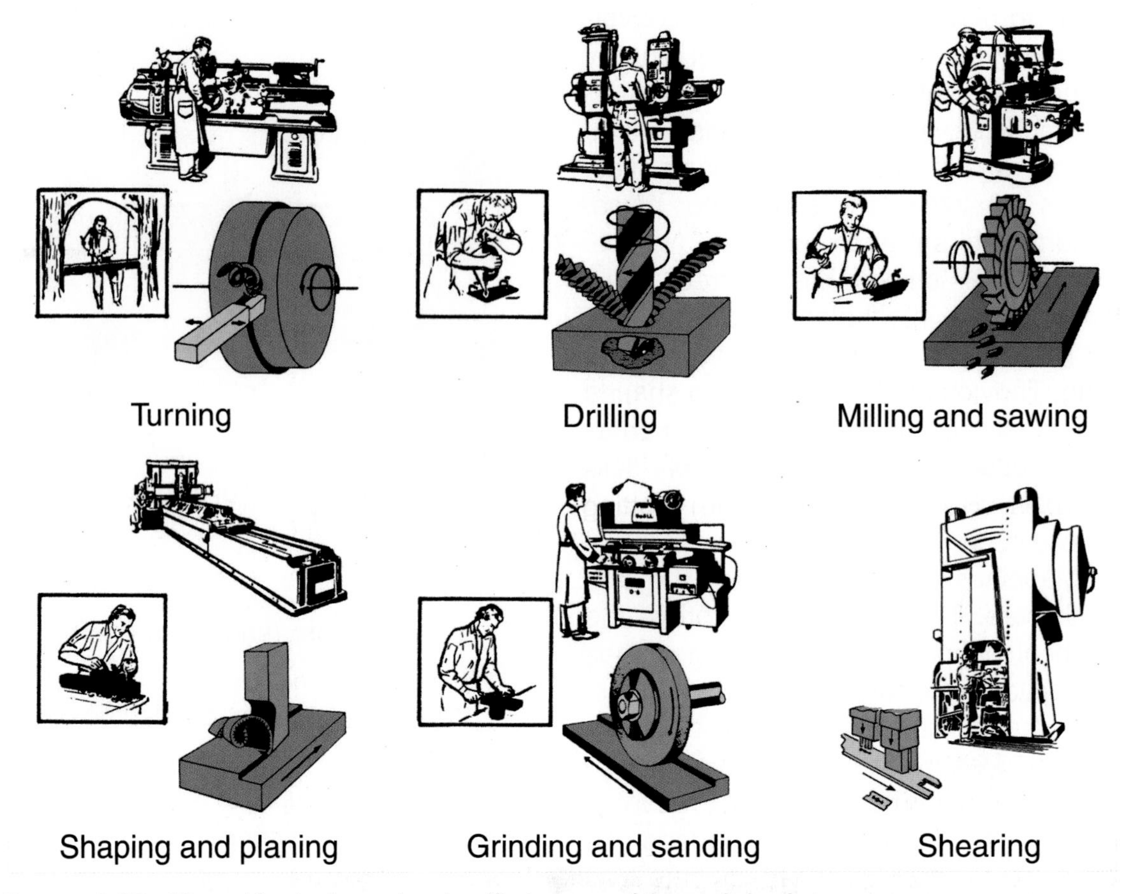

Figure 4-28. All machine tools can be classified as one of these six basic types.

These machines use a stationary tool. The material to be shaped is rotated around an axis. The stationary tool is fed into the work. The cut is created by slowly moving the tool along or into the rotating work.

Drilling Machines

The first drill was probably invented over 40,000 years ago. You learned earlier that it was no more than a pointed stone on a shaft. Today, clamping a drill bit in a chuck (toolholder) is a more common way to produce holes. The chuck is rotated and pushed downward. This causes the rotating drill to feed into the work and cut a hole. The most common ***drilling machine*** is the drill press. See Figure 4-29.

Milling and Sawing Machines

Milling and sawing machines use either the straight or circular saw blades discussed earlier. A motor makes the blade

Figure 4-29. A drill press is the most common drilling machine.

move. The work is fed into the moving blade, creating a cut.

Many of these machines use rotating circular blades or cutters. The most common machines of this type are the milling machine, table saw, and radial saw. The woodworking surfacer, jointer, shaper, and router use the same cutting action.

Power hacksaws, scroll (jig) saws, and saber saws use reciprocating straight blades. The band saw uses a straight blade that has been welded into a loop. The blade travels around two wheels. This produces a linear (straight-line) cutting motion at the workpiece.

Planing and Shaping Machines

Planing and shaping machines are generally limited to cutting metals. Both machines use a single-point tool. The shaper moves the tool into the work to produce the cut. The metal planer moves the work into the tool. Both machines produce a flat cut on the surface of the work.

Grinding and Sanding Machines

Grinding and sanding machines use abrasives to cut materials from the workpiece. The abrasives can be bonded into wheels or onto a backing for sheets, discs, and belts. Generally, the work is moved against the moving abrasive. Grinders and sheet, disc, and belt sanders are the common machines in this group.

Shearing Machines

Shearing machines slice materials into parts. They use opposed edges to cut the workpiece. The material is placed between the cutting edges (knives or blades). One edge is moved down, forcing the material against the second edge. As more force is applied, the material is cut. A pair of scissors uses a shearing action. Common shearing machines are sheet metal shears, punch presses, and paper cutters.

Safety and Technological Devices

People have been developing and using tools since humans first walked on the earth. Early tools were simple and crude. Today, tools range from the common hammer to sophisticated computers. Each tool or machine should be used with care so the operator and other people are not injured.

Learning to Use a Device

Each technological device has its own safety and use considerations. These can be presented in a few sentences. On the other hand, whole books can be devoted to the issue. One simple rule, however, keeps most tool users out of trouble. This rule is "Know your tool." Some simple procedures can help you obey this rule. These approaches are the following:

- **Read the owner's manual.** This booklet tells the new owner about the features of the product and how to install or set up the device. The manual presents proper and safe procedures for use of the product.
- **Search for instructions.** Use the Internet, books, and how-to magazines to learn how other people use the tool or device properly.
- **Ask for instructions.** Seek advice from teachers, skilled workers, and knowledgeable friends and neighbors on ways to use the tool or device safely.

When in doubt about the proper use or safety procedures, do not use the tool or device.

Simple Safety Rules for Hand and Power Tools

You should always have safety in mind when you are using tools. See Figure 4-30. Ask yourself the following questions:

- Is this the right tool for the job?
- Is the tool in proper working condition?
- How should the tool be properly operated?
- Are there any broken or worn parts?
- Are safety guards and switches in place and working?
- Do I have the proper protective equipment (such as safety glasses, protective clothing, or a face mask)?
- Is the work area free of hazards?
- Is the work surface at a proper height?
- Are other people out of the way?

There are some specific questions that should be in your mind as you use common hand and power tools:

- Are electric tools properly plugged in and grounded?
- Are small workpieces properly clamped to a work surface and held in a vise?
- Are the cutting tools or blades sharp?
- Are hammerheads held tightly on their handles?
- Are screwdriver points free of wear?
- Do wrenches grip the nut or bolt properly?
- Are the tools properly stored after use?

For portable electric tools, remember the following rules:

- Do not carry the tool by its cord.
- Do not yank on the cord to disconnect it.
- Disconnect all tools when cutters are being changed.
- Disconnect all electric tools when they are not in use.
- Do not use the tools in wet conditions unless they are specifically designed for such conditions.

Figure 4-30. Note the safety devices shown.

Tool Storage

Tools and equipment must be regularly inspected and maintained so they remain safe and reliable. Always follow the guidelines in the owner's manuals for inspecting, maintaining, storing, and servicing tools and materials. Care should always be taken to properly store tools and materials. They should be kept in an organized fashion in a place where they are easily accessible. Some tools have sharp points that can injure people if not stored carefully. Other tools and materials must be protected from the environment. For example, computer discs should be protected from dust and dirt when stored. If they are not, there is a risk of losing important data.

Summary

Technology is the use of knowledge and action to extend human potential. A major component of technological action is the use of tools. Tools are devices humans have developed to do jobs. They can be classified as measuring, cutting, drilling, gripping, pounding, and polishing tools.

Tools are used by hand. Handwork can be fun. This work, however, is slow. The demand for more efficient farming, products, communication media, energy, medical care, buildings, and transportation devices gave rise to machines. Humans found that some basic mechanisms could be combined with tools to create these machines. They joined the lever, wheel and axle, pulley, inclined plane, wedge, and screw with the basic tools. From these advancements came machine tools. People developed turning, drilling, milling and sawing, shaping and planing, grinding and sanding, and shearing machines. These machines can be used to make all other machines. From these come agricultural, communication, construction, energy conversion, manufacturing, medical, and transportation machines.

STEM Connections

Mathematics

Examine a complex tool or simple woodworking or metalworking machine. Identify a place where mechanical advantage is used. Calculate and graph the mechanical advantage using several different forces applied.

Science

Examine a complex tool or simple woodworking or metalworking machine. Identify as many uses of simple machines (mechanisms) as possible. List and describe them.

Curricular Connections

Social Studies

Trace the development of a tool or machine over time. Start with early developments in places such as Egypt and Rome.

Activities

1. List five common tools you use at home.
   A. Identify the categories of tools in which they belong.
   B. Identify any mechanisms they use.
   C. Describe how they work.

2. List three machines you have seen or used.
   A. Classify them as one of the six types of machines.
   B. Describe the task each machine performs.
3. Build a simple item using tools.
   A. List each tool you use.
   B. Group the tools by the families of tools to which they belong.
4. Select a tool from your classroom lab. Prepare a presentation for the class reviewing the proper safety, inspection, maintenance, and storage procedures for the tool. Include a handout with your presentation.

Test Your Knowledge

Do not write in this book. Place your answers to this test on a separate sheet of paper.

1. Why are humans different from other species of living things?
   A. They can adjust their behaviors to different situations.
   B. They can design, make, and use tools.
   C. In order to control their environments, they have to depend on tools.
   D. All of the above.
2. Devices people develop and use to complete tasks are called _____.
3. A(n) _____ is a basic device controlling or adding power to a tool.
4. Give an example of each of the six primary tools.
5. A lever producing large movement at one end when a force is applied to the other end is being used as a(n) _____.
6. A pulley can be used to change the direction of force, multiply force, and multiply distance. True or false?
7. Explain each of the six mechanisms.
8. Which of the following are major elements of a machine tool? You can choose more than one answer.
   A. Frame.
   B. Tool.
   C. Electric motor.
   D. Feed control.
   E. The part being made.
9. Summarize the qualities of each of the six major classes of machine tools.
10. List two reasons that proper tool storage is important.

Reading Organizer

Create a detailed outline based on what you've read about the different types of tools and machines.

Example:

I. Tools
   A. Measuring tools
   B. Cutting tools
      1. Sawing tools

CHAPTER 4 ACTIVITY A

TOOLS AND MATERIALS AS RESOURCES

INTRODUCTION

You have read about tools and technology. Also, your teacher has told you about tools. Now, you are ready to use this knowledge to work. Your teacher will show you how to use tools to extend your ability to do a job.

In this activity, you will have common hand tools to help you build a simple game. You will be using common measuring, cutting, drilling, pounding, and polishing tools to make a tic-tac-toe board. See **Figure 4A-1.**

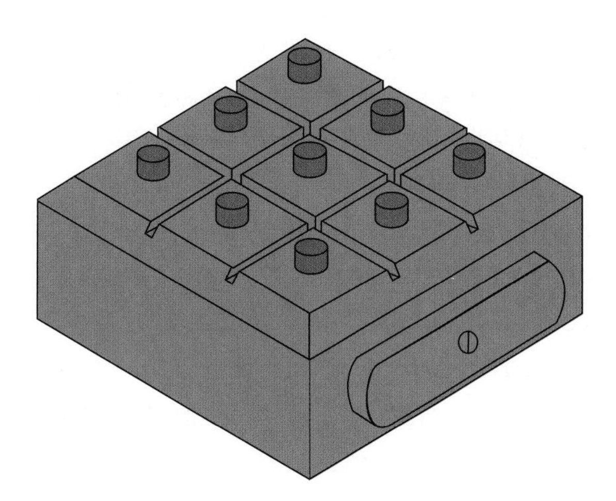

Figure 4A-1. Tic-tac-toe.

EQUIPMENT AND SUPPLIES

- 2 × 4 (1 1/2″ × 3 1/2″) construction lumber.
- 1/4″ × 3/4″ wood strips.
- 3/8″ dowels.
- 3/4″ × No. 6 flathead wood screws.
- A steel rule.
- A try or combination square.
- A crosscut or backsaw.
- A miter box and handsaw.
- A hand drill.

- A brace.
- A 1/2″ auger bit.
- 1/16″, 9/64″, and 13/32″ twist drills.
- A countersink bit.
- A flat wood file or rasp.
- A block or smooth plane.
- Abrasive paper and sanding blocks.
- A screwdriver.
- A scratch awl or center punch.
- A hammer or mallet.

Safety Rules

- Cutting tools have sharp edges. Never carry pointed tools in your pockets.
- Always cut and chisel away from your body or the hand holding the part.
- Always carry cutting tools with sharp edges pointing down.
- Use each tool for its proper purpose.
- Mushroomed heads on chisels or other tools that are struck with a hammer or mallet are dangerous. Bits of material can fly off and strike you.
- Never use tools with loose handles.
- When using a saw, keep your free hand away from the saw blade.
- Never rub your fingers across cutting tools.
- Remove slivers immediately and sterilize the wound.
- Cuts should be allowed to bleed freely for a short time. Bandage properly.
- Keep working surfaces free of scrap and unnecessary tools.
- Sweep up scrap and debris from floors and discard in a waste bin.
- If jigs or fixtures have sharp edges, wear protective gloves. Never wear gloves, however, when operating power equipment.

Procedure

The steps of the procedure for the tic-tac-toe game are numbered from 1 to 43. You do not, however, have to complete the steps in order. You can make the pegs, peg-storage hole cover, and game board in any order. Each number is used only once so you and your teacher can easily refer to a specific step without confusion.

Preparing to Make the Product

1. Study the drawings and bill of materials for the tic-tac-toe game. See **Figure 4A-2** and **Figure 4A-3.**
2. Read the procedure for making the game.
3. Carefully watch your teacher demonstrate how to make the game.

Quantity	Description	Size	Material
1	Game board	1 1/2″ x 3 1/2″ x 3 1/2″	Spruce or hemlock
1	Peg cover	1/4″ x 3/4″ x 3 1/2″	Pine
8	Pegs	3/8″ dia. x 5/8″	Birch dowel
1	Screw	3/4″ x No. 6 flathead	Plated steel

Figure 4A-2. The bill of materials.

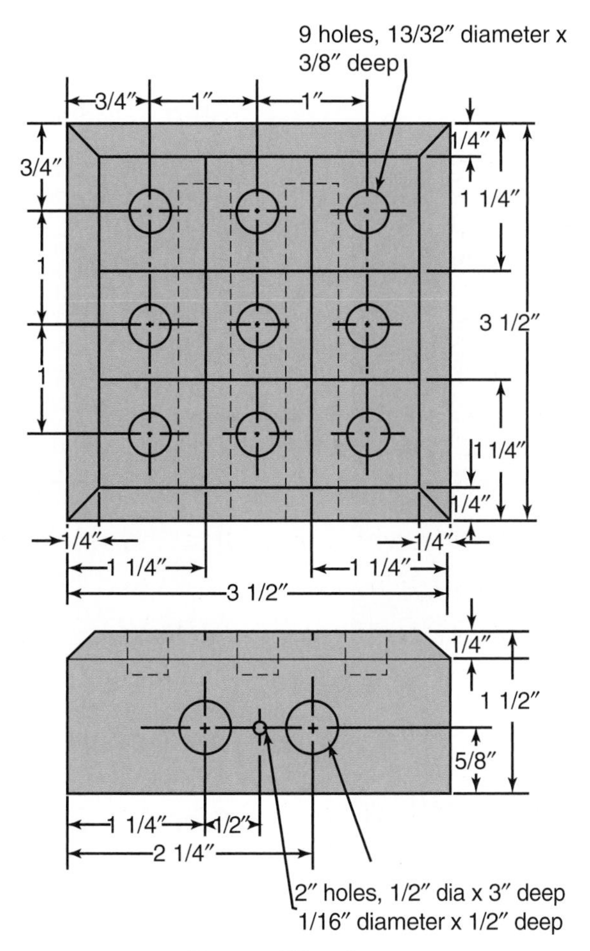

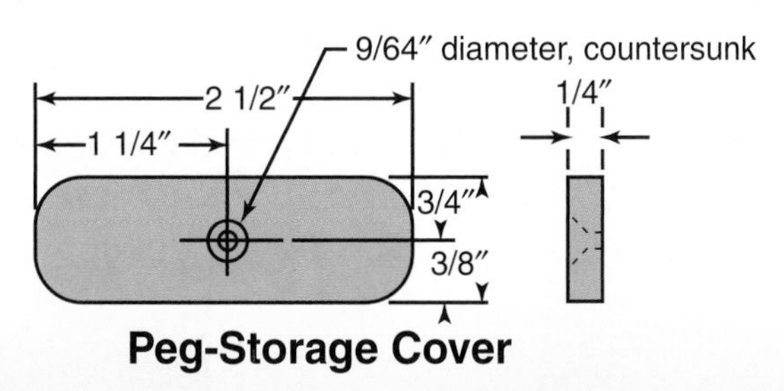

Figure 4A-3. Working drawings.

Making the Game Board

To select and lay out the material, do the following:

4. Select a length of 2 × 4 construction lumber. (Note: The actual size is 1 1/2″ thick × 3 1/2″ wide.)
5. Lay out a line 3/8″ from one end. See **Figure 4A-4.**
6. Lay out a line 3 1/2″ from the first line.

To cut out the game board, complete the following steps:

7. Cut the end off to the outside of the 3/8″ line to square the end of the board.
8. Cut off the game part, barely leaving your line on the part.

Step 1: **Draw lines 1 1/4″ apart.**

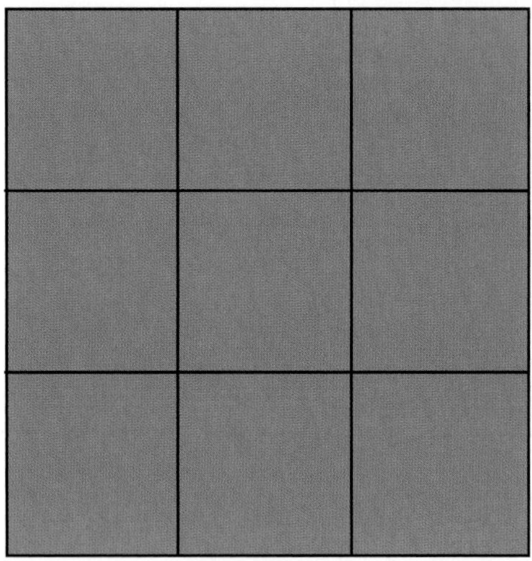

Top View

Step 3: **Locate and mark the peg holes.**

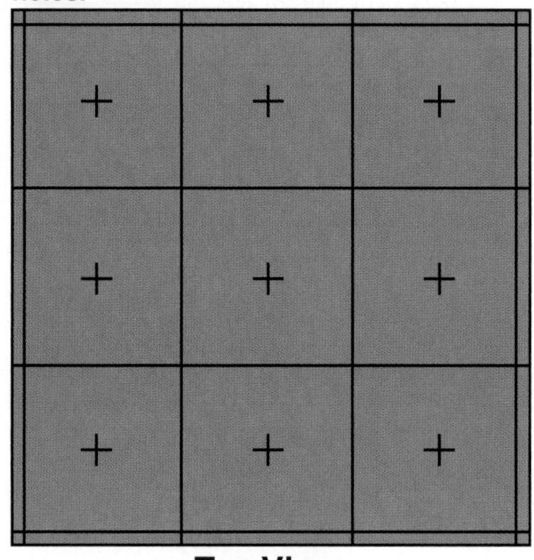

Top View

Step 2: **Draw lines 1/4″ from the edges and the ends.**

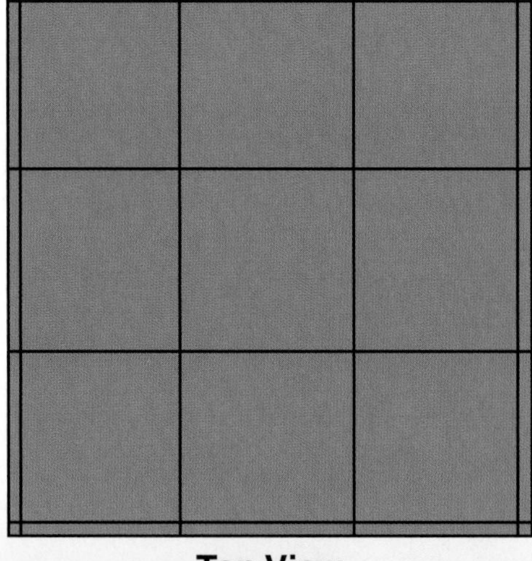

Top View

Step 4: **Locate and mark the peg-storage and cover pivot holes.**

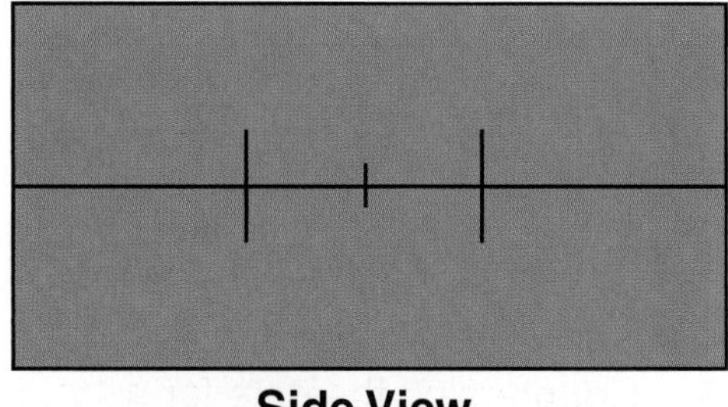

Side View

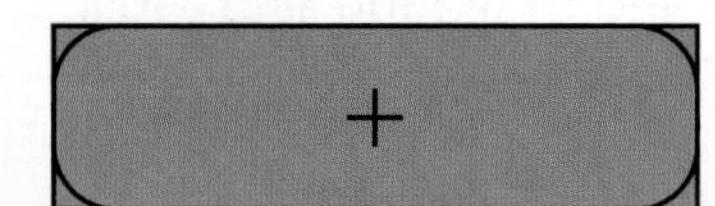

Figure 4A-4. How the game board should be laid out.

To lay out the game board, do the following:

9. Draw lines 1 1/4″ in from the edges and ends of the block.
10. Draw lines 1/4″ in from the edges and ends.
11. Draw lines 1/4″ down from the face on the edges and ends.
12. Locate and mark the nine peg holes.
13. Locate and mark the peg-storage holes on one edge.
14. Locate and mark the peg-cover pivot-screw anchor hole.

To produce the game board, complete the following steps:

15. Saw kerfs (shallow slots) about 1/8″ deep on the four 1 1/4″ lines.
16. Drill the nine 13/32″ peg holes 3/8″ deep.
17. Drill the two 1/2″ peg-storage holes 3″ deep.
18. Drill the 1/16″ pivot-screw hole.
19. File and plane the 1/4″ × 1/4″ chamfers around the top of the block.
20. Sand all surfaces.

Making the Pegs

To select and lay out the materials, do the following:

21. Select a length of 3/8″ dowel.
22. Check the end to see that it is square.

To produce the pegs, complete the following steps:

23. Set a stop block on the miter saw for a 5/8″ cut.
24. Cut the end of the dowel square, if necessary.
25. Cut eight pieces of dowel, 5/8″ long.
26. Sand and lightly break (round) the ends of the pegs.

Making the Peg-Storage Hole Cover

To select and lay out the materials, do the following:

27. Select a length of 1/4″ × 3/4″ pine.
28. Draw a line 1/4″ from the end.
29. Draw a line 2 1/2″ from the first line.
30. Locate and mark the pivot-screw hole.
31. Lay out the radius on each end.

To produce the peg-storage hole cover, complete the following steps:

32. Cut the end off to the outside of the 1/4″ line to square the end of the board.
33. Cut off the hole cover, barely leaving your line on the part.
34. Drill a 9/64″ pivot-screw hole.
35. Countersink the hole for a No. 6 flathead screw.
36. Sand or file the end radii.
37. Sand all surfaces.

Finishing and Assembling the Game

To apply finish to the parts, do the following:

38. Stain four pegs a dark color and let them dry.
39. Apply a surface finish to the board, peg-storage hole cover, and pegs.
40. Allow all finishes to dry properly.

To assemble the product, complete the following steps:

41. Place the dark pegs in one storage hole and the light pegs in the other hole.
42. Make the screw hole for the peg-storage cover. This hole should be over the anchor hole in the game board.
43. Attach the cover with a 3/4″ × No. 6 flathead wood screw.

Challenging Your Learning

Make a chart similar to the one shown. See Figure 4A-5. List the steps in the procedure in which you used each type of tool.

Type of Tool	Procedure Step
Measuring tool	
Cutting tool	
Drilling tool	
Gripping tool	
Pounding tool	
Polishing tool	

Figure 4A-5. A list of where you used each tool.

CHAPTER 4 ACTIVITY B

TOOLS AS A RESOURCE

THE CHALLENGE

Develop a device (tool) that separates marbles according to their diameters.

INTRODUCTION

Since the earliest times, human beings have developed tools to do work. Early tools included stone implements and weapons. Now we have complex tools that do work for us. Machines cut materials with ease and accuracy. Vehicles whisk us to faraway places. Computers process information and control machines. People developed each of these devices to help other people. In this challenge, you are faced with a problem. You have a container holding three different sizes of marbles. These marbles need to be sorted by size (their diameters). See Figure 4B-1. Measuring each marble individually would take too much time. You have decided to use technology to help you, and you plan to build a device to sort the marbles.

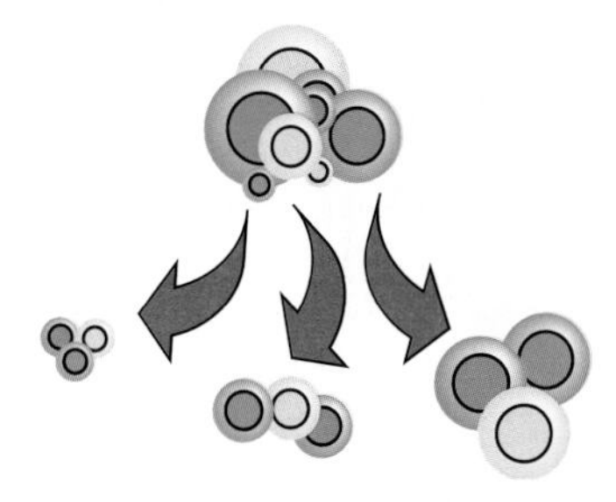

Figure 4B-1. The marbles need to be sorted by the size of their diameters.

MATERIALS

Develop your technological device using any or all of the following materials:

- 1/2″ × 1″ wood strips.
- 1/4″ plywood.
- White glue.
- 3/4″ brads.
- 3/8″ dowels.
- Masking tape.
- Paper clips.
- Poster board.
- String.

CHALLENGING YOUR LEARNING

How did your device work? Describe two ways you could improve it.

CHAPTER 5

Materials and Technology

Did You Know?

- The word *plastic* refers to new substances produced by chemical means. The first such material to be manufactured was Parkesine. The British inventor Alexander Parkes developed it. He won a bronze medal for his invention at the International Exhibition of 1862 in London.
- Celluloid was developed in the late 1800s. The solid form was used as imitation horn for billiard balls. Celluloid sheets were used for men's collars and photographic film.
- Leo H. Baekeland, a U.S. inventor, introduced a new class of large molecule plastics in 1909. He developed Bakelite® plastic. This plastic is a hard, chemically resistant thermosetting plastic.

Objectives

The information given in this chapter will help you do the following:

- Give examples of natural materials.
- Give examples of the three major types of materials used in technological systems.
- Give examples of the four major types of engineering materials.
- Compare renewable resources and exhaustible resources.
- Explain the three major ways to process raw materials.
- Explain what an industrial material is.
- Summarize the major properties of materials.
- Explain why material must be properly stored.

Key Words

These words are used in this chapter. Do you know what they mean?

acoustical property
alloy
ceramic
chemical processing
chemical property
composite
ductility
electrical property
engineering material
exhaustible resource
hazardous material
hazardous waste
magnetic property
mechanical processing
mechanical property
metal
natural resource
optical property
physical property
plastic
polymer
property
renewable resource
thermal conductivity
thermal expansion
thermal processing
thermal property

Preparing to Read

Look carefully for the main ideas as you read this chapter. Look for the details that support each of the main ideas. Use the Reading Organizer at the end of the chapter to organize the main and supporting points.

Humans have always used materials as they have adapted to their world. Over time, there have been four basic phases of material use. See **Figure 5-1.** First, people used the materials they could find. They searched for rocks having the right shape for spearheads. Carefully selected

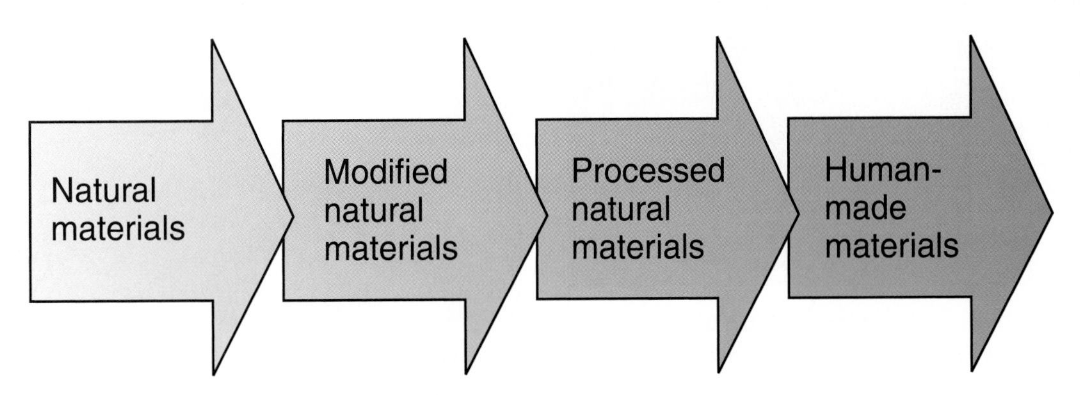

Figure 5-1. The development of materials has gone through four stages.

tree branches became digging sticks (the early plow). Rocks, limbs, and tree bark became houses. During the next phase, people started forming natural materials to fit their needs. Stones were chipped to make arrowheads. See Figure 5-2. Natural copper was shaped to make metal tools and utensils. Trees were cut into boards. As human technological knowledge advanced, a third phase of material use appeared. Combining old materials developed new materials. Iron, coke (a coal product), and limestone were converted into steel. Copper and zinc were combined to form the alloy we call *brass*. Thin strips of wood were glued together to form plywood. Finally, people created synthetic materials. In the late 1800s and thereafter, plastics were developed. These new materials were created to fit special product needs. Plastics were developed for a wide range of products. Special plastics became fibers for new textiles.

Figure 5-2. Early Native Americans, such as these shown in this diorama, were masters at using natural materials.

Today, as in past years, we live in a world of materials. Everything around us is made of materials. See Figure 5-3. Some of these materials appear in nature. We call these *natural materials*, or ***natural resources***.

Trees, grass, soil, minerals, coal, petroleum, and many other materials are found naturally on Earth. See Figure 5-4. Many of these materials have limited use to us in their natural states. Rocks can be used to make retaining walls. Plants can form windbreaks. Trees give us shade on a hot day. Grass and other plants make parks look nicer. Stones can be used for walkways.

Most objects we use, however, are made of natural materials that have been modified. People have processed them into new forms and compositions. The desks, chairs, and tables in your school are made from materials that were once in their natural states. The tabletops might be made from wood. The tables might have metal legs that were once an ore. The chairs might have plastic seats made from petroleum. In this chapter, you will explore the roles materials play in technology.

Figure 5-3. Things made from materials surround us. (Lexington Homes)

Trees

Minerals

Figure 5-4. Natural materials include things growing or found on Earth. Trees grow in soil and provide wood. Minerals are found underground. (Weyerhaeuser Co. and Amoco Corp.)

Types of Materials

A material is any substance out of which a useful item can be made. Not all materials are alike, however. All materials do not fulfill the same purposes. For this discussion, three major types of materials used in technological systems are explored.

The first type of material provides energy to operate technological systems. This type is the input to energy conversion systems. This input powers engines, turbines, and other energy converters. Some of these materials are petroleum products, such as fuel oil, gasoline, kerosene, and diesel fuel. Falling water and wind are also sources of energy, as are uranium, wood, grain converted into ethyl alcohol, and coal.

The second type of material includes process-supporting materials. These are the liquids and gases supporting natural and technological processes. They include the air we breathe and water we drink. These materials also include the water and fertilizers promoting plant growth on farms and forests. This type of material also cools industrial processes, lubricates moving parts, and provides the raw materials for products and thousands of other uses.

A third type of material is called *industrial material*, or ***engineering material***. This group is the main focus of this chapter. Engineering materials have rigid structures. They withstand outside forces. Their shapes are hard to change. Most of us know these materials as solids. Engineering materials form the primary structures for all products having set forms. See Figure 5-5.

Typical products made of engineering materials are bicycles, tennis rackets, cola bottles, garbage bags, microwave ovens, and T-shirts. These products all share one characteristic. They hold their shapes over time. Products made of engineering materials have basic structures that remain constant. Soap powders, toothpaste, lipstick, cherry cola, and pizza are different. All these products are made from materials without structure. They must be placed in or on containers to hold their shapes.

Steel

Lumber

Figure 5-5. The steel on the left and the lumber on the right are examples of solids, or engineering materials.

Engineering Materials

Engineering materials are the building blocks for many products. They can be grouped into four major categories:

- Metals.
- Ceramics.
- Polymers.
- Composites.

Industries and individuals use each of these materials. Also, each type of material is very different from the others.

Metals

Metals are inorganic materials. This means they were never living things. Metals, as a group, have similar internal structures. Their molecules are arranged in a boxlike framework. These material structures are called *crystals*. They combine to form the grains of metal. See Figure 5-6. The crystal framework produces a rigid, uniform material.

These materials are seldom used in their pure forms. We do not often use pure gold, silver, copper, aluminum, or iron. Most metals we use are mixtures of two or more different base materials. Iron is combined with carbon and other elements to make steel. Aluminum is often combined with

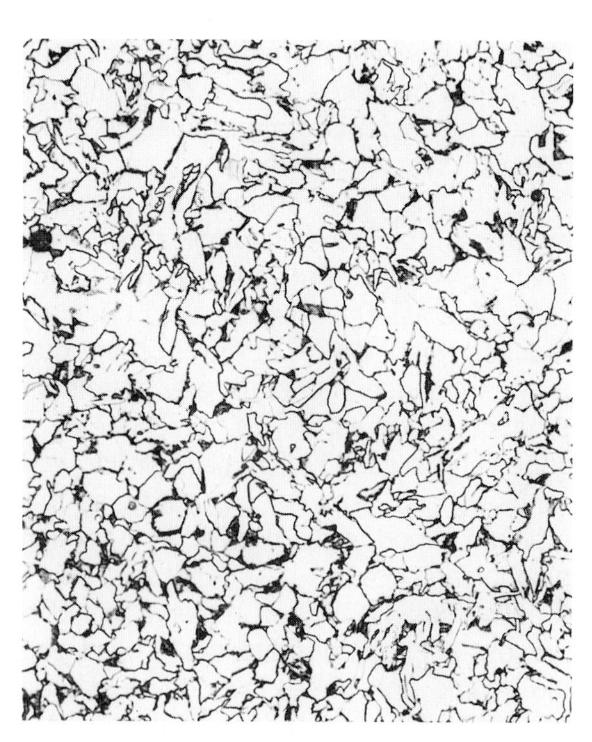

Figure 5-6. How a section of metal looks when magnified 1000 times. (Bethlehem Steel Corp.)

copper, silicone, and magnesium to make more useful materials. These mixtures of two or more elements are called ***alloys***. Bronze is an alloy of copper and tin. Sterling silver is an alloy of silver and copper. Stainless steel is a nickel-chromium alloy of steel.

Ceramics

Ceramics are one of the oldest materials humans have used to make products. See Figure 5-7. Ceramics, similar to metals, are inorganic materials. They are also made of crystals.

Ceramic materials are widely used because they are very stable. Heat, moisture, and chemicals do not affect them. Ceramic materials are stiff and brittle. They do not bend. These materials break under stress. There are three major types of ceramic materials:

- **Clay-based materials.** These materials are substances made up of crystals that a glass matrix (binder) holds in place. Typically, clay-based ceramics are used to make bricks, decorative and drain tiles, dinnerware (earthenware and china), and sanitary ware (sinks and toilets).
- **Refractories.** These are crystalline materials held together without binders. They can withstand high temperatures. Firebrick (used to line furnaces and fireplaces) and the space shuttle reentry tiles are made of refractory materials.
- **Glass.** This is an amorphous material (without a regular structural pattern) commonly made from silica sand. Glass is a very thick liquid appearing to be a solid under normal conditions. This material is used for windows, containers, and many other products.

Polymers

Polymers are organic materials. This means they are formed from once-living materials. The molecules in the material form chainlike structures. Most polymers are made from natural gas and petroleum. Nylon and polyethylene are examples of polymers made from these base materials. See Figure 5-8. Wood and other cellulose fibers can also be used to make polymer materials. Rayon is a cellulose-based polymer.

These materials can be natural or human made. A typical natural polymer is natural rubber. Synthetic (human-made)

Figure 5-7. A Pueblo Indian from the southwest United States made this ceramic pot.

Figure 5-8. These hot air balloons are made from nylon, a plastic. (Jack Klasey)

Career Highlight

Inspectors

The Job: Inspectors measure products against standards to guarantee the quality of the goods a company produces. They might check products by sight, sound, feel, smell, or taste. Inspectors might check the dimensions, color, weight, and textures of materials and products. They check products, tag problems, reject defective items, and send rejected parts for rework. Inspectors maintain inspection records and prepare periodic reports on product quality.

Working Conditions: Working conditions depend on the company and its type of production. Work is generally done while the product is being manufactured. The inspection station is often located directly in the production facility.

Education and Training: Inspectors need mechanical aptitude, mathematical ability, communication skills, mechanical dexterity, and good hand-eye coordination. Most inspectors must have a high school diploma. Generally, they receive on-the-job training in blueprint reading, inspection device use, record-keeping techniques, and report preparation. Inspectors who do complex inspecting tasks are often experienced workers.

Career Cluster: Manufacturing

Career Pathway: Quality Assurance

Manufacturing

polymers are often called ***plastics***. They are materials designed and produced to meet specific needs. Polymers can be grouped into three categories:

- **Thermosets.** These are materials that form rigid shapes under heat or pressure. After a part or product is made, the structure remains rigid. Once set, the material does not soften when heat or water is applied. Cases for television sets, furniture components, and many automotive parts are made from thermoset plastics.
- **Thermoplastics.** These are materials that soften under heat and become rigid when cooled. They can be heated and reshaped a number of times. Product packages are commonly made from thermoplastic materials.
- **Elastomers.** These are materials that can be stretched. They rapidly return, however, to their original shape. To be called an *elastomer*, the material must withstand being stretched at least twice its length. Rubber bands and balloons are common products made from elastomers.

Composites

Composites are combinations of materials. Each material, however, retains its original properties. Mixing cement, sand, gravel, and water forms a composite. We call this composite *concrete*. The cement binds the sand and gravel together. Water starts the curing action of the cement. The sand and gravel, however, have not changed. If you look at a piece of broken concrete, you can see the sand and gravel. This is different from an alloy. You cannot see the copper and zinc particles in a piece of brass.

Likewise, wood chips are glued and pressed together to form particleboard and wafer board. Thin wood sheets, called *veneer*, are glued and pressed together to form plywood. Wood fibers can be heated and pressed under high pressure to make hardboard. All these wood products are composites.

Still another common composite is made with glass fibers and plastic binders. The fibers are coated with a liquid plastic. When the plastic cures, we have a product called *fiberglass*. This product is widely used for automobile, aircraft, boat, and truck parts; bathroom-shower enclosures and bathtubs; fast-food serving trays; and sporting goods.

Obtaining Material Resources

All materials we use can be traced back to Earth. As you have learned, they can be living things. They can be minerals and other elements found in the ground, sea, or air. These materials must be located before they can be removed from the earth through such processes as harvesting, drilling, and mining. Few materials occur in nature in usable forms. Most must be changed into new forms before they can be used as inputs in the manufacturing process. All materials can be grouped into two categories:

- Renewable material resources.
- Exhaustible material resources.

Renewable Resources

Renewable resources are living things. They sprout from seeds or shoots. These resources can also be born. Over time, they grow and mature. Finally, they die. See **Figure 5-9.** These materials might go through their life cycles without human care. They might grow in a forest, on a mountainside, or on a plain. Renewable resources might be part of what we call *nature*. People engaged in farming (growing food and fibers) and forestry (growing trees) might, however, produce these materials. Commonly used renewable material resources are trees, grains, animals, fish, fruits, and vegetables.

Figure 5-9. Young trees grow next to a stand of mature trees. Why should mature trees be cut, rather than younger ones? (Weyerhaeuser Co.)

Locating Renewable Resources

Some renewable materials must be located before they can be harvested. These are the materials growing wild in nature. Foresters search forests to locate mature trees. They also select the correct species. See Figure 5-10. Some trees have value for lumber. Others are used to make paper and particleboard. Still others are good only for firewood. In fact, some trees have little or no commercial value. The western juniper and the pin oak are two examples. You might want them in a yard for shade and appearance. These trees, however, produce little valuable lumber or wood fibers. Fish, also, must be found before they are harvested. Many people would tell you that finding a tree is easier than finding fish.

Growing and Harvesting Renewable Resources

Most people are familiar with farming. Farmers plant seeds in the spring and care for the crop during the growing season. Fertilizer is applied to the soil to increase plant growth. Weeds and pests are controlled by mechanical means (cultivating) and chemical means (herbicides and pesticides). At the end of the growing season, the crop is harvested and placed in storage or used immediately. See Figure 5-11.

Figure 5-11. Farmers harvest mature crops, such as this grain, before the crops spoil. (Deere and Company)

Figure 5-10. Foresters carefully select the trees to be harvested. (©iStockphoto.com/fstop123)

Technology Explained

geothermal energy: energy derived from the natural heat present in the earth.

People are actively seeking new sources of energy. One of these sources is actually very old. This source is the natural heat present in the earth. The earth's magma, heating the earth's crust and water below, develops this heat. The effect of this heating can be seen in geothermal areas, such as Yellowstone National Park. See **Figure A.** Eruptions of volcanoes, such as Mount Saint Helen's, are another example of the earth's geothermal energy.

One of the ways to use this energy is with a geothermal heat pump in a home. See **Figure B.** To use this type of system, two wells are drilled. Warm water is brought to the heat pump. The unit extracts heat from the water in the winter. The pump can add heat to the water (cool the home) in the summer. The altered water is then pumped back into the ground. This type of unit works anywhere. The pump uses the difference in temperature between the water and outside air.

Another use of geothermal energy is in power plants. See **Figure C.** These pumps work much differently than residential heat pumps do. First, the earth's magma heats water to at least 350°F. A well and pipes bring the hot water to the power plant. Some of the water is allowed to enter a set of steam separators. Here, a part of the water is allowed to flash into steam. The remaining hot water is flashed into steam in a second separator.

The pressurized steam is moved to a turbine through pipes. Here, it turns the turbine connected to an electric generator. The output of the generator is fed into the power grid through wires.

A condenser pulls the cooled steam from the turbines and allows it to change back into water. A gas-emissions treatment unit removes sulfide gas. The hot water is allowed to cool in a cooling tower. The water is then returned to the geothermal reservoir through an injection well.

Similar to most technological systems, there are some drawbacks to geothermal power. The water contains chemicals that can deposit on the walls of the pipes. This restricts the flow of the hot and cold water. Also, the plants are more expensive to build than conventional power plants are.

Figure A. A geyser at Yellowstone National Park.

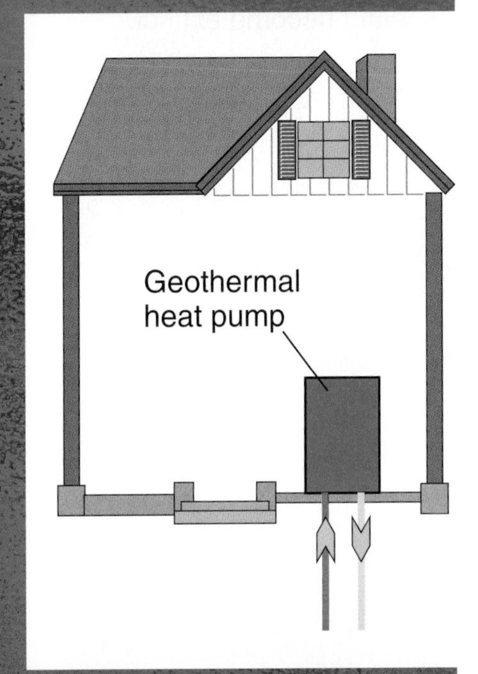

Figure B. A residential geothermal heating unit.

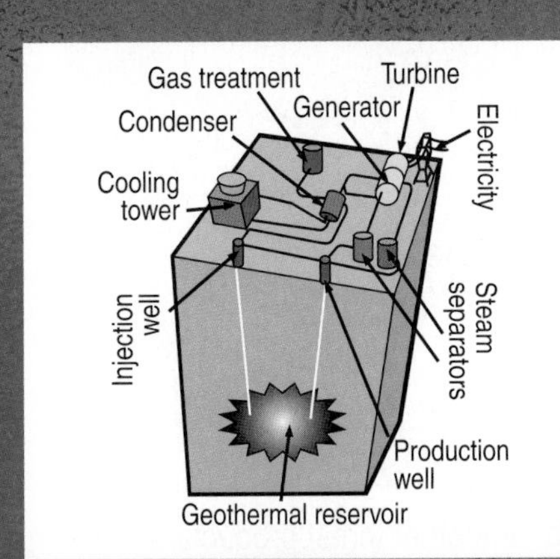

Figure C. The parts and operation of a geothermal power plant.

In a similar manner, ranchers raise cattle and sheep. Adult livestock give birth to their young each year. The young animals are cared for and fed. They grow to be young adults. Some are saved to raise additional young animals. The majority is harvested. These animals are butchered for their meat, hides, and other parts. Additionally, shearing adult sheep harvests wool.

Farming and ranching, however, are not the only ways people grow renewable resources. Many wood-product companies operate large tree nurseries. They grow trees to be planted in forests or on tree farms. The newly planted trees are carefully tended. They receive fertilizer and often are pruned to increase growth. The goal is to grow the maximum amount of wood fiber in the least time. Similarly, fish are now being raised on farms. A large industry has emerged to grow salmon and catfish for sale. See **Figure 5-12.**

The key for all renewable resources is to harvest them when they are ripe (ready to be used). See **Figure 5-13.** Cutting trees that are too small is wasteful. Also, letting a tree mature and die wastes valuable resources.

Likewise, catching young fish breaks nature's cycle. Later, we pay a price. There will be too few mature fish to provide food. Entire species might become extinct.

Figure 5-12. A salmon farm. (©iStockphoto.com/CW03070)

Figure 5-13. Mature trees are cut to produce logs for lumber, plywood, and other forest products. (Weyerhaeuser Co.)

Exhaustible Resources

Many resources are exhaustible. There are only so many of these resources on Earth. When ***exhaustible resources*** are used, they are gone and can never be replaced. This type of resource provides a unique challenge. People must use each resource very wisely. Wasted resources are gone forever. Changes in habits and attitudes do not bring them back. New resources cannot be grown.

Common exhaustible resources include petroleum, natural gas, coal, mineral ores (such as iron, copper, and aluminum), farm soil, clay, and chemical deposits (such as salt and sulfur). See Figure 5-14. Obtaining exhaustible materials involves two major steps:

- Finding deposits of the resource.
- Extracting the raw material.

Locating Exhaustible Resources

Finding the resource requires special skill, training, and a certain degree of luck. The challenge is often the job of geologists. See Figure 5-15. Geologists study Earth to determine possible locations of the materials. They might use aerial maps and satellite pictures. Seismic studies are also used. These involve firing explosives buried in the ground. The resulting shock waves are read on special equipment. "Pictures" of the rock layers below the

Figure 5-15. This geologist searches for natural resources located in the earth. (Amoco Corp.)

Figure 5-14. Modern farming practices help protect soil from erosion. (Deere and Company)

Career Highlight

Machinists

The Job: Machinists produce precision metal parts using lathes, milling machines, machining centers, and other machine tools. They use their skills with machine tools and their knowledge of manufacturing processes and metals to make precision parts. Machinists use blueprints or written specifications to determine the sizes and shapes of the parts. They then select and use the correct machines and cutting tools to produce the parts.

Working Conditions: Most machinists work in relatively clean and well-lit shops. Working with machine tools, however, always presents certain dangers. Machinists must stand for long periods of time and sometimes lift heavy workpieces. Commonly, machinists work 40-hour weeks, with some overtime during peak production periods.

Education and Training: Machinists receive their training through on-the-job training, apprentice programs, vocational-technical schools, or community colleges.

Career Cluster: Manufacturing

Career Pathway: Production

Manufacturing

surface are obtained. These pictures are charts and graphs the equipment makes. They are read to locate promising sites for exploration.

Extracting Renewable Resources

Once a body of ores, minerals, or petroleum is located, it must be extracted. Holes are often drilled to get core samples of the rock below the surface. The samples provide additional evidence of the presence of the resource. If the samples show promise that the resource is there, extraction begins.

Two major methods of extraction are used. See Figure 5-16. Wells are drilled to remove liquid resources. Mines are dug to remove solid resources. Some might be big holes dug into Earth's surface. These holes are called *open pit mines*. Others are tunnels extending into the sides of mountains or vertically into Earth. See Figure 5-17. These tunnels are called *underground mines* or *shaft mines*.

Processing Raw Materials

Most raw materials have little value to people in their own rights. We cannot use them as they come from the ground. These materials must be processed into something more useful. Often the resource found is mixed with other materials. In this condition, it has limited value. Iron ore is of little

Extracting Natural Resources

Mining

Drilling

Figure 5-16. Mining and drilling extract natural resources. (Caterpillar, Inc. and Amoco Corp.)

Figure 5-17. The entrance of an old gold mine in eastern Oregon.

use to us in its natural state. Likewise, a pile of harvested trees cannot be used easily. Processing raw material falls into three general categories. See Figure 5-18:

- Mechanical processing.
- Thermal processing.
- Chemical processing.

Mechanical Processing

Mechanical processing uses machines to cut, slice, crush, or grind material into new forms. The size and shape of the natural resources are changed. Grain is ground into flour. Rocks are crushed into

Figure 5-18. There are three types of processing used to convert natural resources into usable materials. (Inland Steel Company)

Figure 5-19. Saws cut logs into lumber. Sawing is a mechanical means for changing material. (Weyerhaeuser Co.)

gravel. Animal carcasses are cut into meat. Trees are cut into lumber, veneer, and chips. See Figure 5-19.

Thermal Processing

Often, we rely on heat to process materials. The materials are melted or softened so they can be changed into more useful materials. This is called ***thermal processing***. Iron ore, coke, and limestone are heated together in a blast furnace. This action produces pig iron. Later, this material is thermally processed into steel. See Figure 5-20.

Most metal ores are processed using heat. They are first crushed into manageable pieces. Heat is then used to melt the metal away from the impurities. Likewise, heat can be used to fuse materials. Silica sand is heated to high temperatures to melt and fuse it into glass.

Chemical Processing

Some materials are altered using chemical action. This is called ***chemical processing***. Gold is removed from its ore using an arsenic chemical process. Natural gas changes into plastic when chemicals react with it. Aluminum is produced using a combination of electricity and chemicals. This is called an *electrochemical process*.

Figure 5-20. This worker is controlling a steelmaking (thermal) process.

Products of Processing

The products of processing actions are industrial materials. This type of material comes in a standard form or size. The product is called *standard stock* or *industrial stock*. Each of these materials needs further processing to be useful. Flour mixes with other materials to make bread. Lumber must be cut and shaped into useful products. This further processing is called *manufacturing*. Manufacturing is studied in Chapter 23.

Figure 5-21. Each material used to build and equip this kitchen was chosen to meet special requirements.

Properties of Materials

There are hundreds of materials available. They can be used in various ways to meet human needs and wants. For each job (product or use), a material must be selected. See Figure 5-21. It would be unwise to use wood for a heat shield in a welding booth. Even at a low temperature, it would burn. Steel would be a poor choice for a baseball bat. This material is too heavy (dense).

Understanding material properties helps people select the right material for the job. A material property is a characteristic the material has. See Figure 5-22. These ***properties*** can be divided into seven groups:

- Physical.
- Mechanical.
- Chemical.

Properties of Materials

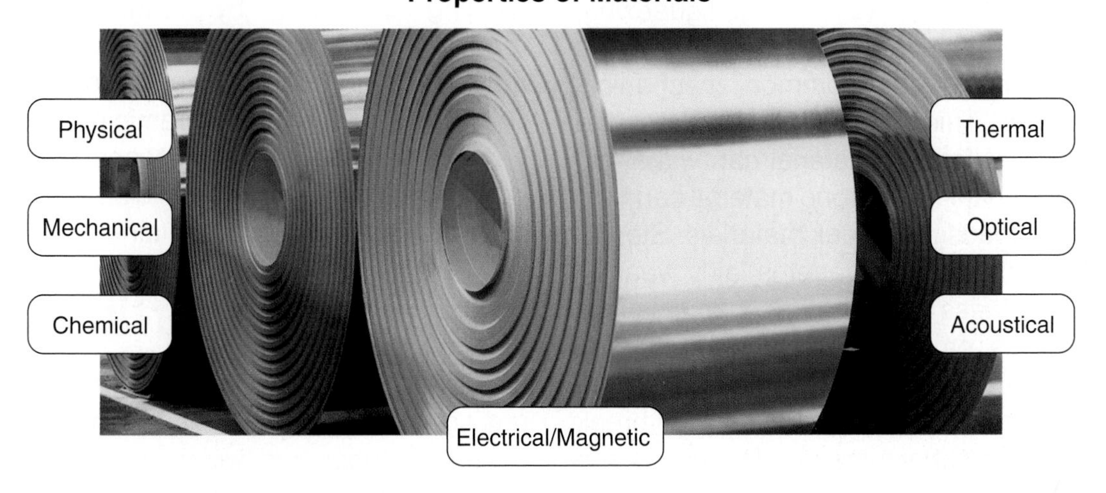

Figure 5-22. The properties of materials. (LTV Steel Company)

- Thermal.
- Electrical and magnetic.
- Optical.
- Acoustical.

Physical Properties

Physical properties describe the basic features of a material. One of these features is density. This is a comparison of weight to size. Density is often given in pounds per square foot.

Another physical property is moisture content. This property tells us the amount of water in the material. Lumber used for furniture should have a moisture content of 6%–8%. Construction lumber has a moisture content in the 12%–16% range.

A third physical property is surface smoothness. This property affects a material's use. A very smooth material makes a poor floor. This material becomes too slippery when wet. Similarly, a rough material is a poor bearing surface. The parts would not slide easily. Abrasive materials have rough surfaces. Glass has a smooth surface.

Mechanical Properties

Mechanical properties affect how a material reacts to mechanical force and loads. See Figure 5-23. They describe how a material reacts to twisting, squeezing, and pulling forces.

Strength is a common mechanical property. This property measures the amount of force a material can withstand before breaking. A strong material can hold heavier loads than weak materials. Steel is very strong. Writing paper is fairly weak.

Hardness is another mechanical property. This property measures the resistance to denting and scratching. Often, hard materials are also brittle. They break easier than softer materials. Diamonds and glass are both hard materials. They shatter (break) if they receive a sharp blow.

Figure 5-23. These windows are manufactured to withstand many mechanical forces. (Andersen Corp.)

A third mechanical property is ***ductility***. This measures the ability of a material to be shaped with force. A ductile material can be hammered or rolled into a new shape. Many aluminum and copper alloys are very ductile. They are easy to form into complex shapes.

Chemical Properties

Chemical properties describe a material's reaction to chemicals. These chemicals might be of any kind. They might be water, some other liquid, or acids in the air.

These chemicals might cause corrosion. They might react with the material, causing unwanted by-products. Water reacts with steel to form rust. When copper reacts with chemicals, it begins to tarnish. See Figure 5-24. Most ceramic materials, however, resist chemical actions. That is why ceramics are used for dishes, bathroom fixtures, and food containers.

Thermal Properties

Thermal properties are a material's reaction to heat. Most materials expand (become longer and wider) when heated. This property is called ***thermal expansion***. The property can be used in many ways.

Figure 5-24. The Statue of Liberty's green color is a result of copper tarnishing in the weather. (Derek Jensen)

For example, gently heating a metal jar lid might make it easier to remove. The metal lid expands more than the glass jar does.

Many materials allow heat to move through them. The measure of this property is called ***thermal conductivity***. The higher the conductivity is, the easier heat moves through the material. If you heat one end of a metal rod, the other end quickly becomes hot, too. Most metals have high thermal conductivity. Wood and most plastics, however, have low thermal conductivity. If you heat one end of a piece of wood or plastic, it catches fire before the other end gets warm. This explains why many kettles are made of metal, while their handles are plastic. See Figure 5-25. The metal conducts the heat quickly to the food. The handle insulates the heat from the cook's hand.

Electrical and Magnetic Properties

Electrical properties describe the material's reaction to electrical current. Materials that easily carry the current are called *conductors*. Other materials that resist the flow of electricity are called

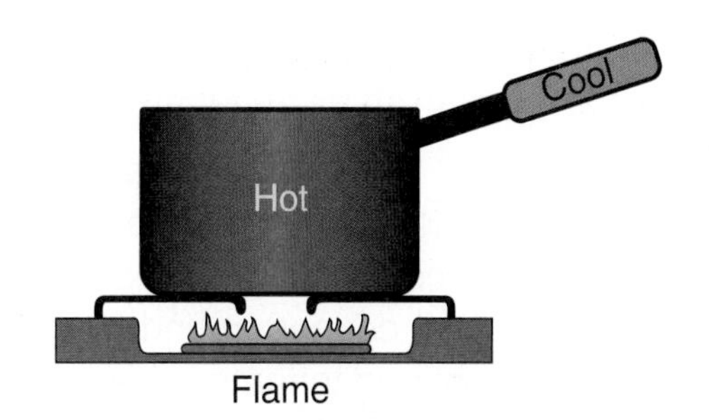

Figure 5-25. A material's thermal properties can be advantageous when designing a product.

insulators. Metals are generally good conductors. Most polymers and ceramics are good insulators. Plastic-insulated copper wire is used to carry electricity in homes and other buildings.

Magnetic properties describe the material's reaction to magnetic forces. Some materials easily are magnetized. Other materials will never become magnets. Most iron alloys (steel) can be magnetized. Almost all other materials are nonmagnetic.

Optical Properties

Optical properties govern the material's reaction to light. Some materials let light pass through them easily. They are called *transparent*. Other materials stop all light. They are called *opaque*. Still other materials let some, but not all, light through. They are called *translucent*.

Another optical property relates to reflecting light. Each material absorbs part of the light striking it. Other parts of the light spectrum are reflected. The part reflected reacts with the human eye. We see the material as having color. Light-absorbing and -reflecting properties are used in photographic filters. These filters allow certain colors of light to pass through and reflect the other colors.

Acoustical Properties

The ***acoustical properties*** of a material determine its reaction to sound waves. Materials that absorb the waves are sound insulators, or deadeners. See Figure 5-26. Other materials reflect sound. Some materials are sound transmitters and carry sound. Smooth, hard materials generally reflect sound. Softer, uneven material absorbs sound. A room having a lot of hard material in it causes sound to echo. That is why most auditoriums have a lot of soft material in them. Curtains, insulating tiles, and curved walls improve the sound quality of a room.

Figure 5-26. This acoustical foam is used to deaden sound in a recording studio. (©iStockphoto.com/sumnersgraphicsinc)

Working with Hazardous Materials and Wastes

Hazardous materials surround us at school, work, and home. Cleaning materials, finishing materials, art and science supplies, lawn and garden chemicals, and gasoline are only a few of the materials we encounter. Follow these guidelines to prevent accidents:

- Read the Material Safety Data Sheet (MSDS) before using any chemical or material for the first time.
- Follow the instructions in the MSDS on how to use, store, and dispose of the chemical.

- Store flammables in an approved container or flammables cabinet.
- Wear the appropriate personal protective equipment (such as gloves, goggles, and respiratory equipment) when using a hazardous substance.
- Keep emergency phone numbers posted.

You must be able to identify and classify hazardous materials and wastes. Be familiar with all hazardous materials in the lab or shop area. Review the MSDSs periodically to refresh your memory and also to make sure you are aware of any new materials.

If you encounter a substance and you cannot identify it immediately, assume it is hazardous. Notify your teacher immediately. Look for nearby containers, and then review the MSDS. When disposing of ***hazardous wastes***, also follow the requirements specified in the MSDS. Use the proper protective equipment (also specified in the MSDS).

The Occupational Safety and Health Administration (OSHA) requires employers to keep a list of all hazardous chemicals and materials used on their premises and maintain a file of MSDSs on the chemicals. Employers also must train their employees to use the chemicals and respond properly should an accident occur. The MSDS includes the following information:

- Chemical name and symbol.
- Manufacturer's contact information.
- Flammability and reactivity data (vapor pressure and flash point information).
- Health hazards.
- Overexposure limits and effects.
- Spill-cleanup procedures.
- Personal-protection information.

Material Storage

All materials must be handled and stored properly so they remain safe and effective. See Figure 5-27. Both raw and processed materials must be appropriately handled and stored to avoid damaging them or wasting the effort it takes to process them. During the manufacturing process, materials usually need to be transported several times and often need to be stored between processing steps.

Prior to being processed, raw materials are sometimes sent to storage. Here, they must be stored correctly so they are not exposed to contaminants. Care must be taken to prevent the materials from rotting, rusting, or deforming in any way.

Figure 5-27. A common sight on many large farms is grain-storage silos. These silos contain and protect the crop from the weather. (©iStockphoto.com/Fertnig)

Think Green

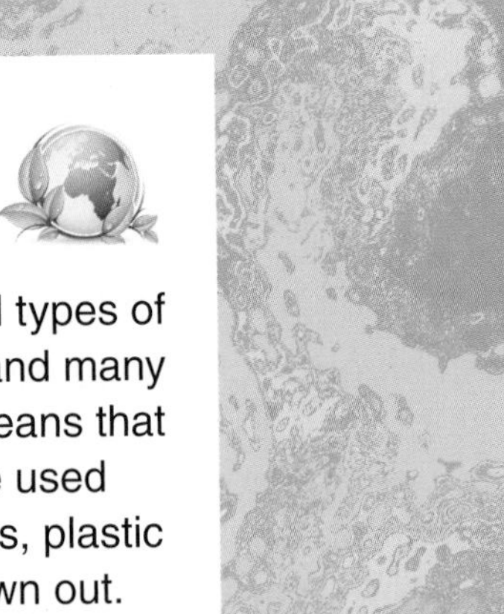

Recycling

Recycling is probably not a new concept to you. Several types of materials have been considered recyclable for a long time, and many communities participate in recycling programs. Recycling means that the useful parts of used materials may be reprocessed to be used again. Everyday items, such as glass bottles, aluminum cans, plastic packaging, or paper, can be recycled rather than being thrown out. Simply throwing out these materials means raw materials must be used to create new products. Also, some of these materials do not decompose well, and may sit in garbage dumps for a very long time. Recycling will help eliminate waste and reduce our depletion of raw materials.

Many products are processed to meet quality specifications. Storage facilities have a responsibility to make sure the quality is maintained throughout the storage period. Processed materials usually must be handled and stored even more carefully than raw materials. Finished products are often packaged with protective material, which helps protect them from damage while they are being handled. It is especially important to use care when handling finished products.

Summary

Humans have used materials throughout history. Materials are key to our existence. They are used for almost everything we need and want. Materials are found in our clothing and shelters. They become the products and fuels we use. Materials are the food and fibers sustaining our lives.

All materials are either renewable or exhaustible. Renewable resources can be grown and harvested. Proper management will give us a constant supply of these resources. Exhaustible resources have a limited supply. Once used, they cannot be replaced. They require careful use, if future generations are to have them, too.

All materials must be grown or located. They then must be harvested or extracted. Finally, they are processed into industrial or standard materials.

Engineering materials are an important group of materials. They have set structures. These materials are the rigid materials used in many durable products.

Metals, ceramics, polymers, and composites are engineering materials. Each type of material has its own set of properties. These properties include the physical, mechanical, thermal, chemical, electrical and magnetic, optical, and acoustical qualities of the material. The combination of properties in a material dictates its range of use. Materials are a basic input to technological actions as systems. They are used in agricultural, energy conversion, information processing and communication, construction, manufacturing, medical, and transportation technologies.

STEM Connections

Integrated STEM Curriculum

Mathematics

Research the properties of several materials in reference books or on the Internet. Develop a set of graphs comparing the properties of these materials.

Science

Look at a pencil. List the properties the materials in it have. Describe why these properties are important.

Curricular Connections

Social Studies

Select a type of material, such as steel, plywood, or nylon. Trace its historical development:

- List the important events in its development.
- Indicate where and when these developments took place.
- List any inventors contributing to the development.

Activities

1. Use magazines to develop a picture set showing at least three uses of each of the following engineering materials:
   A. Metals.
   B. Ceramics.
   C. Plastics.
   D. Composites.
2. Complete an experiment comparing a property, such as strength, for two different materials.
3. Gather a group of 5 to 10 different materials. Arrange the materials in the kit first by hardness, then by smoothness, and finally, by density.
4. Select a material from your classroom lab. Prepare a presentation for the class reviewing the proper handling and storage procedures for the material. Include a handout with your presentation.

Test Your Knowledge

Do not write in this book. Place your answers to this test on a separate sheet of paper.

1. Paraphrase the definition of *natural material.*
2. Which of the following materials are natural? You can choose more than one answer.
   A. Plastic.
   B. Trees.
   C. Rocks.
   D. Petroleum.
3. Give an example of a material used to power engines, turbines, and other types of converters.

4. What is the name of materials having rigid structures?
5. Indicate which of the following are engineering materials. You can choose more than one answer.
   A. Modeling clay.
   B. Metal.
   C. Glass.
   D. Plastic.
   E. Concrete.
   F. Petroleum.
6. Trees and wheat are examples of _____ resources.
7. Resources that cannot be replaced once they are used are called _____.
8. Match the processes on the left with the descriptions on the right. Give one more example of each type of processing.

_____ Producing plastic.	A. Mechanical processing.
_____ Making lumber.	B. Chemical processing.
_____ Producing steel.	C. Thermal processing.

9. What is an industrial material?
10. Name at least three products of primary processing activities.
11. Explain what the term *property* means, in the context of this chapter.
12. List the seven properties of materials and define each.
13. Name three things that can happen to materials that are not properly stored.

Reading Organizer

Draw a bubble diagram for each main idea in the chapter. Make each of the main ideas the central bubble, while using details in smaller bubbles to surround the main points. An example from this chapter is shown as an example.

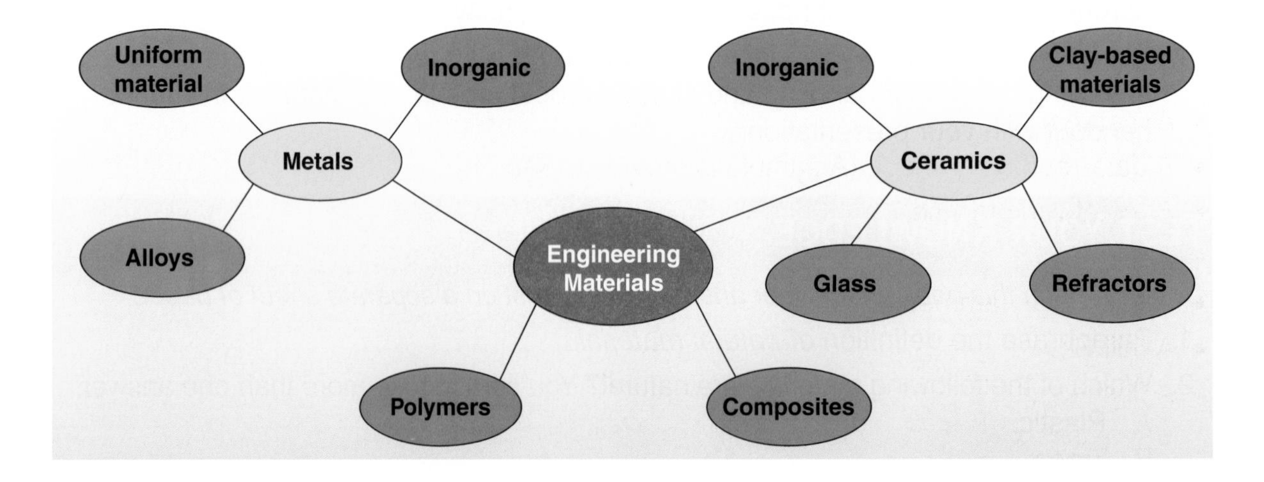

CHAPTER 5 ACTIVITY A

MATERIALS AS A RESOURCE

INTRODUCTION

The world is made of materials. Science tells us that materials are found in three states: gases, liquids, and solids. Solid materials are often called *industrial materials*. They can be grouped as metals, ceramics, polymers, and composites.

Each material has a unique set of properties. These properties tell us how a material acts or performs under certain conditions. As discussed in this section, there are seven major material properties:

- Physical.
- Mechanical.
- Chemical.
- Thermal.
- Electrical and magnetic.
- Optical.
- Acoustical.

In this activity, you are able to test your knowledge of materials and their properties.

EQUIPMENT AND SUPPLIES

- A material kit.
- A data-recording sheet. (A sample is provided. See **Figure 5A-1.**)
- A metalworking vise.
- Pliers.
- A ball-peen hammer.
- A mill file.
- A flashlight.

Data-Recording Sheet Material Identification and Property Testing						
Group:					Date:	
Specimen Number	**1**	**2**	**3**	**4**	**5**	**6**
Name of material						
Type of material						
Density						
Hardness						
Ductility						
Light reflectivity						
Torsion strength						

Figure 5A-1. A sample data-recording sheet. (Do not write in the book.) Fill in the sheet given to you, as materials are tested. On the back of the sheet, list products for which each material is suited. Remember, manufacturers select a material for its properties.

SAFETY

Review with your teacher any safety rules concerning the use of tools. If not properly handled, these tools can cause injuries.

PROCEDURE

Your teacher will divide the class into groups of three to five students. Each group should do the following:

1. Obtain a material kit containing a set of six materials. You need two samples of each material that are the same size. The materials should also be numbered.
2. Obtain a data-recording sheet.
3. Carefully look at each material. Write down the name of each material (such as copper, wood, or steel) on the chart.
4. List under its name whether the material is a metal, ceramic, polymer, or composite.
5. Select one sample of each material.
6. Arrange the samples by weight. Since they are the same size, this shows which are denser than the others.
7. Put a *1* in the column of the heaviest sample in the "Density" row, a *2* in the column of the next heaviest, and so on.
8. Test the hardness of each sample. Take two file strokes on the surface of each sample. See **Figure 5A-2.** The harder the material is, the less the file cuts.

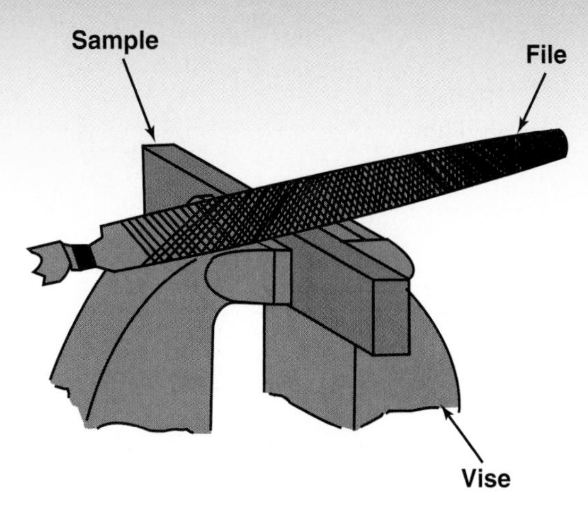

Figure 5A-2. How to test a material for hardness.

SAFETY

Use a file with a handle fitted over the tang. Be careful not to run a hand or finger over filed edges. The edges might be sharp.

9. Mark a *1* in the column of the hardest material in the "Hardness" row, a *2* in the column of the next hardest, and so on.
10. Test the ductility of each sample. Place one end of the sample in the vise. Grip the other end of the sample with a pair of pliers. Bend the sample back and forth three times. See Figure 5A-3. Note how easy it is to bend each sample. Also look for any cracking or breaking along the bend line.
11. Place a *1* in the column of the material that was easiest to bend and did not crack in the "Ductility" row. This is the most ductile sample. Place a *2* in the column of the next most ductile sample, and so on.
12. Take the second set of material samples.

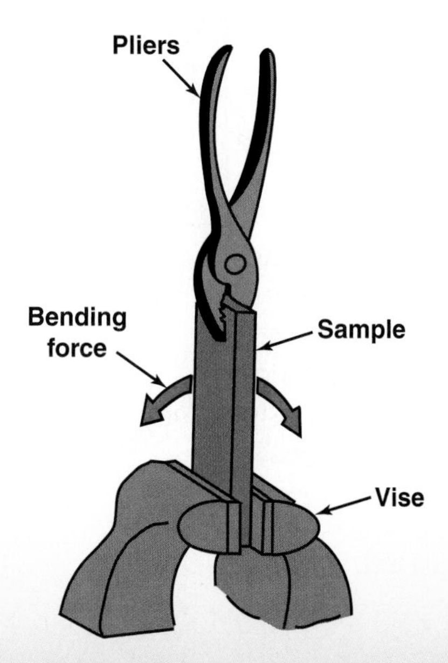

Figure 5A-3. Testing the ductility of a material.

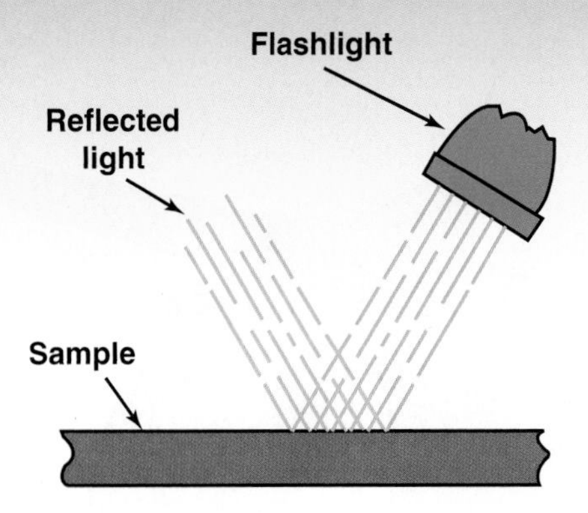

Figure 5A-4. Use a flashlight to test a material's ability to reflect light.

13. Test the light-reflecting property of each material. See Figure 5A-4. Shine the flashlight on each sample. A material that shines is reflecting light.
14. Place a *1* in the column of the material reflecting the most light in the "Light reflectivity" row. Place a *2* in the column of the next best reflector, and so on.
15. Test the torsion strength of each material. Place one end of the sample in the vise. See Figure 5A-5. Grip the other end of the material with a pair of pliers. Try to twist the material. The material that is the hardest to twist has the highest torsion strength.

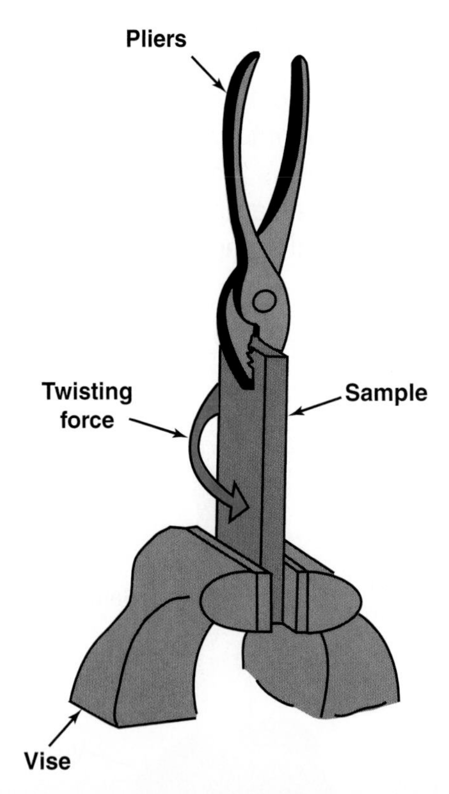

Figure 5A-5. Performing a torsion test.

Safety

Pliers can pinch fingers hard enough to draw blood or raise blisters. Keep your hands away from the jaws. Do not use pliers with damaged or worn jaws. A slip can cause an injury.

16. Place a *1* in the column of the material with the best torsion strength in the "Torsion strength" row, a *2* in the column of the next best, and so on.

Challenging Your Learning

Look around the room you are in. Identify three materials that are used in some item in the room. List two properties each material has that make it appropriate for its use.

CHAPTER 6
Energy and Technology

Did You Know?

- *Geo-* means "earth." *Thermal* means "heat." *Geothermal* means "earth heat."
- *Hydro* means "water." *Hydroelectric* means "making electricity using waterpower."
- Some 2000 years ago, the Greeks used mirrors to focus the Sun's rays on Roman ships, causing them to catch fire.

Objectives

The information given in this chapter will help you do the following:

- Explain the differences between natural uses and technological uses of energy.
- Explain how energy and work are related.
- Compare energy and power.
- Give examples of changing energy from one form to another.
- Summarize the different forms of energy.
- Give examples of several sources of energy.

- Summarize the three most common types of energy supplies.
- Give examples of alternate energy sources.

Key Words

These words are used in this chapter. Do you know what they mean?

bioenergy
biomass
chemical energy
coal
electrical energy
fission
fossil fuel
fusion
heat energy
hydroelectric
kinetic energy
mechanical energy
natural gas
nonrenewable resource
nuclear energy
peat
petroleum
potential energy
power
radiant energy
solar energy
thermal energy
tidal energy
wave energy
wind energy
work

Preparing to Read

As you read this chapter, make a list of examples of objects that require the use of energy to function properly. Think about what type of energy they use and all the alternative ways those examples could be powered. Use the Reading Organizer at the end of the chapter to categorize your examples.

About a half million years ago, early humans learned to make fire. They collected and burned wood. With the heat generated, they cooked food, warmed their homes, and made tools. Much later, ancient Egyptians developed the first sails to power their boats on the Nile River and Mediterranean Sea. Still later, the waterwheel was developed to raise water into irrigation ditches and power crude machines. In recent history, the steam engine was developed. This engine sparked the Industrial Revolution. The steam engine powered factories and transportation vehicles. Today, we are connected into a worldwide communication system. We are developing new ways to move people and cargo around the world and into space. New manufacturing processes are being used. Cutting edge medical treatments are available. Energy, in one form or another, is a key part of all these technological developments.

People use the word *energy* almost every day. They talk about the energy it takes to get out of bed. People say they don't have the energy to run a mile. They worry about the energy crisis the cost of foreign oil is causing. Not all people, however, understand the role of energy in their lives. Energy is one of the most fundamental parts of our universe.

This resource is used in natural and technological systems. In natural settings, our bodies transform food into energy to do work. See Figure 6-1. We use energy as we walk and run. As we think, read, and write, we use energy. Likewise, animals use energy as they go about their daily activities. They use energy to gallop, fly, build burrows, and do other natural activities. Also, energy is released as snow melts, waves crash onto rocks, and earthquakes shake the ground around us. See Figure 6-2.

Technological activities also use energy to do work, using several processes. Energy powers our cars, trains, and planes

Figure 6-1. Human activity requires energy we get from food.

Figure 6-2. These waves crashing on rocks contain a great amount of energy.

and lights our homes, stores, and offices. Medical systems use energy as illnesses are diagnosed and treated. Energy is used to plant, cultivate, and harvest crops. This resource heats and lights buildings and cooks food. Energy provides entertainment through sound systems and television sets. In factories and offices, it powers machines. There could not be technology or industry without energy. See Figure 6-3.

Figure 6-3. Industrial processes require large amounts of energy.

Energy and Work

Energy is essential for meeting human needs and wants. This resource is the foundation for raising standards of living. Over time, people have used more energy. Today, each person uses more than 100 times the energy early humans used. In fact, from 1950 to 2010, American energy use has more than tripled. See Figure 6-4. There is no reason to believe that this growth will not continue. The world's population continues to grow. Many developing nations are using more and more energy as their industries grow and standards of living increase. This increase in demand for energy is greatly offsetting energy-conservation measures in developed nations.

So what is energy? Simply defined, energy is the ability to do work. Energy provides a force. All forms of energy are associated with some type of motion. ***Work*** is the use of force to create movement. This use of force can be calculated using a simple formula: work equals the force required, times the distance traveled.

Energy is either moving or stored. Stored energy is called ***potential energy***. This energy is waiting to be used when needed. Moving energy is called ***kinetic energy***. This energy is the energy of action. To see the difference between potential energy and kinetic energy, think of a pencil. If the pencil is placed on a desk, it has potential energy. The energy is stored because of the pencil's position above the floor. If the pencil falls to the floor, it has kinetic energy. The pencil has the energy of motion. Picking it up and placing it on the desk restores the potential energy. The kinetic energy the person moving the pencil uses is stored as potential energy.

Likewise, think of a rubber band. If you stretch it, your kinetic energy (moving hands) is changed into potential (stored) energy in the rubber band. As it slips from your grip, the potential energy is changed

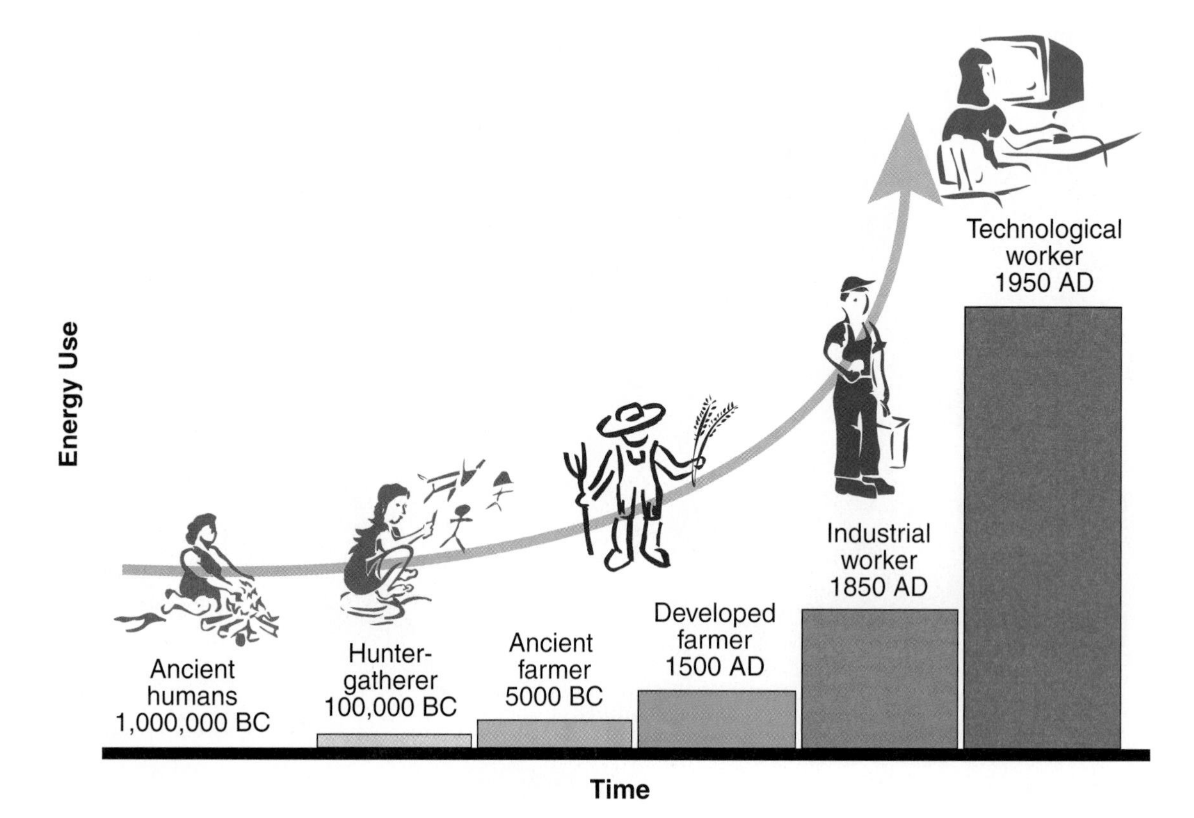

Figure 6-4. Our energy use has changed over time.

into kinetic energy as the rubber band moves through the air.

Therefore, we can say kinetic energy is the energy an object has because of its motion. Potential energy is stored energy that depends on the relative position of an object. See **Figure 6-5.**

Energy and Power

Power and *energy* are two words people often confuse. They are not the same thing. As we said, energy is the ability to do work. ***Power***, on the other hand, is the amount of work done in a set period of time, or the rate at which work is done. The amount of work done is the result of using energy for a definite time span. The rate at which energy is changed from one form to another or moved from one place to another is also called *power*.

Suppose you pedal your bicycle for one mile. That is work. You have used energy to create this movement. The energy you used can be calculated. Typically, one of two measurements is used to measure the energy consumed. The first unit of measure is called a *British thermal unit (Btu)*.

A Btu is defined as the amount of ***heat energy*** needed to raise the temperature of one pound of water by 1°F, at sea level. One Btu equals the energy in one kitchen match. As you can see, it is a fairly small unit of measurement. Often, we measure energy by thousands of Btu. The energy in an average candy bar equals 1000 Btu. The Btu is the measurement unit used in the United States. In countries using the metric system, energy is measured in newton-meters, or joules (pronounced *jewels*). Also, scientists around the world use this unit of measurement. One Btu equals 1000 joules.

Potential Energy

Kinetic Energy

Figure 6-5. The water in the reservoir on the top has potential energy. The water rushing through this power plant has kinetic energy.

Figure 6-6. Can you explain energy, power, and work, in terms of these gondola operators?

When you measure the energy it takes you to ride a bicycle one mile, you are calculating *power.* As you just read, power is energy used over time. See Figure 6-6. Power can be calculated with a simple formula: power equals the distance traveled, times the force required, divided by the time.

Figure 6-7. These wind generators change mechanical power into electrical power.

The Nature of Energy

Energy is everywhere. People and machines use it. Energy is used to move things, light areas, lift weight, and warm spaces.

One form of energy can be changed into another. See Figure 6-7. Stored chemical energy in a flashlight battery can be converted into light energy. Your body uses the stored chemical energy in food to do work. A television set changes ***electrical energy*** into light and sound energy. An automobile engine changes the chemical energy in gasoline into heat and mechanical energy to power the car.

Energy cannot be created, however, nor can it be destroyed. As we change its form, it can be wasted. For example, the lightbulb is an energy converter. A lightbulb's purpose is to change electrical energy into light energy. In doing this task, not all the electrical energy ends up as light energy. Some of it becomes heat energy. The electrical energy that becomes heat energy is not destroyed. This energy can be considered wasted, however. The heat energy is not in a form that is useful in the situation, nor is it wanted.

Forms of Energy

Not all energy is alike. See Figure 6-8. Energy appears in several different forms:

- **Light energy, or *radiant energy.*** This is the energy in the atomic motion present in sunlight and fire.
- **Heat energy.** This is the energy in the increased activity of molecules found in matter whose temperature has been raised.

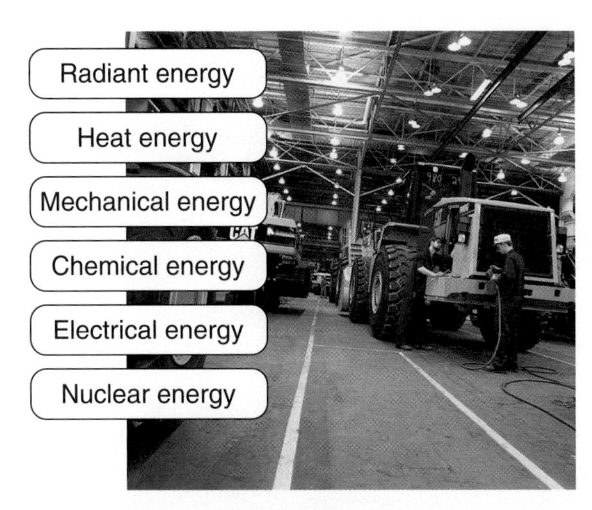

Figure 6-8. How many forms of energy can you see in this picture? (Caterpillar, Inc.)

- ***Mechanical energy.*** This is the energy in the movements that humans, animals, moving water and air, and machinery produce.
- ***Chemical energy.*** This is the energy in reactions between substances such as carbon (coal) and oxygen.
- **Electrical energy.** This is the energy in the movement of electrons in matter, caused by lightning, batteries, and generators.
- ***Nuclear energy.*** This is the energy from the splitting of atoms.

Types of Energy Sources

Technological systems need available, easy-to-control energy. When the energy is controlled, it can deliver power when and where it is needed. Energy sources were limited for early humans, who relied on natural sources. People burned wood to cook food and keep warm. They used water and wind to do work for them.

Still, these energy sources sometimes failed. Rivers were low during droughts. Waterwheels did not always have enough waterpower to drive them. Sailing ships could not move when winds did not blow. Windmills sometimes stopped turning. Over time, people developed energy sources they could control. They moved from dependence on wind power and waterpower.

Primary and Secondary Energy Resources

Energy resources used today can be considered to be either primary or secondary. Primary energy resources are in the form of natural resources. These include resources such as wood, coal, petroleum, natural gas, uranium, wind, water, and sunlight.

The Sun is key to much of the energy we use. This body is a huge energy "factory." Every day, the Sun gives Earth enormous amounts of heat and light. See Figure 6-9. In 20 days, Earth receives as much energy from sunlight as is stored in all coal, oil, and natural gas reserves in the world. Even so, only 13% of all ***solar energy*** leaving the Sun actually reaches Earth.

Some of the sunshine striking Earth creates kinetic energy. Another large share of it gets stored as potential energy. Solar activity accounts for wind power, water-power, and fossil fuels.

Wind, a type of kinetic energy, comes from uneven heating of the atmosphere by the Sun. Heated air rises and causes cold, heavier air to move in under it. The resulting movement is called *wind*. This movement is often strong enough to drive windmills and electric generators.

The Sun also works as a giant water pump. Radiant energy heats water until it vaporizes. The water vapors rise into the sky, where they form clouds. When cooled, the vapors become a liquid again and fall as rain. The rain becomes rivers that can power hydroelectric power plants.

Figure 6-9. The Sun's energy acts as a huge water pump for Earth. This energy heats water. The water turns to vapor. The vapors rise to form clouds. In the clouds, the water vapor condenses to make rain. (Natural Gas Supply Assoc.)

Much of our solar energy becomes stored in plant life. This energy remains stored until something releases it. Animals and humans eat plants for energy and growth. Burning releases the energy in wood and other plants. Plants and animals decay and become coal and petroleum. Burning these materials also releases energy.

Many primary energy resources are changed into more usable forms. These altered forms are secondary energy resources. Examples of secondary energy resources include electricity, gasoline, and propane.

Renewable and Nonrenewable Energy Resources

Primary energy sources can be either renewable resources or ***nonrenewable resources***. Renewable resources are inexhaustible energy sources. Natural or human actions can replace them. These resources include solar energy, wind energy, wave energy, biomass (wood or crops, such as sugar), geothermal energy, and hydropower.

Nonrenewable energy resources have limited quantities. When they are used up, they cannot be replaced. They are gone forever. Typical nonrenewable energy sources include fossil fuels (coal, oil, and natural gas) and uranium. Today, the fossil fuel resources provide slightly less than 80% of our energy.

Common Energy Supplies

Developed countries, such as those in North America and Western Europe, use many different sources to meet their growing demand for energy. The sources used in each location are influenced by the natural resources available and the resources that can be transported from other areas in the world. These common resources are often called *primary energy sources*. Other sources can be called *alternate energy sources*, which are discussed later in this chapter. The most used and practical primary energy sources for many countries are the following:

- Fossil fuels.
- Waterpower.
- Nuclear energy.

Fossil Fuels

All fossil fuels were formed millions of years ago, long before the time of the dinosaurs. For this reason, these fuels are called ***fossil fuels***. The age they were formed is called the *Carboniferous period.* This period gets its name from carbon, the basic element in coal and other fossil fuels.

Fossil fuels are the remains of once-living matter. They are the products of dead plants or animals. A process called *decay* formed them. When living matter decays, it reacts with oxygen. Fossil fuels, however, were formed under conditions where there was very little oxygen. Therefore, the dead matter did not fully decay. The partially decayed matter became a material high in carbon.

The carbon in fossil fuels makes them burn easily. Burning is rapid oxidation and requires large quantities of oxygen. In the process of burning, large amounts of heat energy are released. Fossil fuel deposits can be found in several forms. These include peat, coal, petroleum, and natural gas.

Peat

Peat is decayed plant matter that sank in swamps. This matter is made up of moss, reeds, and trees. Other plant matter and soil covered and pressed on it. The decaying matter compressed into solid material. This matter became somewhat similar to coal. In some parts of the world, it heats homes. Also, gardeners use peat as a soil additive. See Figure 6-10.

Coal

Coal is also decayed plant matter. See Figure 6-11. Coal started out the same way peat did. The weight of soil and rock, however, greatly compacted the coal. Coal became a hard, black, rocklike substance. This substance is mined in deep mines or in strip mines closer to the surface.

Industry and homes once widely used coal. Coal provided heat for powering steam locomotives. See Figure 6-12. Coal-fired stoves and furnaces were common in homes, factories, and public buildings. Today, coal is a major fuel in electricity-generating plants. Also, it is used in steel-making and as a base for many other products.

Figure 6-10. Peat that has been cut for fuel. (Vapo Oy)

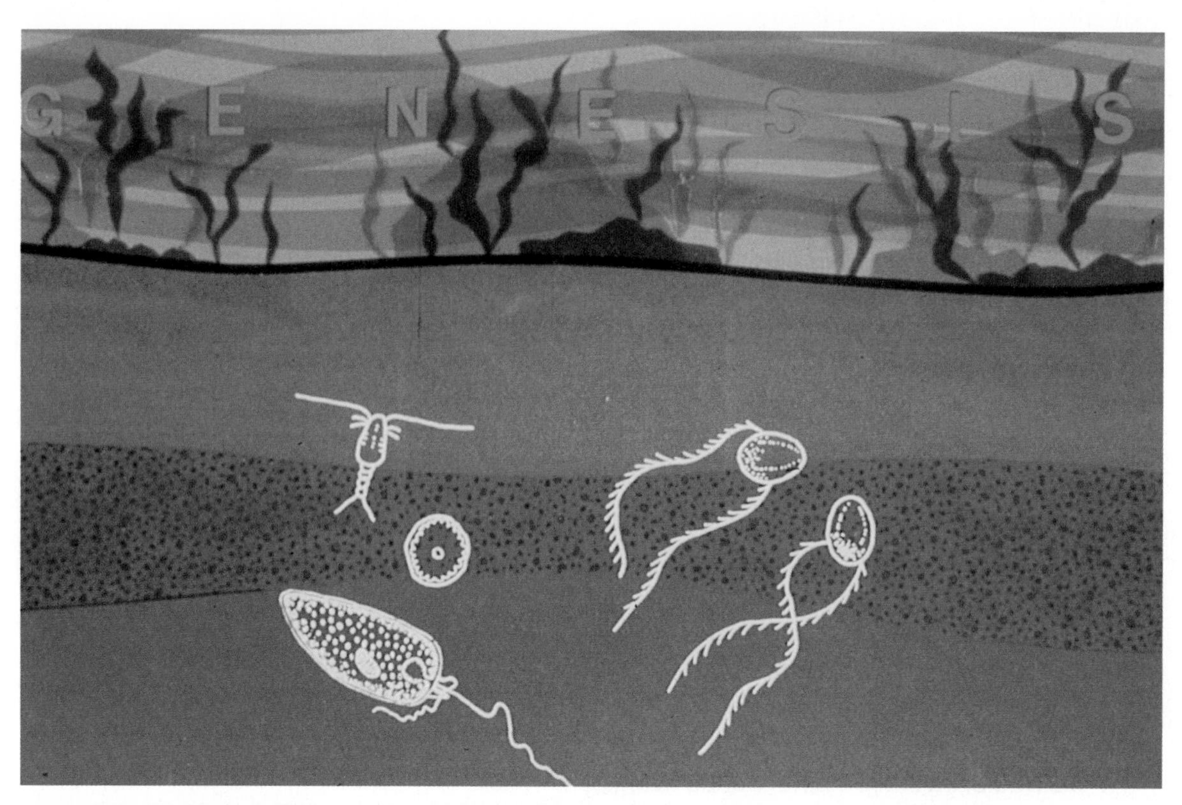

Figure 6-11. Fossil fuels are formed from plant and animal life that existed millions of years ago. (Standard Oil of California)

Figure 6-12. This wood-burning steam locomotive can also burn coal.

Figure A. Livestock waste can be a major source of alternative energy. (Tillamook County Creamery Assoc.)

Technology Explained

biogas: a gas produced by processing animal and plant waste.

As petroleum supplies become scarcer, people are searching for new energy sources. One of these is called *biogas*. This is gas produced by processing biological waste.

A dairy cow can produce over 100 lbs. of manure per day. A large dairy farm with a thousand cows has to deal with over 50 tons of manure per day. See **Figure A.** This waste must be disposed of without harming the environment. One way to do this is to convert it into a gas that can be used to generate electricity.

A process called *anaerobic generation* produces biogas. This process has been used in Europe for a number of years. In the United States, it is starting to be a source of alternate energy.

Anaerobic generation starts in the dairy barn. See **Figure B.** Here, the manure is scraped into a collection pit. A low-volume pump moves the manure to an outside tank. From there, the waste is deposited into a digester. This is an airtight, aboveground, concrete tank. A heat exchanger maintains the temperature at about 95°F. This is the temperature at which anaerobic bacteria work best.

Over the course of two to four weeks, the bacteria work on the manure. They break the waste into smaller particles. These particles are primarily organic acids, hydrogen, and sugar. At the same time, another group of organisms is at work. These organisms are archaebacteria, which change the acids and hydrogen into a gas. This biogas is a mixture of about 40% carbon dioxide and 60% methane.

The gases rise to the top of the tank and are collected. They are fed into a natural gas engine connected to an electric generator. When the gases are burned in the engine, they provide the energy to drive the generator. The electric output is fed into the electric grid (power lines) by using a transformer on a power pole.

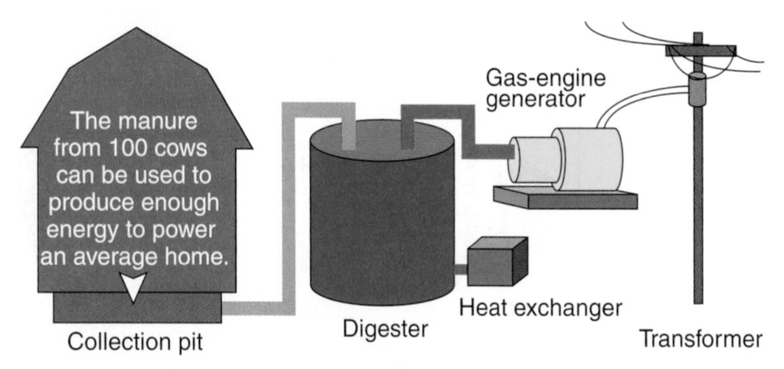

Figure B. The biogas-production system.

Petroleum

Gradually, petroleum products and natural gas replaced coal as the primary technological fuel. ***Petroleum*** is a thick liquid thought to be the decayed remains of plant and animal life. Similar to coal, it became buried and partially decayed. Petroleum is pumped from pockets, or wells, deep below ground.

This liquid requires a refining process that uses heat to separate the petroleum into more usable products. These products include liquids such as gasoline, diesel fuel, kerosene, and fuel oil. See Figure 6-13.

Natural Gas

Natural gas is lighter than air. This gas is mostly made up of a gas called *methane*. Natural gas is a simple chemical compound made up of carbon and hydrogen atoms.

This gas is usually found underground near petroleum. Similar to petroleum, natural gas is pumped from below ground. Natural gas is used for many of the same purposes for which we use petroleum. This gas is, however, most often used to heat buildings and produce electric power. Natural gas is the key ingredient for making many plastics.

The process that produced fossil fuels is called *fossilization*. Fossilization still goes on. We, however, are using these fuels 100,000 times faster than they are being replenished. One day, fossil fuels will be gone. Therefore, we call them *nonrenewable*, or *exhaustible*.

Waterpower

When it rains, the water striking the earth becomes streams. These streams run together to form rivers that flow toward the oceans. This moving or falling water has kinetic energy that can be used to do work. For early humans, this was a primary source of power. In early times, moving

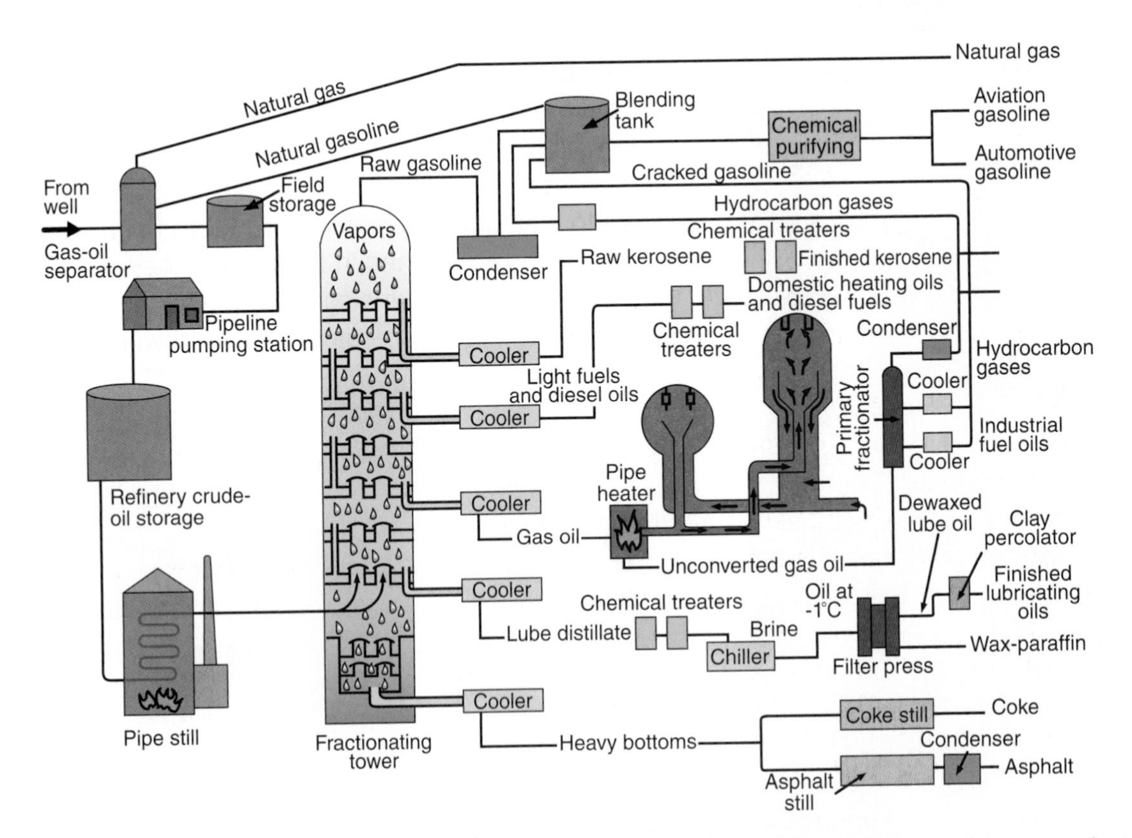

Figure 6-13. When petroleum is refined, many different products are produced. Some of these products are used as fuels for their energy.

Career Highlight

Petroleum Engineers

The Job: Petroleum engineers search for oil or natural gas reserves. They work with other specialists to select drilling methods to be used, monitor drilling activities, develop enhanced recovery methods, and oversee other production operations. These engineers also design equipment and processes for oil and gas recovery operations.

Working Conditions: They might work in offices. These engineers spend a considerable amount of time outdoors at oil and gas exploration and production sites. Most petroleum engineers work in Texas, Louisiana, Oklahoma, Alaska, or California; on offshore sites; or in other oil-producing countries.

Education and Training: A bachelor's degree in engineering is required for almost all entry-level petroleum-engineering jobs. The degree program generally involves general engineering classes, a concentration in petroleum engineering, and a number of mathematics and science classes.

Career Cluster: Science, Technology, Engineering & Mathematics

Career Pathway: Engineering and Technology

Science, Technology, Engineering & Mathematics

water turned wooden wheels called *waterwheels*. See Figure 6-14. The turning wheels created mechanical energy that turned big stone wheels to grind wheat and corn into flour.

Today, we use moving water to make electricity. This action takes place in hydroelectric power plants. *Hydro* means "water." ***Hydroelectric*** means "making electricity from waterpower." To make electricity from water, dams are built across rivers. See Figure 6-15. These dams hold river water, forming a reservoir behind the dam. The water in the reservoir flows through pipes into turbines that turn electric generators.

Unlike fossil fuels, hydropower is a renewable energy source. This means it will always be available. Rainfall constantly replaces water in rivers. The Sun's heat evaporates surface water. This water falls back to Earth as snow, rain, and dew. This cycle occurs endlessly.

Nuclear Energy

Nuclear energy is the energy found in atoms. Scientists learned how to unlock this energy in recent times. In the 1940s, they discovered a process that causes the atom's nucleus to be split apart. In turn, this causes other atoms to split. Each splitting is called a *reaction*. Once started, the reactions continue in a process called ***fission***. See Figure 6-16. With fission, a tremendous amount of heat and light energy is released. This energy, when let out slowly, can be harnessed to generate electricity. The great heat produced is usually

Figure 6-14. A restored waterwheel used to power a flour mill in the nineteenth century. (Zureks)

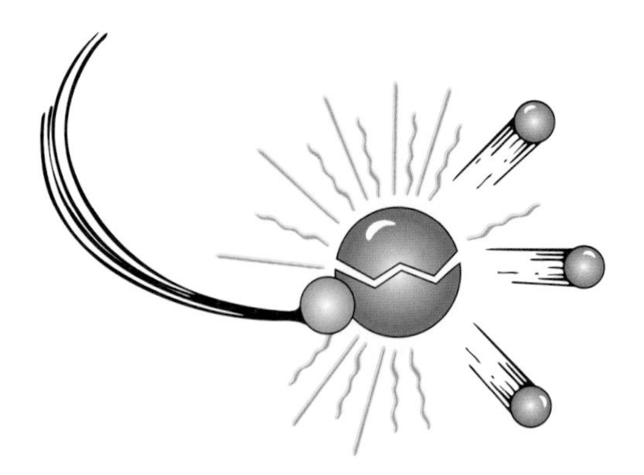

Figure 6-16. Nuclear energy is released when atoms are split. (Westinghouse Electric Corp.)

transferred to water. The water vaporizes into steam, which spins a turbine. An electric generator, coupled with the turbine, produces electric power.

A second type of nuclear reaction is called ***fusion***. This type of reaction causes parts of hydrogen atoms to fuse (join). Similar to fission, fusion releases huge amounts of energy.

The Sun releases energy through the fusion of hydrogen and helium atoms. Our

Figure 6-15. This dam on the Sacramento River in California is used for flood control and electricity generation.

THINK GREEN

Compact Fluorescent Lamps (CFLs)

Compact fluorescent lamps, or CFLs, may be used to replace incandescent bulbs. CFLs use gas and ballasts to light. Because they do not use filaments, CFLs do not create excess heat in addition to light. CFLs use much less energy to light a room than incandescent bulbs. Because of the low energy usage, they also last much longer than conventional bulbs. Using a little less energy for your lightbulbs not only helps you save money, but it is also helpful in that less electricity may need to be generated overall.

technology has been able to start fusion reactions. So far, it has been unable to keep them going.

ALTERNATE ENERGY SOURCES

People are becoming increasingly concerned about the use of nonrenewable energy resources. They are turning to alternate energy sources to reduce our primary dependence on fossil fuels. Individuals, government agencies, and private companies are studying these sources. Large-scale, efficient methods to harness and use alternate energy sources are being explored. Similar to waterpower, these sources will always be available. We will never run out of them. The more important alternate energy sources are solar, wind, ocean, geothermal, and biomass energy.

Solar Energy

The Sun has been an energy source since the beginning of time. Plants use the Sun's light to make food in a process called *photosynthesis*. Early in history, people learned to use solar energy. This energy was used to dry food and warm buildings. When you hang a wet swimsuit outside to dry, you are using solar energy to do work. Today, the Sun is being used to heat water for homes and businesses. Campers can use solar energy to cook food. The two major commercial uses for solar-energy technology are to heat water and produce electricity. See Figure 6-17.

Figure 6-17. The solar cells on the top of this sign convert solar energy into electricity.

Wind Energy

Wind is an indirect use of solar energy. This movement can be used to do work. A sailboat uses ***wind energy*** to push it through the water. Ranchers use wind energy to pump water for their cattle. In Holland, windmills have been used for centuries to pump water from low-lying areas. Wind, similar to the Sun, is free and increasingly harnessed for electricity production. See Figure 6-18.

Figure 6-18. This windmill pumps water for livestock in Australia.

Ocean Energy

Energy is present in the ocean. The ocean can produce two types of energy. These are thermal energy, from the Sun's heat, and mechanical energy, from the tides and waves.

Oceans cover more than 70% of Earth's surface. This makes the oceans huge solar collectors. The Sun's heat primarily warms the surface water. This water is warmer than the water at greater depths. The temperature difference creates ***thermal energy***, which can be used to make electricity.

Wave and tidal energy are other types of ocean energy. See Figure 6-19. Unlike thermal energy, they use the mechanical energy of moving water. Mechanical ocean energy is quite different from thermal ocean energy. The Sun is the ultimate source for thermal energy. The gravitational pull of the Moon, however, primarily drives tides. The wind primarily drives waves.

Figure 6-19. Ocean waves and swells are sources of energy that can be used to produce electricity. (Mila Zinkova)

Tidal energy uses the differences between high tide and low tide. ***Wave energy*** uses the forces present in the coming and going of waves near the shore. Both of these sources can power electricity-generating equipment.

Geothermal Energy

Not all renewable energy resources, however, come from the Sun. Geothermal energy uses the natural heat below Earth's surface. Earth is a molten mass of hot material covered by a thick crust. Deep under the surface, water sometimes makes its way close to the hot rock and turns into hot water or steam. See Figure 6-20. The hot water can reach temperatures of more than 300°F. This water is at a temperature above its boiling point. When this hot water comes up through a crack in Earth, we call it a *geyser*, or *hot spring*. Drilling a well can also tap this water. This hot water from below the ground can warm buildings, keep roads clear of ice, and power electricity-generating plants. See Figure 6-21.

Figure 6-20. Geysers can be used to harness geothermal energy.

Biomass Energy

Biomass includes all the living organisms in an area. These organisms are matter people often treat as garbage. Biomass includes materials such as yard waste, dead trees, and tree branches. These materials can be paper products and other household waste. Biomass includes animal waste and by-products. Also, it can be housing rubble, leftover crops, and bark and sawdust from lumber mills.

These materials can be used as an energy source (fuel). Also, biomass can be recycled and made into other products, such as paper and fertilizer. Some biomass is made into compost (decayed plant or food products) that helps plants grow. Biomass can be used for ***bioenergy***, which is defined as the energy from organic matter. Bioenergy is used in three ways:

- **Biochemicals.** Biomass can be converted into chemicals to generate electricity.
- **Biofuels.** Biomass can be converted into liquid fuels for transportation.
- **Biopower.** Biomass can be burned directly to generate electricity.

Figure 6-21. Geothermal energy from Earth can be used as an energy source. This energy is seen as a geyser. (Aminoil UAS, Inc.)

SUMMARY

Energy is key to our modern way of life, and it heats our homes, provides us with light, powers our entertainment devices, and makes our vehicles move. We use energy in the forms of radiant, or light; heat; mechanical; chemical; electrical; and nuclear energy. Fossil fuels—peat, coal, petroleum, and natural gas—primarily provide this energy. These fuels, along with nuclear-energy resources, are exhaustible. They will not always be available. To conserve these energy resources, people are turning to inexhaustible sources. These sources include water (hydro), solar, wind, ocean, geothermal, and biomass energy sources.

STEM CONNECTIONS

Science

Draw an illustration describing the geology of petroleum or a coal deposit.

Science

Prepare a display explaining the science behind geothermal or thermal ocean energy.

Mathematics

Use the Internet or other resources to discover the energy use for several countries. Prepare a graph showing the relationship between the types of energy and countries selected.

CURRICULAR CONNECTIONS

Language Arts

Trace the roots of several words used in discussing energy. Such words might include *energy*, *geothermal*, *hydroelectric*, and *biomass*.

Social Studies

Develop a map highlighting the distributions of energy reserves (such as fossil fuels and uranium) in the United States or world.

Activities

1. Visit the local electric company and find out how their electricity is generated. Prepare a poster to show the processes it uses to make electric power.
2. Make a model of a petroleum deposit and a well used to extract the resource.
3. List all the ways you use energy. Indicate the type of energy used with each activity. Describe the probable source of that energy.

Test Your Knowledge

Do not write in this book. Place your answers to this test on a separate sheet of paper.

1. Summarize the different ways energy is used in natural and technological systems.
2. The ability to do work is called _____.
3. Using force to create motion is called _____.
4. The amount of work done in a period of time is called _____.
5. List one way energy is wasted when it is transferred from one form to another.
6. Label the six forms of energy.
   A. The energy present in sunlight.
   B. The energy in reactions between substances.
   C. The energy caused by batteries.
   D. The energy moving water produces.
   E. The energy found in matter whose temperature has increased.
   F. The energy caused by atoms splitting.
7. Paraphrase the difference between primary energy resources and secondary energy resources.
8. List the four major types of fossil fuels.
9. Another name for waterpower is _____.
10. Energy from natural hot water or steam from Earth is called _____.
11. Energy from organic matter is called _____.

Reading Organizer

Copy the following chart to a sheet of paper. In the Object column, list at least three examples of everyday objects that require the use of energy. In the remaining columns, explain what type of energy is used by your examples and give alternative ways of powering the examples.

Object	**Current Energy Source**	**Alternative Energy Source**
Example: Outdoor lamps	Electrical energy	Light energy

CHAPTER 6 ACTIVITY A

ENERGY AS A RESOURCE

INTRODUCTION

People have always looked for ways to do work more easily. They want to do jobs efficiently and quickly. Therefore, humans have invented tools and machines. One of the most important inventions is the electric motor. We can find such motors in use almost everywhere. Motors power machines in factories, run refrigerators and freezers, move air through furnaces and air conditioners, and move the hands on wall clocks. Think of your home. How many uses of motors can you list?

This activity lets you build a simple motor. With the help of your teacher, you can see how the motor converts energy into power. A motor changes electric energy into mechanical motion. See Figure 6A-1.

Device	Energy Input	Energy Output
Electric motor	Electric	Mechanical
Electric generator	Mechanical	Electric
Battery	Chemical	Electric
Electric light	Electric	Light and heat
Oil burner	Chemical	Heat and light
Windmill	Mechanical (linear)	Mechanical (rotary)

Figure 6A-1. Ways of converting energy from one form to another. Can you add to the list? Try to do this by listing five energy-converting devices you see every day.

Equipment and Supplies

- Materials listed on the bill of materials. See Figure 6A-2.

Part Number	Quantity	Description	Size	Material
1	1	Base	1/2″ x 4″ x 6″	Plywood
2	2	Bearings*	14-gauge x 10″	Copper wire
3	1	Coil**	18-gauge x 72″	Copper wire
4	4	Magnets	1/4″ x 2″ diameter	Speaker magnets
5	4	Screws	1/2″ x No. 6	Sheet metal
6	2	Connecting wire**	18-gauge x 12″	Copper wire

***These two parts are made from one 20″ piece of 14-gauge wire.**
****These parts are made from one 96″ inch length of 18-gauge wire.**

Figure 6A-2. The bill of materials for making an electric motor.

- 72″ of 18-gauge magnet wire.
- 20″ of 14-gauge, solid, uninsulated copper wire.
- Thread to tie the coil together.
- A coil-winding jig. See Figure 6A-3.

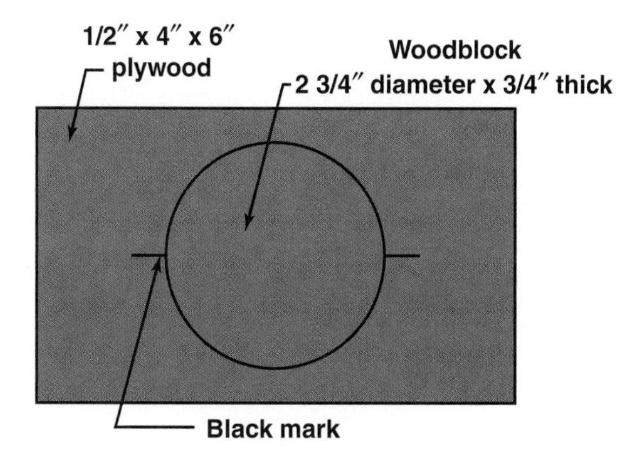

Figure 6A-3. A coil-winding jig. About five are required.

- A bearing-bending jig. See Figure 6A-4.

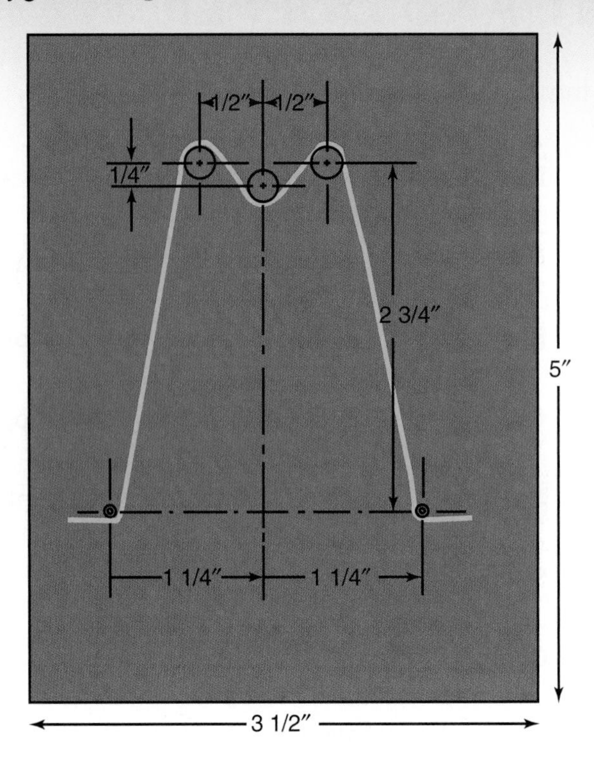

Figure 6A-4. A bearing jig.

- A 1 1/2- or 6-volt battery.
- A scratch awl.
- A hammer or mallet.
- Wire cutters.
- A screwdriver.
- A utility knife.
- Abrasive paper.

PROCEDURE

Preparing to Make the Motor

1. Gather the materials needed to make the motor. See Figure 6A-5:

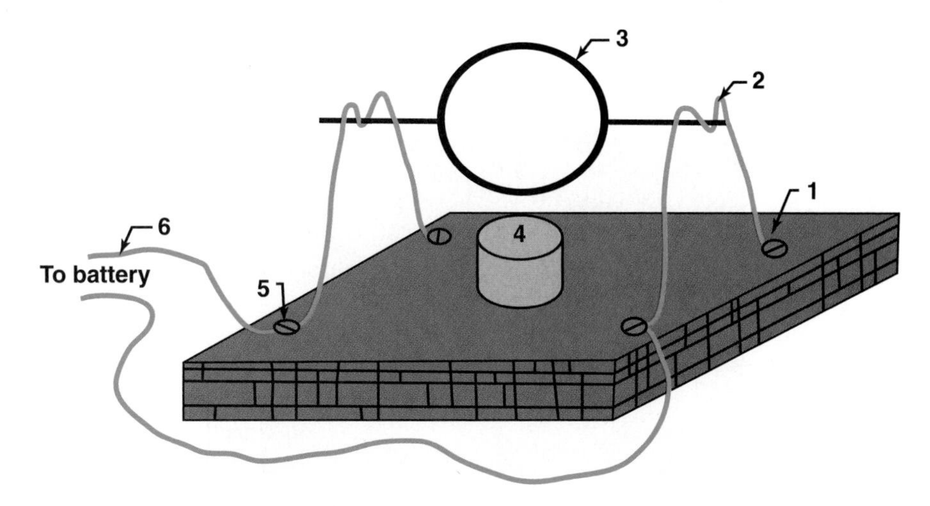

Figure 6A-5. An electronic motor. Refer to Figure 6A-2 for part names.

- A 1/2″ × 4″ × 6″′ plywood base.
- One piece of 18-gauge copper wire, 96″ long (72″ for the coil and 24″ for the two connecting wires).
- One piece of 14-gauge wire, 20″ long.
- Four 1/2″ × No. 6 pan-head sheet metal screws.
- Four 1/4″ × 1″ magnets (or a substitute your teacher suggests).

2. Each group of four to six students at a workbench should secure the following:
   - A coil-winding jig.
   - A bearing-bending jig.
   - A battery.
   - A scratch awl.
   - A hammer or mallet.
   - A screwdriver.
   - A utility knife.
   - Wire cutters.

Making the Motor

In this part of the activity, you will make three important parts of your motor. These parts are (1) a base to hold the components of the motor, (2) a coil that changes electromagnetic energy into mechanical energy, and (3) two bearings that will allow the coil to turn freely.

Making the Base

See Figure 6A-6.

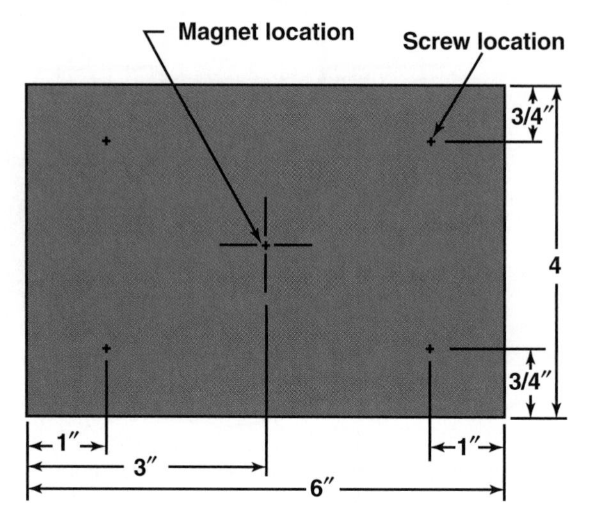

Figure 6A-6. A motor base.

1. Lay out the location of the screw holes and magnets.
2. Start the screw holes with the scratch awl.
3. Prepare for attaching the magnet. (Your teacher will show you how. The method varies with the type of magnet used.)
4. Sand the edges and ends to remove sharp edges and splinters.
5. Start a screw in each of the four holes. Do not tighten the screws at this point.

Making the Coil

See Figure 6A-7.

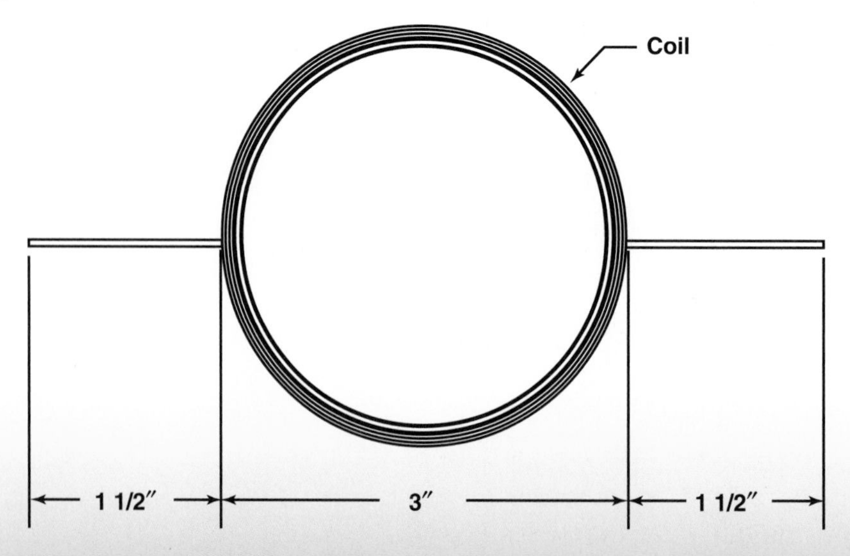

Figure 6A-7. The completed coil for the motor.

1. Cut about 20″ off the 18-gauge wire and lay it aside.
2. Mark 1 1/2″ from one end of the long wire.
3. Using the utility knife, carefully remove the insulation from the end of the wire up to the mark. Be careful to cut away from any part of the body.
4. Place the wire in the coil-winding jig. Leave the 1 1/2″ clean end of the wire extending along one of the black marks.
5. Carefully wind the wire around the center core.
6. Stop winding the coil when you cannot make another turn and leave 1 1/2″ on the opposite black mark.
7. Extend the wire along the black mark.
8. Cut the excess wire, leaving 1 1/2″ along the mark.
9. Remove the coil from the jig.
10. Tie thread at four places to hold the coil together.
11. Remove the insulation from the second end of the coil.

Making the Bearing

See Figure 6A-8.

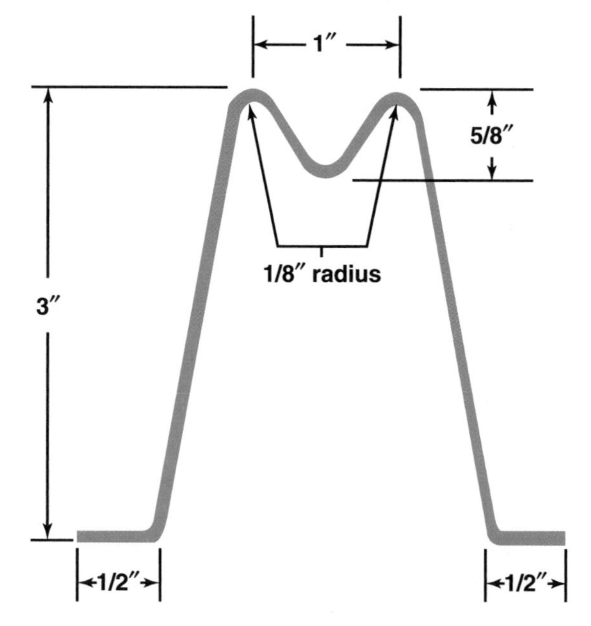

Figure 6A-8. A motor bearing.

1. Cut the bearing wire in half.
2. Use the bearing-bending jig to form two bearings.
3. Cut off the extra wire at the ends of the bearings.
4. Carefully remove the insulation from one foot of each bearing.

Assembling the Motor

1. Cut the 20″ piece of 18-gauge wire in half.
2. Remove the insulation from each end of the wires.
3. Place one bearing under the screws at one end of the base.
4. Tighten the screw on the still-insulated end.

5. Place the end of one of the connecting wires under the other screw.
6. Tighten the screw on the bearing and connecting wires.
7. Repeat steps 3–6 for the other bearing and connecting wires.
8. Attach the magnets.
9. Place the coil between the bearings.
10. Connect the motor to the battery.
11. Test the motor operation.
12. With the help of your teacher, describe how the motor converts electrical energy into mechanical motion.

Challenging Your Learning

Would your motor be more efficient if you had more winding on the coil? If so, why?

CHAPTER 6 ACTIVITY B

ENERGY AND MACHINES AS A RESOURCES

THE CHALLENGE

Develop a device (tool) that uses mechanical energy to lift a load. See Figure 6B-1.

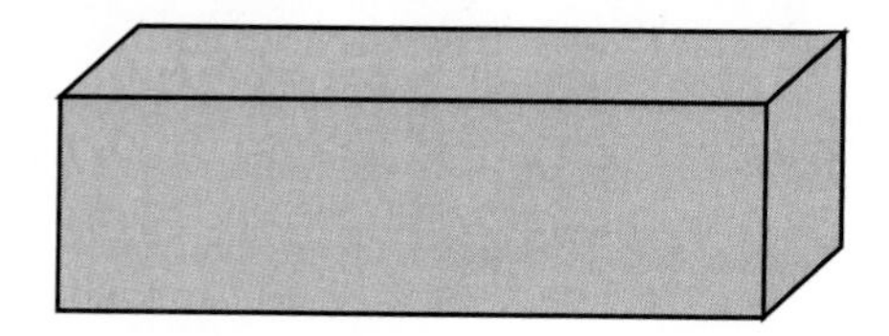

Figure 6B-1. A load.

INTRODUCTION

Average individuals are not particularly strong. They cannot lift loads equal to their own weight. Compare this to the strength of common ants, which can carry loads many times their own weight. People have learned to compensate for this weakness. They have developed technological devices to lift loads and move items. These devices are all based on six simple machines: the lever, inclined plane, wheel and axle, wedge, screw, and pulley. For this challenge, you have a load (a brick) that must be lifted 18″ off the floor. Design and build a technological device that lifts the load. This device must use less force than the brick weighs.

MATERIALS

Develop your technological device using any or all of the following materials:

- 1″ brads.
- 1/2″ × 1″ wood strips.
- 3/8″ dowels.
- Heavy string.
- Pulleys.
- Scales.
- Wire.
- Woodblocks.

CHALLENGING YOUR LEARNING

List one problem or issue that caused your device to work less than effectively. Sketch and describe how you could improve the design to overcome this issue.

Chapter 7
Information and Technology

Did You Know?

Americans have thousands of sources of information. In the United States alone, there are over 1500 newspapers; 11,500 periodicals (magazines and journals); over 220 million personal computers; more than 230 million mobile-phone subscribers; about 1700 television stations; and nearly 14,000 radio stations.

Objectives

The information given in this chapter will help you do the following:

- Explain how people use knowledge.
- Compare data, information, and knowledge.
- Paraphrase the three beliefs that are the foundation of the scientific method.
- Summarize technological knowledge.
- Explain the term *Information Age*.
- Give examples of the three types of research.

Key Words

These words are used in this chapter. Do you know what they mean?

data
descriptive research
experimental research
historical research
information

Preparing to Read

As you read this chapter, outline the details of information and technology. Use the Reading Organizer at the end of the chapter to organize your thoughts.

Humans are unique because they have advanced brains. These brains allow humans to think, plan, and process information. These processes contribute to the ability to know, which is a basic input to technology. Knowledge lets people develop and use technological products and systems. See **Figure 7-1.** Knowledge allows them to be inventive and creative. In Chapter 1, you learned there are four types of knowledge:

- **Scientific knowledge.** This explains the laws and principles governing the universe.
- **Humanities knowledge.** This explains how people have formed societies, developed values, and expressed themselves through art and music.
- **Descriptive knowledge.** This explains how people have used words and numbers to describe objects and events.
- **Technological knowledge.** This describes how people use tools and materials to produce products and systems.

Figure 7-1. Knowledge allows people to develop and use technology. (©iStockphoto.com/Vasko)

People who develop technological products and systems use all four of these types of knowledge. Designers use materials science knowledge to select materials for specific applications. They use physics knowledge as they design mechanical structures and electrical circuits. These people use chemical knowledge as they develop manufacturing and energy-conversion systems. Biological knowledge helps them develop agricultural and medical devices. Geologists use their scientific knowledge of Earth to find resources. See **Figure 7-2.**

Figure 7-2. This worker is using geological knowledge in a search for oil deposits. (©iStockphoto.com/westphalia)

Likewise, designers use humanities knowledge in their work. Knowledge about personal, social, and religious values is used. This knowledge allows designers to develop products people want and value. Historical knowledge is used in restoring significant buildings. See **Figure 7-3.** Knowledge of gender roles in society is also used. In some societies, certain tasks are considered "men's" work. Other tasks are considered "women's" work. In other societies, males and females might do the same work. Knowing these distinctions allows designers to match products to body sizes, shapes, and strengths. Knowledge of style and design tastes allows designers to create products people think are attractive.

Descriptive knowledge is used to tell other people about design ideas. Words,

Figure 7-3. Historical knowledge was used in recreating this historical western fort.

Career Highlight

Home-Appliance Repair Technicians

The Job: Home-appliance repair technicians install and repair household appliances, such as refrigerators, dishwashers, washing machines, ranges, and ovens. They repair equipment by inspecting appliances, diagnosing the problems, replacing defective parts, adjusting components, and testing the repairs. These technicians use service manuals and troubleshooting guides to help perform their tasks.

Working Conditions: They repair and install major appliances by making service calls to customers' homes. Home-appliance repair technicians carry their tools and commonly used parts in service trucks. Most technicians work independently, with little supervision. The average technician works a standard 40-hour week.

Education and Training: Most employers require a high school diploma for home-appliance repair technicians. Often, these technicians receive their appliance-repair training at a trade school or community college. Appliance manufacturers train some technicians.

Career Cluster: Manufacturing

Career Pathway: Maintenance, Installation & Repair

mathematical formulas, and drawings are used in the process. See Figure 7-4. They communicate sizes, shapes, and relationships. Also, descriptive knowledge is used in preparing production and operation directions. Descriptive knowledge is used in preparing documents to tell suppliers of material and energy needs.

Finally, technological knowledge is at the heart of technological enterprises. This knowledge of tools and materials is vital in developing, producing, and using devices. Technological knowledge lets people select appropriate tools and materials. This knowledge allows these tools and materials to be used in making products and systems. See Figure 7-5. Technological knowledge lets people select and use technological devices properly. This knowledge allows people to assess and control technological actions.

In actual practice, knowledge is not used in isolated areas. We don't say, "Now, I'm going to use scientific knowledge. I will then use technological knowledge as I work on this problem." People might learn things in specific areas, such as science or technology. This approach might help the learning process. In daily life, however, we use knowledge as a whole. We apply scientific, humanities, descriptive, and technological knowledge together. To deal with problems and opportunities, we use knowledge as an integrated mass. Knowledge is a resource in our daily lives.

Figure 7-4. This architectural drawing contains descriptive knowledge about a house that is being built.

Data, Information, and Knowledge

Directly related to these types of knowledge are data and information. People often use the three words *data*, *information*, and *knowledge* as if they describe the same thing. This is wrong. The terms describe different things.

Data is individual facts, statistics, and ideas. These facts and ideas are not sorted or arranged in any manner. They are simply an accumulation—a collection of facts, numbers, and ideas.

By contrast, ***information*** is data that has been sorted and arranged. Information is organized facts and opinions people

Figure 7-5. Technological knowledge allows people to use machines and materials to make products. (©iStockphoto.com/thelinke)

receive during daily life. These facts and opinions might come directly from other people, electronic media (radio and television), the Internet, books and magazines, or many other sources. Information is data that has been put in a condition useful to people or machines (computer systems). See Figure 7-6. Information is the product of direct human action or computer processing. Using computer systems to change data into information is called *data processing*.

Finally, knowledge is information humans can apply to situations. See Figure 7-7. When information is organized in a logical way, it is called a *body of knowledge*. This knowledge can be acquired by systematic study. Knowledge is a body of useful information a person has accumulated over time. A body of knowledge is the result of schooling and life experiences. People have general knowledge. They understand facts and principles about many things. People might know how the universe operates or how devices work. They might understand how people react in various situations. This general acquaintance with the world is useful to all people. Also, people have specific knowledge. They can apply information to do a task or job. For example, they might have design knowledge or machine-operation knowledge. They can use this knowledge to develop new products or operate machines to make something.

To put these three terms into context, think of going to an airport. There are thousands of flights each day. Each flight goes from an origin to a destination. A list of all these flights represents data. This list is a massive list. The data is of little use to the average person. Sorting this data into groups by origin city and destination city changes the data into information. Arranging the data further by time of departure makes the information more useful. Being able to use this information to select an appropriate flight is knowledge.

Figure 7-6. This architect is reviewing a set of drawings that provide information about a building under construction. (©iStockphoto.com/WillSelarep)

Figure 7-7. This worker is using knowledge to control a steelmaking process. (Inland Steel Company)

Scientific Knowledge

As you learned earlier, scientific knowledge explains how the natural world operates. This knowledge includes the laws and principles governing natural interactions. Scientific information is gathered using several basic methods. One of the most important of these is the scientific method. The scientific method starts with a person who believes the following ideas:

- Everything that happens in nature can be understood. You have to ask the right questions and then do the right experiments.

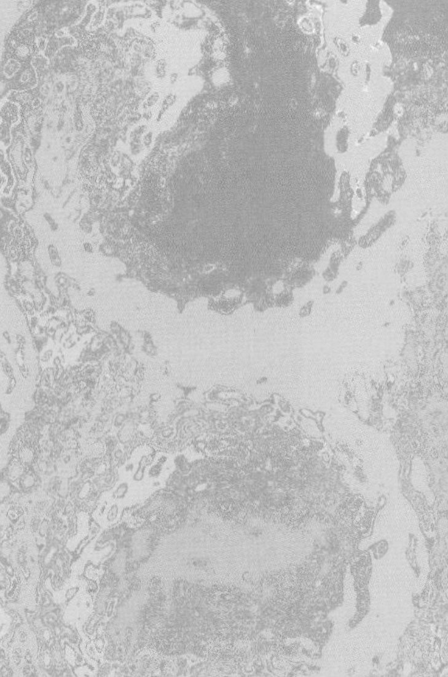

Think Green

Electronic Media Waste

You may think the transition to a more digital world has helped reduce environmental issues, like paper waste or saving on consumable resources. What you may not realize is that the equipment used by technology, which is always changing and being upgraded, is also harmful to the environment. When computers or cell phones are thrown out, they become *e-waste*. The toxic chemicals found in e-waste can contain lead and mercury. Also, human exposure to these toxins can also lead to neurological damage and cancer. Be sure to check with your local recycling programs before throwing out old equipment when you upgrade.

- Nature is always the same. Time or distance makes no difference. A scientist in one place working on an experiment should get the same results as another scientist anywhere else at any time.
- There is a relationship between a cause and its effect. A specific cause produces a specific effect. Scientists design experiments to produce given effects. They then change the conditions one by one. This way, they can find which conditions are producing the effects.

Starting with these ideas, scientists develop hypotheses. They state what they think will happen if certain actions take place. Scientists then design and conduct experiments to test these hypotheses. See Figure 7-8. They collect and analyze data. Finally, they draw conclusions about the causes and effects shown during the experiments. The experiments are designed to produce useful information. Scientists seek answers to questions puzzling them. Experiments must often be repeated to prove the results are not accidental. There is no end to the kinds of experiments scientists can conduct. For example, scientists might launch a space probe to photograph a distant star. They might work in a laboratory studying molds or bacteria.

Technological Knowledge

Technological knowledge is the knowledge of action. This knowledge includes how to do or make something. Technological knowledge is knowing how to use tools and other resources to make

Figure 7-8. These laboratory technicians are conducting a scientific investigation. (©iStockphoto.com/LajosRepasi)

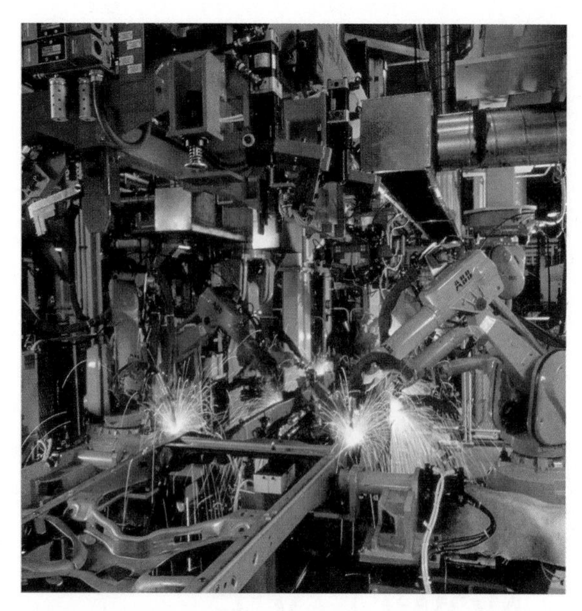

Figure 7-9. Technological knowledge is used when people make products. (Dana Corporation)

products and systems. See Figure 7-9. Technological knowledge involves using information to control and modify the natural and human-made environments.

This knowledge is gained by trying something. Technological knowledge is gained through purposeful action. This knowledge involves building something, trying it out, evaluating its operation, modifying it, and trying it out again. Technological knowledge is gained through the processes of creating, building, testing, and evaluating devices and systems.

Without technological knowledge, agriculture could not feed the people of the world. Technological knowledge prevents food from spoiling before it can be delivered to people. We also use this knowledge to locate minerals and petroleum. To process these materials into usable products, we use technological knowledge. We need technological knowledge to harness atoms and electrons for power.

The Information Age

We are becoming overwhelmed with information. Some people call the period of time we are living in the *Information Age*. This period started with the development of printing with moveable type by Johannes Gutenberg in 1450. The age was hastened by many inventions (technological advancements). These include lithography (a method of printing) in 1798, Bell's telephone in 1876, Edison's sound recorder in 1877, the first successful movie system in 1891, Marconi's radio in 1896, the first workable television in 1926, and the tape recorder in 1935. The advent of the computer after World War II, however, brought the age into being. Today, we have more information available than anyone can read or use. People might be drowning in information. They risk having information overload. The challenge is to gather the information you need and ignore what is not helpful.

Gaining Information and Knowledge

You can gain new technological information by reading about it, listening to the radio or television, having someone show it to you, or developing and testing your own designs. The foundation to all these approaches is research. Research involves seeking and discovering information.

To gather technological information, a person must have a focused approach. See Figure 7-10. This person should do the following:

- Determine the type of information needed to address the technological problem or opportunity.

Gathering Technological Information

Determine type needed

Identify possible sources

Gather

Sort and arrange

Apply

Figure 7-10. The steps in gathering technological information. (AT&T Network Systems)

Career Highlight

Braodcast Engineers

The Job: Broadcast engineers set up, operate, and maintain the equipment needed to broadcast radio and television programs. They are responsible for equipment controlling the signal strength, clarity, sound, and color of the broadcasts. These engineers also operate control panels to allow the director to switch from one camera to another, from film to live media, or from network to local inputs.

Working Conditions: At smaller local stations, broadcast engineers might perform a wide variety of tasks. At larger stations, they perform more specialized duties. They generally work indoors. Those who work with news programs might work outdoors. Engineers generally work 40-hour weeks. They might, however, have weekend or evening shifts.

Education and Training: Some broadcast engineers have electronics, computer-networking, or broadcast-technology training from a postsecondary school. Many learn their skills through on-the-job training from experienced engineers and supervisors. They often begin their careers in small stations and move on to larger ones as they gain knowledge and experience.

Career Cluster: Arts, Audio-Video Technology & Communications

Career Pathway: Journalism and Broadcasting

research can be one of three types. These types are historical, descriptive, and experimental.

Historical research gathers information already existing. This research describes how other people have solved similar problems. See Figure 7-11. Historical information might include existing designs or processes used in the past. This information might come from examining products on the market or reading operating manuals for equipment. Sources of historical information include books, magazines, museums, films, videotapes, photographs, drawings, and models.

Descriptive research gathers information by measuring and describing products and events. This research describes something as it is. See Figure 7-12. For example, it might describe how people are doing a job. Descriptive research might describe the physical attributes (such as size and shape) of products. Henry Dreyfuss conducted a pioneer study of this type. He and his research colleagues measured large numbers of men, women, and children. Dreyfuss analyzed and categorized the data he collected. He then published his findings in great detail. For example, he provided information about the size of

Figure 7-12. Descriptive research gathers information about existing conditions. This worker is gathering information about inventories. (Datastream)

Figure 7-11. Historical research describes how other people have solved similar problems.

men's, women's, and children's hands. This information is valuable for glove makers. Likewise, the data on human-torso sizes is important to clothing-manufacturing companies. The data on the width of people is useful for aircraft and stadium-seating designers.

Experimental research is typical of the research scientists conduct. See Figure 7-13. This approach structures activities so changes or improvements can be measured. For example, a team might carefully study how people are doing a job. The job might be modified to see if the production improves. The quality of the product, the time required to do the job, or the amount of scrap and waste might be compared.

Historical research describes what was possible. Descriptive research describes what is possible. Experimental research describes what can be possible. See Figure 7-14.

Figure 7-13. Scientists often conduct experimental research.

SUMMARY

Technological information and knowledge have helped people control and modify the environments around them. They have helped people build better ways of life. This information and knowledge have given people better food, clothing, and shelter. They have made movement from one place to another easier. Technological information and knowledge have increased the standard of living and life expectancy of the U.S. population.

Types of Research

Historical
What was?

Descriptive
What is?

Experimental
What can be?

Figure 7-14. The focus of each type of research.

STEM Connections

Science

Read an owners' manual for an electrical appliance or a tool. Describe the scientific information included in it. Look for examples of physical, chemical, and biological information.

Mathematics

Measure the size of the hands of your classmates—the length of each finger, width of the hand across the knuckles, and length of the hand from the heel to the tip of the third finger. Average these measurements for boys, girls, and the class as a whole. Provide a drawing of the average female and male hands.

Curricular Connections

Social Studies

Research the effects of a specific invention on the Information Age. Use inventions aiding in communication, such as the radio, television, or printing press.

Activities

1. Read a current-events magazine or newspaper article. Indicate examples of descriptive, humanities, scientific, and technological knowledge, using colored pencils or highlighters.
2. Develop a time line of inventions that have led to the Information Age. You might want to develop a time line for a specific strand. Typical strands are the written word, broadcasting, or recording.

Test Your Knowledge

Do not write in this book. Place your answers to this test on a separate sheet of paper.

1. Knowledge about mechanical structures and electricity is _____ knowledge.
2. Knowledge communicating size and shape is _____ knowledge.
3. Knowledge about using products and devices is _____ knowledge.
4. Facts arranged in a useful order are called _____.
5. Information that can be used to solve a problem is called _____.
6. Rewrite, in your own words, the basic beliefs underlying the scientific method.
7. Why is technological knowledge essential to our existence?
8. What are three sources of information?
9. Research telling us the condition of something is called _____.
10. Research answering the question "What can be?" is called _____.

Reading Organizer

On a separate sheet of paper, create a detailed outline based on what you've read about information and technology.

Example:

I. Types of knowledge
   A. Scientific
   B. Humanities

CHAPTER 7 ACTIVITY A

INFORMATION AS A RESOURCE

INTRODUCTION

Information is used to understand technology and solve technological problems. These facts come from many sources and can be obtained by talking to people or observing their actions. For most researchers, the printed word has been their greatest resource. This resource can be found in the form of books, magazines, technical reports, and product catalogs. Today, computers and the Internet allow researchers from every field to share information.

This activity lets you gather and apply some technical information. You will read about how a computer modem and home-to-home e-mail on the Internet works. Diagrams and illustrations have been included. You will be asked to do the following:

- Outline the information.
- Prepare a brief report.
- Illustrate the report.

PROCEDURE

1. Study the materials included with this activity.
2. Take notes on the important information.
3. Provide a brief outline for the information, using the following headings:
   A. Historical background.
   B. System of operation.
   C. Technological importance of the invention.
4. Write a brief report with illustrations (sketches).

THE HISTORY OF COMPUTERS

Computers are machines that perform calculations and process information with amazing speed. The first computers were very large and often took up a whole room. They were very expensive to design and maintain. Only the government, big businesses, and universities could afford them. Today, you find computers almost everywhere. More likely than not, you have one in your own home.

These machines help people find and organize information, create and test models, and solve problems. They even help us to communicate with each other. Charles Babbage designed the true ancestor of the modern computer in the 1830s. This steam-operated machine was called the *Analytical Engine*. The Analytical Engine was never completed. This machine, however, was designed to perform calculations with a mechanical calculating unit controlled by punched cards. These punched cards were the basis of the card-handling machines Dr. Herman Hollerith developed in the 1880s.

In 1944, Professor Howard Aiken completed his Automatic Sequence Controlled Calculator (ASCC). This machine was over 50′ long and 8′ tall! ASCC took 0.3 seconds to add or subtract, 4 seconds to multiply, and 12 seconds to divide. In 1946, Dr. John Mauchly and J. Presper Eckert completed the first electronic digital computer. This machine was called the *Electronic Numerical Integrator and Calculator (ENIAC)*. ENIAC could perform as much work in one hour as ASCC could in a week. Technological advances in the 1950s and 1960s led to the computers you see today. These advances include more reliable transistors, more accurate analog systems, and digital computers with miniaturized integrated circuits.

How a Dial-Up (Telephone) Computer Modem Works

Computers process information expressed as ones and zeroes. When the machines exchange information over telephone lines or cable systems, however, the data must be converted, or modulated, into analog, or wave, form. After the data is sent and received, it is converted back into digital form at the other end. This is the job of the *modem*. The name comes from the jobs a modem performs: modulation and demodulation.

Modems can be inside your computer's hard drive or added on to your system similar to the way a video game is added on to your TV. In order to use a modem, you have to tell your computer to send information to a certain destination (phone number). In this case, let's pretend you are sending your friend some information on your favorite TV show.

After giving your computer the phone number, the computer orders its modem to open a phone line and dial the number. When your friend's modem answers the phone, your modem announces it would like to make contact. The modem announces this with a hailing tone. This is very similar to when you pick up the phone, hear the dial tone, dial, and then hear the ringing. As the user, you can hear the tones on your computer as the connection is being made.

Your modem and your friend's modem exchange messages after the connection is made. They agree on how fast the information will be sent and how it will be packaged. The modems also agree on how the information will be checked for errors.

The modem begins transmission. Your computer, the sender, feeds its modem digital data. The modem turns the data into a high-speed series of tones and puts it in the phone line. Your friend's modem, the receiver, hears the tones and converts them back into digital form. After the transmission has been completed, your computer instructs its modem to break the connection, or you could say, hang up the phone. See Figure 7A-1.

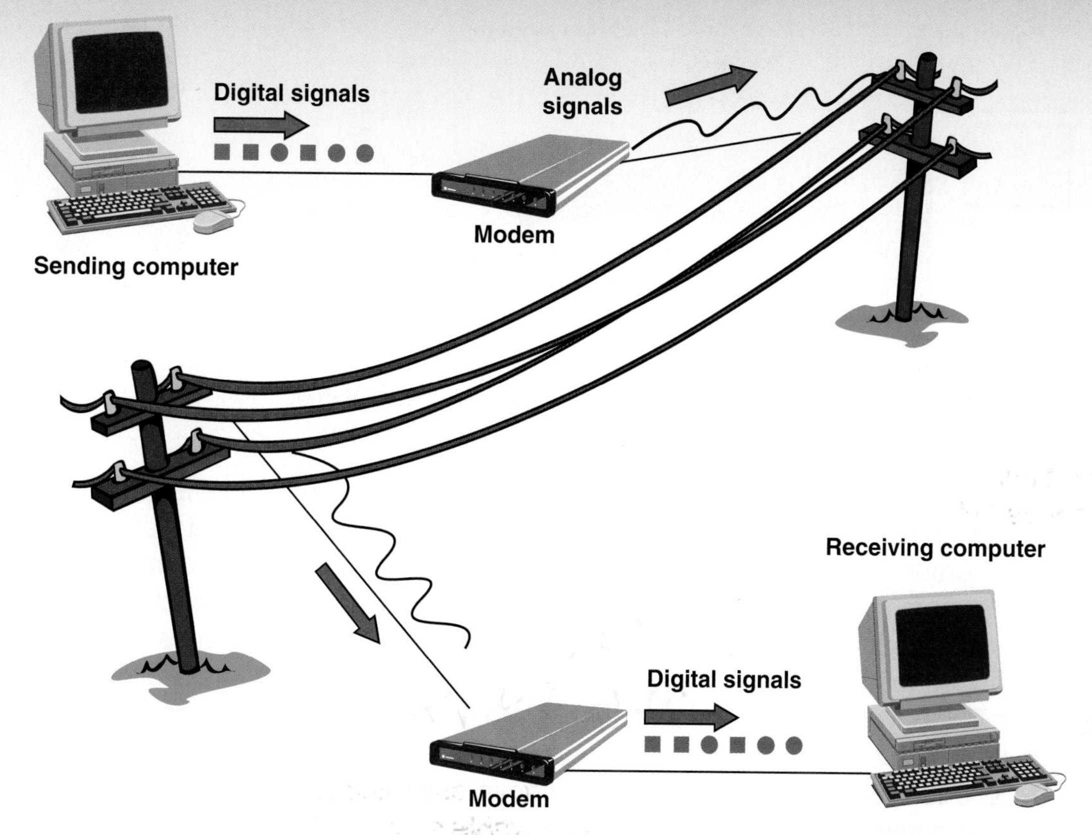

Figure 7A-1. How a computer modem transmits signals from one computer to another.

How a Cable Modem Works

When people first started connecting to the Internet at home, they used dial-up modems. These modems allowed them connection through the telephone system they already had. Dial-up systems, however, send and receive information fairly slowly. This fault encouraged designers to develop a new system that uses the cable television system.

A cable television system has a large number of channels used for television programming. The system still has space (bandwidth), however, to be used for other services. The most common other service is high-speed Internet access.

A cable Internet connection is, most commonly, made through a cable modem. A modem is usually a separate device placed close to a PC. The modem is then connected to the cable system with a standard coaxial cable and to the computer with an Ethernet interface. See Figure 7A-2. There are other ways to connect to the Internet via cable. These include a modem built into the PC and a set-top box, which connects both the television and the computer to the cable system.

There are two major advantages of a cable modem system. First, it is much faster than a dial-up system. These systems are also generally faster than most digital subscriber line (DSL) services, which many phone companies offer as an alternative to dial-up systems. Pages can be uploaded (sent) or downloaded (received) faster on cable systems. Most cable modems have faster download speeds, however, than upload speeds. Second, the cable model is always connected and ready to use. There is no need to dial a special number to be connected to the Internet.

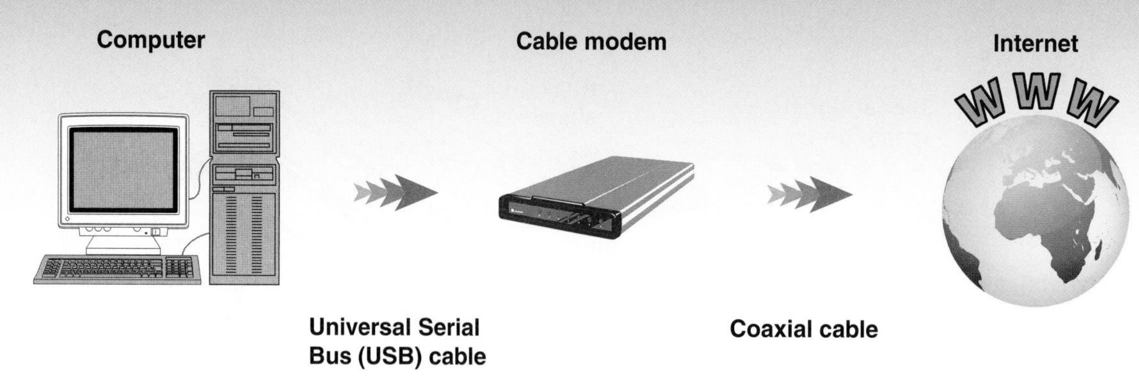

Figure 7A-2. The role of the cable model in connecting a PC to the Internet.

Cable Internet systems use the same type of signal that carries television broadcasts to the home. The Internet signal is a compressed electronic stream of data that travels on a dedicated portion of the cable line. A cable modem decodes information coming to it from the cable system. Cable modems have several important parts. First, each modem has a tuner to capture the data stream from the cable system. Second, it has a demodulator, which changes the signal so it can be recognized by the third part, the analog-to-digital converter. The converter converts the signal so the computer can understand the data.

Although cable systems are fast, they have one major disadvantage. If the data being sent and received is not encrypted (converted into a secret code), other people who share the network can access the information. This is a drawback if you have data or pictures that you do not want to share widely.

How Home-to-Home E-Mail on the Internet Works

If you live in New York and want to send your cousin in California a letter, you can use the postal system or your computer. The Internet allows you to use your computer to send any kind of information over telephone lines. Refer to page 514.

First, you have to type in the information on your computer. You then have to type in your cousin's Internet address and use the send command. When the send command is given, the message flows across a telephone wire to a computer directly linked to the Internet. Here, it is expressed as a collection of the ones and zeroes of computer language.

The Internet breaks the message into packets. Each of these packets has your cousin's address on it. The packets make their way across the Internet separately. Computers route them onto the least busy pathways. The packets can take different routes. This helps the system use the circuits efficiently and provide alternate routes, in case there is a breakdown somewhere. It might take as little as a fraction of a second to send your letter. Everything depends on the traffic!

Your cousin's computer in California puts the packets back together. This computer then inserts the letter into your cousin's electronic mailbox. When your cousin turns on his computer and looks into his mailbox, he will find your letter. He can read the letter on the computer screen or print it out. See Figure 7A-3.

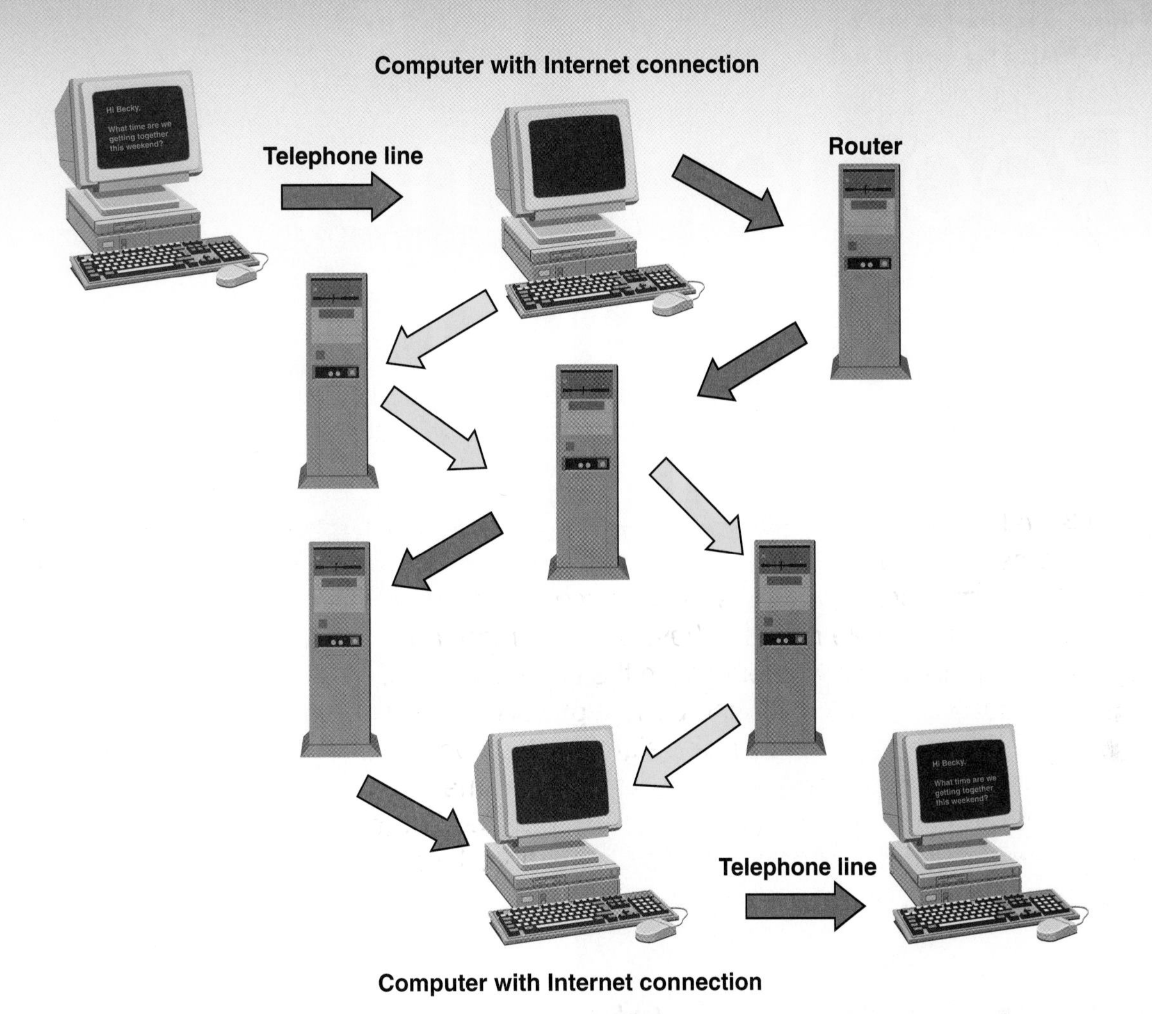

Figure 7A-3. The path information takes to travel from one telephone line to another through the Internet.

CHALLENGING YOUR LEARNING

Explain how a cable model works differently from a telephone-based model. Which one would you choose to use? Why?

CHAPTER 8
People, Time, Money, and Technology

DID YOU KNOW?

- In a recent survey, company leaders thought listening and interpersonal skills are the most important skills managers need.
- Good career decision making involves knowing the following:
  - Your interests, skills, and values.
  - How your personal criteria relate to career options.
  - What those careers involve, in terms of education, experience, and skills.
- According to the U.S. Department of Labor, these are the 10 fastest-growing occupations:
  - Registered nurses.
  - Home health aides.
  - Customer service representatives.
  - Combined food preparation and serving workers.
  - Personal and home-care aides.
  - Retail salespersons.
  - Office clerks.
  - Accountants and auditors.
  - Nursing aides, orderlies, and attendants.
  - Postsecondary teachers.

Objectives

The information given in this chapter will help you do the following:

- Explain the roles of people in technology.
- Compare white-collar jobs and blue-collar jobs.
- Identify the changes in job opportunities that came with the Information Age.
- Paraphrase what a company is.
- Explain how companies and industries are related.
- Give examples of the tasks that managers, engineers, technologists, technicians, and workers carry out.
- Explain the factors to consider in selecting a career path.
- Recall what "climbing the career ladder" means.
- Give examples of time as a technological resource.
- Recall the ways money is used in technological activities.

Key Words

These words are used in this chapter. Do you know what they mean?

career ladder
communication skills
company
critical thinking skills
discrimination
engineer
equality
general job skills
harassment
independent learning skills
industry
information skills
leadership skills
listening skills
manager
problem-solving skills
salary
self-management skills
social skills
specific job skills
teamwork skills
technician
technologist
visualization skills
wage
worker

Preparing to Read

As you read this chapter, make a list of the different jobs available in technology. Then, list the job skills needed for the job and explain why they are necessary.

We are in the greatest growth period in the history of humankind. People are gaining new knowledge at an unbelievable pace. Compared to people in past periods of history, we know much more about our universe and how it was formed. We are developing new technological knowledge daily; learning more about using machines effectively; and rapidly creating new products, systems, and structures. All these advancements are the products of the efforts of people.

These efforts take both time and money. Many people dedicate a part of their day to creating knowledge and technology. Often, they are paid for this time with money. Therefore, it is easy to see that people using their time, which is paid for with money, is key to us living better with new or improved technology.

People and Technology

People are developing new technological knowledge daily. We are learning more about using machines effectively. People are rapidly creating new products, systems, and structures. All these advancements are the products of the efforts of people.

Work and a Changing Society

Our country has already passed through several historical stages. Each of these stages saw great changes in what kinds of work people did. During pioneer days, we were an agricultural society. Growing crops and raising livestock were our main ***industries***. See Figure 8-1. Most people worked on farms and ranches. Some people worked in shops and stores. Other people worked as doctors, lawyers, teachers, and nurses. A small group of people practiced village trades, such as blacksmithing and carpentry. Still other people worked in newly developing industries. Regardless of the type of work they did, most people worked for themselves. They tilled the land, ran the businesses, practiced the trades, and served the professions.

With the Industrial Revolution that started in the late 1800s, work changed. Factories started to sprout up in many parts of the nation. This was particularly true in the Northeast and upper Midwest. Manufacturing became an important industry. Work was concentrated in newly developed factories. Machines were invented to help make products efficiently. People moved off the farms and ranches into expanding cities. Many of these people started to work for other people for the first time. They were trained to run the machines making an expanding array of products.

After World War II, a new age started to develop. It started to become more important for people to exchange information. Many inventions helped lay the foundation for this information exchange. The telephone became more important, as did the telegraph and radio. Books,

Figure 8-1. Early in American history, most people worked on farms.

magazines, and newspapers had become plentiful and inexpensive. Television came into general use during the 1940s and early 1950s. Computers became common in the later part of the twentieth century. They became a key part of information processing and manufacturing process control. See Figure 8-2.

Experts are predicting even greater advances in the twenty-first century. They think more new discoveries will be made during this period than in all of human history. With these changes will come new demands on people. Jobs and communities will change. Whole societies will be different.

Figure 8-2. This is a control center in the printing plant of a newspaper. With computer and electronic controls, the operator is away from the printing press. (©iStockphoto.com/seraficus)

Jobs and Technology

People have always worked. First, they worked to find food and build shelters for themselves and their families. Communities then developed, where people gathered together and shared work. Some people raised crops. Others fished or hunted. Simple trades, such as shoe making and candle making, developed. Society continued to progress over the years. Likewise, the work people did changed. Today, there are literally thousands of different jobs. It is impossible to know about all of them. The various jobs, however, can be grouped into several categories. A common breakdown was developed to reflect the Industrial Era of the nineteenth and twentieth centuries. This breakdown divided jobs into two categories: management and labor. See Figure 8-3.

Managers

Workers

Figure 8-3. During the Industrial Age, employees were thought to be either managers or workers. (Conoco, Inc. and Motorola, Inc.)

Management included all people who were paid ***salaries***. Actually, not all people in the group were managers. They were, however, paid a weekly or monthly salary. This group was often referred to as *white-collar workers*. The name implies these jobs were clean, office-type jobs. White-collar jobs allowed ***workers*** to wear white shirts and blouses. Typical white-collar jobs were engineering, sales, accounting, and supervising.

Labor included people paid hourly ***wages***. They earned money for each hour they worked. This group was often called *blue-collar workers*. This name suggests the work was somewhat dirty. The workers wore colored clothes so the dirt was not as noticeable. Blue-collar jobs were generally divided into three groups:

Figure 8-4. This skilled worker is preparing a model for a museum.

- **Skilled workers.** These were people having extensive training and work experience for a broad job. See Figure 8-4. Often, they served apprenticeships. This training is a combination of classroom and on-the-job instruction. It often takes four or more years to complete an apprenticeship. Typical skilled jobs are carpentry, plumbing, tool- and die making (making manufacturing tools), and mold making (making casting molds for metal, plastic, and glass).
- **Semiskilled workers.** These were people who received specific training for a specific job. See Figure 8-5.

Figure 8-5. This machine operator is considered a semiskilled worker. (Dana Corporation)

Technology Explained

flight controls: a series of moveable control surfaces that can be adjusted to cause an airplane to climb, dive, or turn.

An airplane is the only transportation vehicle that can move in these three distinct ways: go forward, turn side to side, and climb and dive. Airplanes can change direction through a system of flight controls the pilot operates. Flight controls change the lift acting on the wings and other control surfaces.

Lift is what makes an airplane fly. The engine pushes or pulls the airplane through the air. Air flows over the wings, creating lift. If not enough air is flowing over the wings, no lift is produced. The plane loses altitude. The airplane must maintain forward motion to fly.

Pilots make turns by using two types of control surfaces, the rudder and the ailerons. The rudder is a moveable flap on the vertical portion of the tail assembly. The pilot controls the rudder with two pedals. The ailerons are flaps on the rear of each wing. See **Figure A.** The pilot controls the ailerons with a yoke or stick. To make a turn, the pilot, using the pedals, moves the rudder to one side. Using the yoke or stick, the pilot simultaneously lifts the aileron on that side and lowers the one on the other wing. This causes the plane to bank (turn) in the direction the rudder is pointed.

Surfaces on the horizontal stabilizers control climbing and diving. These control surfaces are called *elevators*. When the elevators are tilted downward, the tail is forced upward. The nose is pointed downward, so the plane loses altitude, or dives. When the elevators are tilted upward, the tail is forced down. The nose points up. The plane then gains altitude, or climbs. See **Figure B.**

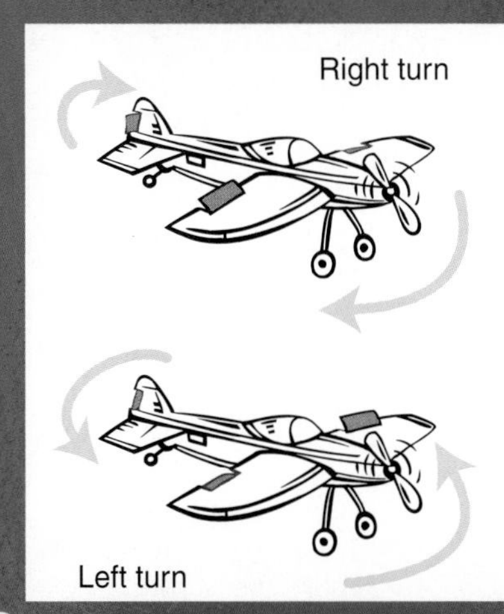

Figure A. An airplane moves to the right when the pilot moves the rudder to the right side, lifts the aileron on that side, and lowers the aileron on the other wing. The airplane turns to the left with the opposite movements.

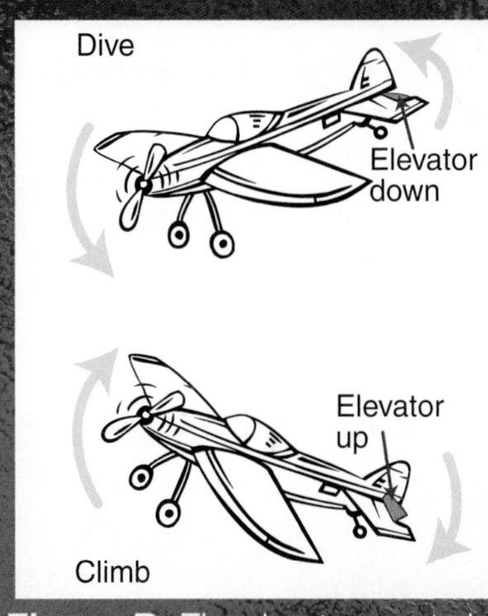

Figure B. The elevators control climbing and diving.

Semiskilled workers had a high degree of skill. This skill, however, was in a narrow range. These workers included machine operators in a manufacturing plant, roofers on a construction site, and truck drivers.

- **Unskilled workers.** These were people doing routine work, requiring little training and experience. They did cleanup, product packing, and other routine jobs.

The use of the terms *white-collar* and *blue-collar* is less meaningful today. These terms were closely related to manufacturing. This sector of employment is shrinking. More manufacturing processes are being automated. See **Figure 8-6.** Fewer workers are needed to do the same amount of work. Also, foreign competition has caused additional manufacturing jobs to be lost.

Figure 8-6. Automation is one reason there are fewer manufacturing jobs than in the past. (Chrysler Corp.)

Jobs in the Information Age

With the Information Age, however, times have changed. The society has changed from one based on hardware (machines and materials) to one based on "thoughtware" (ideas and knowledge). Information processing is more common than material processing. Many production workers work in clean environments. See Figure 8-7. They are knowledge workers, rather than manual laborers. The jobs are more dependent on mental skills than physical skills.

There are fewer jobs for people failing to complete high school. The industrial jobs requiring little education and training are rapidly disappearing or moving overseas. There is a high demand for people having technical and professional training after high school. The days of dropping out of school and getting a fairly good job are gone. In the future, even people with only a high school education might have trouble getting a good job.

Figure 8-7. Many of today's production workers work in clean environments.

Jobs in Technology

There are many jobs directly related to technology. Most of these are associated with businesses or companies. See Figure 8-8. These businesses are economic organizations that change resources into products and services. They are generally organized to make money by providing something consumers want. See Figure 8-9. In brief, a ***company*** is all of the following:

- An economic institution.
- A user of resources (labor, material, energy, information, finances, and time).
- A producer of a product, structure, or system.
- A profit maker.

Companies are present in all technological contexts. They exist in agriculture, communication, construction, energy conversion, manufacturing, medical services, and transportation. The companies within a technology can be grouped into a number of different industries. Farming, food processing, flour milling, and fertilizer industries are part of agriculture. Each industry is composed of a number of companies producing competing products. For example, several companies make cereals or frozen vegetables. The government and universities also provide technology-based employment. Typically, these technology-based careers include managers, engineers, technologists, technicians, and workers.

Figure 8-8. Most technology-related jobs are with companies. (AT&T Network Systems)

Managers are people who organize resources to produce products or services efficiently. See Figure 8-10. Effective managers have special skills they use

Economic institution

that uses

resources

to produce

products or systems

intending to make a

profit

Figure 8-9. A simple definition of *industry.* (Carolina Power and Light)

Figure 8-10. The managers organize the work for the company.

Figure 8-11. This engineer is checking a drawing for a new part.

to organize people to get a job done. Effective managers are not afraid to make decisions. They have great determination and confidence, even when things are not going well. Managers seem to enjoy the challenges of difficult jobs.

Engineers design processes, products, and structures. See Figure 8-11. They apply knowledge of physical and mechanical principles and mathematics to do their work. These people use this knowledge to design roads, bridges, and skyscrapers. They design products and machines. Engineers might direct the work of other engineers and technologists.

The ***technologist*** is a fairly new type of professional person. As engineers became more focused on design, technologists have taken some of their duties. Technologists work closely with engineers to implement the engineers' work. They might be called *industrial technologists*, *manufacturing technologists*, *food*

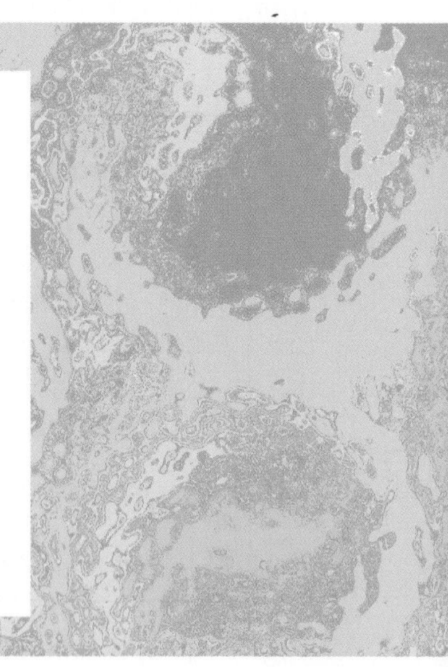

Think Green

Carbon Footprint

A *carbon footprint* is a measurement of how much someone's everyday behaviors can impact the environment. It includes the average amount of carbon dioxide put into the air by energy and gas used at home and in travel, as well as other actions. Several various aspects of technological production contribute to carbon dioxide emissions. Companies are beginning to determine their carbon footprints. Often, simply learning the details of their carbon footprints is enough to induce people to work toward reduction.

Career Highlight

Occupational Health and Safety Specialists

The Job: Occupational health and safety specialists analyze the environments in which people work. They design programs to identify, measure, control, eliminate, and prevent workplace injuries. These specialists deal with the impacts of equipment design on a worker's comfort and safety. They conduct inspections and enforce laws, regulations, and policies related to worker health and safety.

Working Conditions: These specialists might work for a large company or a governmental agency. The job requires on-site observations and actions. During their on-site visits, occupational health and safety specialists might be subjected to unpleasant, stressful, and dangerous working conditions. They might be forced into conflict with the management of the company with which they are working.

Education and Training: Many employers and governmental agencies require occupational health and safety specialists to hold a college degree in safety or a related field. In addition, most occupational health and safety specialists receive classroom and on-the-job training in inspection procedures and the appropriate laws and regulations.

Career Cluster: Business, Management & Administration

Career Pathway: Human Resources Management

technologists, *medical technologists*, or *construction technologists*. Technologists are the major link between engineers and the factory floor or construction site. They help managers and workers make the objects the engineers design.

Technicians are people who know how to operate machines. They diagnose and solve problems at the operational level. Generally speaking, engineers know the theories of how and why something works. Technicians know how to make it work.

Workers are the people who construct and assemble various products. They operate cameras in television studios, printing presses at newspaper publishers, and bulldozers on construction sites. See Figure 8-12. These workers are the people driving buses, flying airplanes, and making cheese.

People and Jobs

How do jobs affect you? Most of us have several roles in life. We have roles in a family. Most of us have citizenship responsibilities to our community, nation, and world. We are workers and consumers of products and services. Most of us will work for or operate businesses.

Goals, Interests, and Traits

You are a special person, and you have special abilities we sometimes call *talents*.

Figure 8-12. This worker is a welder on a construction site.

For example, you might be very good in math or excel in communication skills, such as languages or grammar. These talents should guide you in selecting a career path or job. Where and how you will work depends on the following factors:

- **Personal goals.** Not all of us want the same things out of life. Some want to live and work among friends. For example, you might want to live in the community where you grew up. Perhaps being home every night is important to you. On the other hand, you might prefer traveling, working with new ideas, and making new friends.
- **Interests.** If you have held jobs, you might already know what you prefer. You might like working with others. Perhaps you get along well as part of a team. If this is the case, you should consider a job in which you meet and work with people. You might prefer working alone, using tools, instruments, or computers. This might indicate that you should look for a career working with tools or machines. A job in which you are alone most of the time might be attractive.
- **Ability.** It is good to identify those courses you like and in which you do well. These courses indicate where you will be most successful. An ability to do math is important for accounting and engineering careers. On the other hand, communication careers depend on English and grammar skills. Abilities in several areas might lead to a career in which these skills and interests can be combined. For example, an interest in writing and working with people is important for authors, teachers, and editors. An ability to work with machines and materials is important for auto mechanics, carpenters, computer operators, and printers. See **Figure 8-13.**

Figure 8-13. This person likes to work with tools and machines.

- **Physical traits.** Careers make demands on our mental abilities. They also, however, call on our physical abilities. Certain jobs might require long hours of physical activity. Others depend on above-average strength or good eyesight. Still others need nimble fingers or quick reaction times.

Job Skills

In addition to finding jobs that match personal goals, interests, and traits, a person must have skills. A person has two types of job skills that jobs require. These are general job skills and specific job skills.

General job skills are those skills that can be applied to many different jobs. They are generally developed during elementary and secondary schooling. ***Specific job skills*** are those skills needed to do a particular job. These skills are often developed on the job, through apprenticeships (combinations of on-the-job training and classroom instruction), or at colleges or universities. Let us look at the general job skills you should be developing. These skills are as follows:

- ***Communication skills.*** These skills include the ability to use speech and writing to explain and present ideas clearly.
- ***Listening skills.*** These skills involve the ability to listen and understand what a person is saying.
- ***Information skills.*** These skills include the ability to locate, select, and use information using appropriate technology.
- ***Problem-solving skills.*** These skills involve the ability to identify a problem, find possible solutions, and choose the best solution.
- ***Critical thinking skills.*** These skills include the ability to evaluate different sides of an argument and draw conclusions from this evaluation.
- ***Visualization skills.*** These skills involve the ability to see three-dimensional objects and systems shown by drawings and schematics.
- ***Leadership skills.*** These skills include the ability to influence people to work toward a common goal.
- ***Teamwork skills.*** These skills involve the ability to work with other people to achieve a goal.
- ***Self-management skills.*** These skills include the abilities to accept responsibility for contributing to a goal and manage time effectively to reach the goal.
- ***Independent learning skills.*** These skills involve the ability and desire to develop new knowledge and skills outside formal education settings.
- ***Social skills.*** These skills include the ability to understand and respect what people say, think, and do.

Getting Career Information

Deciding on a career is important. This decision is not similar to any other activity you do in school. Try to learn as much as you can about each kind of work. For each kind, match the job requirements and major responsibilities with your own abilities, interests, and values. Also, pick out the requirements that do not match your characteristics. Decide if there is something you can do to remove the mismatch. If not, that particular career might not be for you.

There are many ways to learn about jobs. The world of technology offers many satisfying and challenging jobs. One of the best ways is to look for books on careers in your school library, guidance office, or public library. Also, the U.S. Department of Labor Web site has occupational information. These sources describe jobs in a number of fields. They also tell you what is needed to get and hold the job of your choice.

The U.S. Department of Labor groups jobs into a number of categories. Each of these groups contains jobs that are somewhat alike. Every job has its own requirements and duties. The table in Figure 8-14 lists the groups and typical jobs.

Employer Expectations

Many general qualities are needed to be successful in the workplace. Behavior required for professional success and advancement includes the following:

Job Classification	Typical Jobs or Occupations
Executive, Administrative, and Managerial	Accountant Public relations manager Construction and building inspector Construction manager Farmer Farm manager Financial manager Health-services manager Hotel manager Industrial production manager Restaurant and food-service manager
Professional and Technical	Aircraft pilot Flight engineer Engineer Engineering technician Drafter Computer programmer Mathematician Food scientist Geologist Lawyer Judicial worker Urban and regional planner Social worker Schoolteacher Librarian Dentist Veterinarian Dental hygienist Broadcast and sound technican Photographer Actor
Marketing and Sales	Cashier Insurance sales agent Manufacturer's sales representative Real estate agent Real estate broker Retail salesperson Securities sales representative Travel agent
Transportation and Material Moving	Bus driver Taxi driver Truck driver
Administrative Support, Including Clerical	Bank teller Communications-equipment operator Computer operator Hotel desk clerk Receptionist Mail clerk Shipping and receiving clerk Mail carrier Bookkeeping clerk Brokerage clerk Statement clerk File clerk Library assistant
Service	Janitor Groundskeeper Pest controller Chef Dental assistant Medical assistant Barber Flight attendant Correctional officer Firefighter
Installations and Repairs, Including Mechanical	Computer repairer Aircraft mechanic Automotive mechanic Musical-instrument tuner Bricklayer Stonemason Carpenter Roofer
Production	Assembler Production supervisor Fishing-vessel supervisor Butcher Inspector Machinist Welder Water-treatment plant operator Printing press operator Textile-machinery operator Upholsterer Dental laboratory technician Photographic-process worker

Figure 8-14. This table lists a sample of jobs with each major job classification. (U.S. Department of Labor)

- **Cooperation.** An employee must cooperate with supervisors, other employees, and customers.
- **Dependability.** A dependable employee is timely, completes all assignments, and sets realistic goals for completing projects. Others trust this type of employee.
- **Work ethic.** Good employees put an honest effort into their work.
- **Respect.** In order to be respected, employees must show respect for others, the company, and themselves.

Today's workplace emphasizes equality. ***Equality*** is the idea that all employees are treated in the same way. To make the workplace a pleasant place to work, negative behaviors, such as discrimination and harassment, are not tolerated. ***Discrimination*** is expressing a biased attitude, act, or behavior toward another person. People sometimes discriminate against individuals based on personal characteristics such as race, gender, age, or religion. ***Harassment*** includes offensive and unwelcome actions against another person. Behaviors such as discrimination and harassment often result in termination of employment.

Teamwork is also important in today's workplace. Solving problems often requires you to work with other people. Teamwork can help get a job done quicker than doing it alone. This type of work allows for more possible solutions to be considered. Working in teams almost always improves the efficiency of developing a design solution. This kind of working, however, takes organization and cooperation among the team members.

A team consists of team members and one or more team leaders. The team members provide their ideas and abilities. Problems are successfully solved using teamwork when all the members and their contributions are involved. The team leaders are responsible for guiding the direction and setting the goals for the team. They also ensure that all the team members contribute to the team's success.

The team leaders guide the rest of the team. The quality of their leadership determines whether the team is a success or a failure. Good leaders have a vision of what must be done and look for ways to reach the team's goals. They also are skilled communicators who are able to encourage the team members to assist each other and cooperate. A leader's job is not always easy. Good leaders, however, are willing to work as long as it takes to achieve success. They can organize and direct the team's activities. Good team leaders are fair and honest. They take responsibility for their own actions and give credit to others when it is due. Finally, good leaders know how to delegate authority. They might even assign leadership roles to other team members to get the job done as efficiently as possible.

Advancing in a Career

Changes in technology are opening up new jobs every day. At the same time, old occupations are being eliminated. People entering the job field today need to be flexible. This means they must be willing to change jobs. Many people make career changes several times during their working lives.

Climbing the ***career ladder*** means working your way up to better jobs, as your skills and knowledge improve. For many people, a beginning job is the first step to a better one. It is wise to choose a job that gives you a chance to move to a better job. See Figure 8-15. Usually people move on to related jobs. Each new job gives added responsibility, more pay, and greater satisfaction. Sometimes, the job promotion and training come from within the company. At other times, there are offers from another company. As you decide on a career, you should think ahead to the chance for promotion.

Figure 8-15. These employees have moved up the career ladder. As they advanced, they were given more authority and responsibilities. (Motorola, Inc.)

It is important to show your superiors you have developed new abilities. Also, you must show interest in taking on more responsibilities. Being on time for work, doing your job with enthusiasm, and responding to job challenges are important. They show your supervisor you are mature and trustworthy.

TIME AND TECHNOLOGY

Time measures duration of an event. In technology, time is a resource. Time measures the duration it takes to make a product, perform a task, or transmit information. See **Figure 8-16.**

To deal with labor and production processes, time is used. In our technological society, people are one of the inputs for producing products and structures. People, however, cannot produce technology without time any more than they can produce technology without knowledge. People are paid by the hour, month, or year for the time spent on the job. In one sense, time is considered renewable. There will always be more time tomorrow, the next day, and the next. In a stricter sense, time wasted or lost is lost forever.

Likewise, time is related to processes. Most tasks have a time limit. Machines can run only so fast or so slow. Materials can be cut only within a range of speed. Plastics take a certain amount of time to melt. A casting must cool for a certain amount of time. Crops have a growing season. X rays must be exposed for a set period of time. Companies must meet time limits in contracts. We call these time limits *deadlines* for completion.

Figure 8-16. For an airline, it is important to control the time it takes to load baggage on an airplane.

Time is an important resource for all technological processes. This resource is also important for every activity in our lives. People need time to perform a task, whether it is manufacturing a product or completing math homework. It is essential to learn how to manage your time wisely because time is such an important input to any process.

Before completing an activity, it is important to develop a work schedule. This requires an assessment of the task and an estimate of how long it will take to complete. If you are writing a report on an emerging technology, you need to plan time to research your information, organize the information, and write the paper. This work schedule might extend over several weeks. If you are growing crops, you need to plan time to prepare the soil, plant the seeds, water and fertilize the plants, and harvest the crops. This work schedule spans several months. In planning your work schedule, it helps to analyze your goals and set your priorities. It also might be beneficial to set a routine for yourself in which you work on this task at the same time every day.

Once your work schedule is developed, you must follow it. Maintaining your schedule is necessary if you wish to complete the task on time. If you are having trouble sticking to your work schedule, you might want to try writing down all your time commitments in a day planner. Make sure your work location is conducive to completing your task. For example, when you sit down to do your homework, make sure you are in a quiet place, away from distractions and interruptions. It is also important to assess your progress as you work and to take short breaks often to relax and rejuvenate yourself.

Learning how to manage your time is essential in order to meet deadlines. Most activities need to be completed within a certain time limit. Companies often have contracts that state the deadlines for completion. Your history essay probably has a due date. Developing and maintaining your work schedule help you meet these deadlines. Prioritizing your tasks, recording your schedule in a planner, and avoiding interruptions can help you finish on time. Depending on the task, you might need to work with others to meet your deadline. This requires additional planning and cooperation. When working with a group, it is often best to divide the work and delegate specific tasks to each group member. When you use these time-management techniques, you should be able to complete your tasks successfully and meet your deadlines.

Money and Technology

It has been said, "It takes money to make money." Technological activities are no exception. People must be paid for their labor. Buildings and machines must be purchased or leased. Material and energy must be bought. In short, all company activities take money. Money is used to purchase resources (materials, machines, and labor), which are used to produce a product or service. The product or service is then exchanged in the marketplace for money.

Businesses obtain the money needed to operate (operating capital) through two basic avenues. The first is equity financing. People provide money in return for a share of ownership (equity) in the company. Often, they are issued shares of stock in the company in return for their investment. They become stockholders in that particular business.

The other source of money is debt financing. People or other businesses (such as banks and insurance companies) might loan the business money to operate. In return, the company pays a set interest rate for the use of the money. Also, over time, the company pays back the original loan (principal). Debt financing is very similar to a home loan or an automobile loan, in that the original loan is paid back with interest.

Summary

We are in a period of history when knowledge is expanding rapidly. In recent years, knowledge has advanced more than it had in all of earlier recorded human history. How you will fit into this changing world depends on your personal goals, interests, abilities, and physical traits. Finding the right kind of work requires serious study. You have many fields from which to choose. Some jobs require skills of managing people. Others demand creative abilities. Still others require skills in making parts and handling tools. Whatever career is best for you, you should consider the opportunities for advancing to more responsible work. This is known as "moving up the career ladder."

STEM Connections

Science

Research science careers in private industries, not-for-profit agencies, universities, and governmental agencies. One source for information is the U.S. Department of Labor Internet site. (Use a search engine and enter "Department of Labor.")

Mathematics

Gather data from the U.S. Department of Labor Internet site. Prepare graphs comparing starting wages or salaries for 10 different jobs.

Curricular Connections

Social Studies

Investigate the types of jobs available in your school. Determine the levels of authority and responsibility each has.

Activities

1. Draw a chart listing your abilities, interests, and values. Include what you are good at doing, as well as what you dislike. Try to be honest and look at yourself as you think others might see you.
2. Find several careers you think you might like. Gather information about the requirements of these careers. Describe the duties, in terms of working with people, information and ideas, and machines.
3. Compare your abilities, likes, and strengths with the career information you have collected. Do not expect a perfect match. The effort will, however, start you on the way to career planning.
4. With two or three other students, create an imaginary business. Decide what product or service you will offer to the public. Write a one-page summary of the company's role in the free enterprise system. Agree on the image you want your company to have. Develop a marketing plan that complies with this image, paying particular attention to ensuring customer satisfaction.
5. Choose a career area that interests you. Research its academic and professional advancement requirements. Prepare a verbal presentation with PowerPoint slides for the class describing the requirements.
6. Develop a schedule for a major assignment in one of your classes. Set deadlines for achieving project milestones. Schedule time to work on the project.

Test Your Knowledge

Do not write in this book. Place your answers to this test on a separate sheet of paper.

1. Why do the major kinds of work people do change over time?
2. Match the statement on the left with the correct type of worker on the right. You will use some answers more than once.

_____	Has a high degree of skill in a narrow range.	A.	Skilled
_____	Requires little or no training.	B.	Semiskilled
_____	Does routine work.	C.	Unskilled
_____	Might serve an apprenticeship.		

3. What is thoughtware?
4. A company is a(n) _____ institution using _____ to produce a(n) _____, with the intent of making a(n) _____.
5. Match the statement on the left with the correct position title on the right. You will use some answers more than once.

_____	Serves as a major link between engineers and the factory floor.	A.	Manager
_____	Organizes resources to produce products.	B.	Engineer
_____	Designs products, systems, and structures.	C.	Technologist
_____	Solves operational problems.	D.	Technician
_____	Often leads a design team.	E.	Worker
_____	Constructs products or structures.		

6. Summarize the four factors that should contribute to your career choice.
7. The path leading to better jobs is called a(n) _____.
8. Time is a valuable resource of technology. True or false?
9. Time is used only to measure human labor. True or false?
10. Money is needed to pay for human labor, materials, machines, and energy. True or false?

Reading Organizer

On a separate sheet of paper, list at least three different jobs in technology, write the job skills related to the job, and explain why the skills are necessary.

Job in Technology	Job Skills	Reason for Skills
Example: Managers	Communication skills	To explain and present ideas clearly
	Listening skills	To understand what employees are saying

New technological knowledge is being developed on a daily basis. The way we use technology has evolved greatly in recent years.

Section 3 Creating Technology

Technology is created to solve problems. The problems can be as simple as creating a device to organize your music collection. In other cases, they can be as complex as developing medicines to help fight illnesses.

People can solve problems using several different methods. They can use the invention process to create a new device. People can change an existing invention, creating an innovation. Also, they can use the design process to create a useful solution.

Many people design products and systems. You might have designed a device to solve a problem of your own. If so, you are a designer. Designers use a number of steps to create new solutions. These steps include activities such as identifying the problem, researching, drawing sketches, making models, testing, and creating final drawings. In the following 10 chapters, you will learn how the design process works.

Technology Headline:

Space Tourism

Planning your next family vacation? Skiing in the Rockies or lying on a beach in Florida sound like wonderful options, but soon a destination that is truly out-of-this-world will be available—outer space! In the near future, average people will become astronauts and experience space travel first-hand. Private companies continue to develop spacecraft that will offer seats—at a significant cost—to anyone who wants to become an amateur astronaut. Rather than book a flight to New York City or Chicago, passengers would book a trip to orbit the Earth!

As the face of space travel rapidly changes, some companies have already begun accepting deposits, reserving seats for future amateur astronauts. Enjoying the wonders of space will no longer be an adventure limited to NASA astronauts. As more private companies begin developing spaceships, it appears as though space travel will become increasingly common. To prepare for this future form of tourism, the United States government proposed laws and regulations for civilian space travel in 2005. While these regulations do not include health screening, they do require companies to train passengers for emergency situations. Once private space flights are fully operational, many plan to hold required sessions of preflight training for their new astronauts. These training sessions will ready one's body for the new sensations of zero gravity, the weightlessness experienced outside Earth's atmosphere. Other training sessions may focus on how to best deal with emergencies in space.

A handful of average citizens have already visited space as part of space tourism programs associated with government-sponsored missions. The cost for such trips typically exceeds several million dollars and such opportunities are rare. The new age of space tourism, however, will offer frequent flights at a much lower cost to passengers—an estimated $200,000 to begin but eventually dropping to around $20,000 as companies become more established—making space travel a more realistic opportunity. This change in price will be made possible in part with the development of spacecraft that is both reliable and reusable. These vehicles will be designed to make repeat trips into space, thereby reducing costs associated with their expensive construction.

Rather than having a specific destination in mind, such as the International Space Station or the Moon, trips would be classified as suborbital flights. This means spacecraft would reach space and orbit the Earth, but would not complete an entire revolution around Earth. The brief trip would last only a few hours.

Perhaps someday there will be a hotel on the Moon or a public space station, but such developments appear to be fixtures of the distant future. For now, passengers will have to be satisfied with trips beyond the Earth's atmosphere and phenomenal views of the blue planet we call home.

CHAPTER 9
Invention and Innovation

Did You Know?

- A patent is the record explaining an inventor's new invention. The U.S. Patent and Trademark Office (USPTO) grants patents. A patent protects the invention from being stolen from the inventor.
- The first patent was granted to Joseph Jenks for creating a better sawmill.
- The USPTO has issued over 7 million patents since it opened in 1790.

Objectives

The information given in this chapter will help you do the following:

- Explain the concept of invention.
- Summarize the history of invention.
- Compare invention and discovery.
- Summarize the three major categories of inventions.
- Explain how invention is a problem-solving process.
- Summarize the steps of the invention process.
- Give examples of some characteristics of inventors.

- Recall the three main reasons for inventing.
- Recall the advantages of invention teams.
- Explain the concept of innovation.
- Explain the concept of adaptation.

Key Words

These words are used in this chapter. Do you know what they mean?

adaptation
challenge
creative thinking
discovery
financial invention
Industrial Revolution
innovation
invention
invention process
leisure invention
patent
problem
problem solving
scientific discovery
social invention
spin-off

Preparing to Read

Look carefully for the main ideas as you read this chapter. Look for the details that support each of the main ideas. Use the Reading Organizer at the end of the chapter to organize the main and supporting points.

Have you ever used a shoestring for something other than to tie your shoes, created a device to help a friend, or even created a new recipe? If you have, you are an inventor. Inventors create new and unique products. These products are called ***inventions***. Inventions are created through the invention process. This process uses imagination and knowledge to turn ideas into devices, products, and systems. See Figure 9-1. Once inventions

Figure 9-1. The products we use today have been developed using imagination and knowledge. (Harris Corp.)

exist, they are often made better through the process of innovation.

INVENTION

The ***invention process*** dates back to the beginning of humanity. When people began to create tools and clothes, they became inventors. Tools and clothes did not exist before people created them. Inventions have followed the materials available at any point in time.

In the Stone Age, axes and tools made of stone and bone were invented. People also invented ways of creating pottery. Next, in the Bronze Age, inventions such as the wheel and the plow were created. Irrigation and writing were also invented in the Bronze Age. The inventions of the Stone and Bronze Ages were important to our world. These inventions were the building blocks for later inventions.

The Iron Age came next. New devices were invented in the Iron Age. One important invention was the screw pump. Archimedes, a Greek inventor, developed this pump. This invention uses a large screw to move water uphill. The screw pump is still used today, although it might not seem to be a major invention anymore. See Figure 9-2. At the time in history when it was invented, however, moving water was a major ***problem***. Archimedes solved that problem by using a screw inside a cylinder. The screw turns and carries the water upward. The cylinder is used to keep the water on the screw. The screw pump is also important because it is one of the first inventions that can be credited to an inventor. Earlier inventions are too old to know who invented them.

There are many early inventions we take for granted. Today, there is paper to write on, compasses to lead our way, and batteries to power our devices. Without these inventions, we would be faced with many of the same problems our ancestors had.

Invention did not stop in the Iron Age. Inventions have been created in every age of time. An important time period for invention is known as the ***Industrial Revolution***. This period lasted from around 1750 to 1850. The Industrial Revolution was a time when many machines and devices were invented. The invention of the steam engine started the revolution. See Figure 9-3.

Figure 9-2. A modern example of a screw pump invented thousands of years ago. (Lakeside Equipment Corporation)

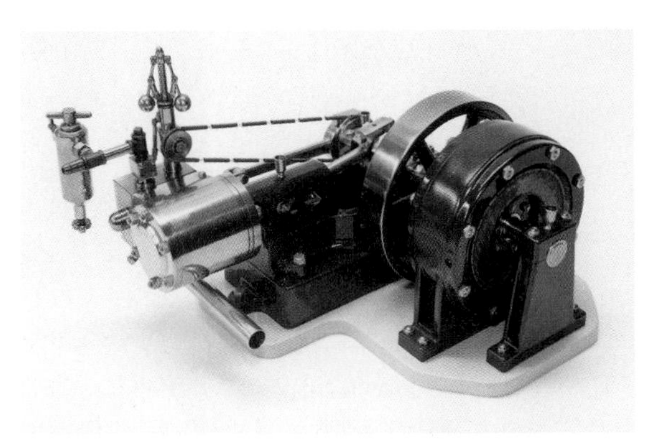

Figure 9-3. Steam engines were used in many applications during the Industrial Revolution. (Jerry E. Howell)

This engine was used in the invention of factories, on the railroad, and in ships. Also, during the Industrial Revolution, the telegraph was invented. Better ways to plant and grow crops were also invented during that time. Many of the inventions of the Industrial Revolution have had lasting effects on the world. Without the Industrial Revolution, there would be no factories to make products. There would be no passenger trains. It would also be much harder to communicate.

Many of the inventions since the Industrial Revolution have been related to information. This recent period of time is known as the *Information Age*. In this age, vacuum tubes and transistors have been invented. These inventions led to the television and computer. The mobile telephone, the Internet, the space shuttle, and MP3 players are all inventions of the Information Age. See Figure 9-4.

Figure 9-4. The MP3 player is just one example of a modern invention.

Over time, inventions have become more complex. The wheel no longer seems to be an important invention. Without the invention of the wheel, however, there would be no automobile. Many things we use and create today come from inventions and discoveries thousands of years old. Inventors often use logical processes of trials and alterations to create useful products and systems. Many of the inventions and innovations we use today have developed gradually over time.

Discovery vs Invention

Discovery is a word often confused with *invention*. You might hear the words used together. It might seem that they are interchangeable. They, however, are not. An invention is a new idea, process, or device. Humans design and create inventions.

Discoveries are different. They are not human made. Discoveries are naturally occurring. A discovery is made when something that naturally occurs is first noticed. Discoveries are often called ***scientific discoveries***. Scientists and researchers are the people who normally make the discoveries. Important discoveries include electricity, vacuums, and Newton's laws of motion. Discoveries are the foundation for scientific knowledge.

Inventions and discoveries are often used together. The Montgolfier brothers, for example, discovered that a balloon filled with hot air rises. That discovery, in 1783, led to their invention of the hot air balloon. Since that time, more discoveries about propane and other gases have been made. These discoveries have led to more inventions related to the hot air balloon. Hot air ballooning is now a safe and fun sport for many people because of these discoveries and inventions. See Figure 9-5.

Figure 9-5. Discoveries and inventions have led to the hobby of hot air ballooning. (Napa Valley Balloons, Inc.)

Inventions are often difficult to create without certain discoveries. Leonardo da Vinci was an inventor who lacked the discoveries he needed. Da Vinci created plans and drawings for bicycles, helicopters, gears, and parachutes. Many of these plans were never built. In his time, the materials he needed to build these inventions did not exist. This was hundreds of years before the materials were discovered or invented.

Categories of Inventions

Unlike discoveries, inventions are human made. Humans create inventions. An invention can be as simple as a paper clip or as complex as a satellite. Inventions can be broken down into several categories. See Figure 9-6:

- **Devices and machines.** These are inventions with moving parts or electrical circuits used to do work. The lightbulb, CD player, digital camera, and mousetrap are all common devices and machines.

Career Highlight

Patent Attorneys

Law, Public Safety, Corrections & Security

The Job: Patent attorneys and agents write and submit patent applications for inventors. They represent the inventor in all communications with the USPTO.

Working Conditions: These attorneys and agents spend a majority of their time researching and preparing patent applications. They can work within large corporations, work in law firms, or be self-employed. Working conditions are similar to other law professions.

Education and Training: Patent attorneys and agents must apply to the USPTO to represent inventors. To become registered with the USPTO, people must pass an examination on patent law and policies. They must also have at least a bachelor's degree in a science or an engineering discipline. Those who meet the requirements and also have a law degree are considered patent attorneys. The people who do not have a law degree are titled patent agents.

Career Cluster: Law, Public Safety, Corrections & Security

Career Pathway: Legal Services

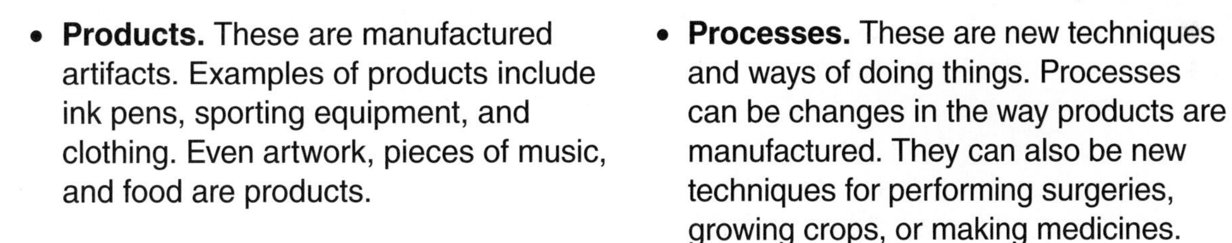

- **Products.** These are manufactured artifacts. Examples of products include ink pens, sporting equipment, and clothing. Even artwork, pieces of music, and food are products.
- **Processes.** These are new techniques and ways of doing things. Processes can be changes in the way products are manufactured. They can also be new techniques for performing surgeries, growing crops, or making medicines.

Devices

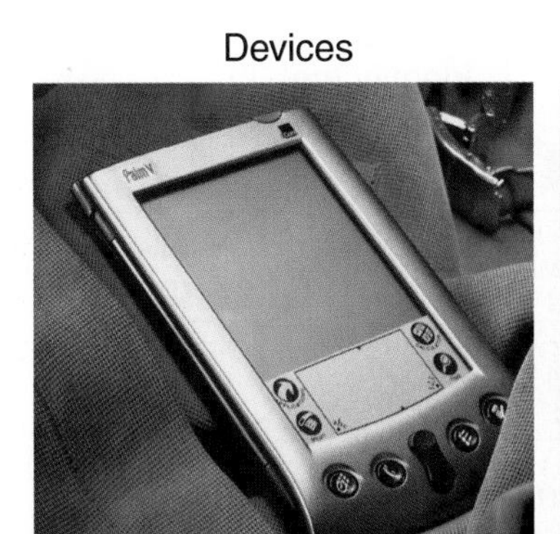

Processes

Products

Figure 9-6. Inventions can be devices, processes, or products. Here are an electronic device known as a *personal digital assistant,* a tire production line, and a medication cart. (Photo by Steven Moeder, courtesy IDEO; Goodyear Tire and Rubber Company; Design Central, design firm—Artromick International, client)

The Invention Process

Most inventions are created through a planned process. A few, however, are created by accident. Charles Goodyear created, perhaps, the most famous accidental invention. Goodyear worked to create rubber that was not sticky or brittle. In 1839, after years of work, he accidentally dropped a piece of rubber on his oven. When the rubber cooled, Goodyear found that the piece was not sticky or brittle. His invention of vulcanized rubber was accidental. Today, we have shoes, tires, and many other products made of his vulcanized rubber. See Figure 9-7.

Accidental inventions are not common. Most inventions are very well planned. Inventing follows a set of steps, and it is a process. The process begins with a problem and ends with a solution. This activity is a problem-solving process.

The invention process is just one type of ***problem solving***. Troubleshooting, experimentation, and research and development are other problem-solving processes. All problem-solving processes are useful. Some technological problems, however, can best be solved through a certain method. Troubleshooting is used to find the cause of a problem in a technological system. This type of problem often calls for some type of specific knowledge. Once the cause of the problem has been recognized, the next steps are to fix the problem and to check if the system works. Experimentation is used to carry out tests on technological products and systems. This process strongly resembles the scientific method, in that both are logical and require tinkering, hypothesizing, studying, fine-tuning, testing, and recording. Research and development is the process used to fix problems. This process is used intensively in business and industry to prepare devices and systems for the marketplace. The invention process creates new products to solve problems.

Most of the inventions people have produced over the course of history were developed using the invention process. See Figure 9-8. This process has five steps:

1. Identify the problem.
2. Collect information.
3. Think creatively.
4. Experiment.
5. Review.

Figure 9-7. Tires are products made from vulcanized rubber. (Goodyear Tire and Rubber Company)

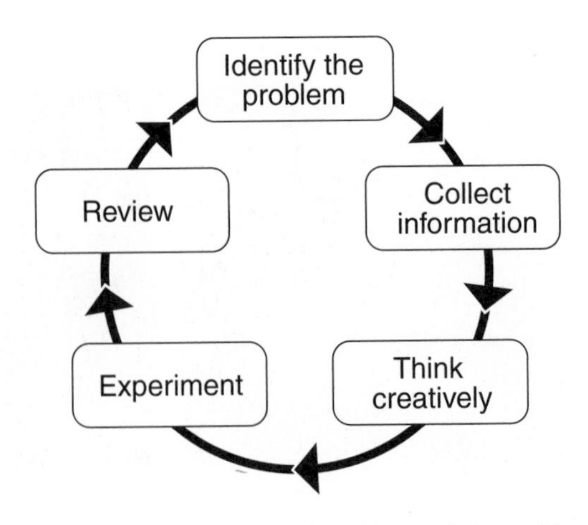

Figure 9-8. The invention process follows this order.

Identifying the Problem

The invention process starts with a problem, or ***challenge***. In this first step, inventors identify the problem. A problem is anything that can be made better through change. For example, the time and effort it takes to mow the lawn are problems. Problems can be dangerous. For example, using electrical appliances around water is a dangerous problem. See Figure 9-9.

Challenges are also problems. A challenge is an obstacle or a goal that needs to be met. An example of a challenge is to safely travel to the planet Mars.

Collecting Information

The next step is to collect information. Inventors research the problem. They learn as much as they can about it. Inventors interview experts. They read books, magazines, and academic journals. Good inventors keep notebooks or personal journals. They write down everything they do and learn.

Thinking Creatively

The third step is the use of ***creative thinking***. Creative thinking requires the human brain. Only humans have the ability to think in creative ways. Animals are unable to invent devices because they do not have creative minds. Animals act on instinct. Humans use creativity to solve problems. See Figure 9-10.

In this step, inventors begin to think of solutions. In the first two steps, they focus only on the problem. The inventors identified and researched the problem. Now they use the results of the research and creativity to design solutions. The inventors come up with as many solutions as possible. They make sketches, drawings, charts, and graphs. At the end of this step, the inventors have a few good designs.

Experimenting

The next step is to experiment. In this step, inventors build inventions. They use the drawings created in the last step. It is important that they know how to use tools and machines. The inventors must build models and test them. See Figure 9-11. This is a trial and error process. Inventions often fail at this step, if they do not work or meet the need. When a solution fails, the inventor fixes it and tries again. This step might take weeks or even years.

Figure 9-9. The danger of using a hair dryer near water was a problem. To solve the problem, manufacturers use a special plug to help prevent shocks.

Figure 9-10. Animals build shelters using instinct. Humans use creativity and technology to build homes. (Wisconsin Department of Natural Resources, Habitat for Humanity International)

Figure 9-11. Inventions are tested to make sure they function correctly. (©iStockphoto.com/Fertnig)

Reviewing

The final step is review. After the invention works, inventors review their notebooks. They make sure the invention solves the problem. It is possible to invent a device that does not solve the problem. If the invention does not solve the problem, the inventor starts over again with step one. If it does solve the problem, the invention process is complete. At this point, inventors might try to patent their inventions. A ***patent*** is a right given to an inventor of a new product, design, or plant. This right protects the inventor from others manufacturing, copying, or selling the invention without the inventor's permission.

Inventors

The people who use the invention process are known as *inventors*. Inventors are curious people of all ages. Both children and adults can invent. Children often make good inventors because they are naturally curious. Alexander Graham Bell and Benjamin Franklin are famous inventors known for their curiosity. Benjamin Franklin was curious about lightning and electricity. He flew a kite in a lightning storm to prove that lightning and electricity are related. Alexander Graham Bell was curious about how sound traveled underwater. He went to a lake and placed his head in the water. His partner then hit two rocks together. They did this at different distances. He found that sound did travel underwater. Bell's and Franklin's experiments led to major inventions because they were curious people.

Inventors look at things in their lives. They wonder how they can make things better. Some inventors think about things such as games and hobbies. See Figure 9-12. James Naismith did just that. He was a physical education teacher who wanted a game his students could play inside during the winter. This teacher created the game of basketball. Milton Bradley was also an inventor of games. He invented games such as The Game of Life® board game.

Creative and imaginative people are often inventors. Creativity is an important part of invention. Inventors use their creativity to see situations in new ways. If inventors were not creative, new devices and systems would not be created. The Wright brothers were creative and imaginative. Orville and Wilber Wright imagined flying. They used their creativity to create the first powered airplane.

Inventors need to have a great deal of knowledge. They must know about the science and mathematics used in their inventions. Samuel Morse invented the telegraph in 1840. See Figure 9-13. He

Figure 9-12. Sports, such as basketball, are invented for fun and recreation.

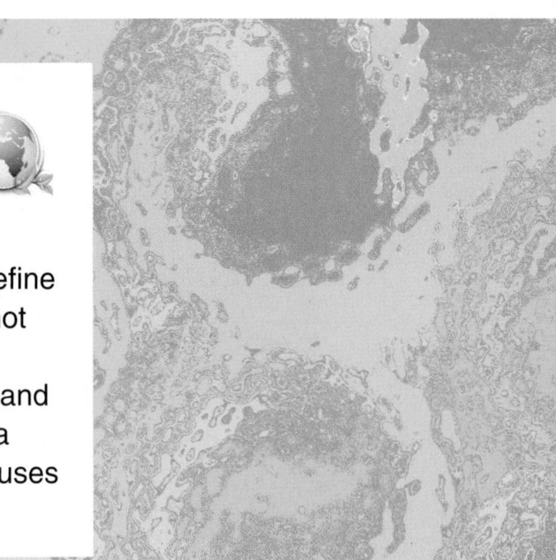

THINK GREEN

Green Materials

Green materials are considered environmentally friendly materials that are alternatives to commonly used materials. To define a material as being green, some criteria are: the material must not be created from exhaustible natural resources; it must emit less carbon dioxide emissions than other materials when processed; and it must either be recyclable or be biodegradable. An example of a green material is recycled paper. It doesn't use new materials, it uses different processing techniques, and it may be recycled again.

Figure 9-13. Samuel Morse's telegraph sped up communication. (Jerry E. Howell)

had to know how to use electricity. Morse also needed technical knowledge. He knew how to use tools to create his invention. Technical knowledge is important for inventors. Many inventors can think of ideas. Some do not have the knowledge of technology to build their inventions.

People who use the invention process must also know about already existing products. James Watt invented the steam engine. He looked at models of steam engines that did not work and took them apart. Watt then discovered why these engines did not work. He built the first working steam engine. Inventors, such as Watt, enjoy taking things apart and finding how things work.

Reasons for Inventing

People invent for three reasons. Some people invent to help themselves and others. The products and systems they develop can be called ***social inventions***. Other people want to make money with their inventions. These inventions are known as ***financial inventions***. There are other people who invent because they find it fun. These devices and processes are called ***leisure inventions***.

Social Inventions

Social inventions are created to help the inventor or other people. They make our lives better and easier. These inventions impact society. Social inventions can be simple, such as a jar opener for the elderly. They can also be complex, such as vaccines for children. The first vaccine was invented to stop smallpox. Edward Jenner invented it to help society. Vaccines are used to control many different diseases. See Figure 9-14. Social inventions can be created to help better the world. Products invented to cause less harm to the environment are social inventions. Safety devices are social inventions. Seat belts, child seats, and safety guards are also social inventions.

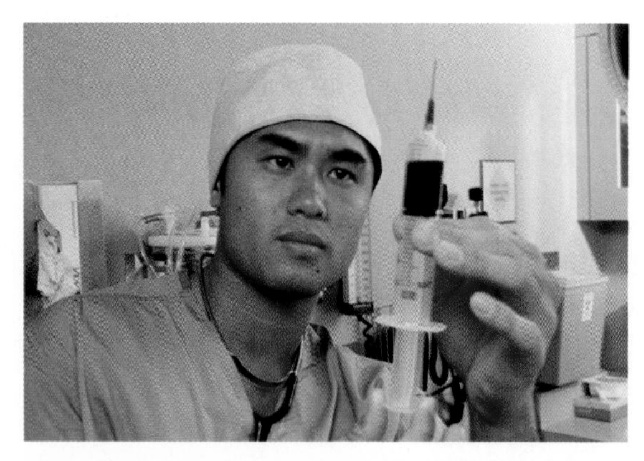

Figure 9-14. Vaccinations are social inventions.

Financial Inventions

Most inventions are created to make money. These are called *financial inventions*. Financial inventions often make things faster or easier to do. Eli Whitney invented the cotton gin to produce cotton faster. The faster the cotton was produced, the more money the farmers made. The computer and computer software are financial inventions. They help businesses operate faster and more efficiently.

Leisure Inventions

Leisure inventions are created for the pleasure of inventing. Inventors who like to tinker create these. These inventors like to invent as a hobby. Leisure inventions include small games and toys. Rube Goldberg was a cartoonist who drew funny leisure inventions. His inventions used many steps to solve simple problems. Leisure inventions are often very creative and are sometimes even patented. See Figure 9-15.

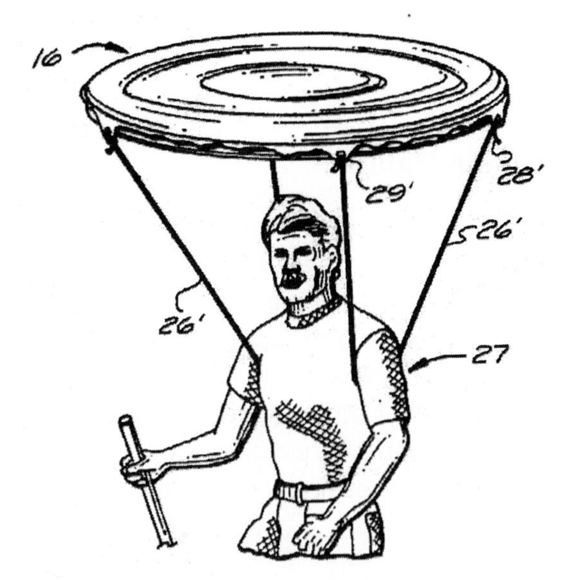

Figure 9-15. This leisure invention of a helium-filled sunshade is patented. (Patent No. 5,076,029, U.S. Patent and Trademark Office)

Invention Teams

Social and financial inventions are usually created in teams. Invention teams are used because each team member has different abilities and knowledge. Some people are better at using tools. Others have more scientific knowledge, while others have more mathematical knowledge. Many invention teams work in research labs. See Figure 9-16. Thomas Edison created the first research lab in the United States. His research lab was built in Menlo Park, New Jersey. Edison's researchers helped invent the phonograph and the electric lightbulb. Research labs can be used to invent many things. For example, they are used to invent new materials, electronic equipment, and

Figure 9-16. The first transistor made from a single material was developed in a research lab. (Lucent Technologies, Inc./Bell Labs)

Technology Explained

high-rise building: a multistory residential or commercial building that has a skeleton frame.

The Great Pyramid of Giza in Egypt is a construction marvel, even today. This type of tall structure, however, was not common throughout history. More common than pyramids were the great cathedrals built in Europe. The cathedrals have high, thick stone walls. This type of wall is called a *load-bearing wall* because it holds all the weight (load) of the roof.

Construction such as this has height limits. Load-bearing walls are not practical in buildings over five stories tall. The walls become too thick and heavy.

In the past, as cities became larger, taller buildings were needed. It took the development of the elevator and the skeleton frame to make tall buildings practical. The elevator allowed people to move quickly between floors without walking up stairs. For example, the Empire State Building's elevators can move up to 1400′ (427 m) per minute. At this speed, people can travel from the lobby to the 80th floor in 45 seconds.

A skeleton frame carries the weight of the building, very similarly to how the human skeleton carries the body. These developments made high-rise buildings, or skyscrapers, possible. The first high-rise building was built in Chicago in 1885. Built for the Home Insurance Company, it was 10 stories tall. The exterior walls provided protection from the weather. These walls, however, did not carry any load. This type of wall is called a *curtain wall* because it merely hangs from the frame.

Today, high-rise buildings use two types of framework. The first type is reinforced concrete. See Figure A. This type of framework is cast on-site. Forms are erected around a network of steel rods. The concrete is poured inside the forms. After the concrete has cured, the forms are removed.

The weight of reinforced concrete limits its use to buildings of moderate height. Taller buildings use steel skeletons. Steel is fabricated into angles and I beams. Steel columns and beams are bolted or riveted together. See Figure B. This allows the beams to expand and contract uniformly with temperature changes.

Once the frame is erected, the floor and roof are installed. The exterior walls are then put in place. See Figure C. These walls can be made of steel or aluminum panels, bricks, concrete blocks, sheets of glass, or other materials.

The title of the world's tallest building passes from skyscraper to skyscraper as new structures are built. The tallest building in the world is currently the Burj Khalifa, built in Dubai, United Arab Emirates in 2009. There are, however, several skyscrapers in both the planning and construction stages that will be taller. The Sears Tower in Chicago is the tallest skyscraper in the United States and stands 1450′ high.

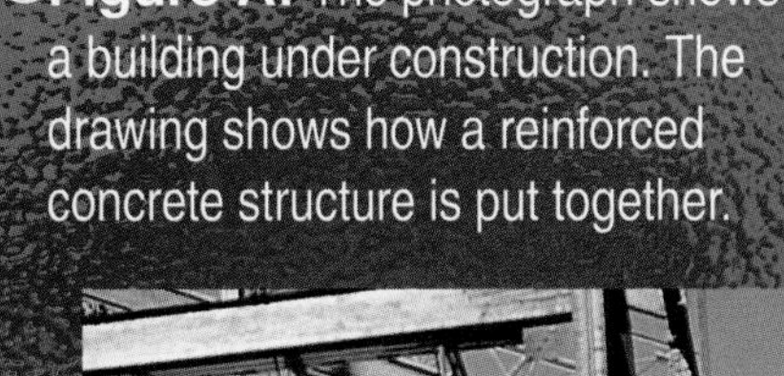

Figure A. The photograph shows a building under construction. The drawing shows how a reinforced concrete structure is put together.

Figure B. A steel-framed building. Notice the different shapes of the steel members.

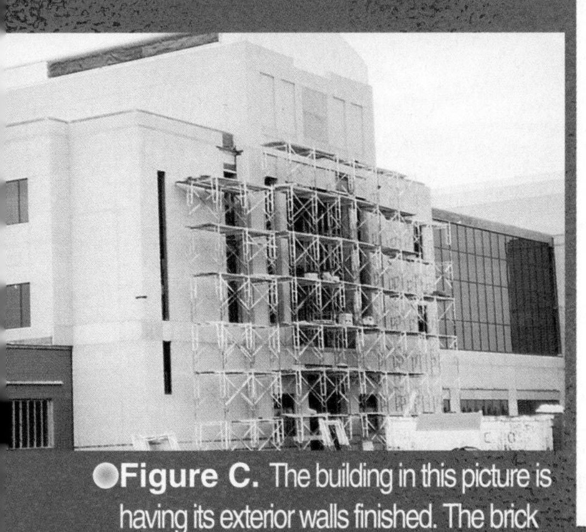

Figure C. The building in this picture is having its exterior walls finished. The brick is in place. The windows are installed.

medications. Research labs can be found in universities, large companies, and government agencies.

Innovation

Inventors do not always invent new devices. They often create ***innovations***. Innovation is the process of altering an existing product or system to improve it. Innovations make inventions better. All technological refinement occurs through the process of innovation. Many of the products created today are innovations. Innovations can do several things for the inventions. They can make inventions work better, be less expensive, or be built with better materials. Devices can even be made from several inventions.

Innovations make inventions more useful. They have been made to improve the speed and capacity of computers. These innovations have made the computer more useful. Computers can do more things faster than they could just a few years ago.

Inventions can be made less expensive with innovation. Henry Ford innovated the way automobiles are built. His mass production system made automobiles cheaper to buy. More people could afford automobiles because of Henry Ford. Many innovations have changed the way automobiles are manufactured today. Today, most automobile production lines are very automated. See Figure 9-17.

Figure 9-17. Current automobile production lines are innovations. (Daimler)

Innovations can be large or small changes to an invention. An inventor might change the material of the invention. The bicycle has gone through many innovations. When the bicycle was invented, it was made of wood. New materials helped to innovate the bicycle. Bicycles have been made of steel, aluminum, and titanium. Some bicycles are even made of carbon fiber. These innovations changed the weight, comfort, and ride of the bicycle. They made bicycles easier to ride.

Putting two inventions together can also make innovations. In-line skates are an innovation. See Figure 9-18. They are a combination of shoes and skateboards. Innovations are important to our society.

Figure 9-18. In-line skates are innovations used for fitness and recreation. (Rollerblade)

One type of innovation is adaptation. ***Adaptations*** are developed when inventions are used for something other than the purposes for which they were intended. Medicines are adapted when they are used to treat ailments besides the one they were meant to cure. The laser is an invention that has been adapted for many different uses. Lasers are used to cut materials, scan grocery items, perform eye surgery, and create interesting light shows. The National Aeronautics and Space Administration (NASA) is known for creating adaptations. NASA has created over 1000 inventions that have been adapted to solve other problems. These adaptations are known as ***spin-offs***. Race-car drivers now use technology from space suits the *Apollo* astronauts wore. Cordless power tools were first used in space. Even the suits firefighters wear are NASA spin-offs. See Figure 9-19.

Spin-offs

Many inventions are adapted to solve problems other than those for which they were intended.

Figure 9-19. The protective clothing firefighters wear is a NASA spin-off from the space suit. (NASA)

Innovations and adaptations make devices, products, and processes easier to use. They help to make inventions better and more useful. Innovations are an important part of invention.

SUMMARY

Invention is a problem-solving process. Humans have used the process of invention for thousands of years. Inventions are new and unique devices meeting our needs. The people who create inventions are known as *inventors.* Inventors are creative and imaginative people. They often work together in research laboratories. Inventors follow the steps of the invention process to create new inventions. They also make changes to older inventions. These changes are called *innovations.* Adaptations are inventions used for new purposes. Inventions, innovations, and adaptations all help to make life easier.

STEM Connections

Science

Choose a scientific discovery (such as electricity, gravity, or friction). Make a list of inventions used to make that discovery. Make a second list of inventions that come from that discovery.

Curricular Connections

Language Arts

Create an invention. Keep an inventor's log. Use the inventor's log to write a simple patent application.

Social Studies

Select an industrialized country (such as the United States, Germany, or England). Research the major inventions developed in that country. Create a display presenting the country's inventions and inventors.

Activities

1. Choose an invention affecting your life (for example, the automobile, telephone, television, or computer). Create a display highlighting the invention. Show the history of the invention and how the invention works.
2. Select a problem in your life or the life of someone in your family. Follow the invention process. Create an invention solving the problem.
3. Research an adaptation, or a spin-off, NASA created. Divide a poster board into two sections. On the first half, list and explain an invention NASA created. On the second half, show how the invention has been adapted and how it is used today.

Test Your Knowledge

Do not write in this book. Place your answers to this test on a separate sheet of paper.

1. Paraphrase the definition of *invention.*
2. Describe why the Industrial Revolution was important to the history of invention.
3. How are discoveries different from inventions?
4. List the three types of inventions.
5. Accidental inventions are the most common type. True or false?
6. Rewrite, in your own words, a definition for the word *problem*, as it relates to the invention process.
7. The invention process is the only problem-solving process. True or false?
8. The invention process has a set of steps. True or false?
9. Summarize three characteristics of inventors.
10. An invention created to help people is a(n) _____ invention.
11. What type of invention is created to increase profits?

12. An invention created for fun is a(n) _____ invention.
13. Today, teams of inventors create most social and financial inventions. True or false?
14. A(n) _____ is an improvement on an invention.
15. Select the phrase best describing an adaptation.
    A. A new and unique product or system created with imagination and knowledge.
    B. An improvement on an already existing invention.
    C. The initial realization of a natural occurrence.
    D. An invention used for something other than the purpose for which it was intended.

Reading Organizer

Draw a bubble diagram for each main idea in the chapter. Make each of the main ideas the central bubble, while using details in smaller bubbles to surround the main points. An example from this chapter is shown.

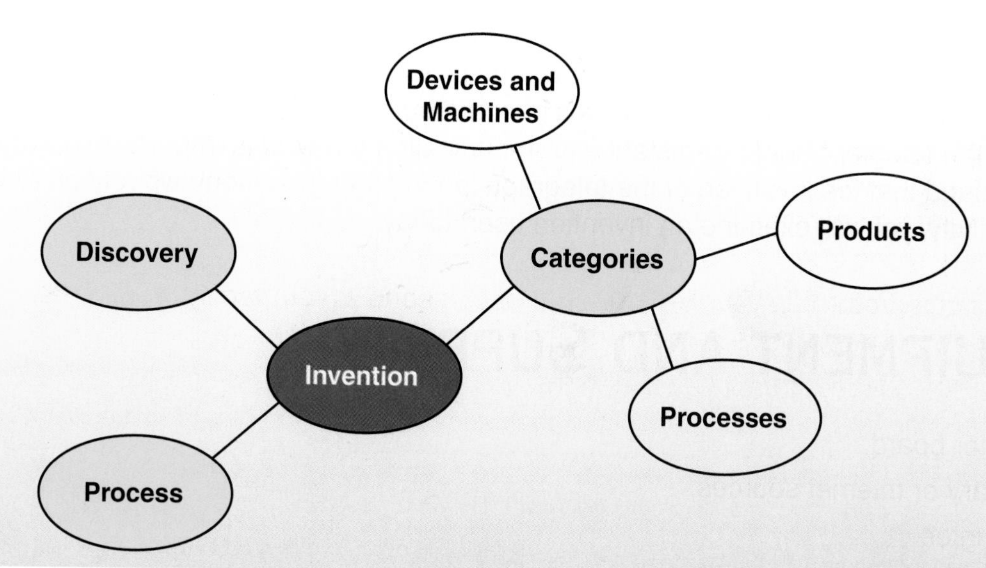

CHAPTER 9 ACTIVITY A

INVENTION AND INNOVATION

THE CHALLENGE

Create a poster board explaining an invention and showing inventions that led up to it and inventions that have come from it.

INTRODUCTION

New inventions and innovations are created every day. You might never encounter some of them. There are many inventions, however, that most of us use all the time. These inventions have become part of our daily lives. Eventually, many of these inventions, as great as they are right now, will become outdated. For example, it is impractical to use the telegraph for long-distance communication today. It is important, however, to understand that the invention of the telegraph led to other inventions we rely on today. In this activity, you will examine an invention used today.

EQUIPMENT AND SUPPLIES

- Poster board.
- Library or Internet sources.
- A printer.
- Markers or colored pencils.
- Tape or glue.

PROCEDURE

1. Select one of the following inventions: the telephone, television, printing press, automobile, airplane, telescope, laser, or single-lens reflex (SLR) camera.
2. Using the library or Internet, research your invention in order to answer the following questions:
   - Who invented the product?
   - When was it invented?
   - Why was it invented?
   - What inventions led up to this invention?
   - How does the invention work?
   - What are the impacts (positive and negative) of the invention?
   - What inventions have come from your invention?
   - Was the invention patented?
3. Create a display (poster board) answering the research questions.
4. Present your display to the rest of the class.

CHALLENGING YOUR LEARNING

Create a model of the invention.

TSA Modular Activity

This activity develops the skills used in TSA's Inventions and Innovations event.

Inventions and Innovations

Activity Overview

In this activity, you will investigate and determine the need for an invention, develop an idea for the invention, and then present your idea using a stand-alone multimedia presentation; a documentation notebook; a model, or prototype; and an oral presentation.

Materials

- Presentation software.
- A three-ring binder with 8 1/2″ × 11″ pages.
- Materials for the model, or prototype (will vary greatly).

Background Information

Selection. Before selecting the theme for your project, consider past inventions and innovations and current needs in each of the major divisions of technology:

- Medicine.
- Agriculture and biotechnology.
- Energy and power.
- Information and communication.
- Transportation.
- Manufacturing.
- Construction.

Use brainstorming techniques to identify several possible inventions and innovations from each area. Select an idea for a final invention to meet an identified need. The invention can be completely new, or it can be an improvement to an existing device, system, or process.

Design. Research issues and gather information about the identified need. Work on the design and details of the final invention. Design and construct a model, or prototype, and visual aids that can be used to enhance the presentations.

Documentation. Determine a format for documenting information in a stand-alone multimedia presentation. Develop a stand-alone multimedia presentation and an accompanying documentation notebook. Prepare an oral presentation further explaining the invention.

GUIDELINES

- The invention should be realistic and have the potential to be workable.
- The multimedia presentation must be between two and four minutes long. This presentation must be self-explanatory. The multimedia presentation should show the development of the idea of the final invention.
- The oral presentation should not be longer than five minutes. The model, or prototype, must be part of this presentation. The goal of this presentation is to convince the audience that the invention is needed and has real potential.
- The following items must be included in your documentation notebook:
  - A cover sheet.
  - The title, or name, of the identified need and a brief description.
  - The title, or name, of the invention and a brief description.
  - Photos of the model, or prototype, of the invention.
  - Relevant information about the stand-alone multimedia presentation.
- The following items must be included in your multimedia presentation:
  - The title of the invention.
  - The identified need and information about it.
  - Information about the design and brainstorming processes.
  - Information about the invention, including the model.
  - An assessment of the invention and its potential for being a workable device, system, or process.

EVALUATION CRITERIA

Your project will be evaluated using the following criteria:

- The effectiveness of the multimedia presentation and documentation notebook to document your work as you investigated and developed an idea for an invention.
- The effectiveness of the oral presentation to convince the audience that the invention is needed, is workable, and has the potential for a return on investment.

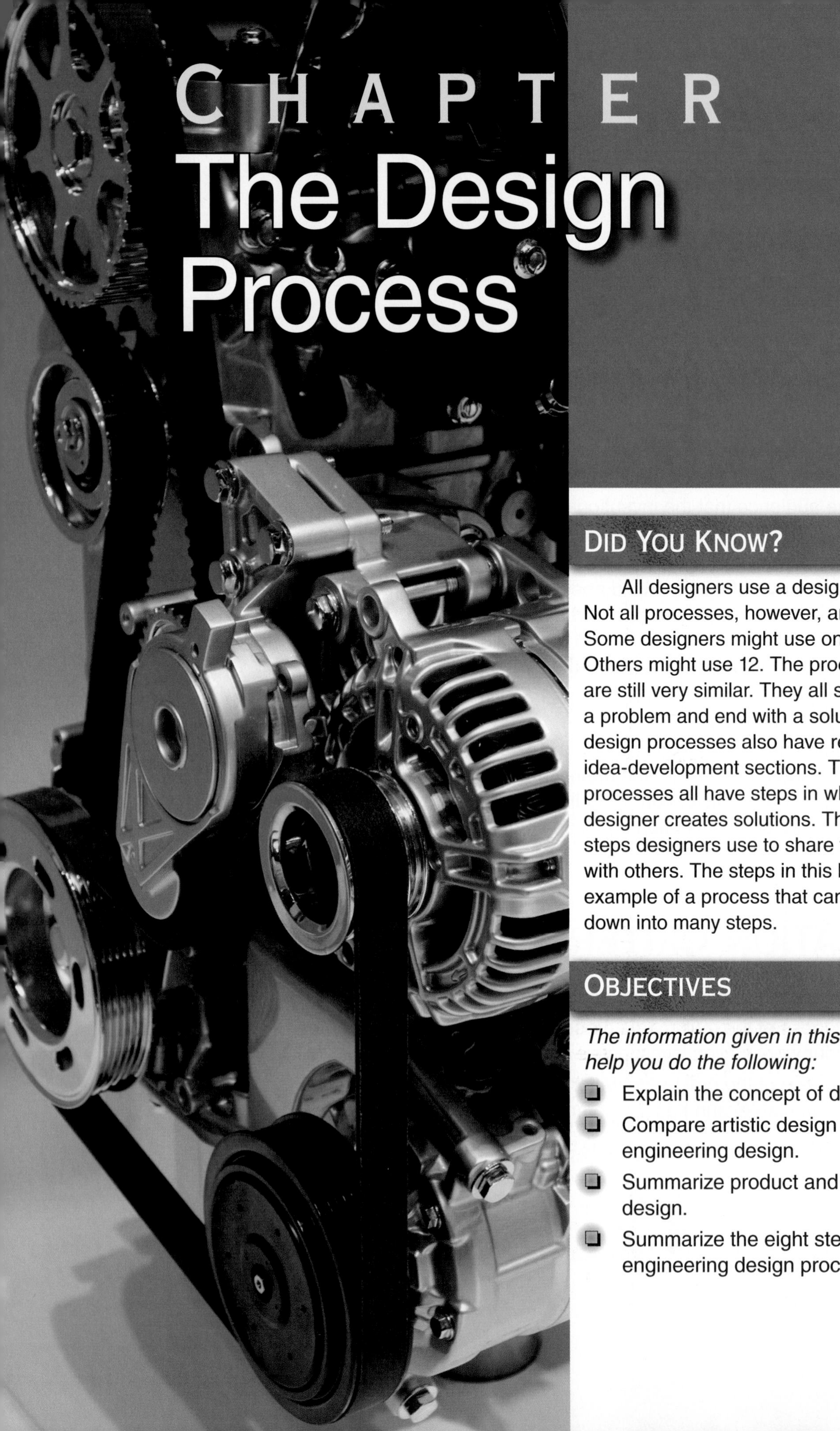

Chapter 10
The Design Process

Did You Know?

All designers use a design process. Not all processes, however, are the same. Some designers might use only 4 steps. Others might use 12. The processes are still very similar. They all start with a problem and end with a solution. All design processes also have research and idea-development sections. The design processes all have steps in which the designer creates solutions. They also have steps designers use to share their solutions with others. The steps in this book are one example of a process that can be broken down into many steps.

Objectives

The information given in this chapter will help you do the following:

- Explain the concept of design.
- Compare artistic design and engineering design.
- Summarize product and system design.
- Summarize the eight steps of the engineering design process.

Key Words

These words are used in this chapter. Do you know what they mean?

artistic design
content
design brief
design process
detailed sketch
engineering design
form
function
mechanical drawing
mock-up
model
product design
prototype
rendering
rough sketch
system design

Preparing to Read

As you read this chapter, outline the details of the design process. Use the Reading Organizer at the end of the chapter to organize your thoughts.

Design affects us every day. Look around the room in which you are sitting. Every item in the room has been designed. The desk you are sitting at, the chair you are sitting on, and the lights in the ceiling have all been designed. Even the clothes you are wearing were designed. If you look through the window, you might see a vehicle, building, or billboard. All these have also been designed.

What Is Design?

We live in a designed world. This means the things we use have been designed. Design is the act of creative planning used to solve a problem. This act results in useful products and systems. A designer generally has a specific problem to solve. Often, the design process takes place in design teams. The members of such a team contribute various types of skills and ideas toward solving the problem. The problem might be to design a physical object, such as a three-bedroom house. The question might be how to create an image, such as an advertisement to sell a product. See Figure 10-1. The problem might even be to grow crops that insects cannot kill. The solution created is usually unique and new.

Design is an action with a purpose. Many different people use the act of design. Engineers design roads and bridges. Architects design buildings. Fashion designers design clothing. See Figure 10-2. Woodworkers design furniture. Artists design paintings and sculptures. Chemists design medicines. Even teachers design lessons.

People use design to help organize their thoughts and actions. The thoughts of designers are the ideas they use to create solutions. Designers get ideas from many places. The actions of a designer

Figure 10-1. Advertisements are designed.

Figure 10-2. Fashion designers use the design process. (©iStockphoto.com/robcruse)

are known as the ***design process***. A designer uses the design process to solve problems. This chapter focuses on types of design and the design process. The steps of the design process are introduced in this chapter. They are explored in more depth in the chapters following.

Types of Design

Design is used in many areas of life. Products, devices, systems, and pieces of art are all designed. All these items can be divided into two types of design:

- Artistic design.
- Engineering design.

Artistic Design

Artistic design is the type of design artists normally do. This design can take many forms. Paintings, sculptures, music, and songs are all forms of artistic design. See Figure 10-3. Artistic design is used to solve problems. The artists usually choose their own problems to solve. They also choose how to solve the problems. The solution might be a piece of pottery, a drawing, a symphony, or a pop song.

This type of design focuses on two aspects of design:

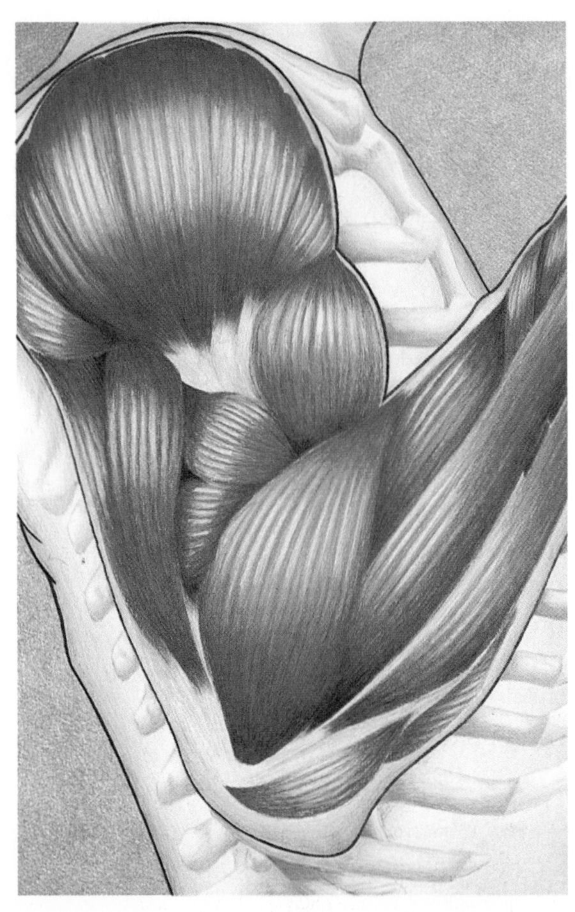

Figure 10-3. Artistic design is different from engineering design. (Mark Chrapla)

- ***Content.*** This is the topic, information, or emotion the artist is trying to communicate. The content is often the problem being solved.
- ***Form.*** This is how the solution looks or sounds. Color, shape, and rhythm are all aspects of form.

Artistic designers have only themselves to please. They decide if their solution has answered their problem.

Engineering Design

Similar to artistic design, engineering design is used to create solutions for problems. Engineering designers, however, are normally given problems to solve. ***Engineering design*** is used to create plans for solutions. These plans can be for buildings, roads, tools, or machines.

Figure 10-4. Engineers use plans to create products.

See Figure 10-4. Many types of people are involved with engineering design. Architects, fashion designers, industrial engineers, and chemists are examples of engineering designers. Their jobs involve two aspects of design:

- **Form.** As in artistic design, this is how the solution looks.
- ***Function.*** This is how the solution works.

Form and function are very important in engineering design. They are important because people other than the designer decide if the solution solves the problem. These consumers decide by buying and using the engineer's product. If the product looks bad or does not work, they will not buy or use the product again. Unlike artistic design, engineering design is successful if it pleases other people. Engineering designers are faced with many types of design. The types can be split into two main categories:

- System design.
- Product design.

System Design

System design involves organizing different parts to solve a problem. By combining different parts, system designers create systems. For example, an electrician combines wires, switches, and outlets to create electrical systems. Audio engineers design entertainment systems. See Figure 10-5. They organize speakers; wires; and audio components, including receivers, DVD players, and televisions. Some designers combine smaller systems to create a larger system. Architects combine electrical, entertainment, heating, cooling, and structural systems to create a house.

Product Design

Product design is the creation of products meeting human needs and wants. Automobiles and clothing are both designed products. Television shows and crops are also designed products. There are many types of product designers. Fashion designers, graphic designers, chemists, and engineers are product designers. Their products are clothing, advertisements, medicines, and machines.

Product designers must consider form and function. Form is the look of the product. Function is the product's usefulness. The

Figure 10-5. The components of an entertainment system.

product must look nice and work well. A chef can be a product designer. See Figure 10-6. Chefs design food to look pleasant and taste good. Industrial designers are product designers who work in many areas of design. They design furniture, computers, and medical and entertainment products. These designers make sure products are easy to use, look nice, and work well.

Designers also make sure products can be manufactured and sold. Toy designers are product designers. They research toys children enjoy so they know if children would want to play with the toy they design.

Steps of the Engineering Design Process

Both system designers and product designers use a design process. The engineering design process is a series of steps to design a new product or system. These steps lead a designer from the problem to the solution. See Figure 10-7. A common design process includes eight steps:

1. Identifying the problem.
2. Researching the problem.
3. Creating solutions to the problem.
4. Selecting and refining the solution.
5. Modeling the solution.
6. Testing the solution.
7. Communicating the solution.
8. Improving the solution.

Figure 10-6. Chefs design food that looks and tastes good.

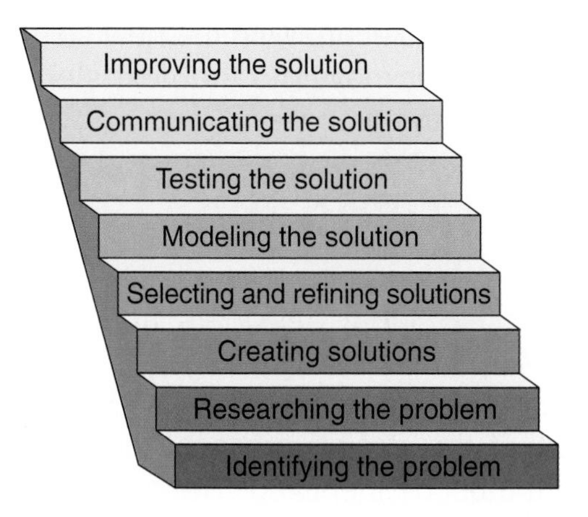

Figure 10-7. The design process uses eight steps.

The design process can be used to solve all kinds of problems. After you recognize and choose a need, want, or problem to solve, this process can result in a solution that can lead to an invention or innovation. The goals of the problem to be solved need to be identified. These goals should specify what the desired result is. To apply the process, you need to do research. You have to then synthesize and study the resulting information you collected.

The eight steps of the design process are usually completed in this sequence. They can, however, be carried out in different orders and repeated as required.

Career Highlight

Industrial Designers

The Job: Industrial designers combine artistic abilities with information about product use, customer needs, material properties, and production methods to create designs that compete in the marketplace. Most industrial designers specialize in an area, such as automobiles, kitchen appliances, or industrial equipment. Designers determine product needs, prepare sketches to show design ideas, and create detailed designs. They are skilled in using drawings, models, and computer simulations to communicate design ideas.

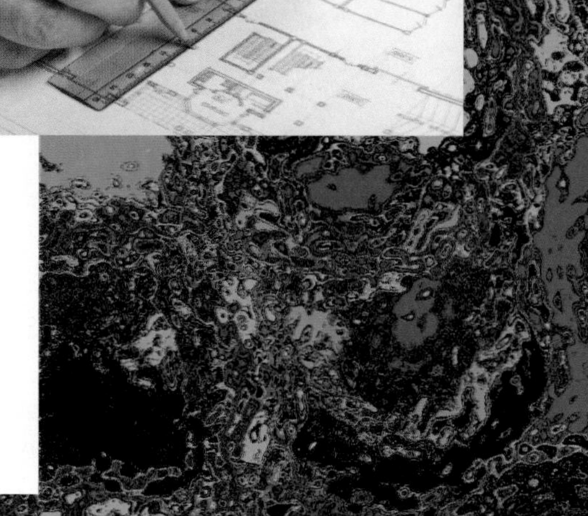

Working Conditions: Many industrial designers work for manufacturing companies, large corporations, or design firms. They generally work standard 40-hour weeks in comfortable design studios. All designers face frustration when their designs are not accepted or are modified extensively.

Education and Training: A bachelor's degree in industrial design or a related field is required for most entry-level positions. In addition, designers must have an eye for color, a feeling for proportion and balance, and an appreciation for detail. They need an ability to prepare sketches and drawings. Designers must have well-developed communication and problem-solving skills.

Career Cluster: Business, Management & Administration

Career Pathway: Marketing and Communications

Each problem is unique and might call for different procedures or for the steps to be completed in a different sequence. For the most part, designers cannot jump forward and skip a step. They can, however, move backward to a previous step. The design process is similar to a steep set of stairs. You must step on each one to move forward. To go backward, you can turn around and jump down to a lower step. The steps in this process can also vary because designers and engineers have their own inclinations and problem-solving approaches. These designers and engineers might want to handle the design process in different ways.

For example, designers might have to go back to a previous step if they get new information throughout the process that changes the problem. They might have to skip back to the first step and identify the problem again. Another example is a designer creating a solution that fails at the testing stage. The designer cannot move forward because the test was not successful. He must, however, move backward in the process. The designer might choose to move two steps back and select a different solution. With the new solution, the designer moves to the next step and continues the process.

Identifying the Problem

All products and systems are designed to meet a human need or want. In the first step of the design process, the need or want must be identified. The need or want becomes the problem the designer tries to solve. Problems can be found in all areas of life. Designers find problems at work, in the home, in vehicles, and even during leisure time. Most designers focus on only one type of problem. For example, some designers focus on problems with landscapes or floor plans. See Figure 10-8. Others specialize in organizational problems.

When designers identify a problem, they write it as a problem statement. The problem statement is very specific and states the exact problem. It is important that the designer focuses on the problem statement and not on the solution. For example, a solution is to design a chair. A problem

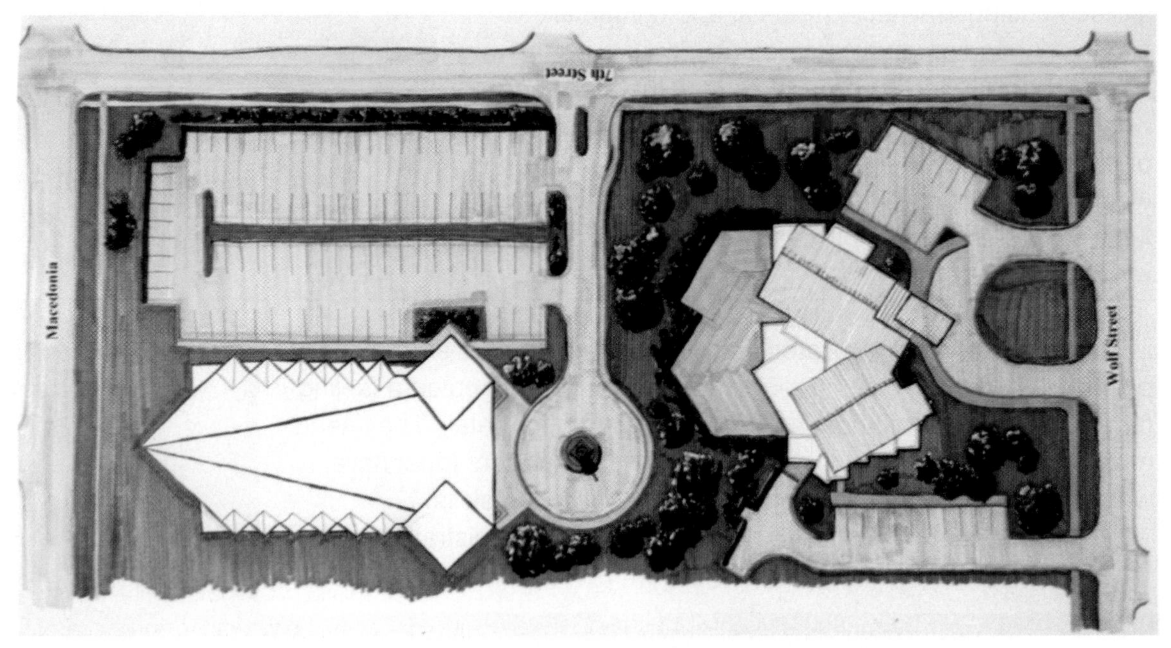

Figure 10-8. Some designers specialize in creating landscape plans. (Keith Nelson)

statement is to design a device that holds a person in a seated position. The solution and problem statement might not seem very different. They are, however, very different. The problem statement is very open and allows the designer the freedom to design many different solutions. See Figure 10-9. The solution statement is very narrow and leaves only one solution.

The problem statement is one part of a document called the ***design brief***. The designer creates the design brief to guide the entire design process. The design brief includes the requirements, or the criteria and constraints, of the design. Criteria identify the desired elements and features of a product or system and usually correlate to the purpose, or function, of these elements. Constraints, such as size and cost, determine the restrictions on a design. The process of identifying the problem and the design brief are discussed further in Chapter 11.

Researching the Problem

Designers create solutions based on research. Since designs are based on problems, the designer must use research to learn about the problem. Designers learn as much as they can about the problem they are solving. Imagine you are a designer creating a better school locker. You need more information before you begin designing and to know the purpose of the lockers. Are the lockers used for books or for physical education supplies? What size do they need to be? What colors do the people who will use the lockers like? What materials are best for lockers? These are just a few of the questions you have to research.

Designers find the answers to many of their questions by using research. They research in books, magazines, and catalogs. See Figure 10-10. Designers also conduct surveys, tests, and Internet searches. They study products that solve similar problems. Research has a process designers follow. Designers begin by listing the problem and finding as much data as possible. In the locker example, designers would gather data such as the average

Figure 10-9. Open-ended design briefs allow creative solutions. (IKEA Home Furnishings)

Figure 10-10. Researchers gather information from sources including books, magazines, and videos. (©iStockphoto.com/simonmcconico)

Technology Explained

sustainable design: the design and creation of products in ways that are less harmful to the environment and the people who use the products than the processes used in designing and creating standard products.

Sustainable design is also known as *green design*. This design is a design movement incorporating a wide view of products and their connections to humans and the environment. Designers who practice sustainable design believe it is important to design products and systems so the products and systems are not harmful to the environment or humans.

In terms of the environment, these designers use and create products that limit the use of nonrenewable resources. For example, sustainable interior designers often use bamboo flooring instead of hardwood because bamboo grows fast and is easily renewable. These designers also attempt to minimize the energy used to create and use their products. Appliances designed to specific energy-saving specifications can be labeled with the ENERGY STAR® logo. See Figure A. Over 2.5 billion ENERGY STAR qualified appliances have been purchased over the last 10 years. The design of products to make them last longer and be reusable and recyclable is another aspect of green design. Products that use recycled materials and the reuse of used products contribute to sustainable design.

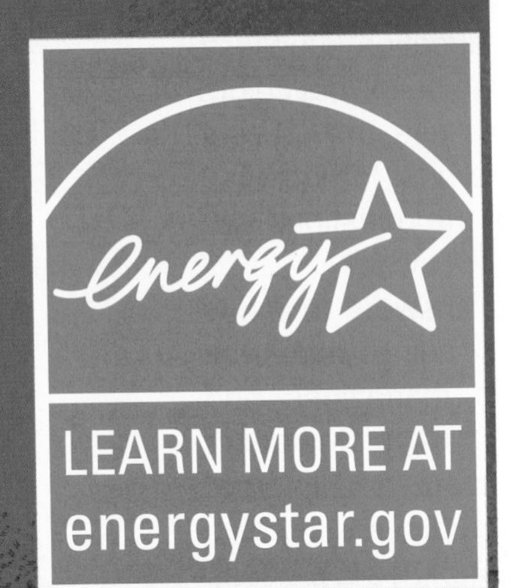

Figure A. The ENERGY STAR logo can be attached to products meeting the guidelines that the U.S. Environmental Protection Agency and the U.S. Department of Energy establish.

Sustainable design not only can help the environment, but also can help humans. Air quality is an issue for sustainable designers that impacts humans. Volatile organic compounds (VOCs) are the gases that are emitted from a number of chemical products and can be harmful to humans. Products such as paints with low or no VOCs have been designed to protect humans and to not impact indoor air quality. Many green designers also believe that products such as bedding, clothing, lotions, and shampoos that contain only natural and organic materials and ingredients are safer and healthier for consumers.

This design has been incorporated in many design industries. From the construction industry and interior design to office-supply and clothing design, sustainable design and "green products" can be found in the marketplace. This type of design and the products that result from it are expected to grow in the coming years, as more and more people see the benefits of sustainable design.

size of lockers. They might also find the percentage of red lockers in all the schools in the state. The designers would then analyze the data and make conclusions. The conclusions are the final results of the research. The designers use the conclusions to design the solution. Chapter 12 focuses on researching problems.

Creating Solutions

In this step, discussed further in Chapter 13, designers begin to create solutions. This begins with developing ideas. The process of developing ideas is called *ideation*. One technique of ideation is brainstorming. Brainstorming is a group problem-solving process in which everyone in the group presents their ideas in an open forum. No one is allowed to criticize anyone else's ideas for any reason. After all the ideas are recorded, the group selects the best ones and further develops them. Questioning and graphic organizers are some other techniques of ideation. The purpose of ideation is to develop as many ideas as possible. The importance is quantity, not quality. Ideas are not judged this early in the process.

As the designers develop ideas, they begin to start sketching. See Figure 10-11. Sketching allows designers to show their ideas to others. It is easier to show someone a design than to try to explain it with words. Sketching at this stage is called *rough sketching*. The purpose of ***rough sketches*** is to get ideas onto paper. This purpose is not to create exact drawings. These sketches are as simple as shapes and outlines. Designers use the rough sketches to select several ideas they will further develop.

When the designers have completed all the rough sketches they can, they review what they have done. They select the sketches they feel have the most promise. The designers then create refined sketches. These sketches are more detailed and complete than the rough sketches. There are three types of sketches used in this step of the design process. Each type has certain advantages and disadvantages. All three are examples of pictorial sketches. Pictorial sketches show the object as it would look to the eye. See Figure 10-12. The pictorial sketches are shaded to make them look more realistic. At the end of this step, the designers should have several good ideas and sketches of possible solutions. The better the ideas are, the better the solution will be.

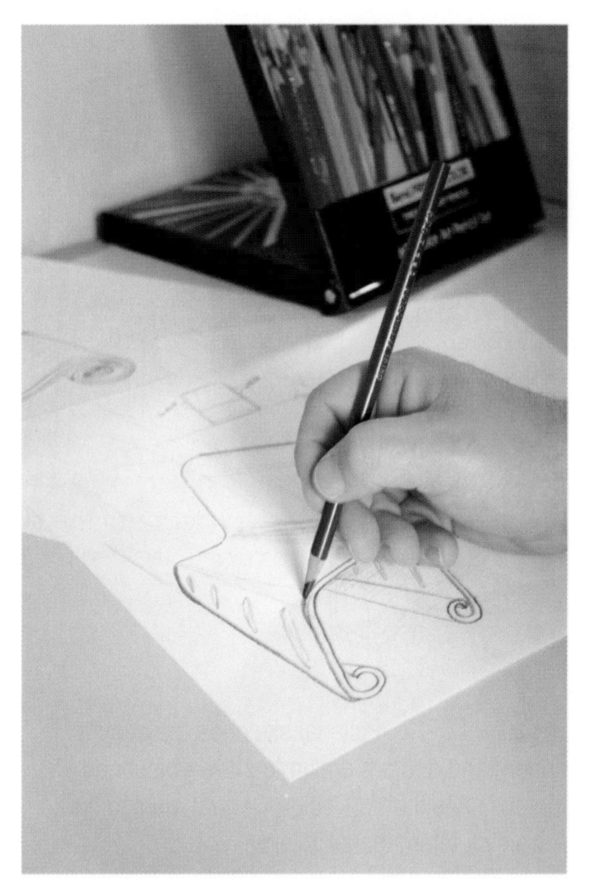

Figure 10-11. Sketches are the first method of putting ideas onto paper.

Selecting and Refining the Solution

The most promising design is selected in this step of the design process. The designer, management, or client selects the design that will be developed into the solution. The selection is based on the aspects of the design. The aspects of design include items such as

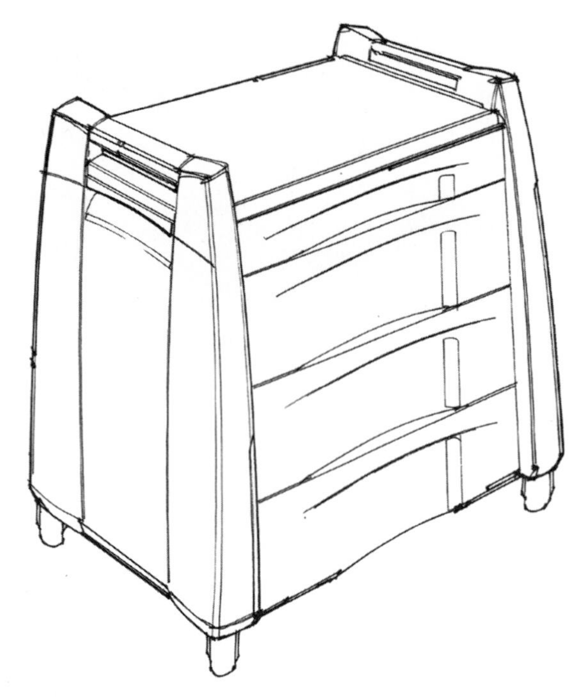

Figure 10-12. Pictorial sketches are easy to understand and are drawn to show objects as the eye sees them. (Design Central, design firm; Artromick International, client)

appearance, function, production, and potential impacts. The individual aspects are listed on a chart. The advantages and disadvantages of each design are added to the chart. The design that seems to be the best solution is then selected. It is possible that one design is not the best solution in all aspects of design. In these cases, trade-offs are considered. A trade-off is made when a designer decides which of the aspects are most important and which are not as important. For example, a less attractive solution might be selected because it is the safest.

Also, the design is refined in this stage. The first refinement is the appearance of the design. Designers use elements and principles of design to create a design with a good appearance. They then create two different types of sketches. The first is called a ***rendering***. A rendering is a sketch shaded and colored in full color. See Figure 10-13. This sketch shows how the design will look in different lights. The second sketch is a detailed sketch. The

Figure 10-13. A rendering shows the object in full color. (Daimler)

detailed sketch is drawn using the sizes of the proposed solution. This sketch also includes dimensions. The dimensions, text, and arrowheads are used to show the size of the object. Selecting and refining the solution are the topics of Chapter 14.

Modeling the Solution

The fifth step in the design process is the creation of a model. ***Models*** can be used to communicate ideas, experiment with concepts, and organize data. Designers make several different types of models. Graphic models, such as tables, charts, and graphs, are used to show information about the design. Mathematical models are equations and formulas. Charts, formulas, drawings, and sketches are some two-dimensional representations of solutions. The models many designers use the most are physical and computer models. Both of these models are used to create three-dimensional objects. Computer models are three-dimensional computer objects. See Figure 10-14. Physical models are actual three-dimensional products.

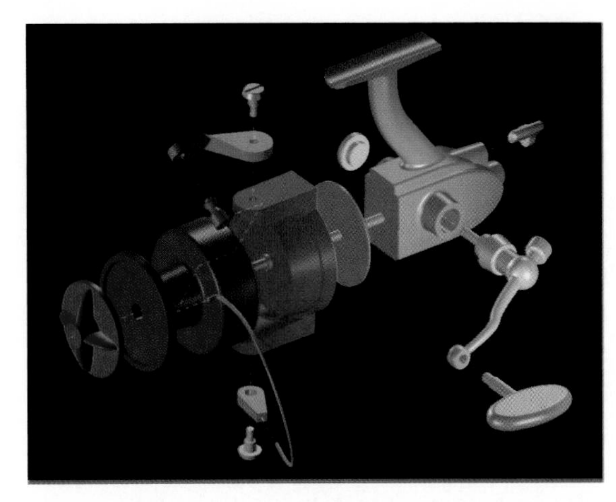

Figure 10-14. The fishing reel is created as a computer model. (Model developed using VX CAD/CAM software, VX Corporation)

Often, physical models are used to help ideas become useful solutions. Creating and testing physical models allow people to actually feel and touch the object. These models can also be looked at in real light and from any angle. Models are particularly vital for the design of large items, such as cars, spacecraft, and airplanes, because it is cheaper to examine a model before the final products and systems are actually made. The two main types of physical models are mock-ups and prototypes. ***Mock-ups*** are appearance models. They show how the object will look in real life. They do not, however, function as the solution will. ***Prototypes*** are physical models that actually function. See Figure 10-15. They are

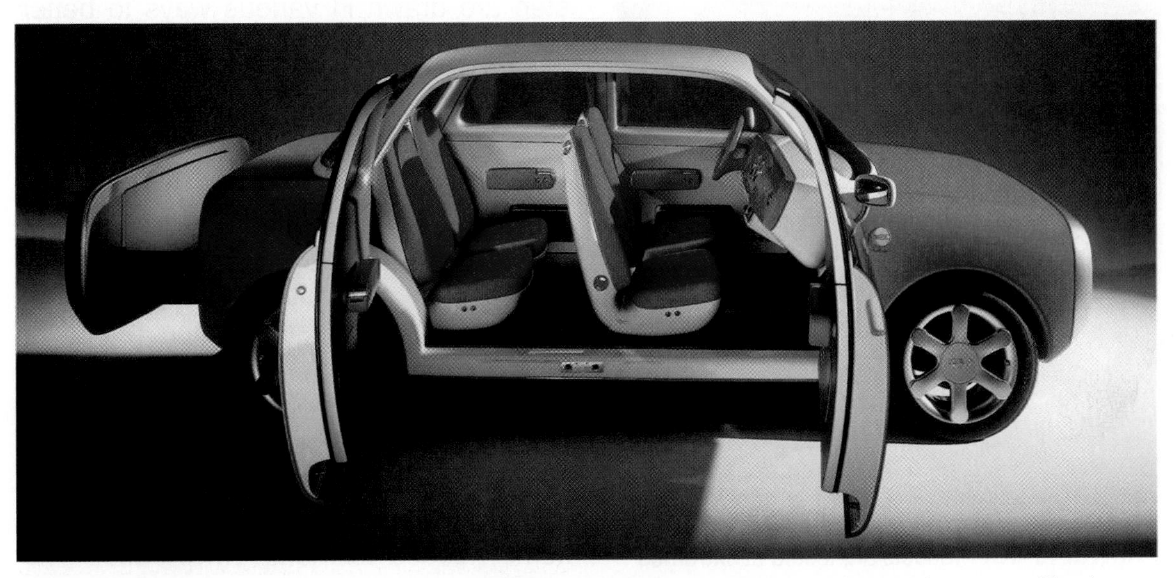

Figure 10-15. A prototype is a working model that can be used to test product performance. (Ford Motor Company)

working models. These models are used to test how well the design will work.

Physical models can be built using many different materials. The materials used depend on the size, shape, and use of the model. Designers and model makers use a process to build models. The models are created by first blocking out the basic shape. The designers and model makers then finely shape the pieces. The pieces are assembled and finished. The models can then be used for testing or a variety of different uses. Models and modeling are discussed further in Chapter 15.

Testing the Solution

The testing stage is the step determining whether or not the solution works. The only good solutions are solutions meeting the design problem. In this step, discussed further in Chapter 16, the designers test the solution to make sure it solves the design problem. The testing stage begins with the designer conducting tests on the model. There are many different tests that can be performed. See Figure 10-16. The designer chooses the test fitting best with the design problem and solution. For example, a solution to a bridge design would be tested for strength. It would not, however, be necessary to test a telephone design for strength. A telephone design would be tested to make sure it fits the users' hands.

Figure 10-16. Products are tested at extremes to make sure they work properly. (Goodyear Tire and Rubber Company)

Designers test the solution using a procedure. They begin by identifying the purpose and type of test they will use. The designers then gather the equipment and supplies and conduct the test. During the test, they gather data. After the test is complete, the designers analyze the data and evaluate the results. The test and evaluation help to convert the designer's idea into a realistic solution by determining if the proposed solution is appropriate for the problem. Evaluation is used to establish how well the design meets the established criteria and constraints and to give guidance for improvement, if needed. Procedures for evaluation vary from visually studying to actually operating and trying the products and systems. Some other strategies include using previous knowledge and asking questions. The designers use the results of the evaluation to make an assessment of the solution and improve the design. The assessment of the design determines whether the solution will be discarded, revised, or sent to the next step. Designs sent to the next step are drawn in various ways to better communicate the solution.

Communicating the Solution

If the solution passes the test, it is ready to be manufactured. Designers, however, cannot just give the manufacturer the model of the solution. They must communicate the solution using a set of documents. The documents include a bill of materials, specification sheets, and drawings. The bill of materials lists all the items needed to make the product. The specification sheets identify the details of the materials and quality of work. The drawings show the sizes of the solution and how the parts fit together. A designer's ideas, processes, and results can be documented in design portfolios or journals or with drawings, sketches, and schematics. The results of the design process can also be communicated in many other ways, such as on a Web page or with a model.

The drawings used are called ***mechanical drawings***. See Figure 10-17. Mechanical drawings are very accurate and precise. They are created using either traditional or computer-aided design (CAD) tools. There are two main classes of drawings: orthographic and pictorial. Orthographic drawings are also known as *multiview drawings*. They use several views to describe the solutions. These drawings are also dimensioned to show the sizes of the pieces. Pictorial drawings are created to look three-dimensional. They are used to show how the solution will look when it is complete. These drawings are easier to understand for most people because they look the most realistic. The sketches, drawings, and all other documents are compiled into a portfolio. When the portfolio is complete and all drawings are finished, they are presented to management or the client for final approval. Communicating solutions is the topic of Chapter 17.

Improving the Solution

Improving solutions, highlighted in Chapter 18, might seem to be the last step of the process. This step is, however, just

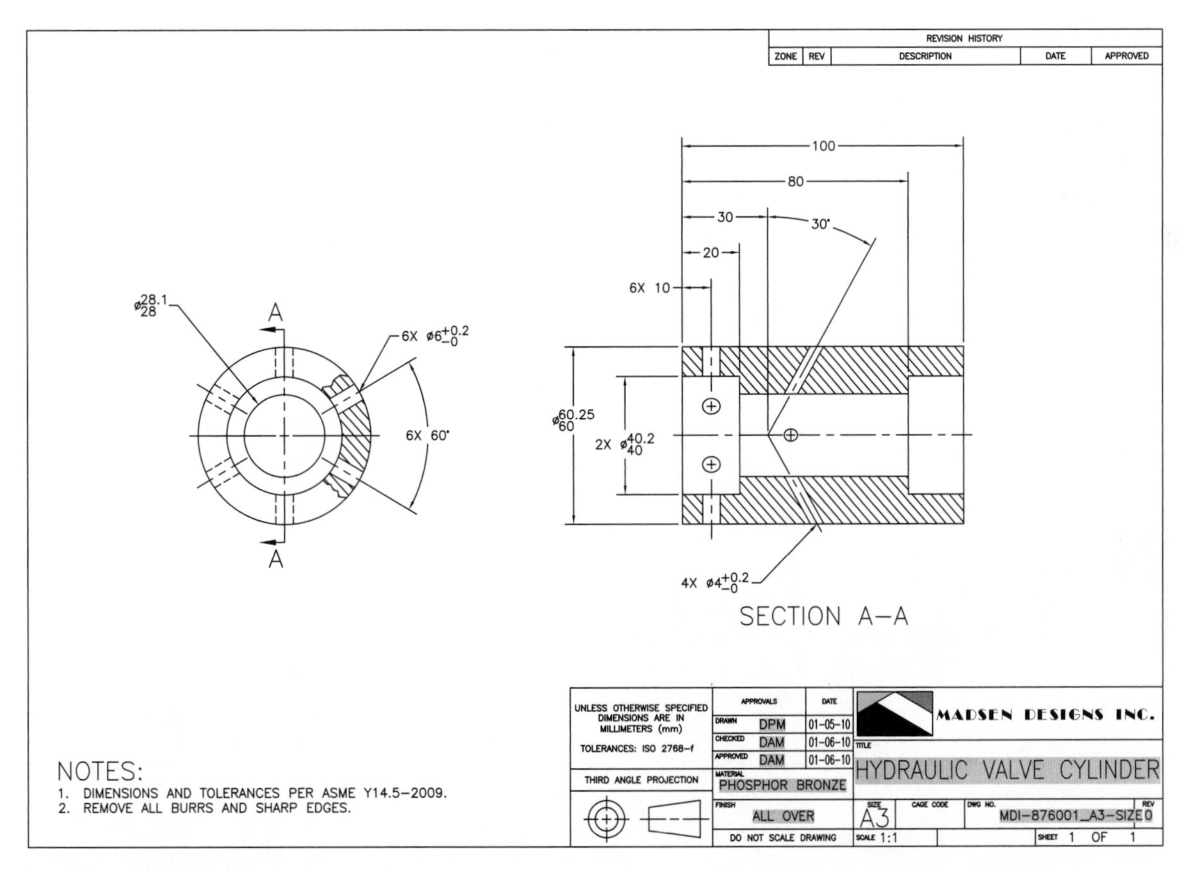

Figure 10-17. Mechanical drawings are accurate technical drawings. (Courtesy David P. Madsen, Madsen Designs Inc.)

the beginning of a new process. Design is never ending because there is no perfect design. All solutions can be improved. The best designs optimize the desired qualities—safety, economy, effectiveness, and dependability—within the specified constraints. A design that is the best solution at one time eventually becomes outdated. All designs build on the imaginative ideas of others.

Designers begin to improve their designs the moment the designs are manufactured. See Figure 10-18. They have several reasons to improve their solutions. These reasons range from profit and competition to function and safety. To improve their designs, designers rely on information. Designers receive information from various sources. Customers and competitors are two of the sources of information. Other sources are government agencies and the company itself.

When designers improve solutions, they go back and redo some steps of the design process. Some improvements require the designer to start again at the very beginning and redefine the problem. Going back only one or two steps can solve other problems. Improving designs is the only way our technology and technological literacy will advance.

Summary

Design is the action of creating a solution for a problem. This action always has a purpose. There are two types of design: artistic and engineering. Artists produce artistic designs for themselves. Engineering designs are produced to solve the problems of others. These designs must have a nice appearance and function well.

Products, systems, and structures are designed using a process. The process for designing identified in this chapter and throughout the book has eight steps. This design process is not, however, the only design process. This process is simply one design process used. All designers use a design process. This process helps reduce the number of product and system failures. The design process also helps ensure that the products and systems meet human needs and solve problems.

Figure 10-18. Designs must be reviewed as they are developed. (©iStockphoto.com/track5)

STEM Connections

Science

Select a type of designer (such as a chemist, an automobile engineer, or an architect). Research and list types of job duties these designers have that include science.

Curricular Connections

Language Arts

Write a letter to a local designer. Inquire about the design process she uses in her own design. Share any responses with your class.

Social Studies

Study how design has changed throughout history. Choose a civilization. Study its designs. Show how and when other civilizations influenced it.

Activities

1. Create a display showing examples of artistic and engineering design. Use the display to explain the differences between the two types of design.
2. Choose an example of a system. Research how it was designed. Explain how the parts work together.
3. Use the eight steps of the design process to create a simple product design.

Test Your Knowledge

Do not write in this book. Place your answers to this test on a separate sheet of paper.

1. Name five items in your classroom that have been designed.
2. Design is an action with a(n) _____.
3. Architects are the only people who use the act of design. True or false?
4. Paraphrase the definition of *artistic design*.
5. _____ design is used to create plans of buildings and roads.
6. Product designers must consider the _____ and _____ of the products they are designing.
7. Give examples of three types of product designers.
8. A mock-up is a working model. True or false?
9. The design process ends with which step?
   A. Testing the solution.
   B. Researching the problem.
   C. Communicating the solution.
   D. None. This process never ends.
10. List and describe the eight design steps.

Reading Organizer

On a separate sheet of paper, create a detailed outline based on what you've read about the design process.

Example:

I. Types of design
   A. Artistic
      1. Content
      2. Form

CHAPTER 10 ACTIVITY A

DESIGNING FOR SPACE

THE CHALLENGE

Design a self-contained space station.

INTRODUCTION

Technology has given us the life we have today. This life is very different from what our grandparents had. Likewise, life for future generations will be very different from life today. People might be living in deep space or undersea settlements. This activity allows you to consider a life in space.

EQUIPMENT AND SUPPLIES

- No. 10 tin cans (large cans found in most school cafeterias).
- A paper-towel tube.
- Five 6″ cardboard or poster board squares.
- 1 1/2″ wide strips of poster board.
- A compass.
- A ruler.
- A pencil.
- Scissors or small tin snips.
- Glue.
- Masking or transparent tape.
- Wallpaper samples or scraps of cloth.

PROCEDURE

Your teacher will divide the class into groups of three or four students. Each group should perform the following steps:

1. Obtain the materials listed above.
2. See Figure 10A-1 to determine the basic structure of the space station.
3. List the basic areas needed for complete, comfortable living. These might include the following areas:

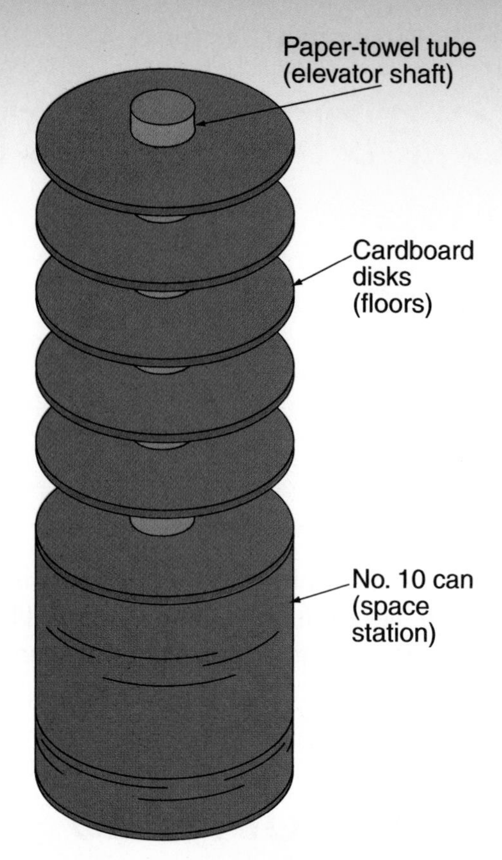

Figure 10A-1. The basic structure of the model space station.

- Living quarters.
- Recreation areas.
- Agriculture, production, or factory areas (the reason for the space station).
- Mechanical support areas (such as areas for heating, air-conditioning, and waste disposal).
- Office or managerial areas.
- Flight-control areas.
- Lab areas.

NOTE

Be sure to consider other types of facilities. This is not a complete list!

4. Assign each basic area to one of the four floors of the space station.

Each member of the group should select one or two floors to develop. All group members should complete the following steps:

5. Cut a 5 7/8″ disk from a square of cardboard or poster board.
6. Cut a 1 1/2″ hole in the center of the disk.
7. Draw a floor plan on the disk. Be sure to consider the wise use of space, ease of movement between different areas, and grouping similar activities together. Remember, 1″ = 8′.
8. Cut walls from the 1 1/2″ strips of poster board.
9. Cut doorways in the wall sections.

10. Attach the walls to the floor.
11. Decorate the walls and floors with wallpaper samples and cloth to represent actual surface treatments.

The group should complete its space station by completing these steps:

12. Cut openings in the elevator shaft (paper tube).
13. Attach the floors to the elevator shaft. See **Figure 10A-2.**
14. Insert the space station into the launch shell (tin can).

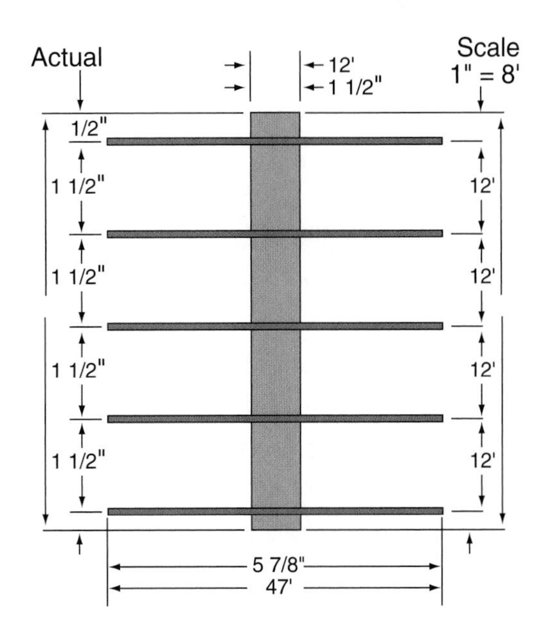

Figure 10A-2. An elevation view of the space station.

Challenging Your Learning

Select an element of the space station that you had to design (such as the living quarters, mechanical support areas, or lab areas). Research that element to determine how it is designed and how it functions on the *International Space Station.*

Safety Rules

- Cutting tools have sharp edges. Never carry pointed tools in your pockets.
- Always cut away from your body or the hand holding the part.
- Always carry cutting tools with sharp edges pointing down.
- Use each tool for its proper purpose.
- Never rub your fingers across cutting tools.

TSA Modular Activity

This activity develops the skills used in TSA's Digital Photography event.

Digital Photography

Activity Overview

In this activity, you will create a digital photo album composed of the following:

- Cover sheet, including an 8″ × 10″ collage of six photographs presenting a single theme.
- A detailed description of the theme being documented by the pictures, including a title.
- Three rough (unedited) photographs of the theme, followed by their edited counterparts.
- A description of each of the three photographs, including an explanation of how the photograph was changed from the rough version to the edited version and a caption explaining who, what, where, when, why, and how the theme is shown.
- Resource page, listing the type of camera and software used.

Materials

- A three-ring binder.
- A digital camera.
- A computer with photo-editing software and a color printer.

Background Information

- **General.** Check your camera's settings before you take pictures. Most often, using the camera's automatic setting allows you to "point and shoot." This means you can concentrate on your subject, instead of worrying about camera controls. To avoid blurry photos, hold the camera firmly with both hands. Press the shutter button with a slow, steady motion. Don't jab the button with your finger. Jabbing causes the camera to shake.

- **Portrait.** Get close to your subject. Fill most of the viewfinder with your subject, instead of the background. Press the shutter button halfway to bring the subject into focus. Watch for a good expression, and then take the photo. Try both people and pet portraits, with and without using the camera's flash. If your subject is against a bright background (such as a window), you probably need to use a flash, unless you want to do a silhouette.
- **Landscape.** If the camera has a zoom lens, use the wide-angle setting to show a broad area. Try to get an interesting object (for example, a rock, bush, or fence) in the foreground to add depth to your picture. The wide-angle lens makes it easier to keep both near and distant subjects in focus.
- **Still life.** Try for a pleasing arrangement of objects that fit a theme (for example, a vase of flowers, pruning shears, and gardening gloves or a baseball and fielder's glove surrounded with related material). Do not use your flash, since that causes harsh shadows. Instead, use room lighting, possibly with a table or desk lamp placed just outside the picture to provide additional light. You can move the lamp around to place shadows where you want them.
- **Action.** Try to shoot action pictures on a brightly lit day, so the camera's automatic mode chooses a fast shutter speed. This prevents blurring. Try to arrange the situation so your subject is moving toward you or at an angle, rather than crossing in front of you. This helps to stop motion. Practice with the camera so you can adjust for shutter lag—the time between pressing the button and the actual shutter operation.
- **Photo editing and enhancement.** Use your computer's photo-editing software to adjust the brightness of your picture (darken or lighten as necessary). If your picture has a color cast, such as a very blue or very yellow appearance, your software should have the ability to remove the cast. Do not be afraid to crop away part of the picture. Many pictures can be greatly improved by eliminating extra background and focusing attention on the subject.

Guidelines

- The photo album can include black-and-white photographs, color photographs, or a combination of these.
- All images must be taken and produced digitally.
- Photographic paper or high-quality glossy paper can be used for photographic prints.
- Images should be enhanced and edited with photo-editing software.

Evaluation Criteria

Your project will be evaluated using the following criteria:

- Composition.
- Photo editing.
- Relation of image to theme.
- Discussion of images.

CHAPTER 11
Identifying Problems

Did You Know?

- Design solutions often create new problems and opportunities.
- Before the gasoline automobile, there was no need for gas stations. Once the gasoline automobile was being used, people had a new problem. Their problem was that they needed to refill their cars with gasoline.
- The sliding keyboard tray is a solution to a problem the use of computers created. Before computers became popular, desk manufacturers did not need a location for a keyboard in their desks.

Objectives

The information given in this chapter will help you do the following:

- Compare needs and wants.
- Compare opportunities and problems.
- Explain the three different types of problems.
- Identify common places to find problems.
- Give examples of problem statements.

- Explain the function of criteria.
- Explain the concept of a constraint.
- Create a design brief.

Key Words

These words are used in this chapter. Do you know what they mean?

constraint
criterion
need
opportunity
personal problem
problem statement
social problem
technological problem
want

Preparing to Read

As you read this chapter, make a list about different problems that resulted in design solutions. Use the Reading Organizer at the end of the chapter to organize your list.

The design process begins by identifying a problem. This is a very important step in the design process. In this step, designers determine what they are designing. This chapter explains how problems are identified.

Needs and Wants

All people have needs and wants. Basic human ***needs*** are the requirements it takes to live. Shelter, clothing, and food are considered to be the basic human needs. Many problems and opportunities can be discovered when looking at human needs. One problem is the need for enough food for the world's population. The need for clean drinking water is another problem. Shelter is a problem found in many communities. See Figure 11-1. Many cities have people who have no homes or shelter. The needs for both shelter and clothing are major problems. Shelter and clothing help to protect people from the elements, such as wind, rain, snow, and heat.

Medication can also be a human need. Many diseases and illnesses require certain medications to treat them. If the basic needs are not met, people will not survive.

Once basic needs are met, people have wants. ***Wants*** are things people desire. They are not needed to survive. Wants, however, help make life better and easier. To many, wants are often mistaken as needs. You will survive without a new skateboard, and you might not need to be in constant contact with your family and friends, using a cell phone. Perhaps you want to have the latest in clothing styles. Likewise, your family might want a certain brand of premium ice cream for dessert.

These strong wants are often the result of advertising. We often think advertising creates demand for only new and different

Figure 11-1. Homes are built to meet the need for shelter. (Habitat for Humanity International)

manufactured products. Advertising is used, however, to compel people to want all types of technological products and services. Think of the advertising you have seen promoting things such as travel to exotic vacation spots, new prescription drugs, the latest exercise equipment, and new breakfast cereals.

Problems and Opportunities

Wants and needs are problems that new products or systems can make better. The design of products, devices, structures, and systems all start with a problem or an opportunity. A problem is a situation that can be made better through an improvement. One example might be that there is no medication to treat a certain disease. See Figure 11-2. A problem could also be that you need to get across town in five minutes. People come across problems every day. Problems often cause difficulty if they are not solved. Arriving at school after the bell rings is a problem. You can see that there will be difficulties if that problem is not solved. If you are not at school in time for the bell, you are tardy.

Opportunities are different from problems. Problems cause people to design new and improved products and systems. ***Opportunities*** use new or existing products and systems in new ways. Crop farming has presented many opportunities over the years. See Figure 11-3. Many farmers found an opportunity in

Figure 11-3. The innovation of older farming methods presents new opportunities. (©iStockphoto.com/Grafissimo)

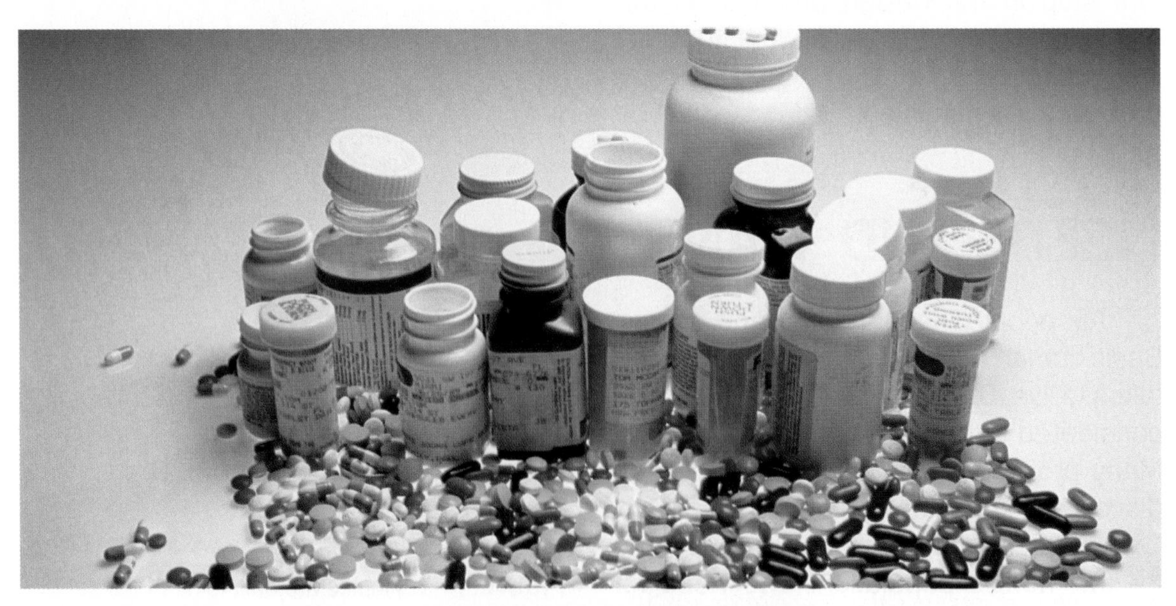

Figure 11-2. Some problems are solved with the design of medications.

no-till farming. Farmers designed the no-till method to increase the yield of crops. In typical farming, the ground is tilled to break it up and remove any excess crop. After it is tilled, the farmers plant new seeds. No-till farming uses the same seed typical farming does. The farmers, however, do not till the ground before planting. They use special planters that cut through the remnants of the last year's crop. This method leaves nutrients in the ground for the new seeds. No-till farming is an opportunity to grow better crops.

Figure 11-4. Disaster-relief workers help solve social problems after natural disasters. (American Red Cross)

Types of Problems

Problems can be found in all areas of life. There are many different kinds of problems. All problems can be placed into one of three categories:

- Social problems.
- Personal problems.
- Technological problems.

Social Problems

Social problems are problems dealing with society. They affect large groups of people. Illiteracy is one example of a social problem. This problem is the inability of people to read or write. Education is used to solve low literacy. Violence and crime are also social problems. The police try to solve these types of social problems. Some social problems are the results of natural disasters, such as hurricanes, floods, and tornados. Governments and organizations such as the Red Cross try to solve many social problems. See Figure 11-4. Environmental concerns are also social problems because they impact the entire society. Many designers try to design products so the products do not cause harm to the environment.

Personal Problems

Personal problems are problems affecting individuals. Emotional problems are personal problems. Low self-esteem is an example of a personal problem. Academic problems are personal problems with schoolwork. Teachers and counselors work to solve academic problems.

Technological Problems

Technological problems are problems that can affect individuals and groups of people. The need for a new building is a technological problem. The building might be a large structure for many people. This new building might also be a small house for one person. The need for electricity to power a machine is another technological problem. See Figure 11-5. These problems can be solved with devices or systems. Designers, engineers, and architects are some of the people who solve these problems. Technology is often used to solve social and personal problems. Life-saving devices and medication are two technologies designed to solve social and personal problems. This book focuses only on how people use technology to solve problems.

Figure 11-5. Electricity distribution plants help spread electricity to homes and businesses.

Recognizing Problems

All technological problems must be recognized before they can be solved. A designer cannot create a solution if there is no problem. Problems are often easy to find. Some designers look at their own lives to find problems. People with disabilities often find problems completing daily tasks. Designers create solutions for these types of problems. See Figure 11-6.

Some people find problems at work. Bette Nesmith found a problem while doing her job. She was a secretary who had to use an electric typewriter for the first time. Nesmith found she had too many errors and did not have a good way to correct them. Her problem led to her creation of Liquid Paper® correction fluid, which is often called *whiteout*.

Other people find problems during leisure time. You might have found a problem while playing a sport. Branch Ricky, a former baseball player, was worried about pitches hitting batters in the head. His problem dealt with safety. Ricky designed the first batting helmets for baseball players. See Figure 11-7. Soon after he designed the helmets, all players were wearing them.

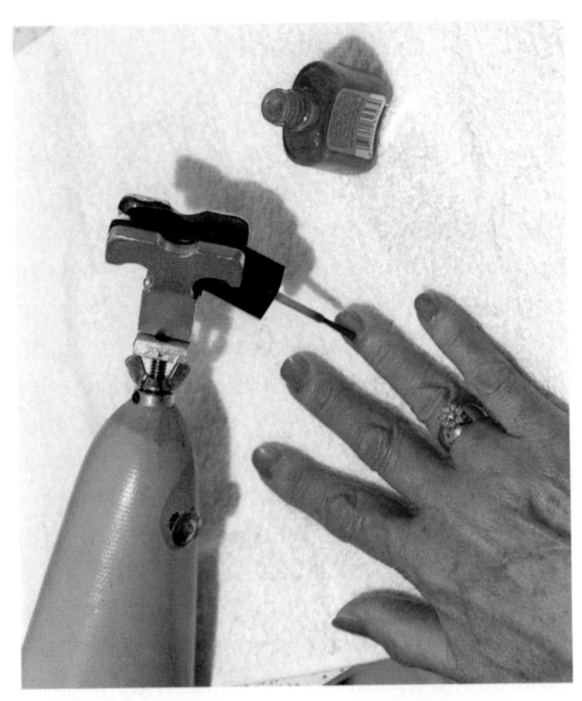

Figure 11-6. Designers create solutions to solve problems for people with handicaps. (NASA)

Technology Explained

universal design: the design of products so all people can use them without adaptations.

Universal design is a design concept stating that as many different people as possible should be able to use a design. The idea of universal design came from the architecture community. Architects used to design adaptations to buildings to accommodate people with disabilities. This was known as *accessible design* or *barrier-free design*. These adaptations were often add-ons to the design, however, and were costly and unsightly. It was determined that a better approach would be to design buildings and products to be more accessible for everyone. By designing these changes into the initial design of the product, costs would be saved. The designs would be more eye appealing and useful to everyone.

There are many examples of universal design in buildings and products all around us. Many commercial buildings have automatic doors that open as customers approach the doors. This makes entering a building easier for all people, whether they are walking, in a wheelchair, or carrying groceries. Door levers, long door handles, rather than doorknobs, are easier for everyone to use because they require less grip and twisting force. See **Figure 11-A.** Softer and larger handgrips on utensils and tools are easier for everyone to use. Signage at international airports, for example, is often universally designed. The signs use icons that are recognizable to everyone, regardless of whether or not you can read the language. Universal design makes products and systems more equitable, easier to use, and more intuitive.

Figure 11-A. Door levers make opening doors easier for all people.

Figure 11-7. Batting helmets were designed to protect baseball players from injury. (American Baseball Company)

Clients are people who hire designers. The designers are paid to solve their clients' problems. Designers must first identify their clients' problems and create a list of the details of each problem. See Figure 11-8. Designers, often working in teams, review the statements from the clients and create problem statements.

Problem Statements

A ***problem statement*** is a clear definition of the problem. For example, a problem might be "to design a device that will evenly spread paint on a wall." This is a clear definition. This definition is an open-ended statement stating a problem, not a solution. A statement such as "design a paint roller" states the solution. This statement is too narrow and makes the conclusion that the only solution is to roll the paint on the wall. It would be hard to develop a new design with a statement to design a paint roller. The first statement allows the freedom for the designer to go in many directions. See Figure 11-9.

Criteria

After the problem statement is written, the designer lists the criteria for the design. ***Criteria*** are elements of the problem needing to be solved. They are some of the requirements, or parameters, placed on the final design. Criteria are listed before the designer starts to develop ideas. They are identified during this first step of the design process. Criteria answer the following questions:

- Who will use the device or system?
- Where will it be used?
- How will it be used?
- What will it do?

Figure 11-8. Designers work with clients to determine the clients' problems. (©iStockphoto.com/MarcusPhoto1)

Figure 11-9. This paint sprayer is one solution to the problem of applying paint. (Eastman Chemical Company)

Career Highlight

Landscapers

Architecture & Construction

The Job: Landscape designers or architects work with clients to design residential and commercial landscapes. Once the design is produced, landscape workers, or landscapers, complete the physical work required to create a landscape. Landscapers install plants, fences, irrigation, and lighting. They also maintain landscapes for clients by mowing lawns, pruning trees and bushes, and adding mulch to plant beds.

Working Conditions: Landscape workers generally work outdoors in a range of weather conditions. Depending on the region of the country, the work can be seasonal. Their work is physically demanding and tiring, as they must do a lot of digging, bending, and machine operation.

Education and Training: Most training for landscape workers is on-the-job training. These workers often learn planting, care, and maintenance from more experienced landscapers or supervisors. As workers progress in their careers, there are opportunities for certification in specific areas, such as tree care or grounds operations.

Career Cluster: Architecture & Construction

Career Pathway: Maintenance/Operations

(Bobcat Company)

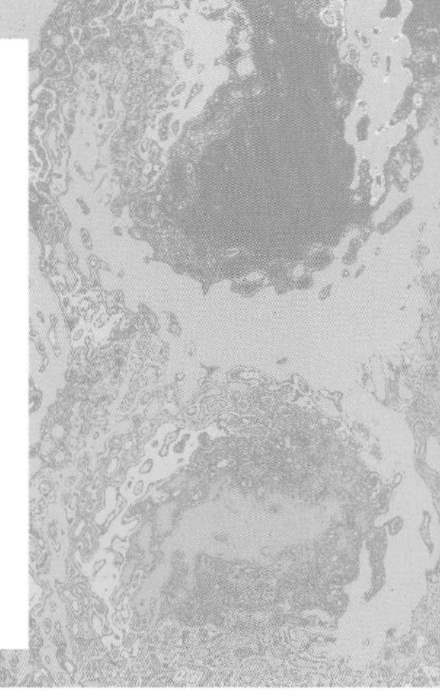

THINK GREEN

Efficient Power Use

There are many ways available to individuals to use less power or energy in everyday life. To be more efficient means to use energy-efficient alternatives to yield the same results as traditional means. We've discussed the example of CFLs as replacements to incandescent bulbs. Another example would be better insulation for your home. With more insulation, it will take less energy to either heat or cool your home. Hybrid vehicles are another example of energy-efficient technology. Hybrid vehicles often use what would otherwise be wasted energy.

- How well will it work?
- How reliable will it be?
- How safe will it be?

These questions state the aspects of use, performance, safety, human factors, and reliability. The list of criteria in the paint-application problem might look similar to this: "Homeowners will use the device. The device must comfortably fit in their hands. This mechanism will be used on the inside walls of their home. The device will spread the paint evenly and thickly enough that only one coat will be needed. This mechanism must also be safe to use and not harmful to the environment."

The criteria are a very important part of identifying the problem. Changing the criteria can change the design of the solution. Imagine if the criteria above were changed to "A professional will use the device on the outside of the house." The solution would be very different.

In some designs, it might be difficult to meet all the criteria. If there is a long list of criteria, the designer might rank the importance of each item. Designers might list the needs versus the wants. They might also assign a number 1 to the most important aspects and a number 2 or 3 to the lesser aspects. It is important that the criteria are accurate. When a solution is complete, the designer evaluates the design against all the criteria. If the design meets the criteria, it is successful.

CONSTRAINTS

After preparing a problem statement and criteria, the designer must begin to list the design constraints. The ***constraints*** are the limitations for the design. A limitation is a boundary. Similar to criteria, constraints are parameters placed on the development of the design. Constraints tell designers how far they can go with a design. See Figure 11-10. There are several categories of constraints providing guidelines for design:

- Technical constraints.
- Financial constraints.
- Legal constraints
- Time constraints.
- Production constraints.

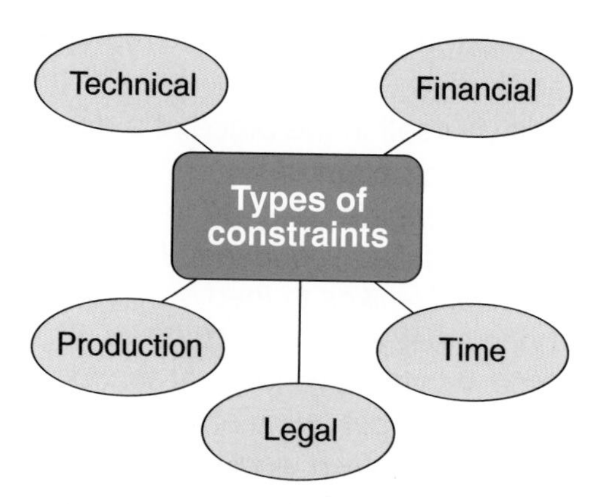

Figure 11-10. Problems can have five types of constraints.

Technical Constraints

Technical constraints are the simplest to understand. They deal with the size and shape of the object and the materials that will be used. Whether a device uses batteries or is hand powered is a technical constraint. An example of technical constraints is "The product must be less than six inches in height and three pounds in weight. This product must also be round and made of plastic."

Financial Constraints

Financial constraints involve the costs of the design. The first cost is the initial cost. The initial cost is the money it takes to design the solution. This is often called the *design budget*. The next cost is how much money it will take to manufacture the product. The last financial constraint is how much the product can make if it is sold, or the product's profitability. In order for a design to be selected, it might have to have a certain level of profitability. Profitability might be a constraint that the client of the design places on the designer.

Legal Constraints

There are also a number of legal constraints. Designers must make sure their designs meet all the government's laws and regulations. They must also check to make sure the design is new and does not break any patent laws.

Time Constraints

Time constraints are limits on the amount of time the design process can take. A time constraint might be that the design process can take only three months. For many products, this is a short time. Other time constraints are how long the product will last. Designers create products to last only a certain amount of time. A lightbulb is a product designed to last a certain amount of hours. See Figure 11-11.

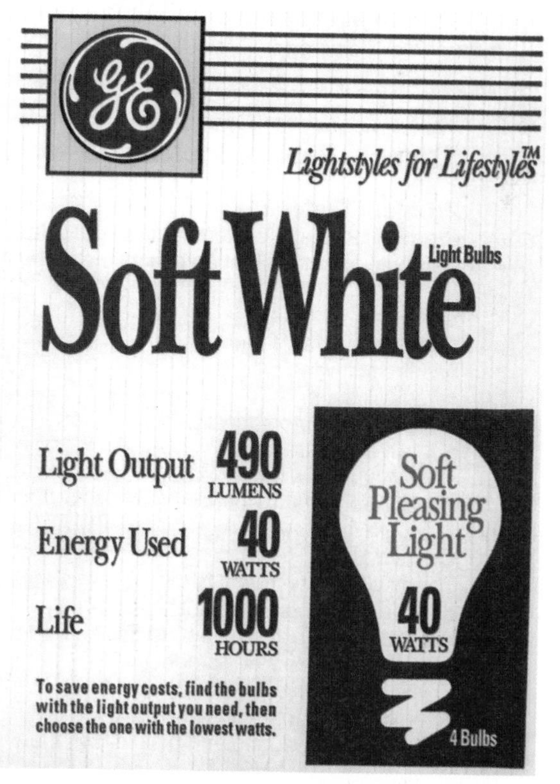

Figure 11-11. Lightbulbs are designed to last a certain amount of time.

Production Constraints

Production constraints are the restrictions on how the device can be built. A designer must look at the available tools and materials. A production constraint might be that the solution must be able to be produced using hand woodworking tools.

Design Briefs

In this first step of the design process, the designer creates a problem statement, criteria, and constraints. The designer puts them together in one document called the *design brief*. The design brief helps the designer completely understand the problem. This document lets the designer see the problem, the problem's requirements, and the problem's limitations together. The design brief is created after a long process. The designer must work through the problem to completely understand it.

Put yourself in the place of a designer. You have a problem:

Your brother is going off to college. He has asked you to feed his fish for him. He likes to feed his fish once a day at exactly 10:00 AM. At 8:00 AM, you have to be at school. You do not get home until 3:00 PM. He leaves in two weeks and will be back to pick up his fish two months after that. He will pay you a total of five dollars a week to feed the fish.

After thinking about the problem, you must create a design brief. The design brief might look similar to the one shown in Figure 11-12.

After writing the design brief, you must review it. See Figure 11-13. You need to ask yourself some of the following questions: Does anything need to be added to the problem statement? Is this statement clear?

Example Design Brief: Fish Feeder

Problem Statement
Create a device or system that will automatically distribute food to fish at an exact time every day.

Criteria
The device or system has the following criteria:
- You will set it.
- It will work automatically.
- It will distribute food.
- Every day, it will work properly.
- It will work at the set time.

Constraints
The device or system has the following constraints:
- It must cost less than $40 to design and build.
- It must be able to be produced with materials on hand.
- It must fit on top of the fish tank.
- It must be designed and built in less than two weeks.

Figure 11-12. Solutions are often designed to fit existing problems. Here is an example design brief for a fish feeder. The problem is clearly stated. The known criteria and constraints are then listed. (All-Glass Aquarium)

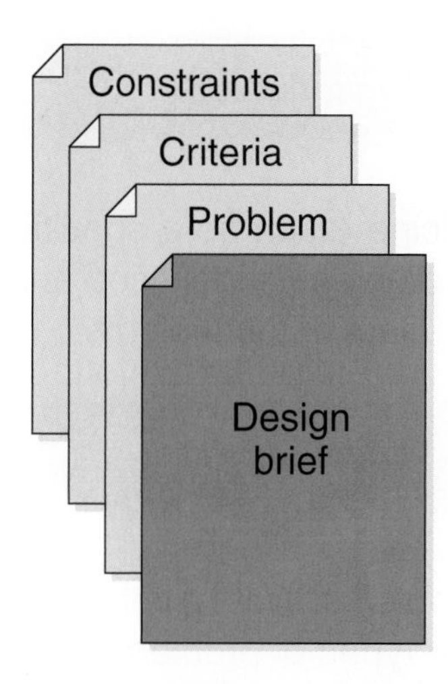

Figure 11-13. A design brief includes the problem, criteria, and constraints of a design.

Does it state the problem? Is the problem open-ended? Are the criteria clear? Was anything left out? Are there more requirements or constraints?

The design brief is complete. Now, you move to step two of the design process. You have to start gathering information to help you solve the problem.

Summary

Identifying the problem is the first step of the design process. This step is very important because it starts the process. Problems can be found in various places. Many problems are found in human needs and wants. Once a problem is recognized, it must be written clearly. Problems are rarely found in clear form. The designer must work to find the true problem.

The true problem is written as a problem statement. A problem statement is a clearly defined problem. Once the problem statement is written, the designer must identify the requirements of the design. The requirements are called *criteria*. The criteria will be used to judge the solution at the end of the process. The designer also lists constraints. Constraints are limitations of the problem. Money, time, and materials are examples of constraints.

After the statement, criteria, and constraints are identified, they are written as a design brief. A design brief is a document clearly describing the problem, requirements, and limitations. When the design brief is complete, step one is finished. The designer, however, might need to come back to step one and make changes later in the process.

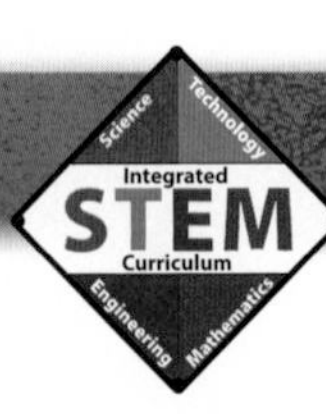

STEM Connections

Science

Research a scientific problem in the area of medicine (such as a specific disease or condition). Create a poster board showing how the problem is identified and treated. Include whether or not the problem is treated with an artifact of technology.

Curricular Connections

Language Arts

Most companies and organizations work to solve problems. Write a letter to managers of a local company or organization. Ask them about the types of problems they work on solving and how they identify problems.

Social Studies

Examine the technological problems of various countries. Compare the problems in other countries to the problems in your own.

Activities

1. Identify and list possible needs and wants of the following types of people:
   A. A baby girl.
   B. A teenage boy.
   C. A middle-aged woman.
   D. An elderly man.
2. Examine the problems existing around you. Create a list of problems you find in the following places:
   A. Your school.
   B. Your home.
   C. Your community or neighborhood.
3. Identify a product you use on a daily basis. Try to imagine the problem that was trying to be solved by the product's creation. Write what you think the design brief (problem, criteria, and constraints) might have included, based on the product.
4. Write a design brief including a problem statement, criteria, and constraints. The problem should be related to your own life.

Test Your Knowledge

Do not write in this book. Place your answers to this test on a separate sheet of paper.

1. A need is a requirement to live. True or false?
2. List four examples of a want.
3. A(n) _______ is a situation that can be made better through an improvement.
4. The use of a new product or system or an existing product or system in a new way is a(n):
   A. problem.
   B. opportunity.
   C. need.
   D. want.
5. Name and describe the three types of problems.
6. Problems can be found only at home. True or false?
7. It is important to write a problem statement clearly. True or false?
8. What is the function of criteria?
9. List and explain the five types of constraints.
10. A design brief includes a problem statement, criteria, and constraints. True or false?

Reading Organizer

Make a chart on a separate sheet of paper. List at least three different problems, the type of those problems, and their solutions.

Problem	Type of Problem	Solution to Problem
Example: Illiteracy	Social problem	Education

CHAPTER 12
Researching Problems

Did You Know?

The glue on Post-it® notes was created while Dr. Spence Silver was researching another product. In 1968, Silver was a scientist researching acrylate adhesives. Through his experimental research, he found an adhesive that was not very sticky. Not until several years later, in 1980, was a use found for it. Art Fry put Silver's research to use. Fry added the adhesive to small pieces of paper and created the Post-it note.

Objectives

The information given in this chapter will help you do the following:

- Explain the concept of research.
- Explain the three types of research.
- Summarize the research process.
- Recall that the problem being researched might be the problem statement from the design brief.
- Give examples of how data is gathered.
- Use graphs to analyze data.
- Explain how conclusions are developed.

Key Words

These words are used in this chapter. Do you know what they mean?

aestheticw
bullet point
conclusion
graph
library research
market
market research
research
research process
scientific method
survey research

Preparing to Read

As you read this chapter, outline the details of the research process. Use the Reading Organizer at the end of the chapter to organize your thoughts.

Have you ever tried to solve a problem, write an essay about a famous person, or design a new device? If you have done any of these things, you have probably done some research. Research is an important part of solving problems, writing biographies, and designing products. See **Figure 12-1.**

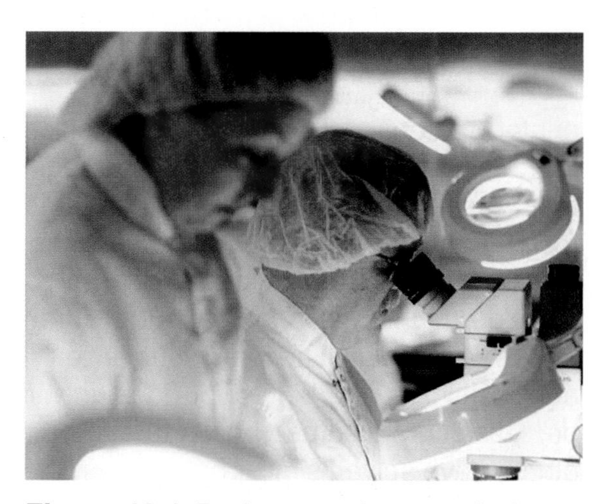

Figure 12-1. Products must be researched before they are manufactured. (NASA)

Researching a problem is the second step of the design process. In the first step, the designer identifies a problem. In the second step, the designer finds as much information about the problem as possible. Take the essay about a famous person, for example. If you are writing a paper about Thomas Jefferson or Eli Whitney, you need to research that person's life. You will be prepared to write about him if you gather a lot of information. Being well-informed will help you develop a good essay.

Similarly, being well-informed helps designers. The more information they can collect, the better their designs will be. If architects are designing an apartment complex, they need to research other apartments. They might talk with people who live in apartments. It is important to remember that, at this step, the designers are only researching the problem. They are not creating a solution. The designers are simply gathering information about the problem. They use the design briefs created in the first step to guide their research. At the end of this step, the designers develop conclusions to help with creating solutions. Creating solutions is the next step of the design process.

In order to design new products and systems, new knowledge must be developed. Without new knowledge, there would be no improvements in the world. New knowledge is gathered using ***research***. Researching is scientifically seeking and discovering facts. This seeking and discovering is used in many different areas. Doctors use research when they are treating patients. When doctors have patients with new symptoms, they read journals and reference books to find a solution. See **Figure 12-2.** Companies use research to sell products. A company selling children's toys uses research to find the things children like. Engineers use research to determine specifications of a product or problem. Designers use research to find data and facts about a problem.

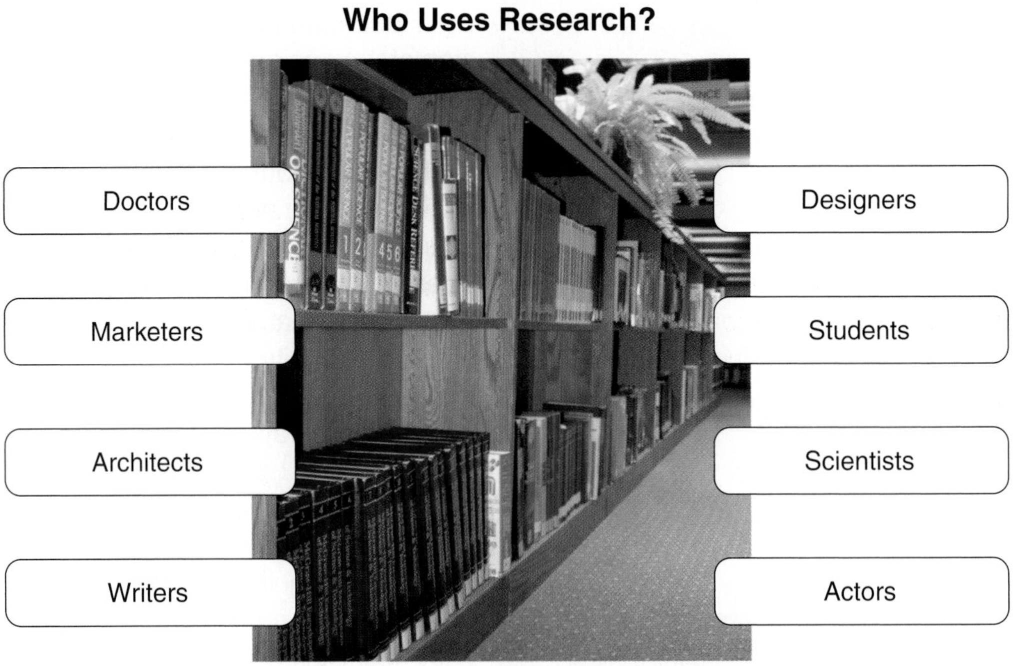

Who Uses Research?

Doctors

Designers

Marketers

Students

Architects

Scientists

Writers

Actors

Figure 12-2. Reference materials are important research tools.

TYPES OF RESEARCH

Finding facts, or researching, can be done in a few different ways. Each way is a different type of research. Researchers choose the type of research best fitting each of their problems. Sometimes, facts about a problem can be found using only a certain type of research. At other times, information can be found using several types of research. There are three major types of research. See Figure 12-3:

- Library research.
- Survey research.
- Experimental research.

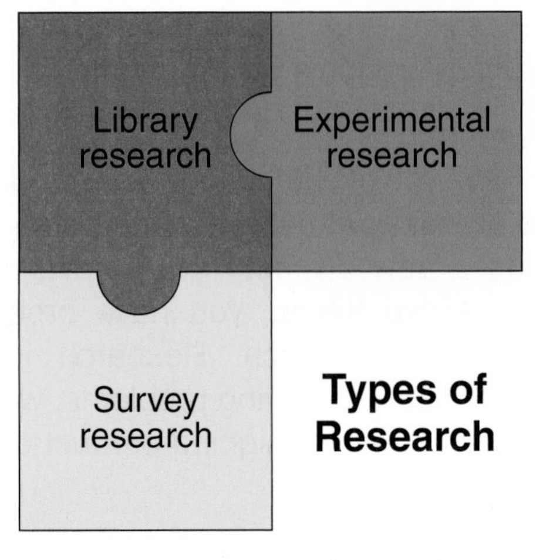

Figure 12-3. Three types of research are used.

Library Research

The first type of research is called ***library research***. Library research is used to get background information on the problem. This type of research uses different forms of printed materials. See Figure 12-4. Books are one example of printed materials. They are a good place for designers and researchers to begin their research. Designers use books to read about the subjects of their designs. Suppose you are asked to design a child's toy. You might read books about children and how they play. The more you can learn about children, the better your design will be. Designers also use books to see how others have solved similar problems in the past. People designing a chair might use books to see how chairs were made over

THINK GREEN

Sustainability

Sustainability is the world's continuing ability to produce natural resources. Resources like water, trees, and petroleum are *exhaustible resources*. If we consume all these resources, they will be gone forever and will not grow back. One way of fighting the loss of these resources is through reducing the use of the resources. Another way of changing the rate at which these resources are being depleted is by finding alternative renewable resources. Individuals, companies, and nonprofit and government organizations are working to encourage sustainability and to prevent the loss of our natural resources.

the last 2000 years. They make notes and record information about the design of chairs.

The use of reference books is also an important part of library research. There are many reference books that are helpful to designers. Books on codes and standards help designers determine the legal requirements of the design. For example, architects use a book on local building codes if they are designing a building.

Technical journals are another reference source. Journals are a good place to find information and data on the technical aspects of a design or product.

Catalogs and magazines are other types of printed material. They are more current than many books. These materials can help a designer learn how other people are currently solving similar problems. Current ideas and trends can be found in catalogs and magazines. Fashion

Figure 12-4. Libraries have large collections of research materials.

designers can use magazines to make sure their designs are in style. Designers often look through the catalogs of their competitors. This helps them stay up-to-date.

The Internet is another form of printed material and library research. See Figure 12-5. The Internet can be used to find news and magazine articles, books, and catalogs. There is a large amount of information on the Internet. One online source of information many designers use is the database of patents. Designers can use the Internet to search for other similar products that have been patented. Researchers and designers, however, must be careful when using the Internet. It is easy for people to put false information on the Internet. Also, it is important that all the information designers use is accurate. Bad information can lead to bad designs. It is important that designers evaluate the quality of their sources of information.

Some other types of related research materials are manuals and experienced people. The information these resources provide can be used to see and understand how things work. This information is helpful in learning how to use a product and determining if the product works well. In addition, a lot of manuals give guidelines on how to troubleshoot a product or system.

Figure 12-5. Internet research is a common form of library research.

Survey Research

The second type of research is ***survey research***. This research is used to find people's reactions to designs and events. Have you ever been stopped in a shopping mall and asked several questions? If you have, you have been part of survey research. The questions probably started with your age and income. This type of information is known as *demographics*. Demographics are certain characteristics such as age, race, income, employment, and marital status. Once researchers have gathered demographic data, they ask if you have ever used a certain product. The researcher then might ask you questions about a product. See Figure 12-6. Those questions are asked to learn your reaction to that product. Researchers can then group the reactions into demographic groups. For example, they might determine that males between the ages of 18 and 25 like a specific feature of a product. They might also find that women in the same age group think differently about the product.

This type of survey research is called ***market research***. The groups of people using certain products are called ***markets***. Market research helps designers determine what types of people (markets) like their product. This research is done in person, as in the shopping mall. Market research is also done on the telephone and through the mail. Have you ever bought a product that had a registration card? Did the card ask questions about why you bought the product? If so, your registration card was also used for market research. See Figure 12-7.

There are companies dedicated to conducting market research. They do not make any products themselves. These companies just survey people about the products of other companies. The survey companies work through the mail and Internet. They send surveys asking about a number of products. These companies

Figure 12-6. Market researchers gather reactions of potential customers to new products.

Product Registration Card

Name______________________________

Address____________________________

City ____________ State ____ Zip______

Name of product______________________

Serial number________________________

Date of purchase___________ Store______

How did you hear about the product?_______

Is the product for home or work?__________

Have you purchased other products from this company?____________________________

If so, which products?____________________

Figure 12-7. Registration cards are used to gather information.

ask if you like certain products. They also ask why you do or do not like the products. The survey companies might even ask about colors or sizes of the products. Once they have surveyed many people, they sell the results to other companies. This information helps designers make good decisions and better products.

Experimental Research

Experimental research uses tests to learn new information. You might think of a scientist doing this type of research. Scientists run tests and experiments to learn about problems such as diseases and illnesses. Designers, however, also use

Technology Explained

Internet cookie: a small text file that a Web site saves on a computer.

Most surveys and research conducted with consumers are done with cooperation from the consumer. When a researcher calls a user of a product, the user agrees to participate in the survey before it is conducted. Technology, however, has changed that. Some companies are able to conduct market research without the consumer's knowledge. These companies use tracking cookies. A tracking cookie is a file that enables companies to track Internet-browsing patterns. See **Figure 12-A.**

Cookies, themselves, are not malicious and are actually quite helpful. They are used when you personalize a Web page or load items into an on-line shopping cart. The cookies allow you to come back to that site and have the Web page still be personalized and the shopping cart still have your items inside. A cookie is a collection of data that a Web server places on a computer when specific Web sites are accessed. When an Internet user visits an online store for the first time, the store's Web server places a cookie on the user's computer. The cookie contains an identification (ID) code that the Web server can link to the user's computer. So, when the user visits the site again, the cookie can use the ID code to recall certain settings or other information. Cookies help companies track the number of visits to Web sites and specific products on Web sites.

On their own, cookies are not dangerous. They cannot search your computer for information, as they are not programs. Cookies also cannot infect a computer with a virus. They cannot provide companies with personal information, such as your name or address. Cookies can, however, provide information on the Web sites you visit. They can be used to track your actions on the Internet. Advertising companies can use this information to target the user with specific ads. For example, if a computer user often visits sporting pages, the advertising company can use that information to send pop-up ads focused on sporting equipment. Most commercial Web sites have privacy statements that can be found on their main pages describing their uses of cookies. The privacy statement states the type of information that is collected and how it is used.

Figure 12-A. Web sites often place cookies on the user's computer while the user searches the Web.

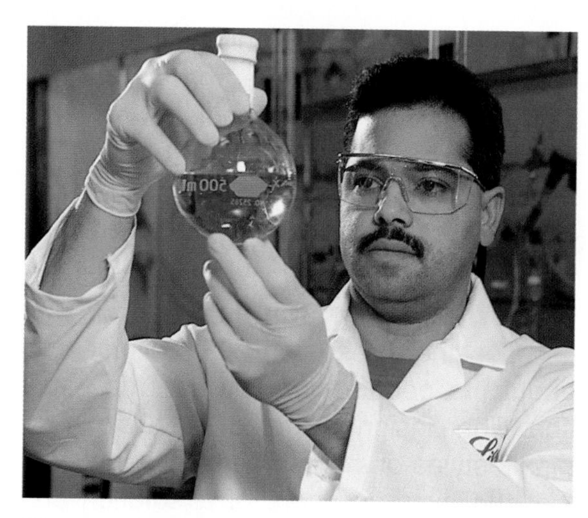

Figure 12-8. Researchers use laboratories to learn new information. (Eli Lilly and Company)

experimental research. See Figure 12-8. Designers conduct tests to learn about materials and procedures. Experimental research can be conducted in laboratories, with the use of computers, or on location with users of the products.

The test, or experiment, starts with a hypothesis. A hypothesis is a hunch about what a test will prove. Designers might have a hunch (hypothesis) that blue telephones sell better than white ones do. They might set up an experiment to prove their thoughts. The designers can test only one variable at a time. A variable is a difference between the items being tested. The color, size, and shape of the telephone are all variables. The designers make sure only the color of the telephone is different. They probably even place the different-colored telephones in the same place in the store. The designers then run the experiment. If more blue telephones are sold, the designers' hypothesis is correct. If the designers' hypothesis is wrong, the designers do more research. They try to determine why they were wrong.

The Research Process

Research is a process. When people research, they follow certain steps. The ***research process*** used is often called the *scientific method*. The ***scientific method*** is a set of steps used to find all the facts about a problem. See Figure 12-9. There are four steps in the scientific method:

1. List the problem.
2. Gather data.
3. Analyze the data.
4. Make a conclusion.

List the Problem

A research project starts by identifying the problem. The problem being researched might be the problem statement from the design brief. The designers might need to review the design brief. They need to make sure they understand the problem.

Gather Data

Once the problem is identified, data is collected. Data is raw facts and figures. These facts and figures can be a group

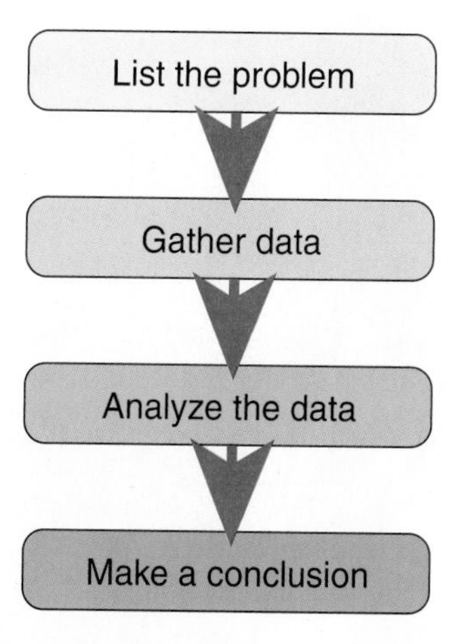

Figure 12-9. The scientific method is a research process.

of numbers or a listing of facts about the problem. Data collected at this point will be analyzed later. These facts and figures will then be used to make ***conclusions*** about the problem.

Data is gathered by designing and using tools, such as data-collection devices for interviews, surveys to be mailed, and computer-based documents on the World Wide Web (WWW). Assessment tools also can include devices designed to perform tests on such things as water quality, air purity, and ground pollution. Once the appropriate instruments are designed, they are used to gather data.

The designers can gather data in several different ways. They can choose library, survey, or experimental research. The designers might use more than one of the types of research. Teams of people are often involved with gathering data. One team might survey people, while another team reads books and catalogs. The more data gathered, the more information the designers have. The data gathered helps the designers make good decisions. See **Figure 12-10.**

Designers use design briefs to guide their research. The criteria from the design briefs help direct the designers. They state who will use the solution and how well it needs to work. The criteria also tell what the solution will do and where it will be used. The designers must research these statements.

Who Will Use the Solution?

Designers research different aspects about the people who will use the solution. The designers gather data showing the lifestyles of the users. They might find that the average users of a solution are married with small children. The designers also research the people's interests. They might identify the hobbies and sports the users enjoy.

Some companies design products for only one market. Other companies create designs for many markets. Baby food is a product designed for one market. The companies designing baby food do not need to research the type of people who use and buy baby food every time they create a new type of food. A company designing shoes, however, might need

Figure 12-10. Teams of designers often work together. (Ford Motor Company)

Career Highlight

Survey Researchers and Market Analysts

The Job: Survey researchers design and utilize surveys as a method of generating information. The surveys might be telephone, online, mail in, or personal surveys. Once the survey data is gathered, market analysts review it. Analysts attempt to understand the data and make recommendations to clients regarding what they find. Companies specializing in market research or large companies or governmental organizations that have a need for survey data and analysis can employ survey researchers and analysts.

Working Conditions: These researchers generally work in a standard office environment. Typically, they work a standard 40-hour week and spend a large amount of time on computers. Survey researchers and analysts are often under pressure to meet deadlines.

Education and Training: Most survey researchers and market analysts hold bachelor's degrees. Courses in marketing, business, statistics, survey design, and psychology are often required. Many of those in this field often have master's degrees in the previous areas. Before being employed as a survey researcher or market analyst, many complete internships in the field.

Career Cluster: Marketing

Career Pathway: Marketing Information Management and Research

to research different users of its product. If the company had always made tennis shoes and decided to design work boots, it would need to research a new market. The people buying tennis shoes might have different lifestyles and interests and might look for different aspects in a shoe than people who buy work boots.

How Well Does the Solution Need to Work?

When designers research how well the solution will work, they study the materials used to build the solution. Materials are important to the design. Some materials are better for certain things than others. See Figure 12-11. Concrete would not be the best material for a spoon. Steel would not make the best backpack. Designers research and gather data on different types of material. The data is related to the materials' properties. Properties are characteristics such as density, smoothness, hardness, and reactions to chemicals and sound.

Designers also research safety. See Figure 12-12. Designs must be safe for people to use. Designers study the safety of other products similar to theirs. They also interview safety experts. Safety

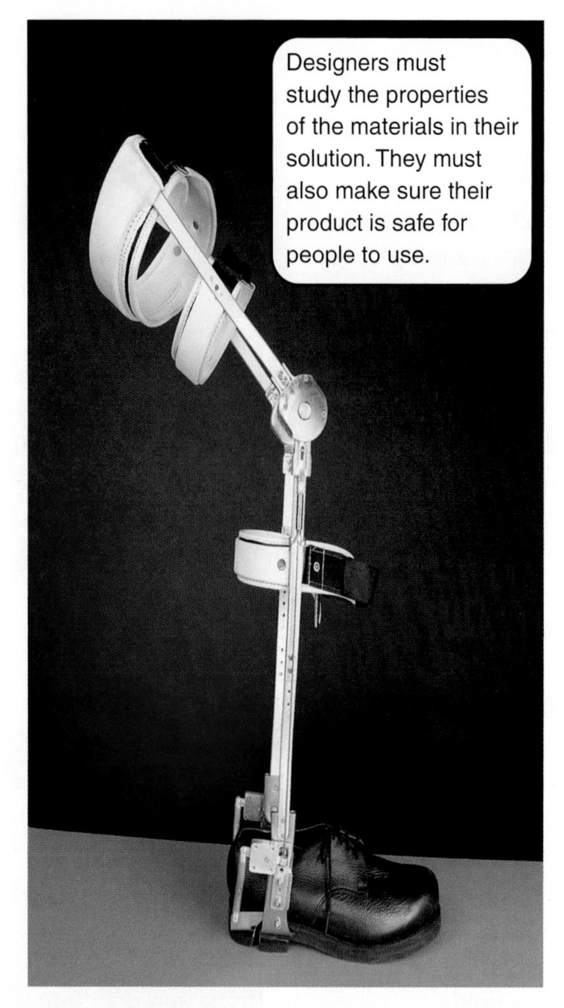

Designers must study the properties of the materials in their solution. They must also make sure their product is safe for people to use.

Figure 12-11. Designers gather information on new materials for use in their products. The materials used in this leg brace are high strength and lightweight. (NASA)

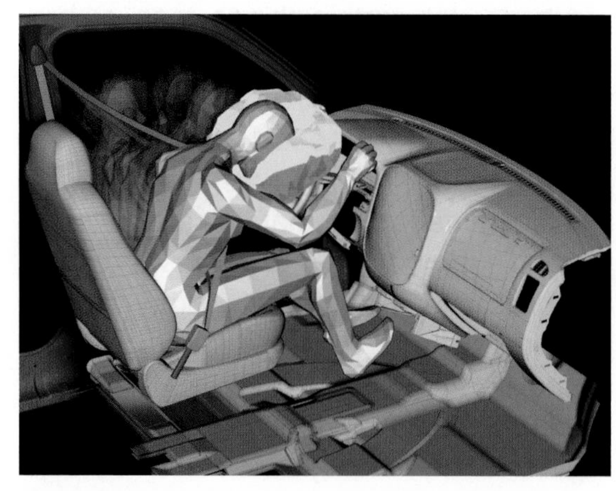

Figure 12-12. The safety of products must be researched. (Ford Motor Company)

includes the effects on people and the environment. The design must be safe for the environment when it is used and discarded. Designers research the safest designs for their solutions.

What Will the Solution Do?

What the solution will do is the function of the solution. When researching the function of the solution, the designers interview and survey people. They collect data showing how people want the product to function and showing the function of other products that solve similar problems. The data collected on the function is in the form of a list of features. In the design of a bulletin board, a feature might be that the solution must hold a piece of paper vertically. Another feature might be that the piece of paper must be read from 20′ away. Features can also be artistic, or ***aesthetic***. Aesthetic features relate to how the design looks. Texture and color are aesthetic features.

Where Will the Solution Be Used?

Research on where the solution will be used determines the size and shape of the solution. If designers are designing a desk chair, it is important to research desks. The designers need to measure or find common sizes and heights of desks. They might also research the average space the users have around their desks. This data would be useful when determining the size of the desk chair. The environment is also important. A chair for an executive of a large corporation would be designed differently than a chair for an elementary school student would be.

What Currently Exists?

An important question that designers need to answer is "What currently exists?" It is easy to get caught up in designing a new product or device. A solution to the problem, however, might already exist. Designers must spend time researching the products that solve their problem and similar problems. If there are solutions to the same problem or similar problems, designers need to determine how well the products function. They research the positive and negative aspects of the existing solutions. The goal of design is not to "reinvent the wheel," but to make new and better solutions.

Analyze the Data

Data alone is hard to understand. Imagine looking at a list of 10 cities, each with 24 numbers next to it. Each number represents the temperature for each hour of the day. This list might not make much sense. Now imagine seeing a map of the United States showing the average temperature of 10 cities. See Figure 12-13. You would probably have no problem understanding the map. The list of numbers and the map both show the temperatures of ten cities. The map, however, is easier to understand because the data (temperatures) have been analyzed. The data has now become information. Information is organized data. Once all the data has been gathered, the designer's job is to analyze it. The designer must convert the data to information. Information has more value to consumers than data because it is easier to understand.

Reviewing data and making graphs are some ways designers analyze data. ***Graphs*** are used to turn numerical data into information. They make data more understandable. A graph is an illustration showing how two or more items compare to each other. Line, bar, and pie graphs are the most common types.

A line graph is often used to show trends. See Figure 12-14. A trend is how something has changed over time. Line graphs use dots and lines to show trends. Dots are placed at intersecting points. The lines connect the dots to make the dots easier to see. A line graph has two axes (x and y). The x-axis is horizontal and usually represents a period of time. The y-axis is vertical and represents the item being studied.

Bar graphs are used to compare different amounts. See Figure 12-15. Bar graphs have an x-axis and a y-axis, similar to line graphs. They can be laid out horizontally or vertically. The most common bar graphs use thick bars of color to show the amounts. Some bar graphs, however, use pictures

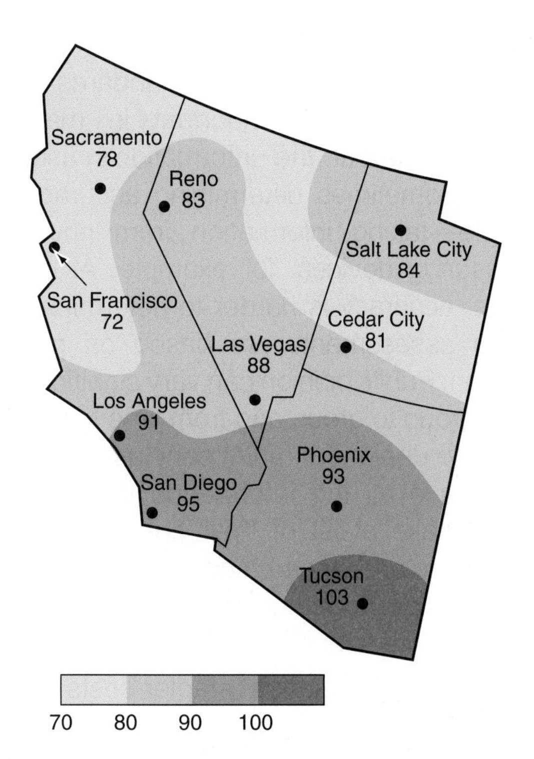

Figure 12-13. Weather maps make complex data easy to understand.

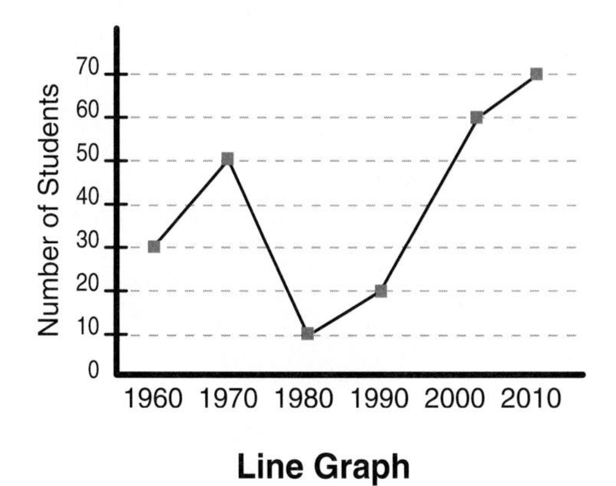

Figure 12-14. Line graphs are used to show data over a period of time.

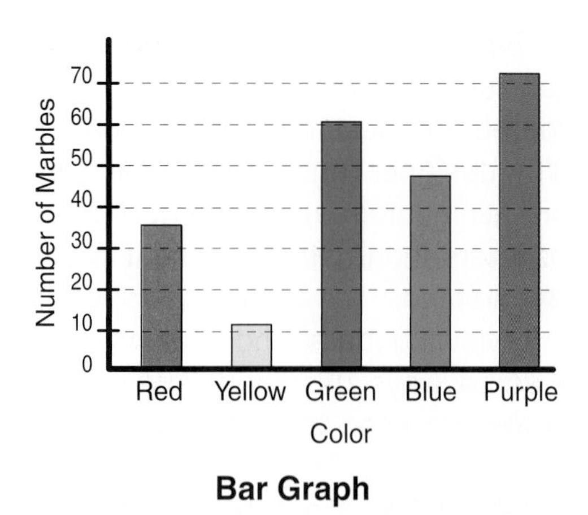

Figure 12-15. Bar graphs use different-colored bars to show data.

instead of bars. A bar graph can be used to show the density of different materials.

Pie graphs are used to show how something has been divided. See Figure 12-16. A pie graph is drawn in the form of a circle. Sections are then divided inside the circle, very similar to cutting a pie. Pie graphs often deal with percentages. The types of materials recycled can be represented in a pie graph.

Graphs work well for numerical data from surveys and experiments. They do not work as well, however, for some data gathered from library research. Much of the data from library research is in the form of notes. Designers take notes while reading printed materials. This data is organized and analyzed differently. One method of analysis is to organize all notes according to the topic of research. The designers rewrite the data as bullet points. A ***bullet point*** is a single piece of information, such as one item from a list. The designers then circle the data they feel is important to their designs. This process has created information from data. The notes were data. Once they are organized and the important parts are circled, they become information.

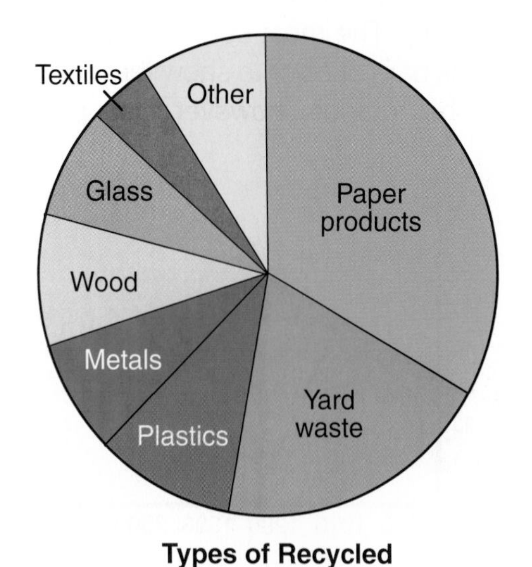

Figure 12-16. Pie graphs are useful for breaking data into pieces.

Make a Conclusion

The last step of research is to make a conclusion. Conclusions are the findings of the research. The conclusion can be made only after data has been gathered and analyzed. The information coming from the data must also be read and reviewed. The accuracy of the information obtained must be interpreted and evaluated. It is important to decide if the information is useful before drawing any conclusions.

Developing specific standards for what is useful is important in making decisions about the information's usefulness. Sometimes determining accuracy is simple—taking information from physical measuring devices, for example. At other times, accuracy is harder to determine, as when assessments are based on public opinion. Public opinion can vary significantly from group to group and from time to time.

The designers make conclusions after they have all the information they need. They make a list of their conclusions. If designers are researching a new design to staple paper together, their findings might be similar to this:

- The average manual stapler costs between $2 and $15.
- The average size of a stapler is 8″ long and 2″ wide.

- Staplers can be either manual or automatic.
- Most people like black staplers.
- Most stores carry three to five different brands of staplers.
- Staplers, on average, can staple 15 sheets of paper.
- The typical stapler punches the staple through the paper and folds the staple over.

The findings are the results of research. Library, survey, and experimental research would have been used to develop these conclusions. The size and cost of the stapler can be generated through library research. Survey research can discover the facts that people like black staplers and that most stores carry three to five brands. The number of pages a stapler can staple can be gathered through experimental research. See Figure 12-17.

Summary

Step two of the design process is researching the problem. The rest of the design is based on the conclusions made in this step. There are three types of research designers can use. Library research is used to develop background information. This research is also useful in finding how others have solved similar problems. Survey research uses questionnaires to find the reactions of consumers. Experimental research is used to conduct tests and find data through experiments.

The research done in this step is a process. This process resembles the scientific method. The research begins by listing the problem identified in the previous step of the design process. The designers then gather raw numbers and facts, or data. They research the who, what, how, and why of the problem. After the data is collected, the designers analyze it. They make graphs and lists of the data. These new lists and graphs are information. The information is then reviewed. The designers report the findings, or conclusions. All the research done in this step will be used to develop ideas and sketches for a new product or system. At the end of this step, the designers might not be satisfied with the research or the design brief. They can always go back and redo either of them.

Figure 12-17. Testing available products is part of experimental research. (©iStockphoto.com/Clicker)

STEM Connections

Science

Choose a scientific advancement (such as the polio vaccine or sheep cloning). Create a poster board describing the experimental research done to create that advancement.

Mathematics

Choose an article from a newspaper with numerical data. Create a graph related to the information.

Curricular Connections

Language Arts

Learn how to use the school or public library to conduct research and how to use card catalogs and search engines.

Activities

1. Give examples of how the following companies would use the three types of research to gather data for the products they design.
   A. A national soft drink company.
   B. A small newspaper company.
   C. An automobile manufacturer.
   D. A fast-food company.
2. Gather data related to a problem in your community. Create a graph informing the public about the problem or potential solutions.
3. Using the fish-feeder design brief in Chapter 11, follow the steps of the research process. Report your conclusions to the class.

Test Your Knowledge

Do not write in this book. Place your answers to this test on a separate sheet of paper.

1. Designers are the only people who conduct research. True or false?
2. _____ research uses printed materials to gather data.
3. _____ research is used to find people's reactions.
4. A hypothesis is a result of research. True or false?
5. Where can you find the problem to be researched?
6. Raw facts and figures are examples of _____.
7. Organized data is called _____.
8. Graphs are used to turn data into information. True or false?
9. The findings of research are known as _____.
10. Explain the four steps of the research process.

Reading Organizer

On a separate sheet of paper, create a detailed outline based on what you have read about the research process.

Example:

I. Types of research
  A. Library research
    1. Books
    2. Reference books

CHAPTER 13
Creating Solutions

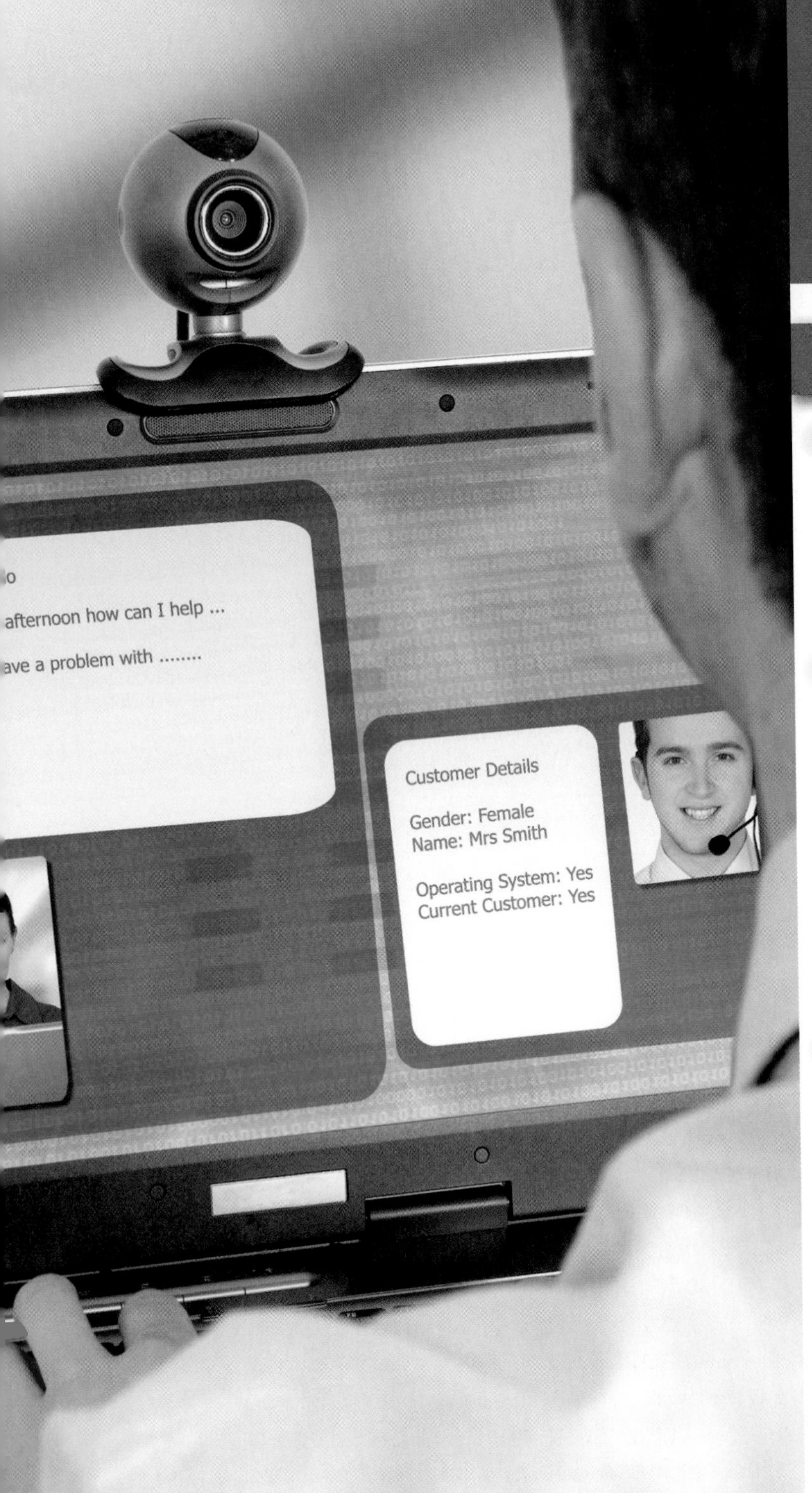

Did You Know?

- Mechanical drawings use symbols that help explain features of the drawings. These symbols are standard so everyone who looks at a drawing can understand it. Different types of drawings use different sets of symbols.
- The symbols used in welding drawings explain how the parts should be welded together. Some common welding symbols include the following:
  - Seam
  - Weld "V" groove
  - Weld all around
- Symbols used in engineering drawings include the following:
  - ∅ Diameter
  - Drill through
  - Countersink
- Some common architectural symbols include the following:
  - Door
  - Window

Objectives

The information given in this chapter will help you do the following:

- Explain the concept of ideation.
- Explain brainstorming, graphic organizers, and questioning.
- Summarize the qualities of a rough sketch.
- Summarize the qualities of a refined sketch.
- Apply the procedure for creating sketches.
- Compare shading and shadowing.
- Create an isometric sketch.
- Create an oblique sketch.
- Create a perspective sketch.

Key Words

These words are used in this chapter. Do you know what they mean?

box
brainstorming
cone
cylinder
elements of design
graphic organizer
ideation
isometric sketch
oblique sketch
orthographic drawing
pictorial drawing
pyramid
questioning
refined sketch
rough sketch
shading
shadowing
sketch
sphere
thumbnail sketch

Preparing to Read

Look carefully for the main ideas as you read this chapter. Look for the details that support each of the main ideas. Use the Reading Organizer at the end of the chapter to organize the main and supporting points.

The design process can be divided into two sections: problem seeking and problem solving. The first two steps of the process are problem seeking. In step one, the problem is identified. A design brief is created. The problem is then researched in step two. Both of these steps deal with seeking information about the problem. Step three is the first step to involve solving the problem. In this step, the designers begin to think of ways the problem can be solved. They develop many ideas and draw sketches for each of the ideas. See **Figure 13-1.** The sketches help to explain the designers' ideas to others. The ideas and sketches developed are then refined, modeled, and tested in the next steps of the design process.

Ideation

Some people might come to this step and think they already know what their final solution is. They might feel that developing a number of solutions is a waste of time. This is the wrong attitude to have. Designers must explore a number of ideas. The more ideas explored, the better the design will be. Designers should generate original and creative ideas. They have to think broadly and develop a wide variety of ideas. If designers use narrow thinking, their designs will not be unique. For example, imagine you are designing a telephone. You should explore many different ideas (broad thinking). One idea might be a phone fitting in your ear, similar

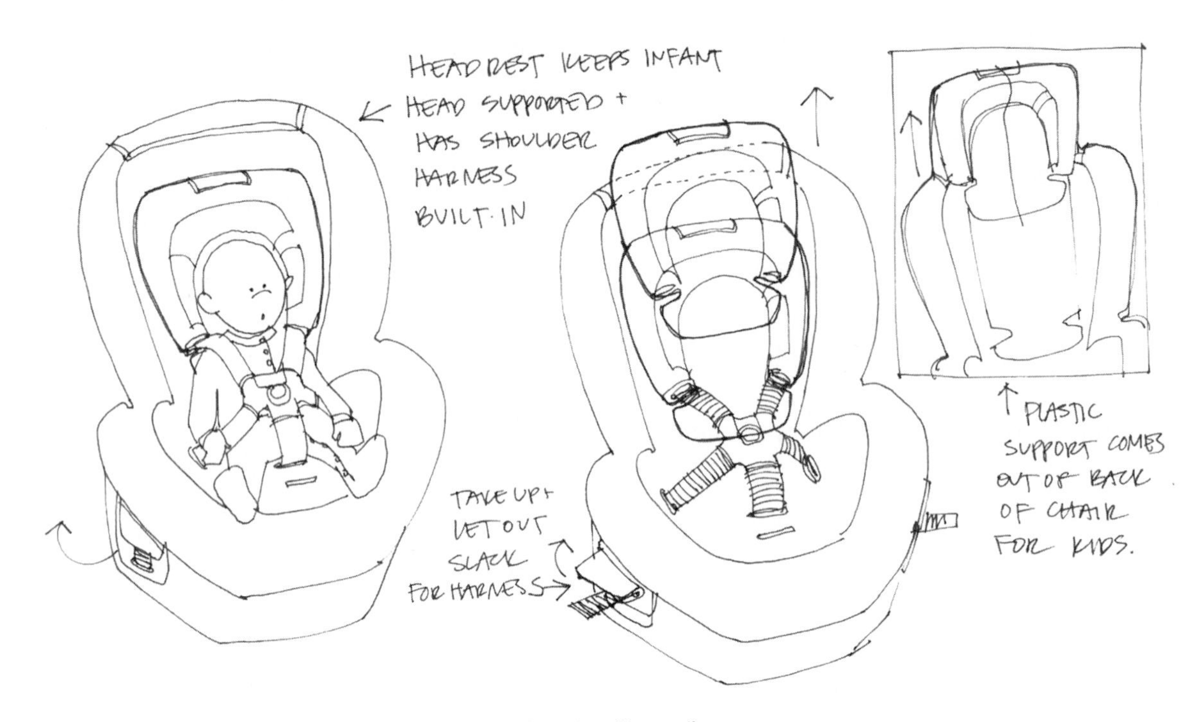

Figure 13-1. Sketches are ideas on paper. (Design Central)

to a hearing aid. Another idea might be a digital phone with a computer video display. See Figure 13-2. These are two different solutions to the design of a telephone. You do not want to create ideas such as a red desk phone and a black desk phone. That is narrow thinking. These ideas are essentially the same thing, with only the color differing. The small changes, such as color, can be made later. It is important that designers explore many ideas. Exploring ideas is called ***ideation***.

Creating a number of new ideas to solve a problem is ideation. The process of ideation is creative and imaginative. See Figure 13-3. Creativity leads to new ideas and solutions. The solutions created at this step should be simple and rough. Later in the process, the solutions will be revised and refined. Ideation begins by reviewing the problem and design brief. Remember, the solutions developed must solve the problem. The solution must also fit the criteria and limitations. After reviewing the design brief, the research conclusions must be reviewed. This review helps the designer understand

Figure 13-2. Broad thinking leads to new solutions to problems. (Motorola, Inc.)

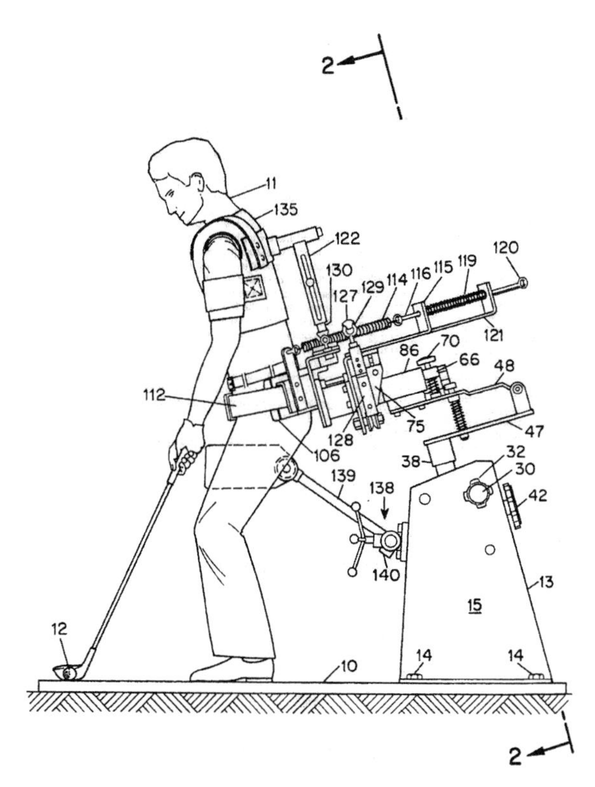

Figure 13-3. Unique products are the results of creativity and imagination. (Patent No. 5,050,855, U.S. Patent and Trademark Office)

what others have done and what people want from the solution.

Ideation is a free-flowing activity. Ideas must be able to flow without others shooting them down. In ideation, all ideas are good ideas. There are no right or wrong answers. It is important to record (sketch or write down) all ideas. Ideas seeming silly or wild at first might make more sense later. If you do not record them, you might forget some good ideas.

The environment you use to create solutions is very important. You must be able to concentrate on the problem. In front of the television at home might not be the best place to focus on creating ideas. Many designers find quiet places to work. They might leave their offices and work outside or in a library. See **Figure 13-4.** The only things needed to create ideas are a pencil and paper. Ideation can be done almost anywhere. When creating ideas, designers often flip through magazines or

Figure 13-4. Designers often work best in quiet areas away from their offices and desks. (©iStockphoto.com/LajosRepasi)

books. They might even listen to inspiring music. Designers often see pictures or hear words that spark ideas for solutions to their own problems.

There are several different methods of ideation. Some people find one works better for them than others do. Three of the main methods are brainstorming, graphic organizers, and questioning.

Brainstorming

Brainstorming is a method used to develop many ideas. This method is useful in small groups for coming up with unusual solutions. The group members are often selected to include people with different backgrounds. It is helpful to have people with different knowledge and experiences because they bring new and different ideas to the activity. Brainstorming is a process that begins with the design problem. The process has several steps.

In the first step, the small group must choose a leader and a recorder. The leader starts and ends the brainstorming and makes sure the group stays on task. The recorder is in charge of recording all the ideas shared. See Figure 13-5.

Figure 13-5. A typical brainstorming session can generate many ideas.

Next, the leader sets a time limit. Brainstorming is done best when a time limit is set. The group members stay more focused if they have only a certain amount of time. At this point, the leader shares the design brief with the group. The leader makes sure everyone understands the problem. The ideas generated must solve the problem. The group leader also shares the research conclusions. It is important that the research is used to come up with ideas.

The members of the small group all focus on the problem. The group leader then asks for ideas. The group members try to list as many creative ideas as possible. Solutions can become very creative when people get ideas from each other. One person might present a new idea. The new idea might spark an idea in your mind. You might never have thought of your idea without the other person's idea. Brainstorming works well when the group members use each other's ideas to add to their own. In brainstorming, as in other forms of ideation, it is important that there is no criticism. If people are criticizing or making fun of ideas, people do not want to share their thoughts. The best thoughts might not be shared if people are afraid to give their ideas.

Throughout the whole session, the recorder makes notes of all the ideas developed. When the time is up, the leader must stop the session. The leader might ask all the group members for one final idea. Once the group is done, the brainstorming is over. The notes will be reviewed later in the design process. The main purpose of brainstorming is to develop a list of ideas. The purpose is not to choose the final solution.

Graphic Organizers

A ***graphic organizer*** is another method of developing ideas. This method is a diagram that helps to organize thoughts. One person or a group of people can create a graphic organizer.

Career Highlight

Architects

The Job: Architects are licensed professionals who are trained to design buildings. They identify a client's needs, develop building concepts to meet these needs, and prepare plans for the building. Architects design the overall look of buildings and develop plans and specifications that construction personnel will use in erecting the structure. They must be able to communicate their vision and plans to clients and construction managers.

Working Conditions: These professionals spend a great deal of their time in offices, consulting with clients and developing drawings. In addition, they spend time visiting construction sites to review projects under construction.

Education and Training: All architects must be licensed before they contract to provide architectural services. Many architecture-school graduates work for another architect, however, while they are in the process of becoming licensed. In most states, obtaining a license requires a degree in architecture from an accredited university program. Most of these programs are five-year bachelor of architecture programs. Typically, architecture programs have courses in architectural history, building design, structures, and construction methods. They also require several professional-practice, mathematics, science, and liberal arts classes.

Career Cluster: Architecture & Construction
Career Pathway: Design/Pre-Construction

To create a graphic organizer, designers begin with a blank sheet of paper. They write the problem in the middle of the sheet and circle it. The designers then create branches from the middle. The branches are the major features or functions of the design. If the design problem is to design a new camera, the word *camera* goes in the middle. See **Figure 13-6.** The branches are features, such as type of camera, loading the film or media, focusing the picture, taking the picture, and changing the settings. Below the branches, the designer lists the possible ways to design the features or functions. Under the "Taking pictures" branch, the solutions might be a button on the top of the camera, a voice-activated control, an automatic timer, and a handheld switch.

The graphic organizer helps to make sure all the features of the problem are considered. This organizer is also helpful because all the solutions are listed on the same sheet of paper. It is easy to see all the ideas you have developed.

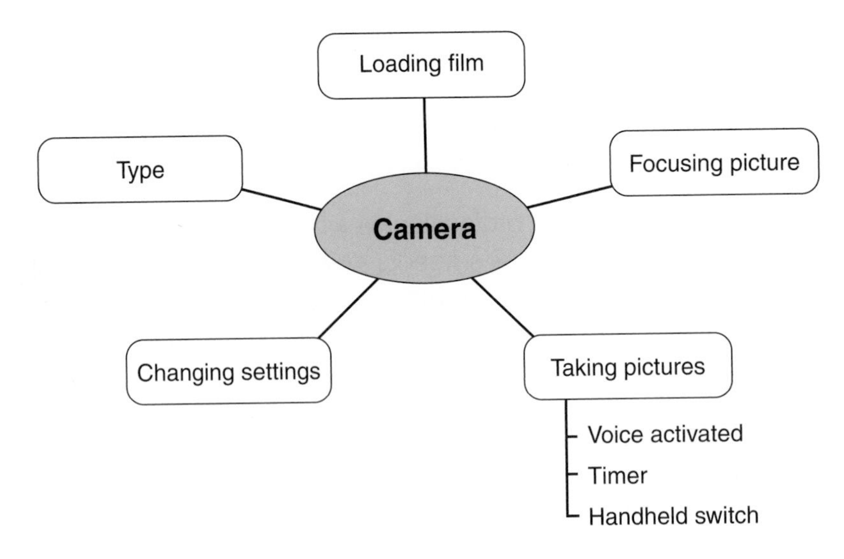

Figure 13-6. Graphic organizers help generate ideas.

Questioning

Questioning is the third method of ideation. This method is done differently from brainstorming and graphic organizing. Questioning is a process in which the designer asks the question "Why?" This process is done while working with existing products. See Figure 13-7. The designer asks why things are done the way they are in the existing product. Imagine you are designing a better way to toast bread. If you were using the questioning method, you would locate a toaster and a loaf of bread. You would use the toaster and ask yourself questions such as the following:

- Why is the bread loaded from the top?
- Why do I push a lever to start the toaster?
- Why does a knob control the darkness scale?

Figure 13-7. Questioning relies on designers to ask questions about ways they use a product.

The key to questioning is that the questions are written. Once the questions are asked, they need to be solved. Designers often find creative solutions to problems because they asked "Why?"

Brainstorming, graphic organizers, and questioning are all useful methods of creating ideas. These approaches to ideation all result in lists of solutions. Lists, however, are not always the best way to describe solutions. Have you ever tried to describe an idea you had using a list? Describing an idea using a list can be very difficult to do. Imagine someone asked you to describe how your bicycle works or how the White House looks. Are these things easy to describe? They would probably be hard to describe without drawing the gears of your bike or the front of the White House.

SKETCHING

Using a ***sketch*** can make describing ideas much easier. The methods of ideation can be very helpful in coming up with ideas. Sketches are useful, however, for describing the ideas to other people. See Figure 13-8. Sketches are tools that help designers communicate their ideas. They are a way to record the designers' thoughts on paper. As thoughts flash into the heads of designers, the designers sketch them so others can see them. Sketching is usually done in two steps:

- The first step is often called *design sketching*, *preliminary sketching*, or *rough sketching*. These sketches are quick and simple sketches.
- The second step is often called *working sketching*, *final sketching*, or *refined sketching*. These sketches are more detailed and require more time to complete.

Figure 13-8. Sketches are used to communicate ideas between one person and another. (Product Development Technologies, PDT)

Rough Sketches

During the design process, designers always have ideas that pop into their heads. This is especially true during the ideation stage. Designers might have finished a brainstorming session and have an idea on the way home or during dinner. For this reason, many designers carry sketch pads and notebooks to record such ideas. When designers forget their sketchbooks, you might see them make sketches on napkins or small pieces of paper. At this point, it is not important what type of paper or writing utensil the designer uses. All that matters is that the designer gets the ideas onto paper. The designers are not concerned with the quality of the sketches. Good designers try to capture as many ideas as they can.

These sketches are called ***rough sketches***. See Figure 13-9. The word *rough* might seem to suggest that the sketches are hard to understand. This, however, is not the case. *Rough* just means the sketches are in the early stage of development. They are not completely thought out. Rough sketches are basic ideas showing shapes and outlines. They do not show a great amount of detail.

Rough sketches can be ideas of a complete solution, such as an entire

Figure 13-9. Rough sketches are early sketches showing the basic shapes and outlines. (©iStockphoto.com/feuers)

bicycle. The bicycle designer creates rough sketches of bicycles of different sizes, shapes, and designs. Size and shape are two ***elements of design*** developed using rough sketches. The sketches can also be ideas of smaller parts of the whole solution. See Figure 13-10. In the bicycle example, the rough sketches can be different ideas for the brakes and gears of the bicycle.

These sketches are also known as ***thumbnail sketches*** because of their size. Most rough sketches are fairly small, with several rough sketches fitting on a sheet of paper. Good designers are able to create rough sketches fairly quickly. These sketches might take beginners a while to create. The time spent on creating rough sketches is not what is most important. The importance of rough sketches is that many ideas are generated.

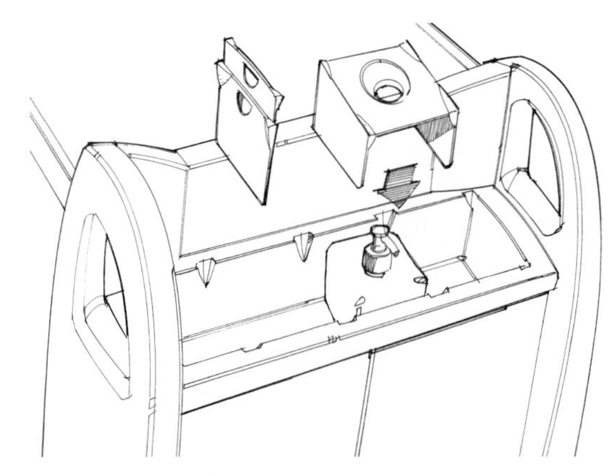

Figure 13-10. Some sketches show solutions of only part of a design. (Design Central, design firm; Artromick International, client)

Refined Sketches

After ideation, the designer should have a large number of rough sketches. These sketches might contain many good ideas. There might be ideas on how to solve the entire problem or how to solve small pieces of the larger problem. The sketches are simply ideas on paper. They make up a library of solutions. The rough sketches need more work done to them before they become proposed solutions to the problem. From the rough sketches, the designer must select the most promising solutions. This is the first time in the entire design process that the ideas can be reviewed. In the brainstorming and rough-sketching stages, all ideas were valid. No ideas were criticized or discarded.

The designer creates new sketches based on the best ideas from the rough sketches. The new sketches are called ***refined sketches***. See Figure 13-11. A refined sketch might focus on one rough sketch. This sketch might, however, combine parts of several sketches to make one new idea. The sketches combine the ideas ideation created. The purpose of the refined sketch is to narrow down the design ideas. The designer creates several different refined sketches. The best solutions will be chosen in the next step of the design process.

The Sketching Process

Sketches are not always easy to create. See Figure 13-12. Sketching is a technique. There is a certain way to sketch. Sketching has several steps:

1. Visualizing the object.
2. Blocking out shapes.
3. Adding an outline.
4. Drawing design features.

Visualizing the Object

The technique of sketching begins with visualizing the object. Visualizing, or seeing, the object might seem to be a simple thing. You might be able to look at an object and sketch it. In rough sketching, however, you might not be able to look at something real. If you are creating a new and innovative product, there is no image at which to look. You might have the idea only in your mind and have to "see" the

Figure 13-11. Refined sketches are more detailed than rough sketches. (Daimler)

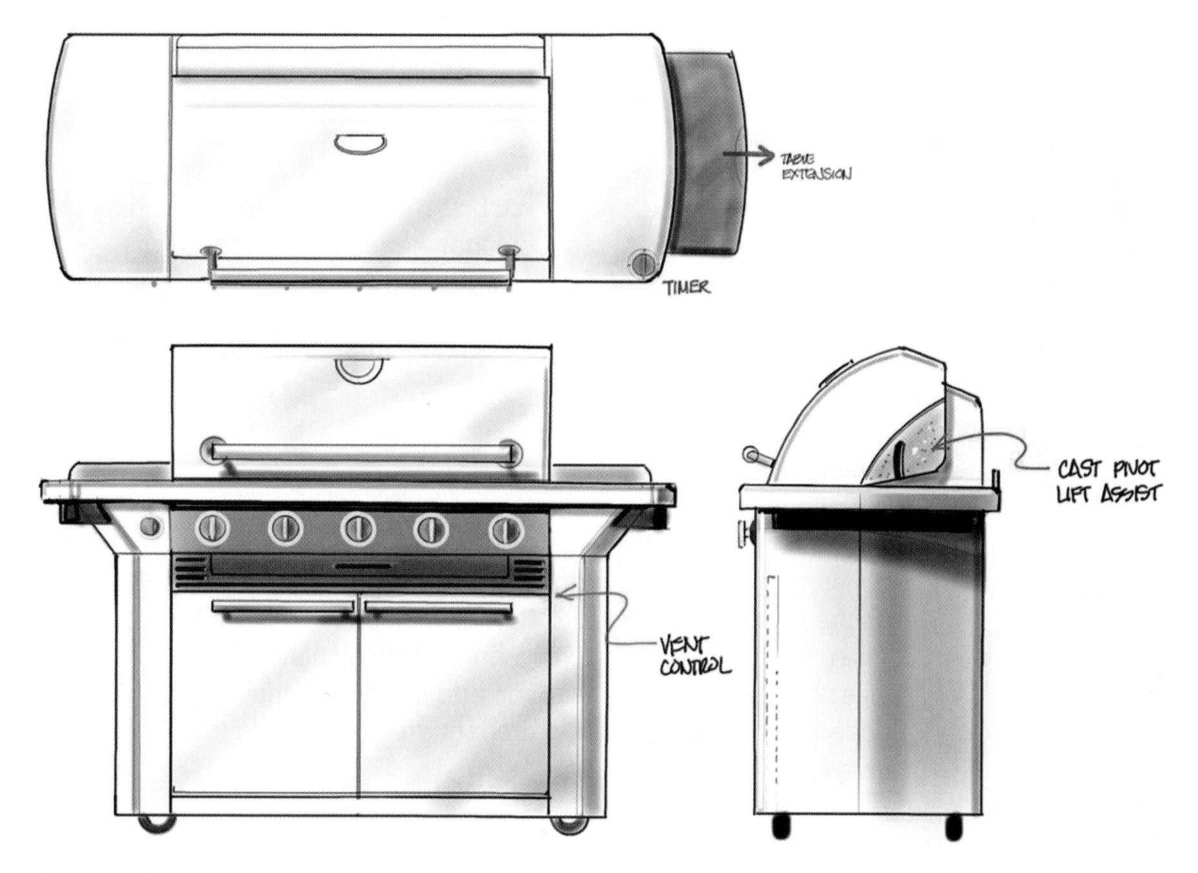

Figure 13-12. Sketches help develop ideas. (Design Central)

Technology Explained

elegant solution: a product meeting a human need in the simplest, most direct way.

Complex products and devices surround us. Many of these products and devices have all kinds of frills and little add-ons designers thought were necessary. Many of these add-ons, however, complicate the designs and make the products hard to use.

The opposite of this is the product meeting a need in the simplest and most effective way. Engineers call these *elegant solutions*. These solutions should be the goal of every designer of technological artifacts.

Think of an adhesive bandage, commonly known as the *Band-Aid® bandage*. This bandage is designed to hold gauze over a small cut or scratch. Can you think of a better solution? Over the years, no one has thought of one. Therefore, the Band-Aid bandage is an elegant solution.

Another commonplace product is the paper clip. See **Figure A.** The paper clip temporarily holds sheets of paper together. Everyone uses paper clips. No one has improved them. Again, this is an elegant solution.

Think of a number of other devices that work so well that improving them is a challenge. These devices include the zipper, safety pin, stapler, ballpoint pen, pipe cleaner, and thong sandal. Small details might change, such as the types of materials used. The basic design, however, remains the same. Elegant solutions do not, however, have to be simple products. There are many examples of complex technological products that elegantly solve problems.

Consider the *Apollo* spacecraft. See **Figure B.** This spacecraft carried three men into space and back with relative ease. The spacecraft's complex systems propelled and navigated the capsule. The spacecraft provided heat, light, and fresh air for the three astronauts inside. The capsule also protected the astronauts from the outside environment.

Elegant solutions are everywhere. Have you ever tried to read and record large volumes of numerical data? This is a time-consuming process and one prone to error. A bar code reader, such as the one shown in **Figure C,** makes this task quick and accurate. The unit shown is the size of a handheld calculator. Despite its small size, the device efficiently reads and processes the information coded in the bars. Bar code readers are widely used to take inventory in retail stores.

Figure 13-A. The paper clip is an example of an elegant solution.

Figure 13-B. The *Apollo* spacecraft is an example of an elegant solution. Complex devices can also be elegant solutions.

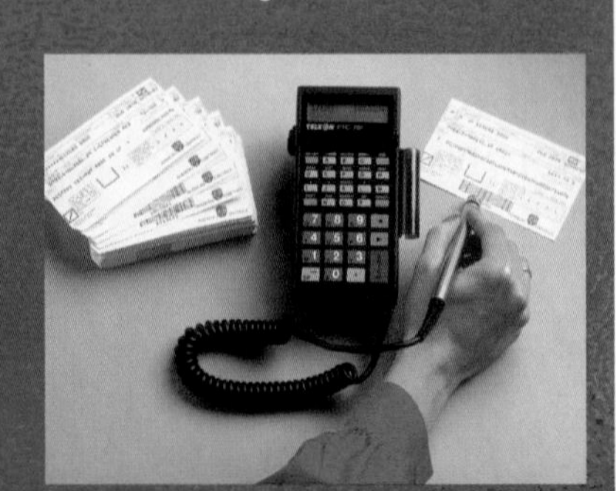

Figure 13-C. The bar code reader is an elegant solution to a modern problem.

object in your imagination. Sketching is often thought of as seeing and thinking with a pencil. Good designers can easily see their ideas and use pencils to recreate them. Seeing your ideas takes practice. The more designers sketch, the better they become.

There are two ways designers see objects. They can see objects in either two or three dimensions. When designers see in two dimensions, they see the object in six different views. They see the front, top, left side, right side, back, and bottom. Sketches made in two-dimensional views are called *orthographic sketches*. See Figure 13-13. ***Orthographic drawings***, which are more precise than sketches, are discussed more in Chapter 17. People who are not trained to read them might find orthographic sketches and drawings hard to understand. At this stage in the design process, it is easier for designers to see ideas in three dimensions. Three dimensions are the easiest because this is how the eye sees things. This is also best because almost all people can understand a drawing in three dimensions. Three-dimensional drawings are called ***pictorial drawings***. The three major types of pictorial drawings are isometric, oblique, and perspective drawings, all of which are discussed later in this chapter.

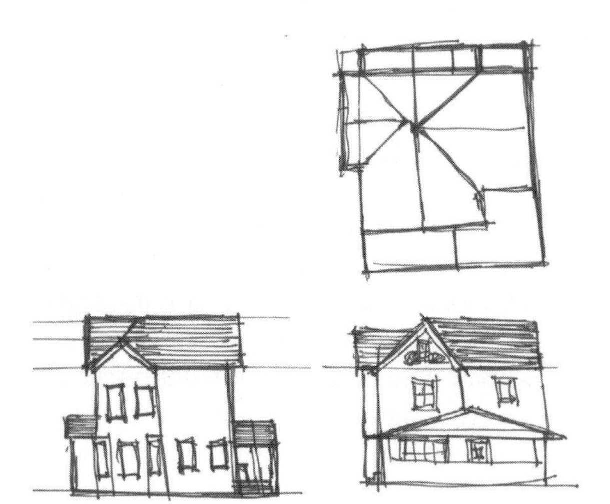

Figure 13-13. Orthographic sketches show each side as a separate view. (Keith Nelson)

Once the designers can see the idea, they begin by breaking down the object. Breaking down the idea requires the designers to divide their ideas into basic shapes. Some designers are able to see the basic shapes easily. Other designers must work harder to see the shapes. All objects, however, can be broken into one or more of these basic shapes. See Figure 13-14:

- ***Box.*** A box can be either a cube or a prism. A cube is a three-dimensional square. All sides of the cube are the same length. A six-sided die is an example of a cube. A prism is very similar to a cube. The difference is that a prism has at least one side that is a different length. A cereal box is a prism.
- ***Cylinder.*** A cylinder is a round shaft. This shaft has one circle on each end, similar to a tube. A soda pop can is an example of a cylinder.
- ***Sphere.*** A sphere is a perfectly round object. Baseballs and basketballs are spheres.
- ***Cone.*** A cone is a shape that is round at one end and comes to a point at the other. An ice-cream cone is an example of a cone.
- ***Pyramid.*** A pyramid is similar to a cone because it comes to a point at one end. The other end of the pyramid, however, is a square. The Egyptian pyramids are examples of this shape.

Blocking Out Shapes

After the object or idea has been broken down into pieces, the sketching begins. The sketching starts by blocking out shapes. Blocking out means drawing light lines where the basic shapes will be. These lines serve as guidelines used to make the sketch. Blocking out is very important for beginners. The more advanced designers are able to block out shapes very quickly.

Shapes in Designs

Box

Cylinder

Sphere

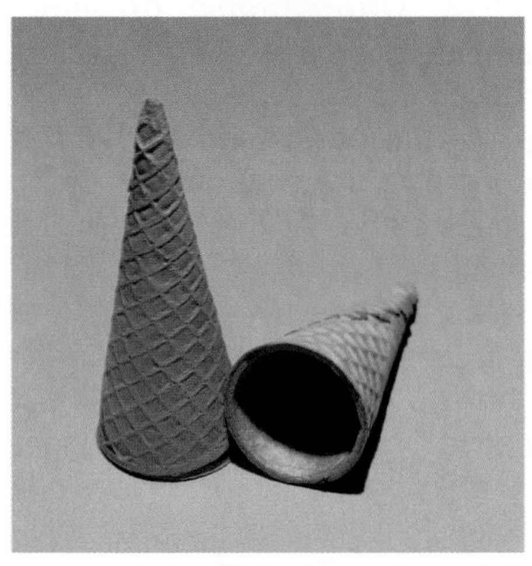

Cone

Figure 13-14. The basic shapes can be seen in many common objects. A file cabinet is a box shape. The shape of a cylinder can be seen in a can. Baseballs are spheres. An ice-cream cone is a cone shape. (Sauder Woodworking, The Coca-Cola Company, American Baseball Company)

Beginners must take their time to make sure the shapes are correct.

To begin blocking out, the designer draws a vertical line. Next, two lines are drawn at roughly 30°, starting from the bottom endpoint of the vertical line. These three lines are called the *isometric axes.* The intersection of the isometric axes forms the front corner of the sketch. The shapes can now be blocked into the sketch. Each shape is blocked in a little differently.

Boxes

The box uses the intersection of the isometric axes as its front corner. See **Figure 13-15.** The vertical line should be drawn to the same height as the object. The line drawn to the left should equal the length. The line to the right should equal the width of the box. Once the lines are drawn, the designer draws a line the same length as the vertical line at the other two endpoints. The designer then connects the three vertical lines. This forms the front and right side of the object. To form the top of the box, the designer draws two additional angled lines, the same length as the other angled lines. The first angled line is drawn from the upper-left corner of the front side

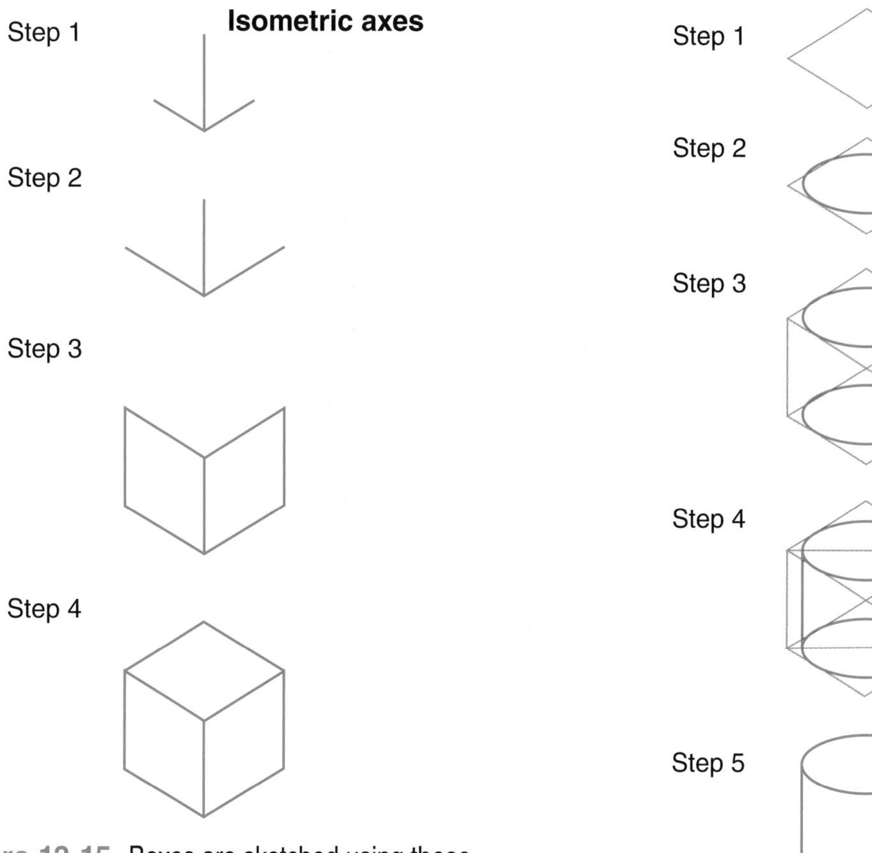

Figure 13-15. Boxes are sketched using these four steps.

Figure 13-16. Cylinders are sketched using these five steps.

of the box. This line should match the line at the top of the right side of the box. The final line should connect the end of the line that was just drawn to the upper corner of the right side of the box. This creates the top of the box. In a cube, all the lines drawn are the exact same size.

Cylinders

The cylinder is drawn using a square at the axes. See Figure 13-16. To begin the cylinder, designers use the angled lines of the axes and draw a square. They then draw an ellipse. An ellipse is an isometric circle. This circle is drawn through the midpoint of each line of the square. This forms the bottom of the cylinder. To make the top of the cylinder, designers draw vertical lines from the corners of the square. They then create a square positioned at the top of the cylinder. The designers draw an ellipse in the top square. To finish the cylinder, they draw a horizontal line from one corner of the bottom square to the other corner. Two vertical lines are then drawn where the line crosses the ellipse. These create the sides of the cylinder.

Spheres

Spheres are drawn by first creating a cube. See Figure 13-17. The cube is created as described previously. Once the cube is drawn, a circle is created inside of it. The circle is drawn so it touches the midpoint of the outside lines of the cube. When the sphere is drawn, it looks similar to a plain circle. In order for the sphere to look three-dimensional, it must be shaded. Shading is discussed later in this chapter.

Cones and Pyramids

The cone and pyramid are drawn using the rectangular box. The box is drawn

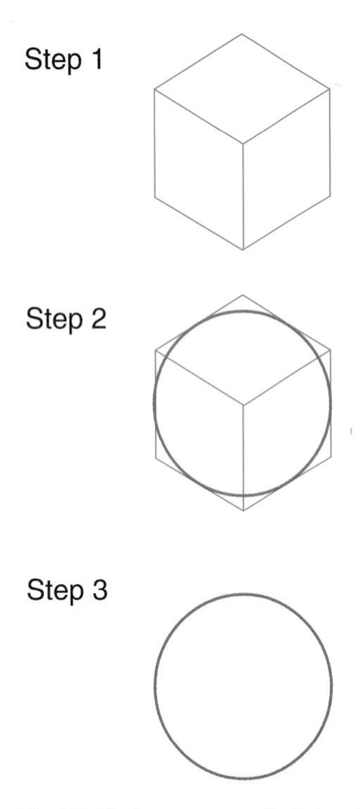

Figure 13-17. Spheres are sketched using these three steps.

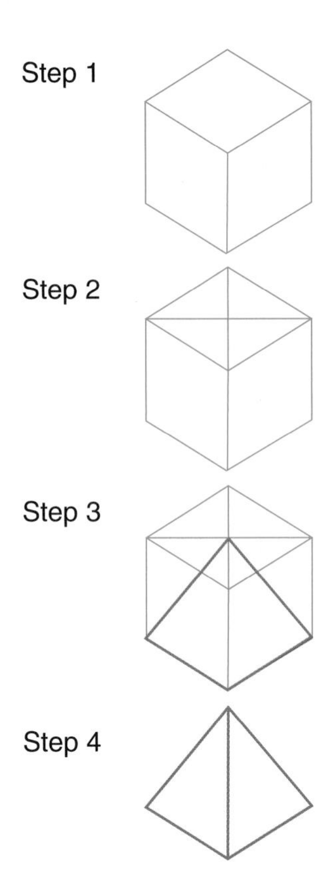

Figure 13-18. Pyramids are sketched using these four steps.

as shown in Figure 13-15. Next, an *X* is drawn on the top of the box. The center of the *X* is the center of the top of the box. The center point is the point where the top of the cone or pyramid will come to. To draw a pyramid, the designer draws lines from the bottom corners of the box to the center point. See Figure 13-18. To draw a cone, an ellipse must be drawn on the bottom of the box, just as when creating a cylinder. See Figure 13-19. Lines are then drawn from the sides of the ellipse to the top center point.

Combinations

Unless the designs are simple products such as ice-cream cones or basketballs, the sketches created will have more than one basic shape in them. Each of these shapes must be blocked out. You have learned how to block out shapes using the isometric axes. Sketches with more than one shape have more than one set of axes. Imagine designing a guitar. See Figure 13-20. You have several boxes making the shapes of cylinders and

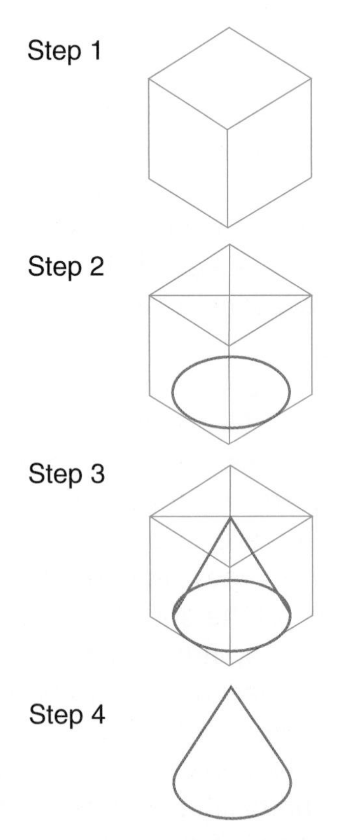

Figure 13-19. Cones are sketched using these four steps.

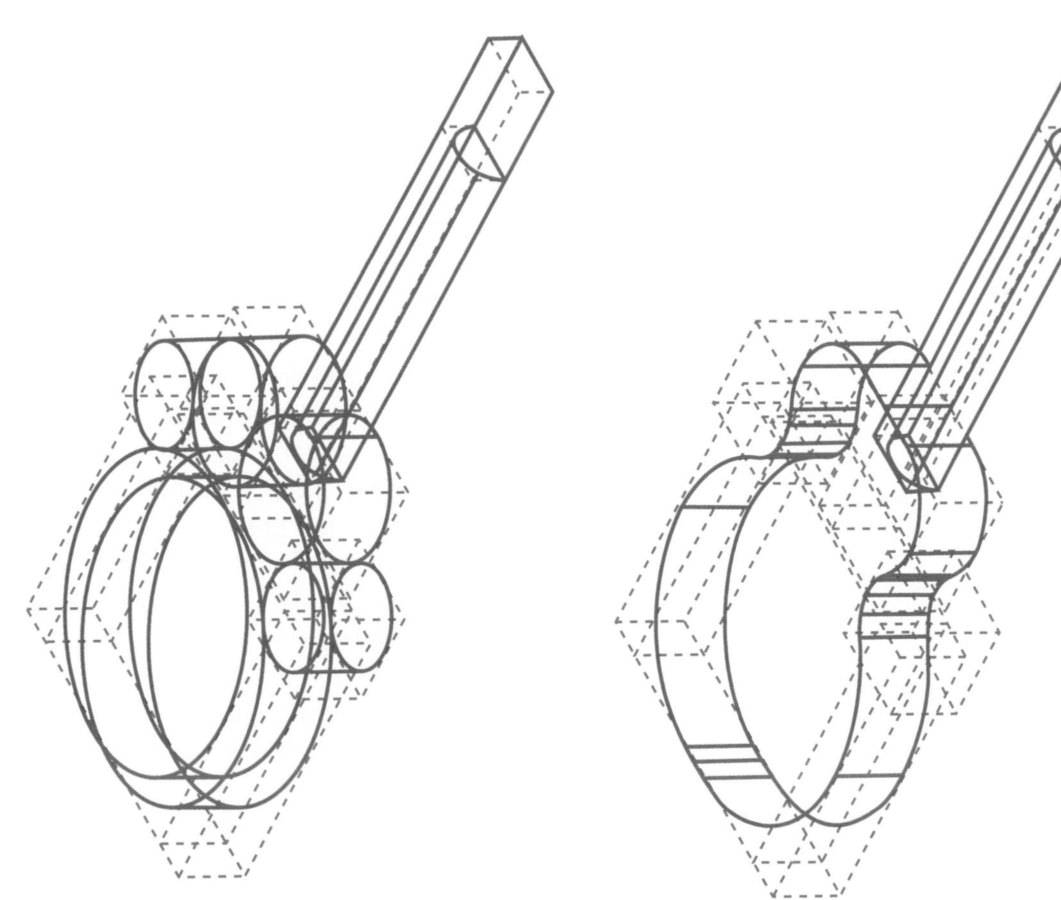

Figure 13-20. A guitar is blocked out using boxes and cylinders.

Figure 13-21. An example of the outline of a guitar.

one long box representing the neck of the guitar. Create the boxes lightly so they can be erased. This sketch might not look very similar to a guitar at this stage. The look of the guitar will come together in the next stage, adding the outline.

Adding an Outline

The blocked-out sketch might not look very similar to the designer's idea. This sketch should be, however, the basic shape and proportion of the idea. Designers must add details to the shapes to make the shapes look the way they want them to look. The first detail added is an outline. Creating the outline might include subtracting parts of the shapes. The guitar in the previous example is several cylinders and boxes. To create the outline, a designer subtracts parts of cylinders to make the guitar rounded. See Figure 13-21. It is important that the blocked-out shapes are drawn lightly because they are guidelines. When the designers add the outline, they draw the lines darker because these lines will not be removed.

Drawing Design Features

The next details to be added are the external features. The external features are objects not included in the outline, but ones that can be seen when looking at the design. The number pad on a telephone is an external feature. These features help in understanding the sketch. The tuning knobs, frets, and sound hole are all external features of a guitar. See Figure 13-22. External features also help to set designs apart from each other. A designer might have a number of sketches for a product that have the same outline. The only differences are the external features. For

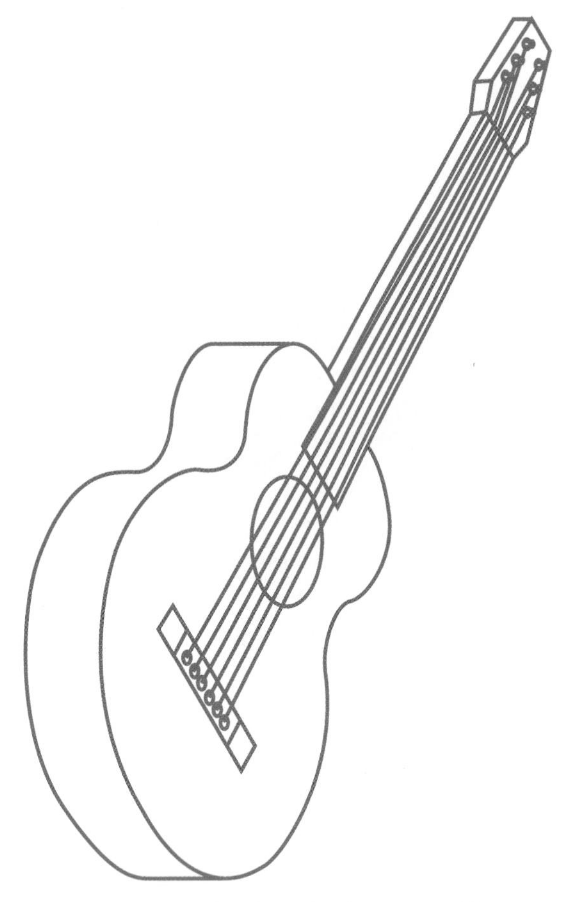

Figure 13-22. A guitar with the external features.

rough sketches, after the external features are added, the sketches are finished. For refined sketches, additional techniques can be used to add more detail and life to the sketches.

Refining Techniques

There are some techniques that can be used on refined sketches to make the sketches appear more realistic. Shading and shadowing are the two most common techniques. Both of these techniques illustrate how the objects will look when placed in light. They add depth and dimension to flat sketches.

Shading

Shading is a technique that shows how light is reflected from an object. This technique relies on a light source. Most designed objects come in contact with either sunlight or interior lighting. Shading helps to show how the object will look in either of these two light sources. This technique also allows the designer to see the shape and form of the object. To begin shading, the designer must determine from which direction the light source is coming. The direction of the light source determines how the object is shaded. The standard is to imagine that the light source is located over the designer's left shoulder.

There are several techniques used in shading. The most common is using different tones to create the shading effects. A tone is a shade of color. Gray is a tone between black and white. In this method of shading, the area of the object closest to the light source is the lightest tone. The area furthest from the light is the darkest. This effect can be created with the edge of a pencil or markers. When using a pencil, designers create the darker areas in one of two ways. They can either press harder in the darker areas or go over these areas several times. When using markers to shade, the designer selects markers of different tones to create light and dark areas. See Figure 13-23.

Other shading techniques use either dots or lines instead of different tones

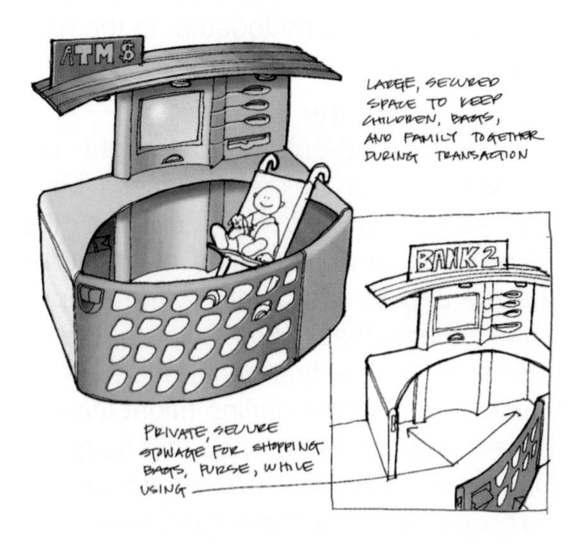

Figure 13-23. Shading and shadowing help designers see the shape and form of the object. (Design Central)

of color. The dot method uses different amounts of dots to show light and dark areas. See Figure 13-24. The more dots an area has, the darker it is. Pens are normally used for dot shading. Line shading is done with different-width lines. The darker areas are shaded with thicker lines. Both of these methods work well for shading a rounded object.

Shadowing

Once the object is shaded, the designer adds the shadow. ***Shadowing*** shows how light is cast around an object. This technique is normally done using a pencil. The shadow is placed on the opposite side of the light source. If the light is coming from the front, left side of the object, the shadow is placed on the back, right side. The shadow should follow the rough shape of the object. Shading and shadowing take a lot of practice to perfect. The best way to practice shading and shadowing is to look at objects around the room and sketch them. Examine the objects to see how the light hits them. Try to add shading and shadows to the sketches.

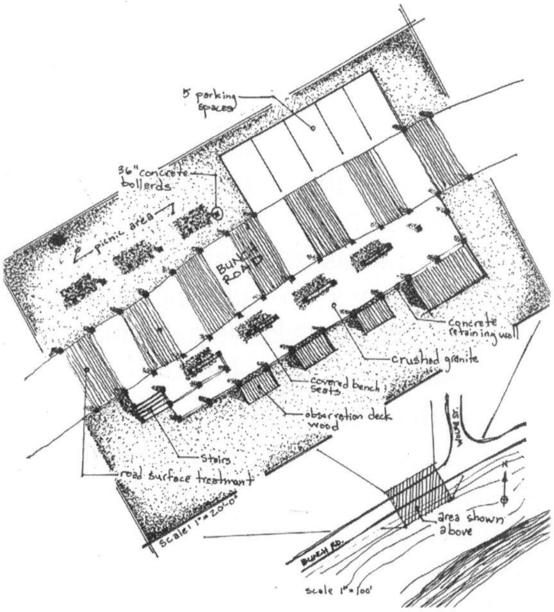

Figure 13-24. Drawings can also be shaded dots. (Keith Nelson)

Types of Sketches

There are several ways sketches can be drawn. The three most common types of sketches are isometric, oblique, and perspective sketches. Any of these types of sketches can be used to create rough or refined sketches. All these types are drawn using the same basic process.

Isometric Sketches

The most popular type of pictorial sketch is an ***isometric sketch***. Isometric drawings show the front, top, and sides of an object, just as the eye sees them. Isometric sketches can be drawn with great detail and accuracy. They can also be drawn in rough form. See Figure 13-25. Isometric sketches use the isometric axes described earlier in this chapter. The sketches are created using different shapes. Experienced designers are able to create isometric sketches quickly and easily.

Figure 13-25. Isometric sketches are angled to show three sides of the object. (Design Central, design firm; Artromick International, client)

Oblique Sketches

Oblique sketches are similar to isometric sketches. See Figure 13-26. In isometric sketches, the front and side are both at 30° angles. In oblique sketches, only the side view is at an angle. The angle is usually 45°. This corner can, however, be any angle.

These sketches show the front view in its true shape. Isometric sketches tend to distort shapes such as circles and arcs. In an isometric sketch, all circles are shown as ellipses. In oblique sketches, circles in the front view are actual circles. For this reason, circular objects are better drawn as oblique sketches. Also, products having one surface that is the most important are drawn in an oblique sketch. For example, the front views of stoves, radios, and televisions are what most people see. Therefore, oblique sketches show these designs better than isometric sketches do.

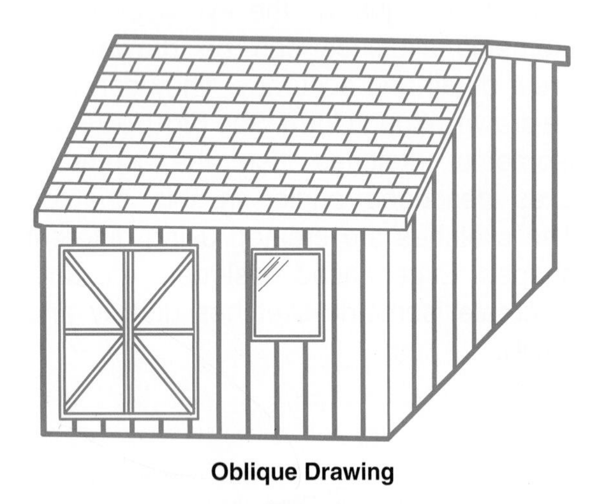

Oblique Drawing

Isometric Drawing

Figure 13-26. Oblique drawings show the front view directly and the side at a 45° angle. Isometric drawings are drawn with both the front and side views at 30° angles.

Oblique sketches are produced using the same steps as isometric sketches. See Figure 13-27. The designer begins by drawing the axes. In oblique drawings, the designer creates the oblique axes. The oblique axes are composed of one vertical line, one horizontal line, and one angled line. There are two types of oblique axes: right and left oblique. In the right oblique, the angled line is drawn back and to the right. In the left oblique, the line is drawn back and to the left.

The designer uses the oblique axes to block out the shapes. The same basic shapes are used in oblique sketches. The box, cone, pyramid, and sphere are all created the same way as in the isometric sketches. The cylinder is the only shape created differently. When a cylinder appears on the front of the object, it is drawn as a circle. If it appears on the side of the object, however, it is drawn as an ellipse.

Once the shapes are blocked out, the designer adds the details. The designer again adds the outline and external features. The last details to be added to refined sketches are shading and shadowing. These details help to make the sketch appear more real.

Perspective Sketches

The third type of pictorial sketch is a perspective sketch. Perspective sketches are often used in making refined sketches. They are used to show how objects look to the eye. Imagine yourself standing in the middle of a road. As you look down the road, the buildings get smaller. It looks

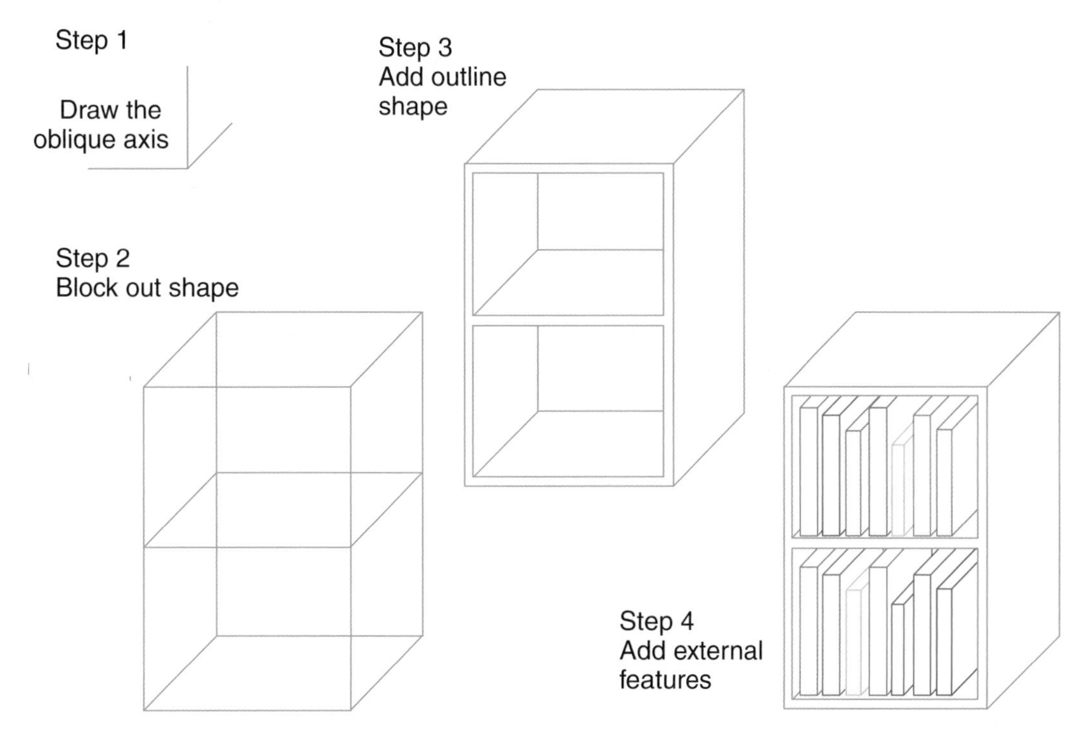

Figure 13-27. Oblique drawings are drawn with these four steps.

as though all the buildings and even the road narrow down to one point. This is how perspective drawings look.

When you look down the road, the road and the buildings seem to come from a single point. This point is called the *vanishing point*. See Figure 13-28. The line where the sky meets the road is the horizon line. Vanishing points are always located on the horizon line. The ground line is the line formed at the front of the sketch along the bottom of the objects.

There are several types of perspective sketches. See Figure 13-29. The most common are one-, two-, and three-point perspective sketches. The two-point perspective sketch is one that has many applications and is used regularly. See Figure 13-30. To create a two-point perspective sketch, the designer follows these steps:

1. Draw the horizon line and the ground line.
2. Add the vanishing points.
3. Sketch a vertical line from the ground line that is the height of the object.
4. Sketch perspective lines from the vanishing points to the top and bottom of the vertical line. These perspective lines form the left and right sides of the sketch.
5. Sketch vertical lines that are the width of the object.
6. Block out the shape.

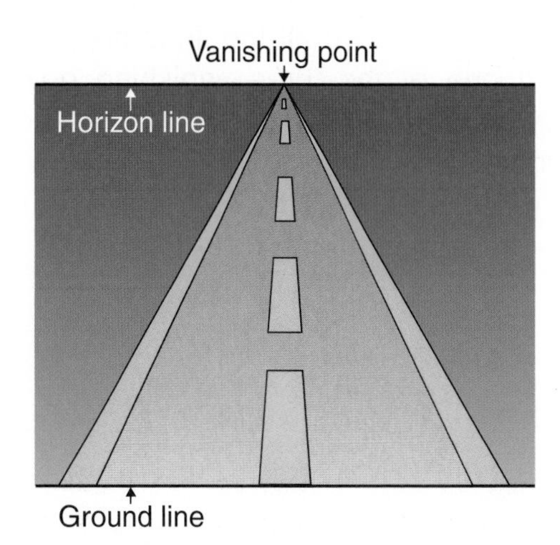

Figure 13-28. Perspective drawings have three basic elements.

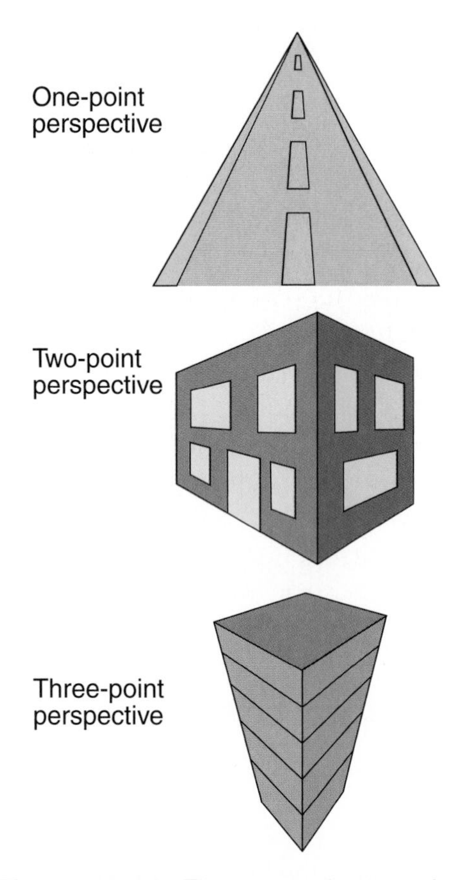

Figure 13-29. The three main types of perspective drawings.

7. Add details.
8. Shade the drawing.

One-point perspective sketches are created almost the same way as the two-point perspectives are created. The difference is that all lines are either vertical or angled to the vanishing point. In the three-point perspective, all the lines are drawn from one of the three vanishing points. There are no vertical lines in a three-point perspective.

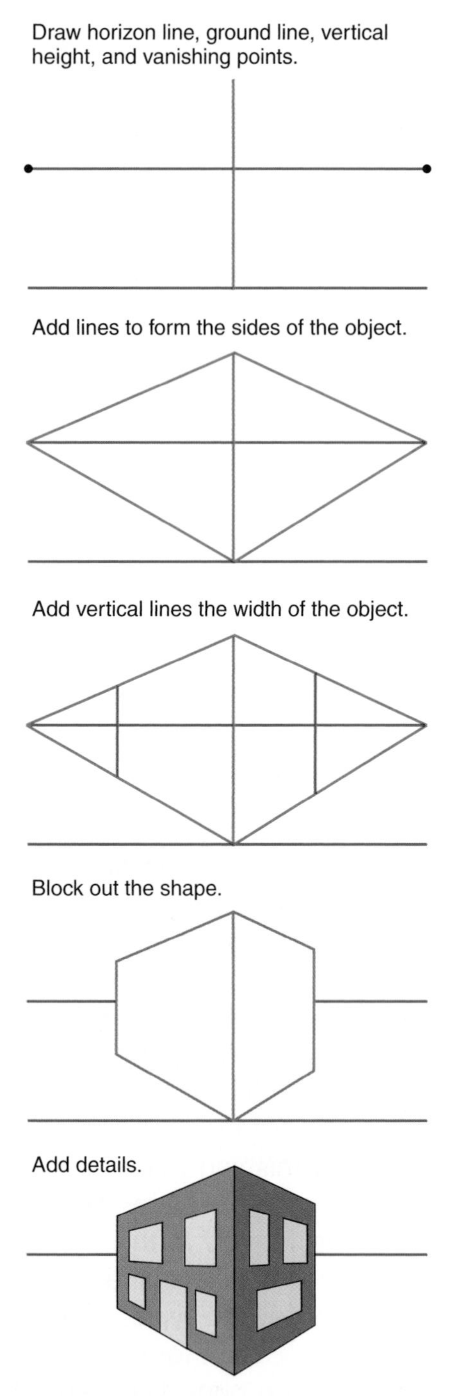

Draw horizon line, ground line, vertical height, and vanishing points.
Add lines to form the sides of the object.
Add vertical lines the width of the object.
Block out the shape.
Add details.

Figure 13-30. Perspective drawings are easy and fun to draw.

Summary

The process of creating solutions begins with ideation. Designers use brainstorming, graphic organizers, and questioning to develop as many ideas as possible. It is important that all ideas are taken seriously during ideation. No idea is wrong when the designers are developing ideas.

While the designers are creating ideas, they begin to sketch. Sketching is the process of putting ideas onto paper. The ideas begin as rough sketches. The rough sketches are simple, quick sketches that help to communicate the designer's ideas. Designers generate as many rough sketches as possible.

They review all their rough sketches. The ideas coming from the review are then drawn as refined sketches. These sketches are more detailed and better than the rough sketches. The refined sketches are the best ideas developed through ideation and rough sketches.

Isometric, oblique, and perspective sketches are all popular pictorial sketches. They are helpful and allow others to understand the thoughts of the designer. These pictorial drawings are all used in creating solutions. The refined sketches will be taken to the next step of the design process.

STEM Connections

Science

Study the difference in mass and volume of different shapes.

Mathematics

Measure several simple objects. Use isometric grid paper to draw an isometric sketch of the different objects.

Curricular Connections

Language Arts

Examine the methods of ideation used in creative writing.

Activities

1. Use a method of ideation to develop ideas about how the layout of your school could be made better.
2. Find and cut out images from magazines and newspapers showing isometric, oblique, and perspective sketches and drawings. Create a poster board showing the different types of pictorial sketches.
3. Use the design brief and the research gathered in Chapters 11 and 12 to develop ideas and create sketches of possible solutions.

Test Your Knowledge

Do not write in this book. Place your answers to this test on a separate sheet of paper.

1. The process of exploring ideas is called _____.
2. A diagram that helps to organize thoughts is known as a(n) _____.
3. A thumbnail sketch is larger and more detailed than a rough sketch. True or false?
4. A refined sketch uses the ideas from ideation, rough sketches, and the elements and principles of design to create a possible solution. True or false?
5. Give two examples of types of pictorial drawings.
6. Draw an isometric sketch of each of the following:
   A. A box.
   B. A cone.
7. The sketching technique that shows how light is reflected from an object is _____.
8. Draw an oblique sketch of a rectangular object (such as a textbook, bookcase, or rectangular building).
9. Paraphrase an explanation of a perspective sketch.

Reading Organizer

Draw a bubble diagram for each main idea in the chapter. Make each of the main ideas the central bubble, while using details in smaller bubbles to surround the main points. An example from this chapter is shown.

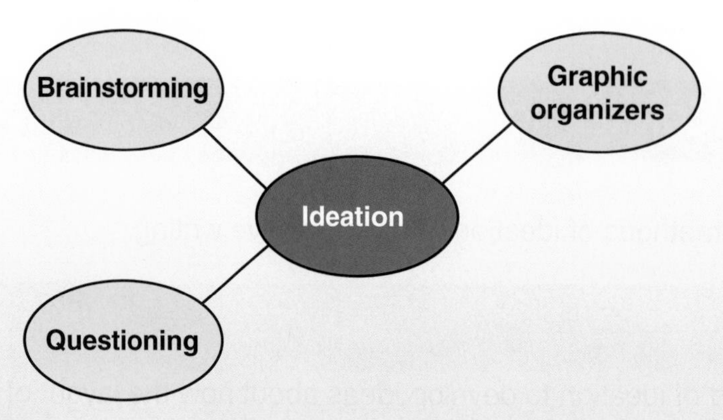

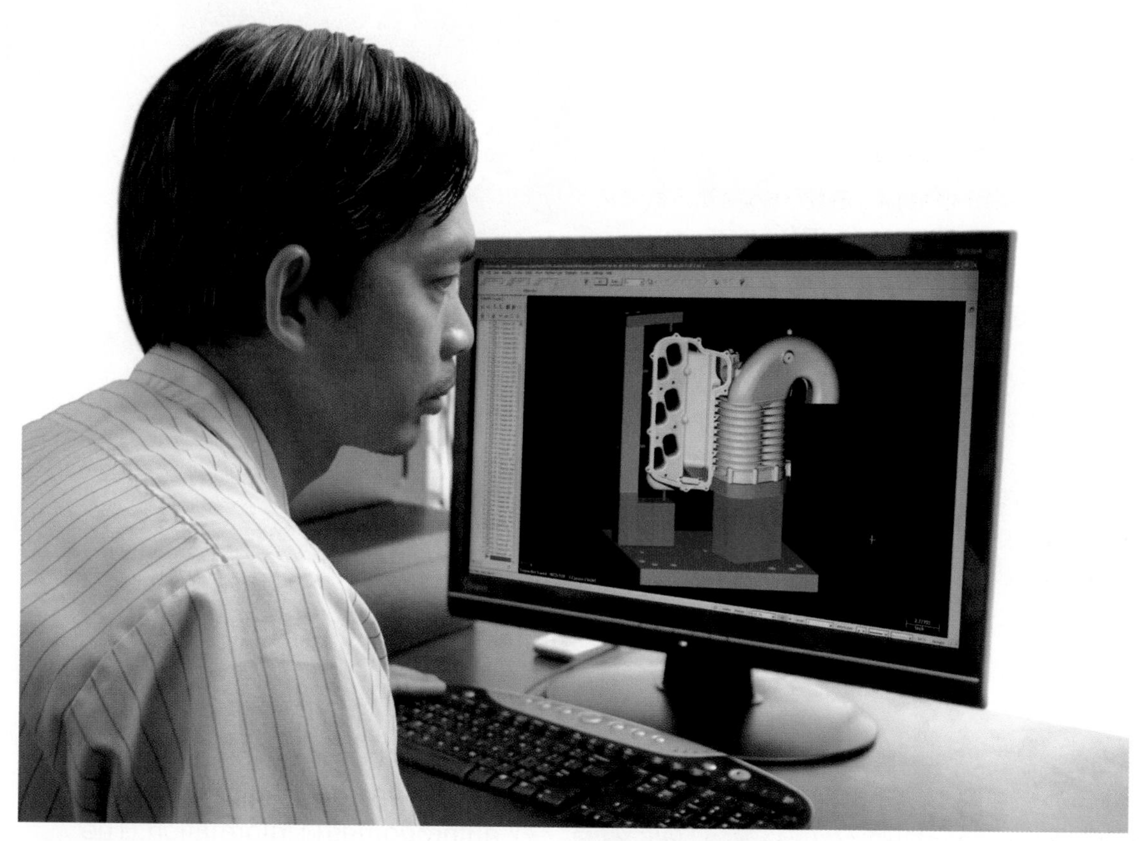

A great deal of designing today is done on computers using specialized software programs in addition to drafting and design knowledge.

CHAPTER 14
Selecting and Refining Solutions

DID YOU KNOW?

- The study of the measurement of the human body is called *anthropometry*. Human factors engineers use anthropometric information. The information is displayed in charts. The engineers use the information to design products that will fit the most people.
- Have you ever wondered how companies know their shirts fit "all" people? The size "one size fits all" was developed using anthropometric information. The shirts are designed to fit the middle 90% of people. There are 5 people in every 100 who are too small. Another 5 out of 100 people are too big. So the shirts do not really fit all people.

OBJECTIVES

The information given in this chapter will help you do the following:

- Compare internal and external selection.
- Explain the seven aspects of design.
- Summarize the purpose of a design sheet.

- Explain how evaluation grids are used.
- Give examples of the six elements of design.
- Summarize the five principles of design.
- Explain the purpose of a rendering.
- Summarize the three types of information provided in a detailed sketch.

Key Words

These words are used in this chapter. Do you know what they mean?

appearance
aspect of design
balance
client
color
contrast
design sheet
design style
dimension
ergonomics
evaluation grid
external selection
finances
human factors engineering
in-house design team
internal selection
line
outsourcing
production
proportion
rhythm
shape
texture
unity
value

Preparing to Read

As you read this chapter, outline the details of the process of sketching design solutions. Use the Reading Organizer at the end of the chapter to organize your thoughts.

The fourth step of the design process is selecting and refining solutions. Before designers can select solutions, they must have completed the previous steps. In those steps, the designers identified a problem. They also created a design brief and researched the problem. The designers then made conclusions. They wrote problem statements and created possible solutions. These designers developed and refined a number of ideas and sketches. See **Figure 14-1.**

Figure 14-1. In the previous step of the design process, designers created sketches. (Product Development Technologies, PDT)

The designers now have several sketches. The sketches are a collection of the most promising solutions. These are good solutions to the problem. The designers have gone through each of the first three steps to reach these solutions. Beginning designers often want to skip steps. Completing each task, however, helps to make sure the design is well planned. Expert designers take the first steps very seriously. They know that if they spend time creating many solutions at first, the final one will be very good. If designers simply create their first ideas, the solutions might not be as successful.

Selecting Solutions

Now it is time for the designers to choose the one solution they feel is the best. This step begins with designers selecting their best idea. This is the second time in the design process a decision must be made.

Types of Selection

The first decision was made when the designer picked the ideas to refine. This decision is made to select the best solution to the problem. There are two ways the best design can be selected. External or internal processes can be used to select the solution.

External Selection

The first way a solution can be selected is by someone outside of the design process. This is known as ***external selection***. Designers often work for ***clients***, or customers, who will buy the designs. In external selection, the clients select the best design. An industrial designer might design a lightbulb for an electric company. In this case, the electric company is the client. This company can select the design it feels best meets its needs.

Other times, management chooses the best design. Many large companies have in-house design teams. An ***in-house design team*** is a group of designers working for the company. These designers specialize in designing products for only their company. Many automobile manufacturers have in-house design teams. These design teams often have a review board that approves their designs.

Designers must present their ideas to the client or review board. See Figure 14-2. The presentation shows all the work they have done, the design brief, research, rough sketches, and refined sketches. The designers use the presentation to inform the clients or managers about the features of the refined designs. After the presentation, the clients or managers make the decision.

The clients or managers choose the most promising design. They choose designs based on what they feel is important for their company. The importance of several criteria can help determine which is the best solution. The amount of money a product might generate, known as *profit*, is one of these criteria. Another is how well a product satisfies the customers' needs. Lastly, clients and managers want to select products that will help their company become leaders in their field. The clients and managers make decisions as representatives of their companies.

Internal Selection

Internal selection is the second way to select the best solution. The design team conducts an internal selection. A designer, or a team of designers, makes the choice. The designers review all the designs, using a set of criteria. They evaluate the designs and select the solution they feel is the best.

This selection has a process that helps to ensure the selection is fair. All the solutions must be evaluated exactly

Figure 14-2. Designers must present their ideas to clients and management. (Design Central, design firm; Artromick International, client)

the same way. The first step of the evaluation process is to identify the ***aspects of design***.

Aspects of Design

There are many aspects of design. The aspects of design are the design criteria. They can be different for every design. A designer creating a soda can examines different aspects from what an engineer designing a space shuttle examines. Aerodynamics, for example, are not important for a soda can. See Figure 14-3. Some of the common aspects designers look at are appearance, function, human factors engineering, production, safety, environmental impacts, and finances. The designers choose the aspects they use to evaluate the design.

Appearance

Appearance is the look of the solution. A ***design style*** often determines the look. Design styles are periods of design in which many of the same appearance features are evident. They can be found in many societies. Queen Anne, Elizabethan, and Victorian are all English design styles. Louis XV and French Regency are design styles from France. American styles include Colonial and Federal styles. Products are often built to match a certain style.

The Art Deco style is an easy design style to recognize. This style was popular during the 1920s and 1930s. The Art Deco style uses shapes and straight lines to produce a certain look. Many kinds of products, from toasters to buildings, have been designed in the Art Deco style at some time. See Figure 14-4. Popular design styles today are the Modern and Contemporary styles. These styles use materials such as metal, wood, and glass to create a clean and up-to-date appearance.

Function

The function of the solution is how the solution works. Solutions can have mechanical, electrical, hydraulic, or pneumatic features. Mechanical solutions use moving parts to function. Gears, pulleys, springs, and other moving objects are

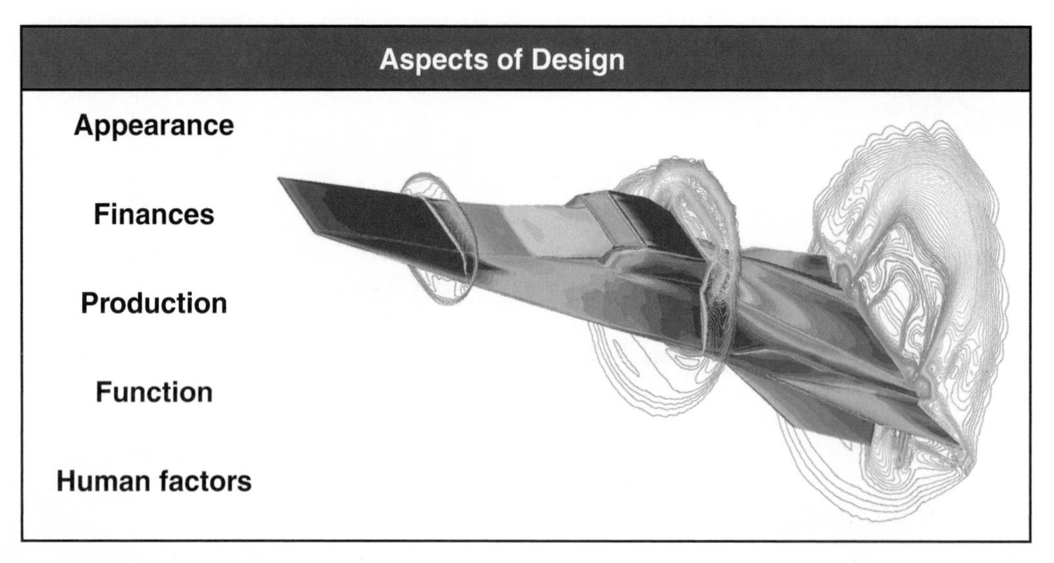

Figure 14-3. Aerodynamics is an important aspect of design for a space shuttle. (NASA)

Figure 14-4. The Art Deco style can be found in many places, including inside buildings. (Photo: Timothy Hursley, Courtesy: Frist Center for the Visual Arts)

used in mechanical solutions. Electrical solutions use the movement of electrons, called *electricity*. Designs using electricity often have batteries, transistors, lights, and speakers.

Hydraulic solutions and pneumatic solutions are very similar to each other. They both use the power of fluids to operate. Both types of solutions have pistons, valves, and gauges. The main difference is the type of fluid used. Pneumatic solutions function using compressed air. Hydraulic solutions use liquids, such as water and oil. See Figure 14-5.

Another aspect of function is how well a product works. It might be difficult at this point in the design process to know how well a product will work. Designers must

Career Highlight

Graphic Designers

The Job: Graphic designers combine the principles and elements of design with the design process to create visual products. The products can be magazine or newspaper layouts, advertisements, corporate logos, or product packaging. These designers are involved in all aspects of the design. Graphic designers create the initial sketches and follow the design through to the final printing. They typically utilize both hand- and computer-sketching tools and techniques.

(©iStockphoto.com/kryczka)

Working Conditions: These designers typically work in offices. The offices can be either corporate offices or private offices. Self-employed, or freelance, graphic designers often work from home. Deadlines often influence the work of a graphic designer. These designers might work longer hours when deadlines are approaching.

Education and Training: Most graphic designers have bachelor's degrees in graphic design or a related art field. Those in graphic design who have associate's degrees often work as assistants to graphic designers. Education or training in business is also important for self-employed or freelance graphic designers.

Career Cluster: Arts, Audio/Video Technology, & Communications
Career Pathway: Visual Arts

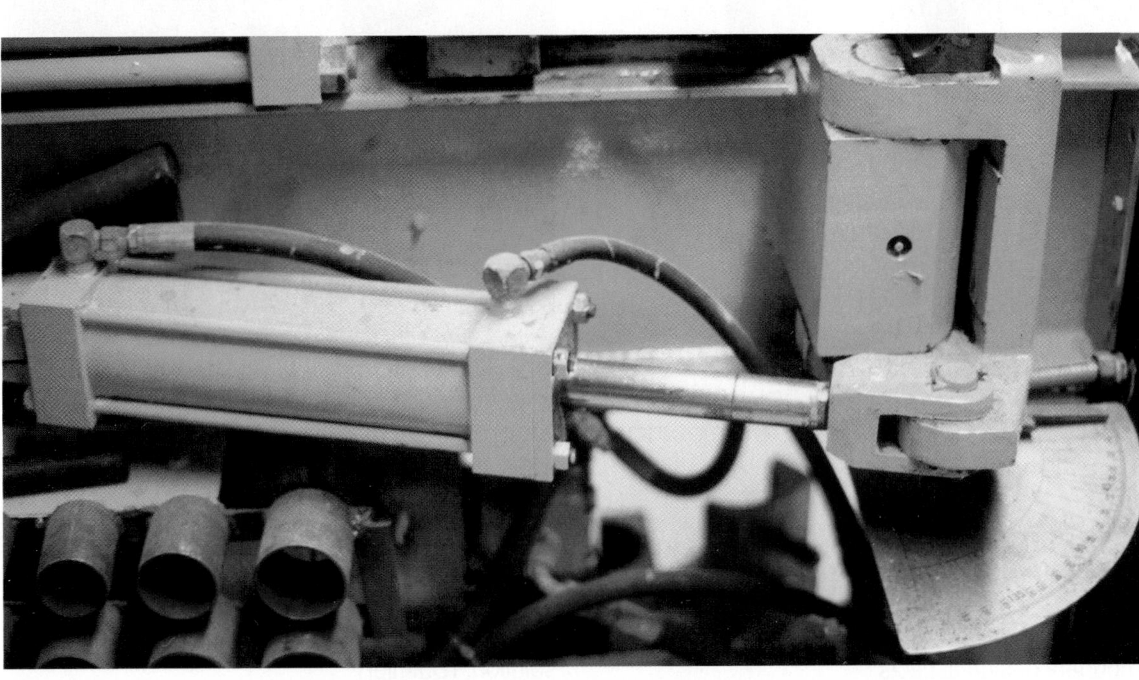

Figure 14-5. This device uses a hydraulic piston to bend exhaust pipes. (WB Automotive)

think about this, however, as they select a solution. They must use their knowledge of the product to consider which solution will work the best. Function might be the most important aspect of design because, if a product does not work, it will not be a useful solution.

Human Factors Engineering

Humans are the users of many design solutions. These products must be designed with the human users in mind. A designer would not create a 50-pound bowling ball. The ball would be too heavy to use. This is called ***human factors engineering***. Human factors engineering is used to make sure people can use the solutions.

This engineering is also known as ***ergonomics***. There are several types of ergonomic issues. The first issue is the size and shape of the solution. The solution must be the right size for people to use. For example, a telephone must fit in a person's hand. A telephone must also be able to be held up to the user's ear. So, the telephone cannot be too big or heavy.

Comfort is an ergonomic issue. Chairs and seats are designed to be comfortable. See Figure 14-6. Many car seats have different settings. These settings help make the seat comfortable. There are adjustments that change the angle and location of the seat and settings that support the user's back.

Body movements are also an ergonomic issue. Humans can move only in certain ways. This is why shirts do not have pockets on the back. People would not be able reach into their pockets very easily if the pockets were on the backs of their shirts. Buildings are also designed with this in mind. It would be very hard to get to the second floor if stair steps were four feet tall.

Figure 14-6. Comfort and ergonomics are important design aspects.

Figure 14-7. Designers must consider the tools and machines that will be used to produce the solution. (Daimler)

Production

Production involves the building of the solution. Designers must think about how the solution will be made. See Figure 14-7. There are several considerations going into the production of a design. First, the designer must select the material. Different designs call for different materials. Is stone the best material for a mechanical pencil? No, plastic would probably be a better choice.

Designers must think about tools and machines. They must determine what is needed to build the solution. Designers must also decide if their company and clients have the abilities to build the product. Some designers do not have the tools, machines, and abilities to build a certain solution. Design companies, or firms, often specialize only in the design of the products. They pay companies specializing in production to produce the products. This is called ***outsourcing***. Outsourcing means design companies send the designs outside of their companies to be built.

Safety

The safety of a product involves the hazards related to using a designed product. The safety of the user and others who come in contact with the product must be thought about. All hazards of some products cannot be removed. In these cases, additional safety devices or warnings can be added to the solution. For example, medications often have warnings about potential side effects. In these cases, the designers know that there can be potential safety issues and notify the consumers.

Environmental Impacts

The impact of the product on the environment is also an aspect of design. When selecting solutions, designers must consider whether or not the product will be harmful to the environment. This aspect should be considered for the entire life of the product. Designers should decide if a product will be harmful in its production, use, or disposal. Some innovative designs might be quite useful. They will be harmful, however, if they are disposed of in a landfill. Designers must think about this aspect of the design. See Figure 14-8.

Figure 14-8. Compact fluorescent lights are more energy efficient than standard lightbulbs. They, however, contain small amounts of mercury. The mercury can be harmful if placed in landfills.

Finances

Finances involve money, including the money both spent and collected. The money spent is known as *expenses*. The main expenses are found in design and production. Design costs are the money spent to pay designers and buy design materials. The production cost is the amount of money used to manufacture the solution. This includes paying workers, buying materials, and using tools and machines.

Figure 14-9. The goal of most designs is to make a profit when consumers buy the product.

Technology Explained

modular design: design that incorporates premade components.

Modular design is a design concept similar to the idea of interchangeable parts. Interchangeable parts are components (such as nuts, bolts, and screws) designed and machined to a standard size. For example, a 1/4″-20 bolt always meshes with a 1/4″-20 nut. The creation of interchangeable parts revolutionized manufacturing. Modular design is the same concept on a larger scale.

Instead of being concerned with individual components (such as bolts and nuts), modular design is concerned with larger assembled components. These components can be modules if they have a standard size and standard connections. For example, a designer creating a car interior does not need to focus on designing every aspect of the car stereo. Most car stereos are a standard size. They also have the same wires attached (such as power, antenna, and speaker wires). Designers can design a place for the stereo module to fit into. They do not, however, have to design the stereo. In this case, modular design is helpful for designers because they do not have to think about every aspect of the design.

Modular design can also be helpful to consumers. This is true especially if a product needs to be upgraded. A computer is an example of modular design. The designer who created the computer case designed spaces for a processor, memory, a hard drive, and a DVD drive. See Figure A. As technology advances, a consumer can remove the memory module and upgrade to faster memory. If a new device is developed to take the place of a DVD drive, a computer user can remove the old drive and replace it with a new drive.

Figure A. The computer technician installs a new hard drive into a slot that has been designed to fit a number of computer components. (©iStockphoto.com/Andrew_Howe)

The money collected is the income. Income is gathered through product sales. See Figure 14-9. The most important figure in a business is profit. Profit is income left after the expenses are paid. This income is the money the company makes.

Design Sheets

Designers use the aspects of design to select the best solution. The aspects must first be identified in each of the refined sketches. Designers can create a ***design sheet*** for each of the promising solutions. See Figure 14-10. The design sheet shows the refined sketch of the design. This sheet also lists the aspects of design. The designers then write descriptions or create sketches for each aspect. They create a design sheet for each of the possible solutions. The design sheet acts as a summary of the different solutions. This sheet helps the designers choose the best solution.

Evaluation Grids

Once the design sheets are complete, the designs are evaluated. The designers read the design sheets and begin selecting the best design. It is often hard, however, to pick the best one. Designers can use an ***evaluation grid*** to help them determine the best solution. See Figure 14-11.

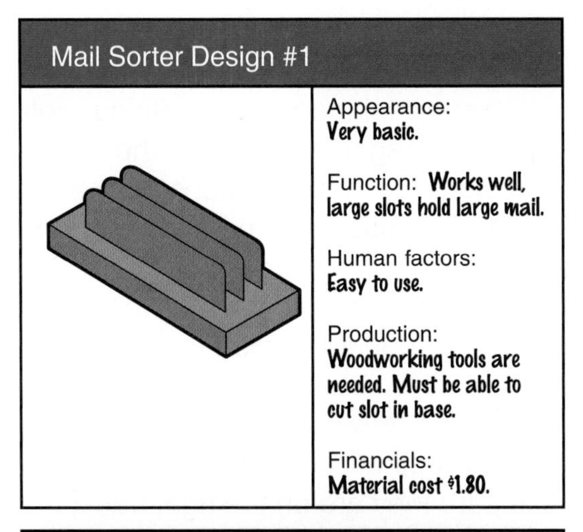

Mail Sorter Design #1
Appearance:
Very basic.
Function: Works well, large slots hold large mail.
Human factors:
Easy to use.
Production:
Woodworking tools are needed. Must be able to cut slot in base.
Financials:
Material cost $1.80.

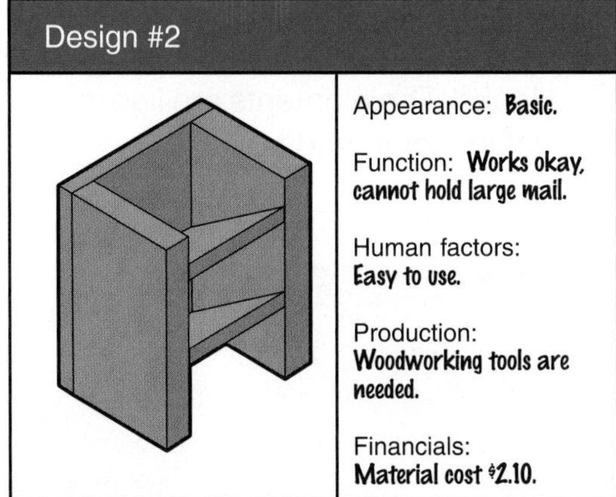

Design #2
Appearance: Basic.
Function: Works okay, cannot hold large mail.
Human factors:
Easy to use.
Production:
Woodworking tools are needed.
Financials:
Material cost $2.10.

Figure 14-10. Design sheets help designers review the aspects of each design.

✓ Above average
– Average
× Below average

	Design #1	Design #2	Design #3
Appearance	×	–	✓
Function	✓	×	✓
Human factors	–	×	×
Production	–	✓	–
Financials	✓	–	×
Design score	1	-1	0

Design score = Number of ✓ – Number of ×

Figure 14-11. Evaluation grids are used to help select the best solution to the design problem.

An evaluation grid is a chart used to record data. The possible solutions are listed across the top of the grid. The aspects of the design are listed along the side. This forms a grid. The grid is used to make an objective choice. An objective evaluation means the designs are evaluated fairly. To make sure the evaluation is objective, the designs must be compared to a standard. The standard can be an old product. For example, if you are creating a new ladder, the old ladder can be the standard. Each new ladder design is compared to the existing ladder design. Many designs, however, are brand new. For these design problems, the designers use the design brief as the standard. They compare the new design to the criteria of the design brief.

The designer fills in the grid by placing symbols in the boxes. The symbols they use can be numbers from 1 to 10 or pluses and minuses. Another set of symbols is the check mark, *X,* and dash. A check mark means the solution is better than the standard. An *X* means the design is below the standard. The design is worse than or does not meet the standard. A dash means the design is even with the standard.

When rating the solutions, the designers might ask themselves questions. The questions can be as follows: Does the design style match the type of product? Will the design function well? Does the design look nice? Will it require a lot of maintenance? (See Figure 14-12.) Will the design be comfortable? Does the design fit the user? Will it be too expensive to buy the right machines? Does the manufacturer have the right tools to build the solution? Will this solution cost too much to produce? Does the manufacturer have the knowledge to work with the nece materials? Will the company ma When the designers have a questions, they fill in the grid.

The designers then calcul For example, they might list the

Figure 14-12. Computers are designed and built to require very little maintenance. (Dell Inc.)

check marks, dashes, and *X*s. They then subtract the number of *X*s from the number of check marks. The result is the design score. The design with the highest score is the best design. If several designs are tied for first place, they can be combined into one design.

Refining Solutions

The solution chosen is never perfect. There are always changes that can be made. Many of the changes are found when the designs are evaluated. Designers might see features in other designs that they like. This is the time when the designers can make those improvements. One area often refined is appearance. Appearance is very important to the sale of the solution. Products that look nice sell better than products that do not. The appearance of a product can be refined using the elements and principles of design.

Elements of Design

When designers are refining solutions, they begin to examine the look of the design. The design must be functional. The solution must work properly and solve the problem. The design should also be pleasant to the senses. The appearance of a product is called the *aesthetics*. Aesthetically pleasing products are nice to see. See Figure 14-13. Designers try to create aesthetically pleasing designs. To make sure they are on the right track, they use the elements of design. The elements of design are tools controlling how an object looks. The elements are line, shape, form, value, color, and texture.

Figure 14-13. Designers use the elements of design to create aesthetically pleasing products. (Design Edge)

Line

Line is the most basic of all the elements. In mathematics, a line is the shortest distance between two points. In design, a line is described as a stretched dot. Lines can be used to create motion. A smooth line on a sports car is meant to show that the car is fast. See Figure 14-14. Lines can also be structural. On a building, lines are often beams and columns holding up the building. Lines can make objects seem bigger or smaller. Vertical stripes on clothing make people look slimmer.

Shape

Shape is the space made by enclosing a line. Shapes have lengths and widths. A shape is a flat, or two-dimensional, element. Shapes can be regular. Street signs are regular shapes that all have different meanings. Shapes can also be irregular. Leaves are irregular shapes.

Form

Forms, similar to shapes, have lengths and widths. Forms also have heights. A form is three-dimensional. Form is also referred to as *volume*. Volume is the space an object takes up. The form of some objects is standardized. Standardized means the object can take up only a certain amount of space. A car stereo has a standardized form. All car stereos must fit in the same space in all cars. Designers of car stereos cannot change the form of their product.

Value

Value is the degree of light and dark in the design. This element of design is the amount of white, black, and grays in between. Value can be seen on an achromatic scale. An achromatic scale shows white at one end, black at the other, and the shades of gray in between. See Figure 14-15. Value can make things stand out and is helpful when creating a pattern.

Color

Color is a property of light. The light reflected off an object determines the color. There are several mixtures of color. When a color is added to white, the color is called a *tint*. When black is added to a color, the result is a shade. Black and white themselves are

Figure 14-14. Sleek lines can give the appearance of a fast vehicle. (Ford Motor Company)

Figure 14-15. The achromatic scale shows the shades of gray between white and black.

not colors. There are different types of color:

- **Primary colors.** These are the three colors that can be mixed to produce any other color. Red, yellow, and blue are the primary colors.
- **Secondary colors.** These are the colors produced when two primary colors are mixed. The secondary colors are orange, green, and violet.
- **Intermediate colors.** These are mixtures of a primary color and a secondary color. Yellow-green, blue-green, blue-violet, red-violet, red-orange, and yellow-orange are the six intermediate colors.

A color wheel is used to show the primary, secondary, and intermediate colors. See Figure 14-16. Colors are used for many different reasons. They can affect our moods. Warm colors—such as red, yellow, and orange—can make us excited and happy. Cool colors—such as blue, violet, and green—can make us feel comfortable and relaxed. Can you guess why most sports cars are warm colors?

Texture

Texture is the last element of design. This element is the surface on an object. See Figure 14-17. Texture can be both the appearance and the feel of the product. For example, the grit of a piece of sandpaper can be both seen and felt. A certain texture is often needed on products. Plates and bowls should have a smooth texture so food can be wiped off easily. Hand tools should be ribbed or have an easy-to-grip texture.

Principles of Design

The design elements are easier to use when the designer also examines the principles of design. The principles of design are the ways the elements are arranged. They control how the elements work together. The principles are balance, contrast, unity, rhythm, and proportion.

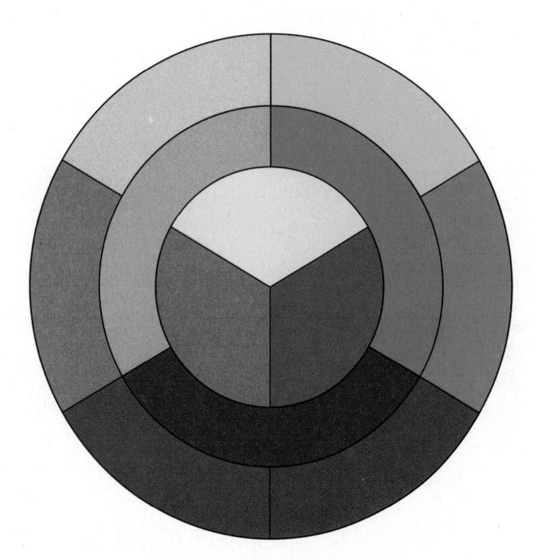

Figure 14-16. The color wheel is based on the three primary colors. All other colors are mixtures of these three.

Figure 14-17. All objects have textures. Some objects can be recognized by seeing or touching the textures.

Balance

Balance is the weight of the elements on each side of the design. There are two types of balance: symmetrical and asymmetrical. Symmetrical, or formal, balance is produced when both sides of the design are the same. See Figure 14-18. If you draw an imaginary line through the exact center of a symmetrical object, the two sides are mirror images. Asymmetrical, or informal, balance is created when the two sides are not the exact same, yet the two sides have an equal amount of elements.

Contrast

Contrast is used to add variety to a design. This principle helps bring attention to a certain part of the design. Value and color are often used in contrast. The use of black and white to make something stand out is the use of contrast. The contrast of black text on white paper makes this book easy to read. Two unlike colors can also be used to bring attention to something.

Unity

Unity is the opposite of contrast. This principle helps the design blend together. The use of either warm colors or cool colors creates unity. Cool colors on the color wheel range from yellow-green to violet. The warm colors are those on the other half of the wheel. In graphic design, using the same typeface throughout a design can create unity.

Rhythm

Rhythm is the effect of motion. This effect is created with repetition of elements. The repetition leads the eye through the design. This movement gives the feeling of motion.

Proportion

Proportion relates to the size and shape of the object. See Figure 14-19. If an object is too tall or too wide, it is said to be out of proportion. The other principles of balance, contrast, and unity help to proportion objects.

Figure 14-18. This lamp is symmetrically balanced. Each side is a mirror image of the other. The red line shows the center of the object. (Fine Art Lamps)

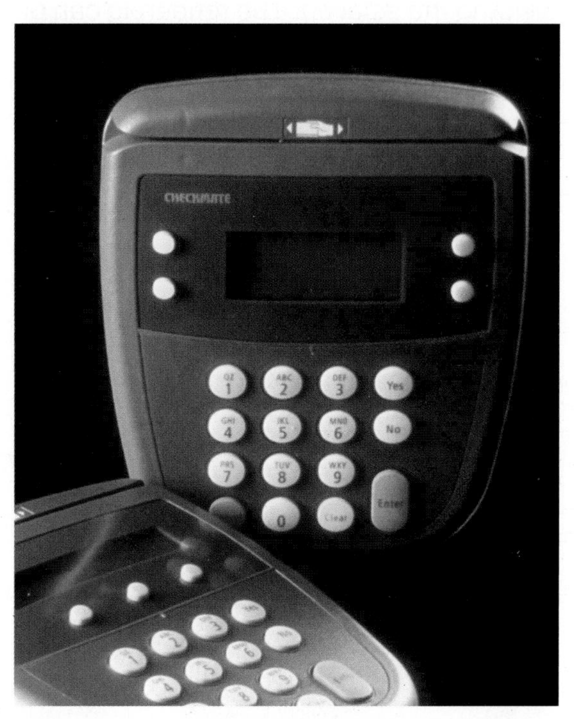

Figure 14-19. The size of the buttons and the shape of the card reader are in proportion. (Design Central, design firm; Ingenico, client)

The elements and principles of design can be found in a number of areas of design. Think of a light show at a pop-music concert. The lighting might have been designed with a number of the elements (such as line, value, and color), as well as principles such as balance, contrast, and rhythm. The same can be said for products from graphic designers, architects, landscape designers, and automobile designers.

Renderings

The overall appearance of the solution is best shown in a rendering. A rendering is a type of sketch. This sketch is the best representation, on paper, of how the final solution will look. See Figure 14-20. Renderings are often shaded with color. A series of renderings showing the solution in different lights is often prepared. This shows how the object will look when it is being used.

These sketches are created by first drawing the solution. The rendering can be drawn as an isometric sketch, an oblique sketch, or a perspective sketch. The color is added in a similar way as in shading. A light source is selected. The lightest area is closest to the light. As the object gets darker, the color used becomes darker. In a rendering, the area closest to the light might be a soft yellow color. The color furthest from the light can then be a dark red. Renderings are very helpful for visualizing the actual object. They are used in many types of design, especially packaging, fashion, and architecture.

Detailed Sketches

Renderings are an excellent way to show the appearance of a design. They do not, however, show the size of the solution. It is important for the designers and clients to know how big the object will be. Detailed sketches are used to show the size of the solution. These sketches use text to describe the object. The text is called ***dimensions***. See Figure 14-21. The dimensions show three different types of information:

- **Size information.** Dimensions are used to describe the size of the object. Size dimensions show the lengths, widths, and depths of rectangular objects. They also list the diameters of circles and the radii of arcs.

Figure 14-20. Renderings are created in color to show how the final design will look. (Keith Nelson)

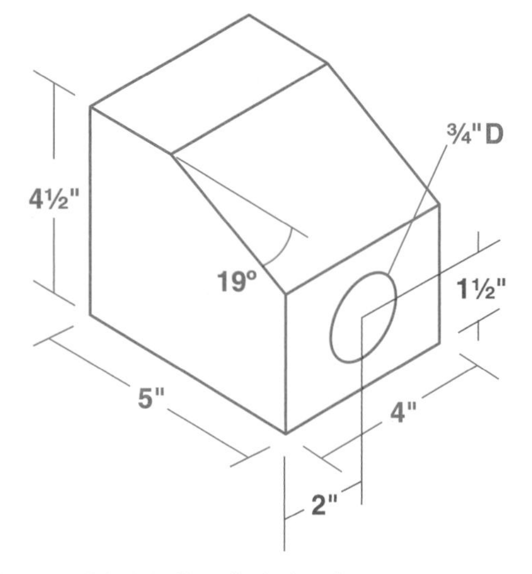

Figure 14-21. Detailed sketches use dimensions to show the size (in red), location (in blue), and shape (in green) of the object.

- **Location information.** This describes the placement of objects. Location dimensions describe where the parts of the object are located. In a floor plan, the location dimensions show the placement of doors and windows.
- **Shape information.** Some dimensions show the shape of the object. Angle measurements show shape information.

The detailed sketches are often pictorial sketches. The designers first create the sketch. They then add the dimensions. The information in detailed sketches is very helpful. The designers will use the detailed sketches in several of the next steps. These sketches will be used to construct models and prototypes. They will also be used to create final drawings of the solution.

Summary

The detailed sketches and renderings are the final outputs of this step. This step involves both selecting and refining a solution. Either the designer or a person outside of the design process makes the selection. Designers use a process to select the best solution. They review the solutions from the last step. These designers identify the aspects of design. They then use a grid to determine the best solution.

Designers improve the appearance of the solution. They use the elements and principles of design. The elements of design are the tools designers use. They help to create products with good aesthetics. When the appearance is finished, the designers create two types of sketches. The first is a rendering showing the solution in full color. The second is a detailed sketch. This sketch has dimensions showing the size, location, and shape of the solution.

The designers will take the work they have done into the next step. The renderings and detailed sketches will help to create models. If the designers feel they do not have a good enough design, they might choose to go back to an earlier step. If they change things at an early stage of design, it might help create a better solution later in the process.

STEM Connections

Mathematics

Calculate human factors. Measure the height and arm span of all the students in the class. Calculate the averages. Create graphs showing the data.

Science

Find examples of the elements and principles of design in nature. Create a display of your findings.

Activities

1. Research design styles in different areas of design (such as fashion, architecture, and furniture). Create a poster board showing the different styles.
2. Gather various examples of a product (for example, flashlights, mechanical pencils, or drinking cups). Develop an evaluation grid for the product. Use the grid to determine the best design.
3. Create a rendering and a detailed sketch of a product idea.

Test Your Knowledge

Do not write in this book. Place your answers to this test on a separate sheet of paper.

1. The designers on the design team do external selection. True or false?
2. Name the seven aspects of design.
3. Another name for human factors engineering is _____.
4. What is the purpose of a design sheet?
5. A(n) _____ is used to record data when selecting the best solution.
6. The element of design made by enclosing a line is called a(n) _____.
7. Value is the amount of white, black, and gray in a design. True or false?
8. Match the principles of design below with the right descriptions:
   The use of two opposite colors to bring attention to something. _____
   The effect of motion. _____
   The weight of elements on each side of a design. _____
   The blending of a design. _____
   A. Balance.
   B. Contrast.
   C. Unity.
   D. Rhythm.
9. A rendering is a black-and-white drawing. True or false?
10. List the three types of information dimensions can show.

Reading Organizer

On a separate sheet of paper, create a detailed outline based on what you have read about the process of selecting design solutions.

Example:

I. Selecting solutions
    A. Types of selection
        1. External

CHAPTER 15
Modeling Solutions

Did You Know?

- The Styrofoam® product is a type of polystyrene foam. Dow Chemicals produces it. This foam was first created in the 1940s and has never been used to create cups or plates. The objects typically called *Styrofoam cups* are another type and brand of foam.
- Many animated movies use computer modeling. The models are created as wireframe models. The frames are colored. It can take hours or even days for a computer to turn the wireframe models into surface models.
- Small figurines and dolls have been found dating back to 13,000 BC. The dolls even had clothing.

Objectives

The information given in this chapter will help you do the following:

- Explain the concept of a model.
- Summarize the three purposes of a model.
- Compare graphic, mathematical, computer, and physical models.

- Give examples of the uses of different modeling materials.
- Recall the types of tools used in modeling.
- Explain the process of model making.

Key Words

These words are used in this chapter. Do you know what they mean?

acetate
acrylic
bristol board
chart
chipboard
clay
computer model
contact paper
corrugated cardboard
double-sided tape
foam
foam-core board
formula
graphic model
hardboard
hot-melt glue
illustration board
machineable wax
matboard
mathematical model
mounting tape
paraffin wax
physical model
plastic cement
plywood
polystyrene
polyurethane
schematic
solid model
spray adhesive
surface model
table
veneer
wax
white glue
wireframe model
wood glue

Preparing to Read

Look carefully for the main ideas as you read this chapter. Look for the details that support each of the main ideas. Use the Reading Organizer at the end of the chapter to organize the main and supporting points.

Have you ever built anything with building blocks? See Figure 15-1. Have you ever played with an action figure or a doll, seen your family tree, or looked at a weather map? If you have done any of these things, you have used a model. Models are an important part of the design process. The fifth step of the process is designing and building a model. The designers use the sketches and renderings from the previous step to construct a model of the solution.

Model:
A representation of...
• A Product
• A System
• A Process
• An Idea

Figure 15-1. People of all ages can use building blocks to build models. (LEGO and the LEGO brick configuration are trademarks of the LEGO Group ©2002 The LEGO Group. The LEGO® trademark and products are used with permission. The LEGO Group does not sponsor or endorse the publication.)

What Is a Model?

A model is a representation of a product, a system, a process, or an idea. When you think of a model, you probably think of a small object resembling a bigger, real-life object. A model car or dollhouse might come to mind. These are both good examples of models. They are not, however, the only examples of models. There are many types of models used in many different professions. Engineers use models to show products and systems, such as roads and bridges. Architects use models to show buildings, monuments, and subdivisions they have designed. See Figure 15-2. Amusement parks use models of people made of wax to entertain us. Weather forecasters use models to show weather patterns. Scientists use models to show small objects, such as atoms and DNA.

Why Are Models Used?

There are three main reasons models are used. Models are used to communicate, experiment, and organize. Some models might even be used for more than one of these reasons.

Communication

Models can be used to communicate. They allow people to see solutions in easy-to-understand ways. Just as sketches are used to visualize what the designer is thinking, models help people to see the designer's thoughts. Models used to communicate are often used when the idea or product is hard to understand. Engineers might create a model showing how a product is built. See Figure 15-3. The model communicates better than the raw data they collect. These models might also be used to get feedback from other people. A designer might make a model of a product to make sure the client approves of the design. Any changes the client wants are less expensive to make at this stage of the design process than after the final product is produced.

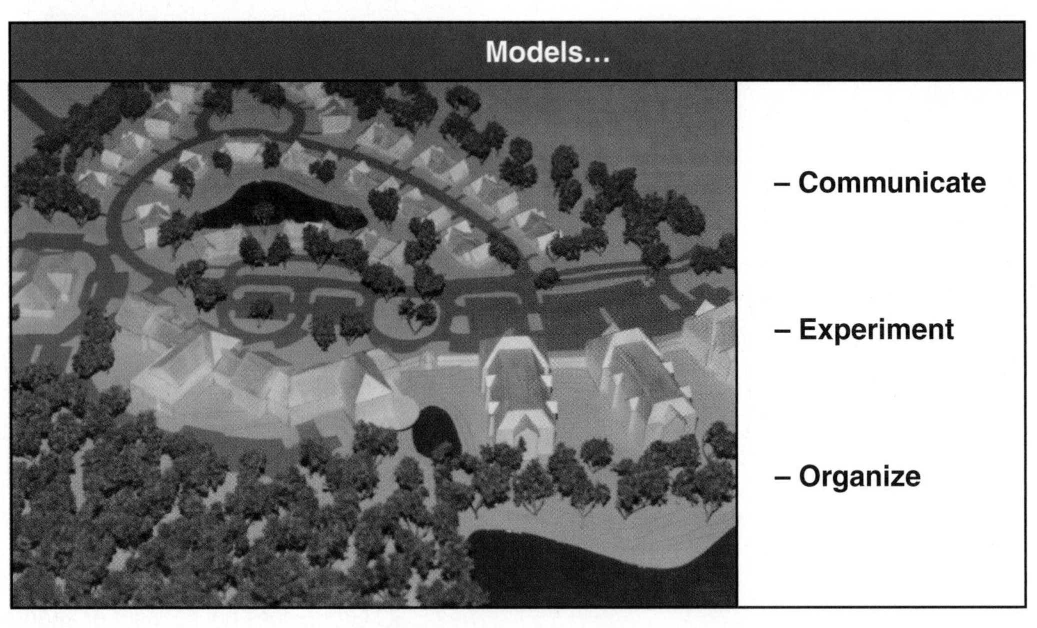

Figure 15-2. City planners and architects use models to show how their designs will actually look. (Custom Modeling & Graphics Studio, Inc.)

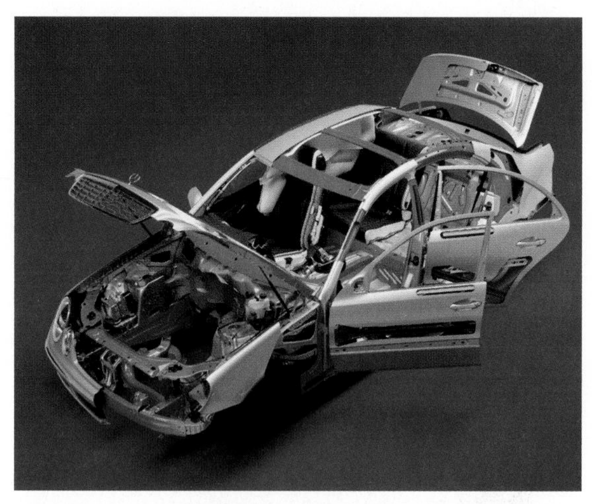

Figure 15-3. Models help communicate the design and construction of solutions. (Daimler)

Experimentation

Models demonstrating how the product will work can be built. These models are used to experiment with the product's operation and use. Models can show that the designers' ideas actually work. They are good ways for designers to test their ideas. For example, a model of a new hinge can be created to determine if the hinge will work correctly. Models can be created to test the size and shape of solutions. They might also be created to experiment with different methods of production. See **Figure 15-4.**

Figure 15-4. These models were produced at different temperatures, as an experiment to determine the best way to manufacture them. (Photo: Rick English, Courtesy: IDEO)

Organization

The last use of models is organization. Organizational models are often used when designers are creating new systems. If manufacturing engineers are designing a new manufacturing line, they might create a model. Their model will contain all the tools, machines, and conveyors used in the manufacturing line. They can use the model to move the objects around to design the plant layout. Companies also use models to show the organization of their business structure. A corporate organizational chart is this type of model.

Types of Models

All models are representations of other objects. Some models are visual representations. Models can also be mathematical or even three-dimensional representations. All models fit into one of the four major types: graphic, mathematical, computer, or physical models.

Graphic Models

Graphic models are visual representations on paper. They are used to organize and communicate information. One example of a graphic model, discussed in Chapter 12, is a graph.

Graphs

Graphs are visual representations of data. Data is gathered in several steps of the design process. Designers collect data during the research step and gather a great amount of data in the testing step. The data can be modeled in a graph.

Charts

A ***chart*** is the second type of graphic model. Charts can be used to model how a set of information is arranged or to highlight a series of events. They can also show sets of data. An organizational chart is used to show how objects are arranged. A family tree is an example of an organizational chart. This chart shows how a family is organized back through history. Organizational charts are created similarly to trees. See Figure 15-5. The chart begins at the top of the tree. There are branches and leaves below. Organizational charts branch out (get larger) as they go down.

A process chart, or flowchart, is used to show a series of events. Many restaurants have process charts informing the cooks of the order in which to prepare the food. Manufacturing plants use flowcharts to plan the operations in a production line. Flowcharts are discussed further in Chapter 23. Charts showing sets of data

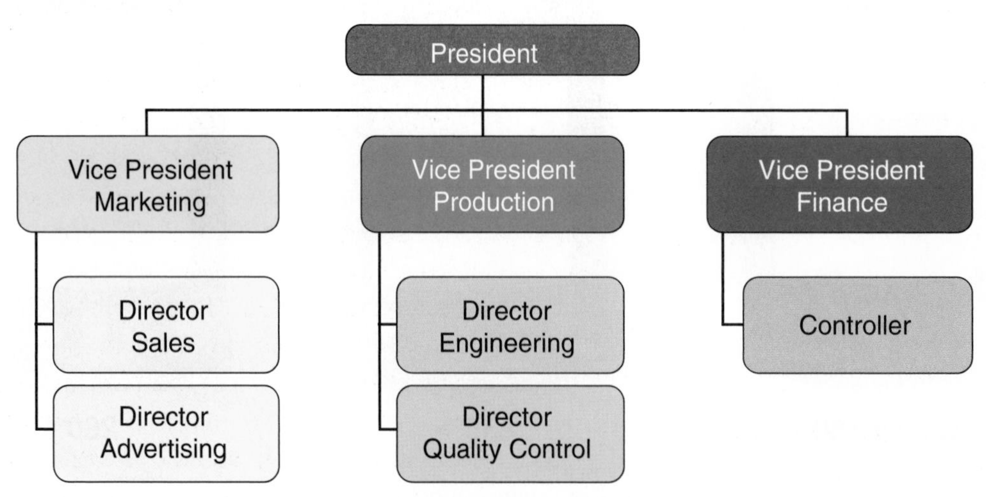

Figure 15-5. Organizational charts are graphic models showing structure and organization.

are known as ***tables***. A table is a graphic model made of rows and columns. Tables are often used to compare sets of data. The evaluation grid discussed in Chapter 14 is an example of a table.

Schematics

The last type of graphic model is a schematic. A ***schematic*** is used to show a process, similar to a flowchart. Schematics, however, use pictures and symbols to show the process, rather than words. They are also known as *diagrams*. One common use of a schematic is to show electrical circuits. See Figure 15-6. In an electrical schematic, the electrical components are described with symbols. The electrical wires are shown as solid lines. Electrical schematics show how the electricity flows through a circuit.

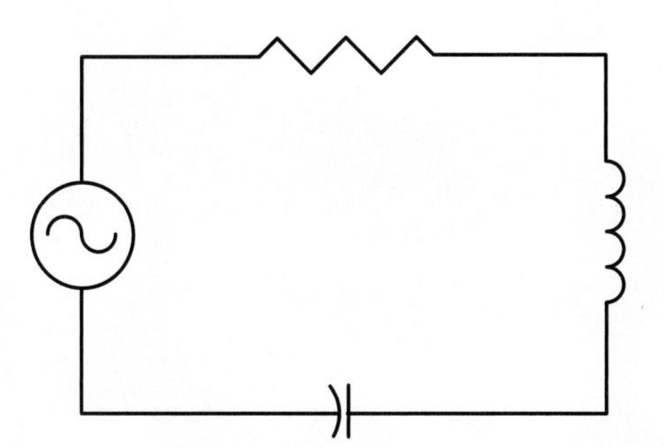

Figure 15-6. Electrical schematics are diagrams showing the layout of electrical systems. (Forrest M. Mimms III)

Mathematical Models

Graphic models are not the only type of model that uses symbols. ***Mathematical models*** also use symbols, along with numbers and letters. See Figure 15-7. Mathematical models use the symbols to represent values in a ***formula***. Formulas are the most common type of mathematical model. These models help us to understand complex math, science, and technology concepts. The concept of momentum can be written as a mathematical model: $p = mv$.

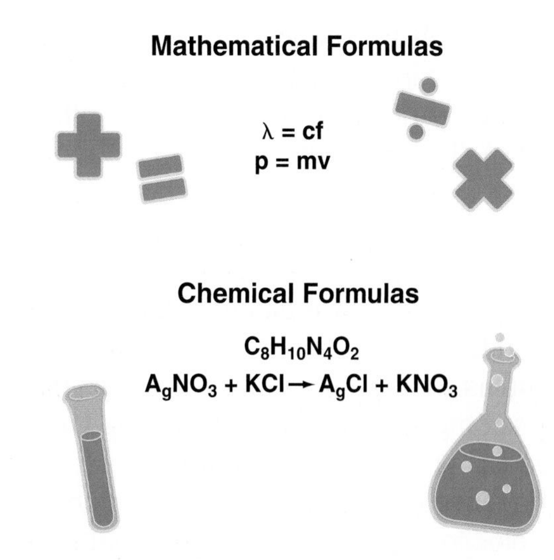

Figure 15-7. Formulas and equations are common mathematical models.

In this formula, *p* stands for momentum. Mass is represented by *m*. Velocity is equal to *v*. The formula for momentum uses letters of the alphabet to represent the concepts. Other formulas use symbols to stand for items in the formula. The formula for the length of a wave (such as a light or sound wave) uses the Greek letter lambda, λ. The formula is $\lambda = cf$. In the formula, the wavelength is λ. The speed of light is *c*. The frequency is *f*. The formulas for momentum and wavelength are used to make it easier to calculate the concepts.

Concepts are not the only use for mathematical models. Mathematical models can represent materials and compounds. These models are often called *chemical formulas*. Chemical formulas use the symbols from the periodic table to represent compounds. For example, the chemical formula for caffeine is $C_8H_{10}N_4O_2$.

Two or more chemical formulas can be used together to show a chemical reaction. Chemical reactions can be found in all aspects of life. One chemical reaction in the area of technology occurs in photography. The first step in making film is creating a light-sensitive emulsion. This can be done by combining a soluble silver salt, such as silver nitrate ($AgNO_3$), with a soluble halide, such as potassium chloride (KCl). This results in a light-sensitive halide, silver chloride (AgCl), and potassium nitrate (KNO_3). The chemical reaction is modeled as $AgNO_3 + KCl \rightarrow AgCl + KNO_3$. The silver chloride is light sensitive. Now the film can be used to take photographs.

Chemical and mathematical formulas are often used along with many other formulas. They can be used to develop large complex mathematical models. These models can be used to determine the aerodynamics of a vehicle, the predictions of an election, and even weather patterns. Weather forecasters rely on mathematical models to determine the weather forecast. NASA researchers use mathematical models to plan the routes rockets and space shuttles take. See **Figure 15-8.**

Computer Models

Computer models are models created using a computer to represent a design, a situation, or an event. See **Figure 15-9.** The main type of computer model is the three-dimensional computer model. This model is created with the use of a CAD system. CAD helps designers create models that look how the real design will

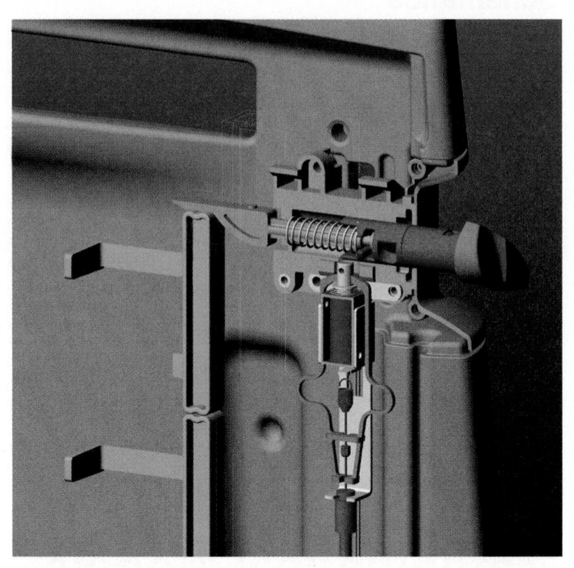

Figure 15-9. This computer model shows the function of a push-button latch without building an actual model. (Design Central, design firm; Artromick International, client)

Figure 15-8. Mathematical models are used in space travel. (NASA)

Career Highlight

Model Makers

The Job: Model makers create three-dimensional replicas of products for companies or designers. Such models can be electronics, cars, mobile devices, and even real estate. Model makers work with the designers of the products to create an exact version of how the designers want the product to look before it actually goes into production.

Working Conditions: These workers can be independent contractors who own their own model-making business. Model makers can also work in design divisions of large companies. They work with small pieces, must understand scales, and must have technical skill and patience.

Education and Training: Model makers can obtain an associate's degree in model making or industrial-design technology. They can also gain a bachelor's degree in a field such as engineering, art, or design. Many other model makers receive on-the-job training.

Career Cluster: Manufacturing

Career Pathway: Production

(©iStockphoto.com/dt03mbb)

look. There are three different types of three-dimensional computer models. Each one has advantages and disadvantages.

Wireframe Models

The first type is a wireframe model. ***Wireframe models*** use lines to represent the edges of objects. See Figure 15-10. Every edge of the design is shown as a line. There is no covering over the model, so the model is see-through. For this reason, wireframe models are not good for showing the appearance of complex objects.

Figure 15-10. Wireframe models show objects as if the objects were made of thin wire.

Surface Models

The next type of three-dimensional computer model is a surface model. ***Surface models*** are very similar to wireframe models, except they have coverings. The covering is known as the *surface*. The colors of the surfaces can be used to show

how a certain design will look. Surface models can be shown in different light and from different angles. They are good for showing the appearance of designs. These models cannot, however, be used to show interior details. The insides of surface models are hollow.

Solid Models

The most complete three-dimensional computer model is a ***solid model***. Solid models have edges and surfaces similar to wireframe and surface models. They also have interior features. Solid models are used to represent the insides and outsides of designs. See Figure 15-11. Solid models are created using objects such as boxes, cones, and spheres. They can be given the properties of different materials and tested to determine the properties of the design. Other computer models are used to predict situations and events, such as the weather and economic activity.

Physical Models

The last type of model is the physical model. ***Physical models*** are actual three-dimensional replicas of the design. See Figure 15-12. Physical models are built so the designers and clients are able to see how the design will look. They are often the most complex type of model and take the most time to complete.

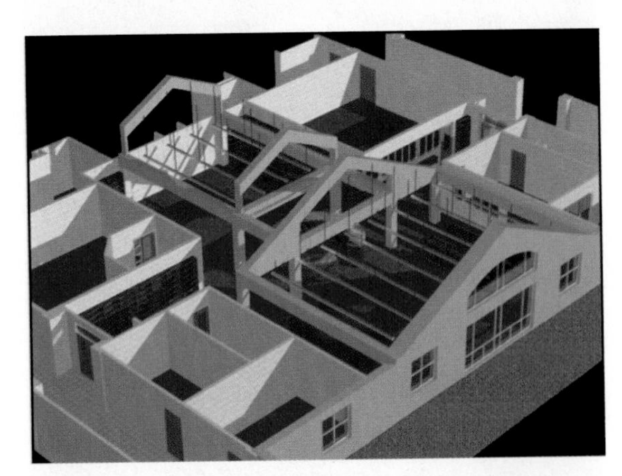

Figure 15-11. Both surface and solid models can be designed with the same colors and textures that the actual solution will contain. (Custom Modeling & Graphics Studio, Inc.)

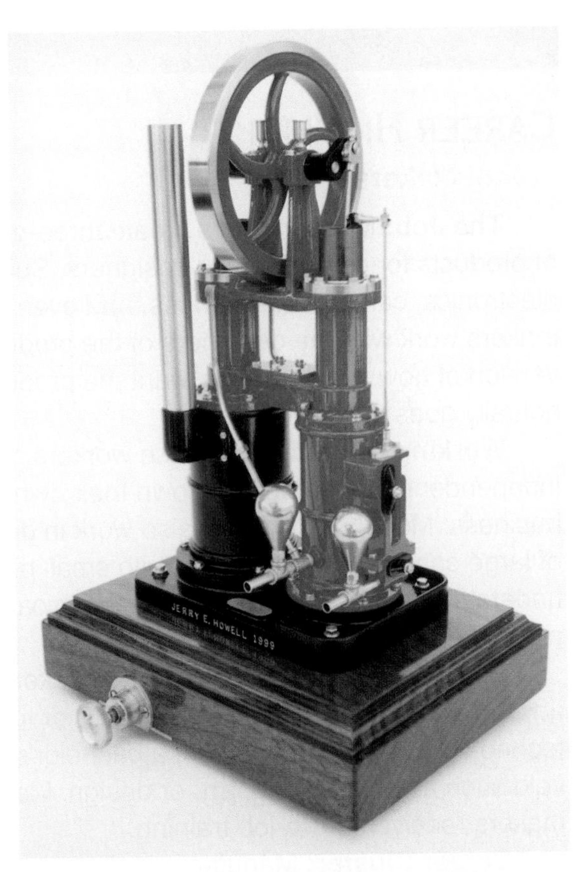

Figure 15-12. Physical models are real, three-dimensional replicas of the design. (Jerry E. Howell)

Different types of designers have many different names for physical models. Architects use the terms *study models* and *presentation models*. Engineers might call physical models *experimental models* or *operational models*. There are also conceptual, plot, and scale models. The two terms best describing all models are *mock-up models* and *prototype models*.

Mock-Ups

A mock-up is a model concerned with the look of the solution. A mock-up model is an appearance model. This model looks similar to the final solution. Mock-ups are helpful in assessing the elements and principles of design found in Chapter 14. Elements such as color, form, and shape can be examined with a mock-up. See Figure 15-13. If designers

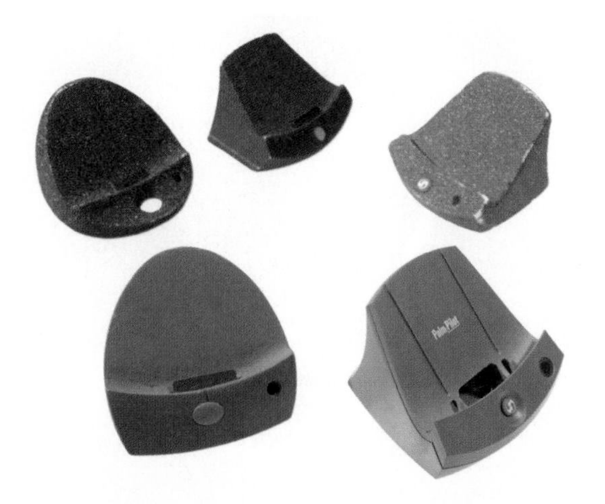

Figure 15-13. Several mock-ups are developed to explore different solutions to the same problem. (Photo: Rick English, Courtesy: IDEO)

Figure 15-14. This architectural mock-up was developed at a 1/4″-to-1′ scale. (©iStockphoto.com/Franck-Boston)

are creating a new design for a computer monitor, they will create a mock-up of the design. The mock-up helps the designers and design team determine whether or not the design has the right look.

In many cases, a mock-up is created at the same size as the final product. This, however, is not always possible. If a product is too big to be created full-size, the model is built on a smaller scale. Architects often build models on a scale of 1/4″ to 1′. In a model built to this scale, everything measuring 1/4″ will be 1′ when the real building is built.

Architects often create mock-ups of buildings to see how the buildings will look. See Figure 15-14. These architects are able to move the mock-ups to different lights to see how the buildings will look at different times of the day. If the mock-up includes other buildings around their design, they can determine if their building will affect any others. They might find that, when a light hits their building, it reflects back on the building next to it. The designers use this information to make changes to their design.

Mock-ups only look similar to the solution. They do not function as the final product will. Mock-ups might not even be built out of the same material as the final product. It would be a waste to build a mock-up of an automobile out of steel. Clay, cardboard, and foam are often used for mock-ups because they are inexpensive and can be formed easily.

Prototypes

Physical models that do function similar to the final product are called *prototypes*. A prototype is the most realistic representation of a design. Prototypes are working models. They are usually full-size and made from the same materials as the final product. Prototypes look, feel, and work very similar to how the final solution will.

These models are used to test the function, operation, and safety of the design. See Figure 15-15. To test aerodynamics, prototypes of automobiles and aircraft are placed in wind tunnels. Prototypes of crops are planted in test fields. To make sure flashlights fit comfortably in a user's hand, prototypes are used. Some prototypes are also called *betas*. Beta versions of software are given to computer users to test their operations.

Prototypes can be either comprehensive prototypes or focus prototypes. Comprehensive prototypes examine the

Figure 15-15. This engine prototype is being tested to make sure it functions correctly. (Boeing)

entire design. A comprehensive prototype of a computer has all the components in the design (such as the processor, memory, disk drives, and video card). A focus prototype examines only one part of the design. See **Figure 15-16.** A common focus prototype on an automobile is the door latch. A focus prototype of the door

Figure 15-16. This prototype allows designers to test grill doors, as well as examine the overall appearance. (Design Central)

latch might be created to test how the latch functions. If the designers are concerned about only one object, such as the door latch, it is unnecessary to build a prototype of the entire design.

It is common for designers to discover unexpected results when they are building prototypes. They might find things they did not think about in the sketches and drawings. The design might be too big or small. The process might not work the way they expected it to work. The designers might even find that it is impossible to build the design they wanted to or that they need to create different tools or machines to build their design. This is one of the reasons modeling is so important.

Figure 15-17. This model was built for a science fiction movie. The materials used to build it were chosen for their appearance on-screen, rather than for their function. (AMS Phoenix/Jim Dore)

Modeling Materials

Physical models can be made from many different types of materials. The materials used depend on a number of factors. Some of these factors include the use and shape of the model, the budget, the time the designer has to work, and the abilities of the designer.

The use of the model helps determine the best materials to use. The model's use includes who will be viewing the model. See Figure 15-17. Different materials can be used if the client will view the model from a distance, rather than handle it. The shape of the object also determines the materials to use. A model of a skyscraper is best constructed with a flat material, such as foam board. It is much more difficult to use a material such as clay, which is not flat, because the designer has to sculpt the sides until they are flat.

The budget of the modeling process also controls the materials used. The best material might be too expensive to use in the modeling stage. For example, jewelers often make models of rings out of inexpensive materials. Making models out of gold and platinum costs too much.

The time the designer has to make the model is another important issue in determining the materials to be used. Some materials take more time to work with than others. For example, some types of ***foam*** must be coated with a sealer before they can be painted. See Figure 15-18. Other materials can be painted on without the sealer. If there is a shortage of time, certain materials can be used to save time.

The ability of the designer is also an important consideration. Some designers are better at working with certain materials. They might not be able to create high-quality models with other materials. These designers might not even have the right tools and machines to work with some materials.

All these factors go into the decision of the right materials to use. There are hundreds of materials designers use to create models. In this chapter, some of the most common materials are discussed. Modeling materials can be broken down into several categories: sheet materials,

Figure 15-18. This foam needs to be sealed before it is painted.

found objects, sculpting and molding materials, and adhesives.

Sheet Materials

Sheet materials are commonly found as two-dimensional objects. These materials are used because of their flat surfaces. They can be found in most building-material and hardware stores in standard sizes. Sheet materials each have their own characteristics and are very useful for different types of models. They can be divided into four categories: wood, paper, plastic, and metal.

Wood Sheet Materials

Wood sheet materials are used for larger models or for the bases of smaller models. See Figure 15-19. There are three main types of wood sheets: plywood, hardboard, and veneer. Plywood and hardboard are very similar. They can be purchased from a hardware store in 4′ × 8′ sheets. ***Hardboard*** comes in thicknesses of 1/8″ and 1/4″. ***Plywood*** ranges in thickness from 5/16″ to 3/4″. Hardboard and plywood are both attached to other materials using glue, nails, or screws. Both plywood and hardboard can be cut using any type of hand- or power saw. The difference between the two is in how they are manufactured.

Plywood is a material made by gluing thin sheets of wood together. See Figure 15-20. Plywood is not very attractive and is often used as a base for models. This material is normally used in places that will not be seen. Plywood does, however,

Figure 15-19. There are many types of wood sheet materials. (Weyerhaeuser Co.)

Figure 15-20. Plywood is made of thin layers of wood. (Weyerhaeuser Co.)

come in different grades. This material can be purchased to look nice if it is needed.

Hardboard is made by compressing and rolling wood fibers together. This type of sheet does not have a grain as pieces of wood do. Hardboard has a solid dark-brown color. This type of sheet is very hard and stiff and can be used in a number of ways in models.

Veneer is a thin sheet of wood. This sheet is less than 1/8″ thick and comes in many different widths. Veneer is available in various types of wood, including oak, cherry, and maple. This sheet is used to give the appearance of a high-quality wood. When cost is an issue, inexpensive plywood or hardboard can be used for the model. The veneer is then glued on top. This sheet gives the model a very nice appearance.

Softwoods such as pine, balsa, and basswood are often used in modeling. Each of these woods is easy to cut and shape. Balsa and basswood are the easiest with which to work. They are found at hobby stores in many small sizes. These woods are very lightweight and have a nice appearance. Pine is heavier and used for larger models. This wood can be found in different grades. The best grade is known as *appearance grade* and can be used in places where it will be seen. Often, several types of wood sheet materials are used together in the same model. See Figure 15-21.

Figure 15-21. This model was built from various wood sheet materials. (Custom Modeling & Graphics Studio, Inc.)

TECHNOLOGY EXPLAINED

rapid prototyping: computer-controlled technologies used to create models and parts.

Rapid-prototyping systems are different from other model-making processes, such as computer numerical control (CNC) machines, including lathes and mills. CNC machines start with a blank of wax, wood, metal, or plastic and shape it into the model or part. They use a subtractive approach. Rapid-prototyping devices use an additive approach. Rapid-prototyping machines build the model in layers, "from the ground up."

There are many different types of rapid-prototyping technologies. The most common are stereolithography (SLA) and Fused Deposition Modeling (FDM). SLA and FDM use different processes to create the part or model. The first stage in both processes, however, is the same. The computer-software program prepares a solid model for rapid prototyping by slicing it into a number of very thin layers. In SLA systems, the computer information is sent to a laser and scanning system. The laser beam is directed into a vat of liquid resin, one layer at a time. The liquid coming in contact with the laser becomes solid. A plate holding the hardened resin is lowered deeper into the vat so the laser can build more layers on top of the model. Once the laser is finished, the model is raised from the vat and cured.

FDM is similar to SLA in that it builds the model layer by layer. The process, however, is different. FDM devices have a nozzle that streams (or extrudes) melted plastic. See **Figure 15-A.** The nozzle is very accurate and is controlled by a computer system. The plastic is extruded first to form the bottom layer of the model. The nozzle is then raised. The second layer is added to the model. This process continues until the model is complete. See **Figure 15-B.**

Both the SLA and FDM methods of rapid prototyping can be used to produce very detailed parts and models. Each process has its own advantages and disadvantages. SLA is messier and generally more costly than FDM. This process, however, produces parts that have excellent finishes. FDM machines can be used in office settings and are not as costly. Rapid prototyping, as a method of building models, is becoming more and more popular and will continue to do so.

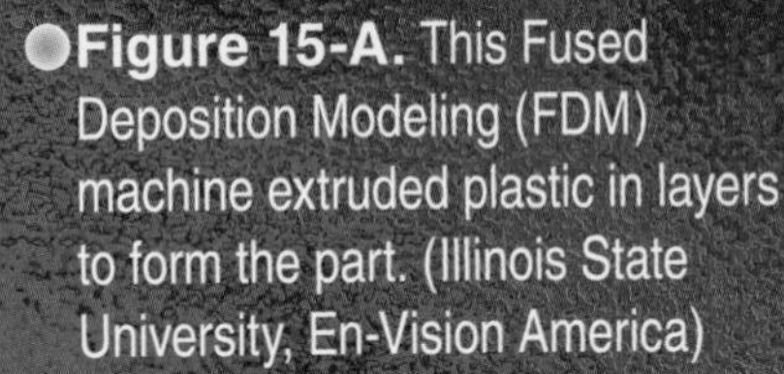

Figure 15-A. This Fused Deposition Modeling (FDM) machine extruded plastic in layers to form the part. (Illinois State University, En-Vision America)

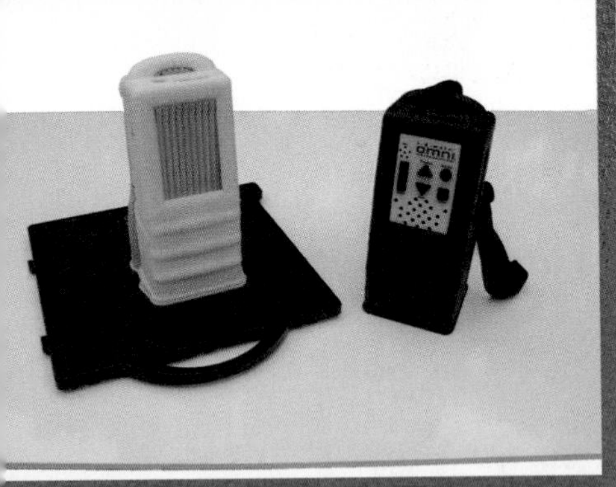

Figure 15-B. The model of the talking bar code scanner was created using FDM. This model is shown with the final product. (Illinois State University, En-Vision America)

Paper Sheet Materials

Paper sheet materials are used for many applications in models. They are often used for walls in architectural models. These materials can also be used as flat panels on industrial models. All the paper sheet materials can be cut with hobby or utility knives. See Figure 15-22. They can be attached to each other with tape or almost any type of glue. There are several different types of paper sheet materials. See Figure 15-23. They are different in size, thickness, color, and flexibility. The paper materials used in modeling are thicker than normal notebook paper. Two of the thinnest paper products used are bristol board and chipboard.

SAFETY

When using a utility knife, always adhere to the following precautions:

- Use sharp blades.
- Wear eye protection.
- Keep your hands out of the path of the blade.
- Cut in a direction away from your body.
- Cap the blade of the knife when it is not in use.

Figure 15-22. Paper sheet materials can easily be cut with a utility knife.

Figure 15-23. Paper sheet materials come in various sizes and thicknesses. Each one has different applications.

Bristol board is similar to common poster board. This board is usually found in sheets 1/16″ thick. Bristol board can easily be curved. This product is not, however, very rigid. Bristol board is smooth and shiny on both sides. This board can be found in different colors. Bristol board is most common, however, in bright white. ***Chipboard*** is a little different from bristol board. This product has a gray color and is produced from two sheets of gray paper pulp. Chipboard can be found in thicknesses ranging from 1/32″ to 1/8″. This product is more rigid than bristol board. Chipboard is not, however, as attractive as bristol board. This product is very common. Chipboard is often used for backers in calendars, notepads, and packages of paper.

The next two types of paperboard materials are illustration board and matboard. Both of these materials are around 1/16″ thick. They are both sturdy and come in various colors. ***Illustration board*** is colored the same throughout the entire board. This board comes in handy because designers do not need to color the edges they cut. This material can easily be drawn on to add features to the model. ***Matboard***, or museum board, comes in many colors and in various textures. The color on matboard is placed only on the top of the board. The inside of matboard is always white.

The two thickest paper sheets are corrugated cardboard and foam-core board. Each of these materials has two outer sheets attached to an inner core of material. ***Corrugated cardboard*** has outer sheets of heavy paper. The inner core is made of heavy paper bent into ridges. ***Foam-core board*** has an inner layer of foam covered on each side with paper. Both materials are very sturdy and come in sizes from 1/8″ to 1/2″. They can be bent if they are scored with a knife first. Corrugated board and chipboard are less expensive than other sheet materials and are used for models in early stages. See **Figure 15-24.** Foam-core board is used for final mock-ups and can be used for some prototypes.

Plastic Sheet Materials

Plastic materials are often used to represent glass. They can also be used to make a model see-through. ***Acrylic*** is a plastic material that comes in sheets and can be purchased at either a hardware store or a hobby store. The most common acrylic sheets are clear. They can, however, be found in various colors. Acrylic is usually too thick—1/8″ to 1/2″—to cut with a knife. Band saws work well for cutting acrylic. See **Figure 15-25.**

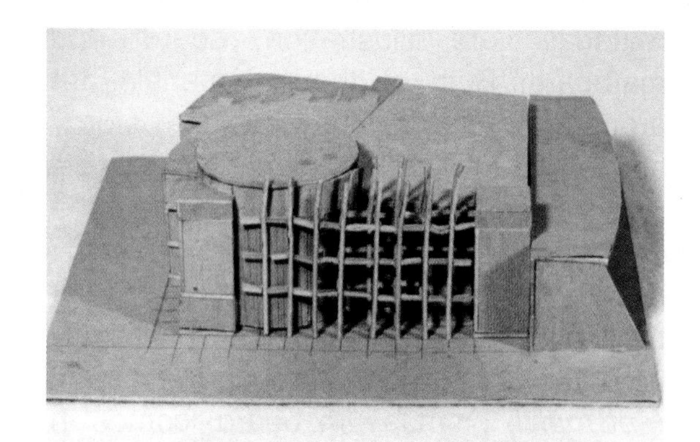

Figure 15-24. Early development models are created with chipboard and corrugated board. (Keith Nelson)

Figure 15-25. Acrylic is best cut with a band saw.

Safety

When you are working with a band saw, always adhere to the following precautions:

- Before cutting your material, check to make sure the blade is tight and tracked between both guides. Set the upper guide to 1/4″ above the material. Allow the blade to get up to speed before you begin cutting the material. Plan the cuts you are going to make. If you are cutting a curve, plan to make relief cuts.
- While you are cutting your material, wear eye protection. Keep your hands at least 2″ from the blade. Place your hands out of the path of the blade. Keep the material flat on the table. Push straight through the blade. Never back the material out while the machine is running.
- After cutting your material, turn the machine off immediately. Use a scrap piece of material to clear away any scraps from near the blade.

Thinner plastics, such as ***acetate***, can be used for windows in architectural models. Acetate can also be used for clear screens in product models. Overhead transparency sheets are a common example of sheets of acetate. Acetate and acrylic can both be secured with plastic cement or tape. ***Contact paper*** is also a plastic sheet material. This paper is a vinyl material that is sticky on one side. The nonsticky side is printed with a color, pattern, or texture. Contact paper is applied to another material to give the model a certain look. For example, contact paper with a brick pattern might be used to simulate a brick walkway.

Metal Sheet Materials

The last type of sheet material is metal. Metal sheets come in a variety of types. Common types of metal used in models are aluminum, copper, and tin. These metals can be found in sheets and are easy to use. They are strong. These metals are, however, also bendable. Thin metals used in models cannot be glued together. Instead, they are soldered together using solder and a soldering iron. See Figure 15-26.

Safety

When you are soldering, always adhere to the following precautions:

- Work in a well-ventilated area.
- Wear eye protection.
- Avoid touching the material shaft or tip of the soldering iron.
- Never leave the soldering iron unattended.
- Use pliers or a vise to hold the workpiece and avoid burns.
- Wash your hands after you are finished.

Found Objects

Objects found or manufactured for other purposes can also be very useful in models. Model makers keep a variety of objects around their shops that can be used to represent things on their models. For example, dowel rods can represent columns on an architectural model. Cork or rubber stoppers can be used as round buttons on a model of a telephone. Other useful objects are film canisters, hobby sticks, matches, straws, milk cartons, wire of different sizes, and fishing line. These

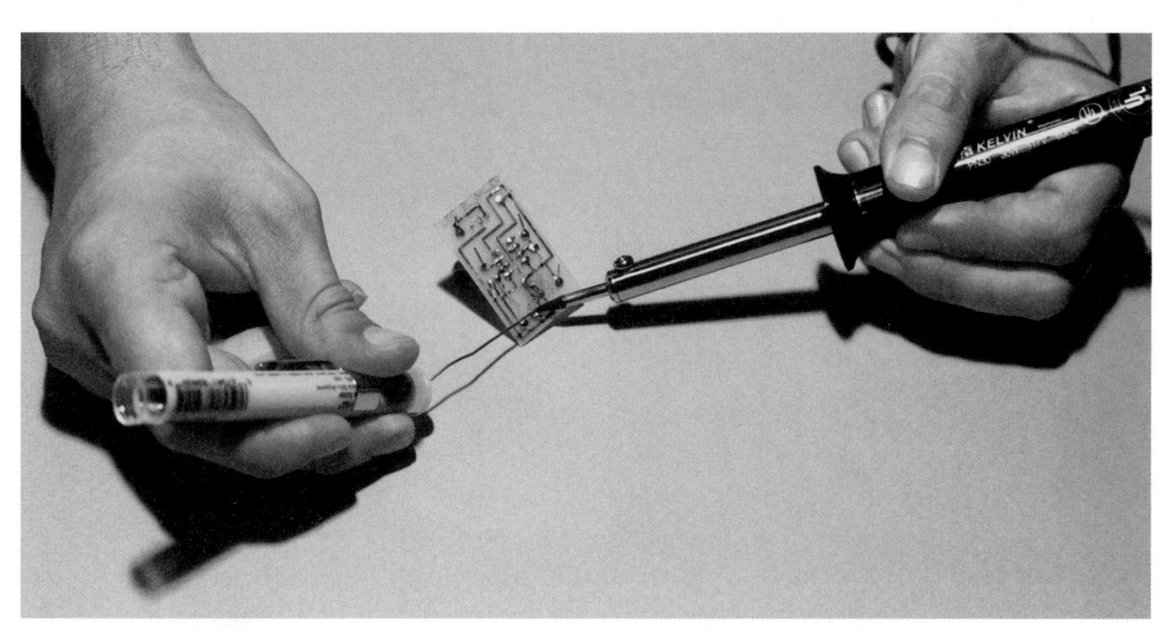

Figure 15-26. Soldering is a technique often used in electronics.

objects help the designers add details to their models.

Sculpting and Molding Materials

Sculpting and molding materials are much different from sheet materials. They are used to create shapes that are not flat. Sculpting and molding materials are used when the solution has curves or rounded areas. They are easier to work into irregular shapes. You might have worked with these types of materials as a child, if you made sand castles with mud or played with Play-Doh® modeling compound.

Clay

One of the most common types of sculpting materials is ***clay***. See Figure 15-27. Clay is a natural material found in the world around us. Today, there are several types of clay used to create models. There are polymer clay, water clay, and oil-based clay. Polymer clay is human made. Different clays have different uses. Some types are workable at room temperature. Others must be heated. Some clay dries out when it is left in the air. Other clay dries when it is placed in an oven. There is even some clay that never hardens.

Clay can be used for two different purposes. This material can be used to create a complete model. This is called *modeling clay*. Models built with modeling clay are solid clay. These models are usually small. Larger models use styling clay. Styling clay is placed on top of a blank. The blank (wood or foam) is used as filler. Large models would be too heavy if they were solid clay. The styling clay gives the appearance of a solid model. This clay, however, is used only on the top layer of the model.

Foam

Foam is another type of sculpting material. See Figure 15-28. Foam is used in many types of models for several reasons. This material is easy to shape. Foam is sturdy enough to be shipped and stored. This material is also available in lumberyards and craft stores. There are two main types of foam: polystyrene and polyurethane.

Polystyrene is the less expensive of the two foams. This foam can be purchased at hardware stores because it is used as a construction material. Polystyrene is used in construction as sheet insulation. Common names for this foam are *bead board* and *blue board* (because of its light blue color). Polystyrene can be cut and shaped with everything from saws to sandpaper. Even a hot wire cutter can be used to cut through

Figure 15-27. This designer is sculpting a clay model. (Daimler)

Figure 15-28. Foam is an excellent material for these early study models. (Photo: Rick English, Courtesy: IDEO)

this foam by melting the foam. Hot glue and many types of model glue, however, also melt the foam. Wood glue and white glue can be used to attach polystyrene pieces together. The pieces can then be brush painted with latex paint. Spray paint melts the foam.

Polyurethane, commonly known as *urethane foam*, is often used for large or thick models because it breaks easily if it is thinly shaped. See Figure 15-29. Polyurethane is expensive and not sold in common places, such as hardware stores. Urethane foam does have its advantages. This foam is more rigid. Urethane foam can be cut more easily with larger tools, such as table and band saws. Any type of glue or paint can be used to finish the urethane. Urethane can also be purchased in different densities. This is important for prototypes that must function a certain way.

Wax

Another widely used sculpting and molding material is ***wax***. Wax is a material that can come from several different sources. Waxes can come from animals, vegetables, or minerals. Mineral waxes, such as paraffin wax, are used most often in modeling. ***Paraffin wax***, made with petroleum, can be carved for small models. This wax can also be heated and cast into a mold. Molds are often made of plaster or silicone rubber. The mold is a reverse of the product. Molds allow material, such as wax, clay, or plaster of paris, to be poured into them and dried. When the material is dry, the mold is opened. The material is in the shape of the product.

Another type of wax, called ***machineable wax***, is often used in modeling. See Figure 15-30. Machineable wax is a very dense material cut and shaped in milling machines and lathes. This wax is used for small models or parts of prototypes. Machineable wax is fairly inexpensive and works well in modeling.

Adhesives

There are many different types of adhesives designers use when they are making

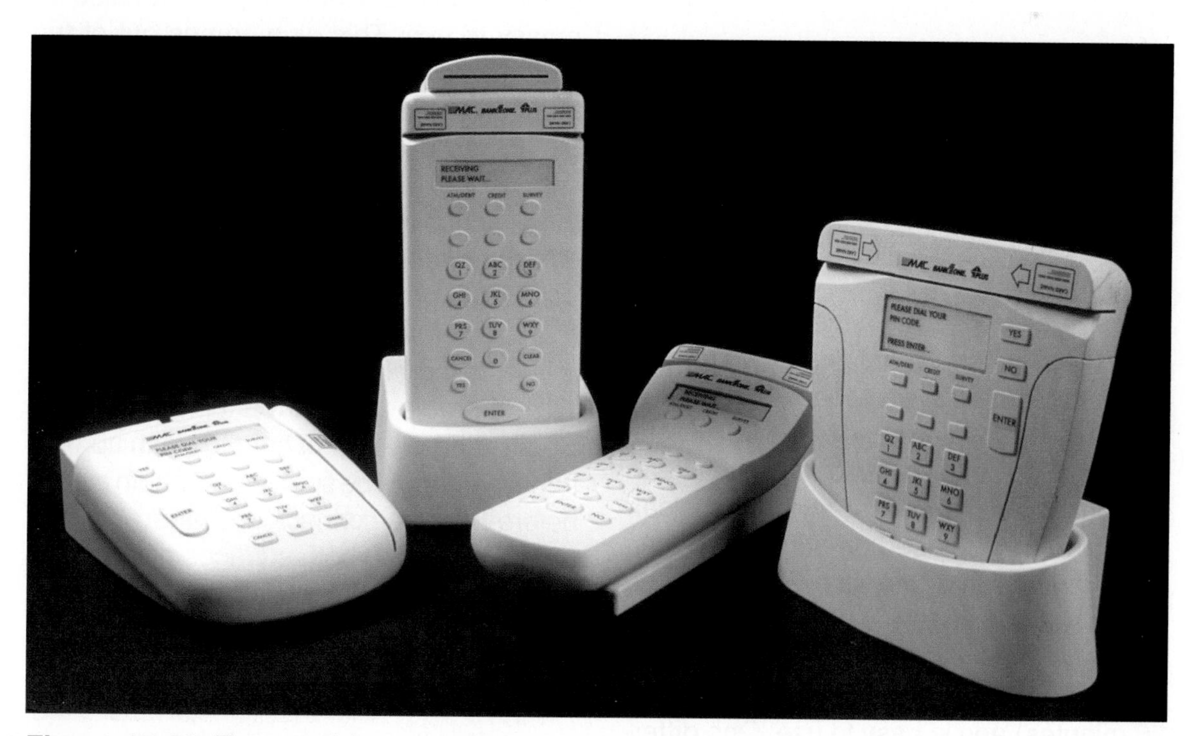

Figure 15-29. These models were made from urethane foam. They were then sanded and painted. (Design Central, design firm; Ingenico, client)

Figure 15-30. Machineable wax comes in different standard sizes.

models. Some of these adhesives have already been discussed. Each one has advantages and disadvantages. All adhesives are best used with certain materials and might not work with other materials. Some can even damage certain materials.

Glue

One of the safest adhesives is ***white glue***. You have probably used white glue for many projects at school and home. White glue is water based and very easy to use. This glue is also easy to clean up. White glue sets fast and dries clear. This glue can be used on paper products, foam, and even some wood. ***Wood glue*** is normally used, however, to glue pieces of wood. This glue is very similar to white glue. This adhesive also dries fast (in 20 to 30 minutes) and is easy to use. One difference is that it is yellow when it dries.

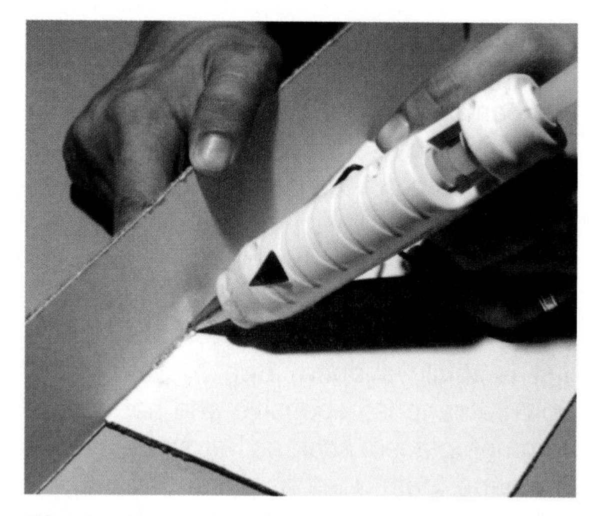

Figure 15-31. Hot-melt glue is one type of adhesive used in modeling.

Hot-melt glue is also used in modeling. See Figure 15-31. Hot-melt glue dries very fast and is clear. This glue can be used on most materials, except certain foams. Hot-melt glue comes in the form of glue sticks. This glue requires a glue gun to heat and apply the glue. A disadvantage of hot-melt glue is that it is not very strong. Even after the glue is dry, the joint that has been glued is flexible. Hot-melt glue can also be messy to use. This glue comes out of the glue gun in an uneven flow and leaves thin strings of glue coming off the model.

Other adhesives commonly used include spray adhesive and plastic cement. Both of these adhesives are very sticky and messy. ***Spray adhesive*** is a type of glue applied by spraying it out of an aerosol can. The area around the model must be prepared before it is sprayed. The overspray from the can will get on the area surrounding the model. Spray adhesive is very hard to clean up. The adhesive should be sprayed onto both pieces before they are connected. ***Plastic cement*** is also applied to both surfaces being glued together. This cement is used on plastic pieces. This adhesive is very strong and dries clear and fast. Both spray adhesive and plastic cement can be dangerous. They have very strong odors and should be used only outside or in well-ventilated areas.

Tape

Tape is another adhesive used in the construction of models. This adhesive should be used only in places hidden from view. Masking, transparent, and duct tapes are used on paper and cardboard models. They can be used to tape corners together. ***Double-sided tape***, or carpet tape, is sticky on both sides. This tape is used to stick one material on top of another. Other types of tapes have special uses. ***Mounting tape*** is made of a thin piece of foam that is sticky on both sides. This tape can be used similarly to carpet tape, when a gap between the materials is needed. Aluminum tape is an adhesive that is tacky on one side and shiny on the other. This tape can be used to give the appearance of metal on a model.

Modeling Tools

Designers build models in special areas designed for model making. They have all the supplies, materials, and tools they need in one place. Designers often specialize in one type of modeling (such as cardboard or foam), so they design their modeling shops to fit their needs for the materials with which they work. For example, a designer specializing in building cardboard models does not need metalworking tools and machines.

The tools used most often in paper and cardboard models are basic tools. Model makers use scissors, utility knives, and hobby knives to cut the material. Hobby knives are very useful because the blades can be changed to make different types of cuts. The designers use rulers and scales to measure and draw straight lines. Power tools, such as a band saw, can be used for making long cuts in cardboard. The band saw can also be used to cut pieces of acrylic for use in the models. Designers using plastics have strip heaters used to bend and shape plastic.

Model makers working with foam have utility and hobby knives. They also use hot wire cutters and small power tools to cut the foam. See **Figure 15-32.** To shape the foam, they use small hand tools, such as files and rasps.

Designers modeling with clay have specialized tools just for working with the clay. Their tools allow them to shape and carve the clay. These designers even have fine finishing tools to make smooth surfaces.

Model makers building wood models often have the most power and machine tools. Some of their machines might be bench-top size (small enough to fit on the top of a table). See **Figure 15-33.** Other machines are full industrial size. Many modeling shops have table saws, band saws, and thickness planers as standard woodworking machines. They also have hand power tools, such as routers, drills, and scroll saws.

The size of the model affects the size of the area the designer needs to work. Some model-making areas are in the corner of the designer's studio. Other designers

Figure 15-32. Most designers have work areas including all the tools they need to build models. (Product Development Technologies, PDT)

Figure 15-33. Model makers often use small power tools to cut, shape, and sand small pieces. (Dremel)

have whole buildings dedicated to making models. They might even have teams of people working on different models.

Making a Model

Modeling is a process using tools, materials, and human knowledge to build a model. Models can range from very large, full-size models of automobiles to small models of jewelry. It does not matter how big or small the models are. They all follow the same basic steps. See **Figure 15-34.** The model-building process has five basic steps to complete:

1. Determine the type of model and material.
2. Rough out the shape.
3. Smooth the shape.
4. Assemble the model.
5. Finish the model.

The first step of the modeling process

Figure 15-34. Models go through a number of steps before they are finished products. (Photo: Rick English, Courtesy: IDEO)

is to determine the type of model you want to build. To do this, the designer examines the purpose of the model. If the model is to show how the solution will look, the designer creates a mock-up. The designer builds a prototype if the model is to test the function of the solution. The type of model helps to determine the materials used to construct it. Paper sheet materials are normally used only for mock-ups. Prototypes, however, are often constructed with plastic.

Once the type of model and materials are selected, the shape is roughed out. In this step, the basic shape of the model is created. Designers rough out shapes in many different ways. An architect might cut out the walls and floor of the structure. Fashion designers cut out the fabric to be used in the model. Clay modelers might cut a foam blank and lay the clay on all sides. Industrial designers cut foam into the rough shapes using the hot wire cutter. Other designers might use a computer numerical control (CNC) milling machine to cut pieces of machineable wax. See Figure 15-35. Laser cutting machines are even becoming popular in model making. They use lasers to cut through material and create nice, clean cuts.

The next step is to smooth and shape the model. In some models, this step can have two separate actions. The first action is to smooth the model. Designers use finishing tools and sandpaper to create a nice, smooth model. In some modeling procedures, this step is done in the same action as shaping the material. Rapid prototyping is a process that does the shaping and smoothing in one step. See Figure 15-36. Rapid prototyping creates a plastic model from a three-dimensional computer model.

Most models have more than one piece. The designer must assemble the pieces to make the whole model. The model is assembled by using adhesives discussed earlier in the chapter. The

Figure 15-35. CNC machines are used to produce models and parts that were first designed on computers.

Figure 15-36. This rapid-prototyping machine allows the designer to create a plastic or wax model very quickly. (Stratasys, Inc.)

designer uses the adhesives carefully to ensure the model looks nice and fits together well. The assembly of the model can include putting the walls of a house model together. Assembling can also be done by sewing pieces of fabric together for a clothing model.

Once the model is assembled, the finishing touches are placed on the model. Small details, such as buttons or knobs, can be added to the model. Designers might draw on many of the details. They might also add model furniture or scale-sized people. Color or decals might be applied to the model. The model can be painted or stained. After the details have been added, the model is complete.

Summary

Models can be found in many places. At different times in our lives, we have all used models. We have seen models in stores, on television, and at school. A model can be any representation of a product, a system, a process, or an idea. Models can be used to communicate ideas, experiment with concepts, or organize information.

There are four types of models we come in contact with every day. Graphic models are used to make visual representations. Mathematical models help us understand mathematical and scientific concepts, such as formulas and chemical reactions. Computer models allow us to construct three-dimensional objects on computers. Physical models are the actual three-dimensional objects representing the final solution.

These models can be either mock-ups (appearance models) or prototypes (working models). They can be built from a variety of materials. The use of certain materials depends on the size, shape, and use of the models. Once the materials are determined, the modeling process begins. Models are created by first blocking out the shape and then finely shaping the pieces. The pieces are assembled and finished. The models are then used in a variety of applications.

STEM Connections

Mathematics

Create a log of the different mathematical models you see and use in one day. Share your log with the rest of the class.

Science

Select a number of different types of adhesives. Conduct an experiment on the properties of the adhesives. Prepare a display comparing their qualities.

Curricular Connections

Social Studies

Create a model of an ancient civilization. Model the buildings and landscape of the time.

Activities

1. Choose a technological device (such as a battery, radio, or computer). Prepare a display highlighting the uses of graphic, mathematical, computer, and physical models to explain your device.
2. Select a company known for innovative design. Write to the company, requesting information about its modeling techniques and asking for images showing the models it constructs.
3. Use the sketches and drawings you have produced in the previous chapters to construct a model.

Test Your Knowledge

Do not write in this book. Place your answers to this test on a separate sheet of paper.

1. A model is a representation of a(n) _____, _____, _____, or _____.
2. The three uses of a model are to _____, _____, and _____.
3. A(n) _____ is a visual representation on paper.
4. Formulas and chemical reactions are which type of model?
5. Summarize each type of computer model below.
   A. Wireframe model.
   B. Surface model.
   C. Solid model.
6. Two types of _____ are mock-ups and prototypes.
7. A mock-up is a full-size, fully functioning model. True or false?

8. Match each type of material below to its description. Give an example of how each can be used in model making.

_____ A material made by gluing thin sheets of wood together.
_____ A single thin sheet of wood.
_____ A material made with paper on the outside and foam on the inside.
_____ A material also known as *museum board.*
_____ A plastic material too thick to cut with a knife.
_____ Two different types include modeling and styling.

A. Plywood.
B. Veneer.
C. Matboard.
D. Foam-core board.
E. Acrylic.
F. Clay.

9. All model makers use the same tools. True or false?
10. Paraphrase the five steps for making a model.

Reading Organizer

Draw a bubble diagram for each main idea in the chapter. Make each of the main ideas the central bubble, while using details in smaller bubbles to surround the main points. An example from this chapter is shown.

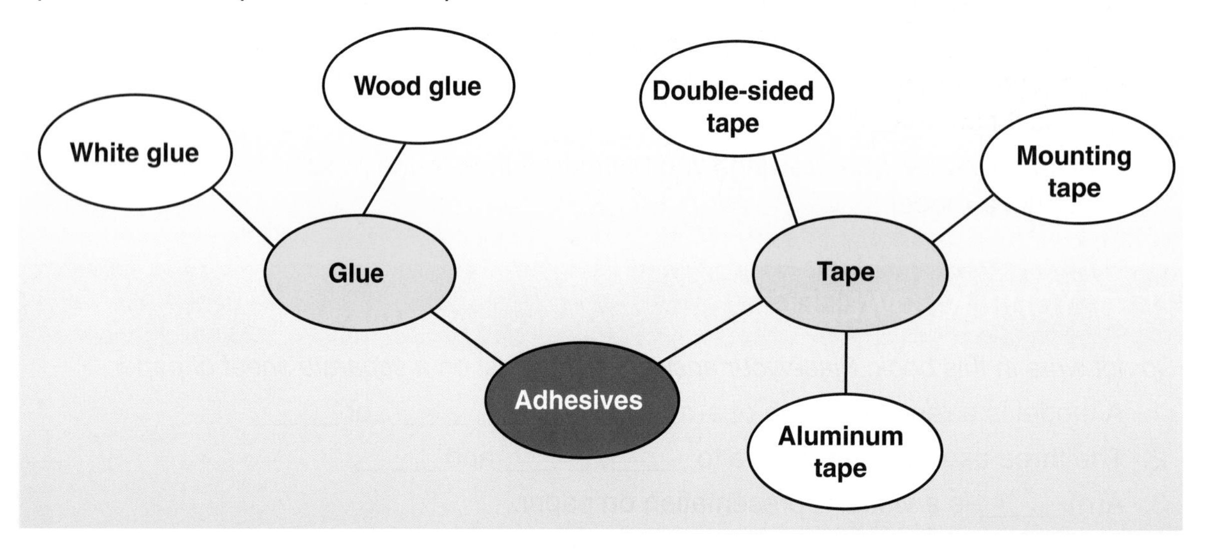

Sketches and renderings are used to create the physical model of a design. Corrugated board may be used in creating an early development model.

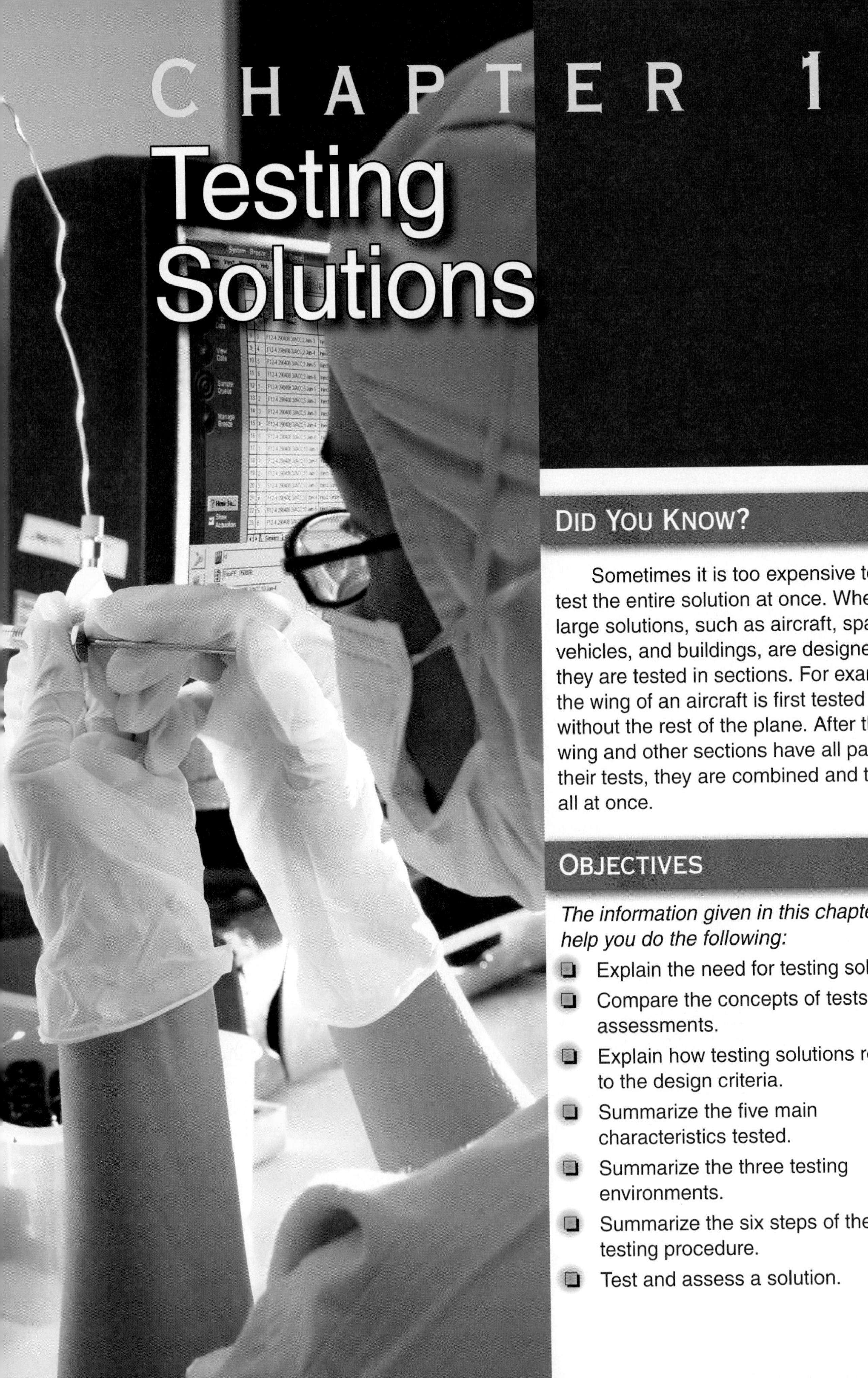

CHAPTER 16
Testing Solutions

Did You Know?

Sometimes it is too expensive to test the entire solution at once. When large solutions, such as aircraft, space vehicles, and buildings, are designed, they are tested in sections. For example, the wing of an aircraft is first tested without the rest of the plane. After the wing and other sections have all passed their tests, they are combined and tested all at once.

Objectives

The information given in this chapter will help you do the following:

- Explain the need for testing solutions.
- Compare the concepts of tests and assessments.
- Explain how testing solutions relates to the design criteria.
- Summarize the five main characteristics tested.
- Summarize the three testing environments.
- Summarize the six steps of the testing procedure.
- Test and assess a solution.

Key Words

These words are used in this chapter. Do you know what they mean?

assessment
computer testing
consumer safety
cost
durability
field test
human engineering
laboratory
material property
performance
reliability
safety
test
trade-off
variable

Preparing to Read

As you read this chapter, choose an existing object and think about the way it was designed. List the ways it was likely tested before it was chosen as a final solution. Give reasons for those tests.

At some point in school, you have probably been given a test. See **Figure 16-1.** The test might have been an aptitude test used to show your abilities. This examination might have been a test of your skills in a game or subject area. The test might even have been a test to assess how well you have learned information. Whatever kind of test it was, it was given for one reason—to measure something about you.

Designers use tests for the very same reason. They want to measure something about their solution. Testing is necessary in design because the solution must solve a problem. Remember, design is not just creating an object. Design is creating a solution to a problem. Designers use tests to determine if their solution solves the problem.

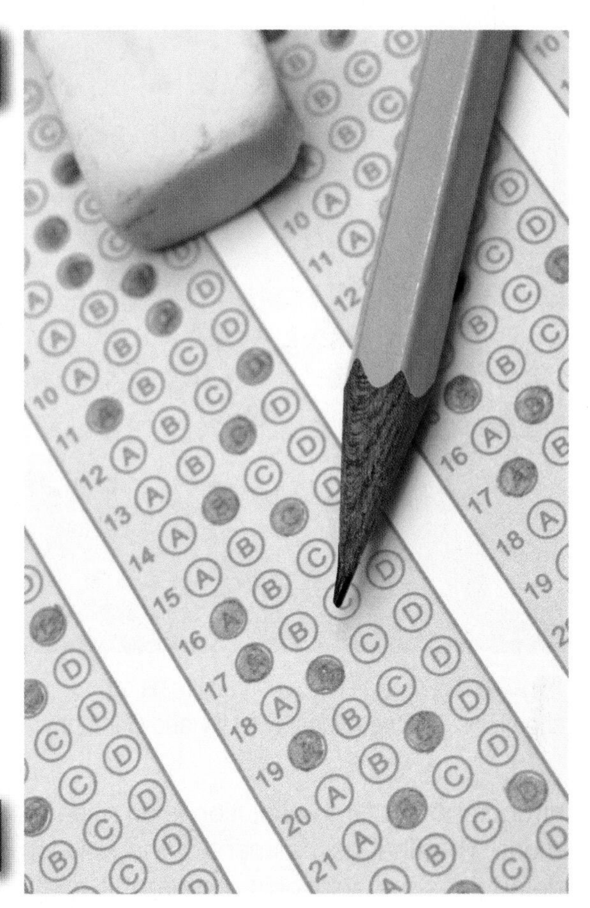

Figure 16-1. A classroom exam is a common example of a test. (©iStockphoto.com/AquaColor)

Tests and Assessments

In this step, designers perform two different tasks. The first task is to test the solution. The second is to assess the solution. The two terms, *test* and *assess*, are often used together. Sometimes, they are even used in place of each other. They are, however, two different concepts. A ***test*** is an experiment or examination. Tests can be very scientific experiments on the properties of a solution. See **Figure 16-2.** A test can also be a written document describing how a person feels about a product. Tests are used to generate results.

These results are then used to assess the solution. An ***assessment*** is a judgment made after reviewing the results of a test. For example, determining that a material is not

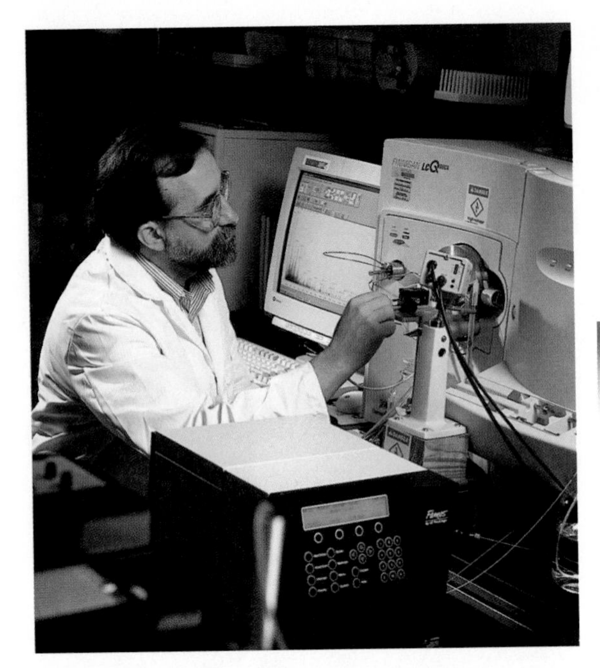

Figure 16-2. Scientists conduct tests on chemicals and medicines. (Eli Lilly and Company)

strong enough for the solution is an assessment. The final assessment of the solution determines the solution's outcome. It might be found that the solution needs to be redesigned. Testing can lead the designers back to an earlier step in the design process. The designers can always go backward in the design process. It is necessary to go back when a solution fails the tests.

USING THE DESIGN CRITERIA

An automobile design might require aerodynamic testing. A solution for a stereo needs to be tested for sound quality. A new recipe requires taste testing. See Figure 16-3. Every design solution requires testing. All solutions, however, do not need the same types of tests. It would be silly to taste test an automobile or test the sound quality of a piece of food. The design criteria determine the tests needed.

Figure 16-3. Food is evaluated in taste tests.

Design Criteria

The solution will do the following:

- ✓ Function as a television antenna.
- ✓ Weigh less than 0.5 lbs.
- ✓ Be placed on top of a TV.
- ✓ Look attractive.
- ✓ Cost less than two dollars to build.

Figure 16-4. The criteria from the first step of the design process should be reviewed to choose the correct tests.

The design criteria were set in the early stages of the design process. See Figure 16-4. The criteria state how well the solution should work and who will use it. They also determine what the solution will do and where it will be used. The tests used in this step of the design process must reflect the criteria. They show whether or not the solution answers the problem and fits the criteria.

If the criteria are written well, they help the designers decide which tests to conduct. If the criteria are not written well, the designers might choose to rewrite the criteria to make sure they are very clear. Since the criteria help evaluate the solution, any rewriting of the criteria must be done before the testing begins.

Testing for Certain Characteristics

The models created during the design process are used for testing. These models give the designers actual objects to test. See Figure 16-5. The designers can test the models for a variety of characteristics. Some of the characteristics tested are similar to the aspects of design. These aspects were the principles used to select the best design (discussed in Chapter 14).

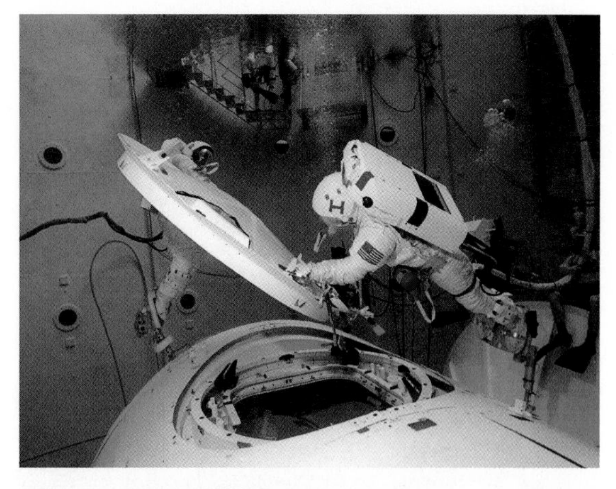

Figure 16-5. NASA astronauts use a prototype of the space station to test maneuvers they will use in space. The model is placed underwater to simulate space. (NASA)

Testing is much different at this step. When the best solution was selected, the designers had only drawings and sketches. The decision was based on how the designers felt the design would function. In this step, the assessment of the solution is based on data and information showing how the solution actually performs. The main characteristics tested are function, human engineering, economics, appearance, and safety.

Function

Function is a characteristic describing how the solution works. This characteristic is tested with two different types of tests. The first type is called a *nondestructive test*. This type tests the function of a product without destroying it. Visual and X-ray tests are nondestructive tests. The second type is a *destructive test*. Destructive tests are often used to test the limits of a product. A fire test is a common

Figure 16-6. Researchers are using a destructive fire test on this glass wall. (Underwriter's Laboratory)

destructive test. See Figure 16-6. When testing the function, researchers conduct tests in three categories: performance, durability, and material properties.

Performance is how well the solution works. This category might test how fast the solution goes or how much weight it holds. Performance can also be the quality of a picture or sound. Civil engineers test performance when they build bridges. They make sure the bridges hold different amounts of weight before they open the bridges to the public. Computer companies test the performance of computer processors before they rate the processors' speed. See Figure 16-7.

Durability is how long the solution will work. All solutions with moving parts eventually wear out. The designers test durability of the solution to determine how long

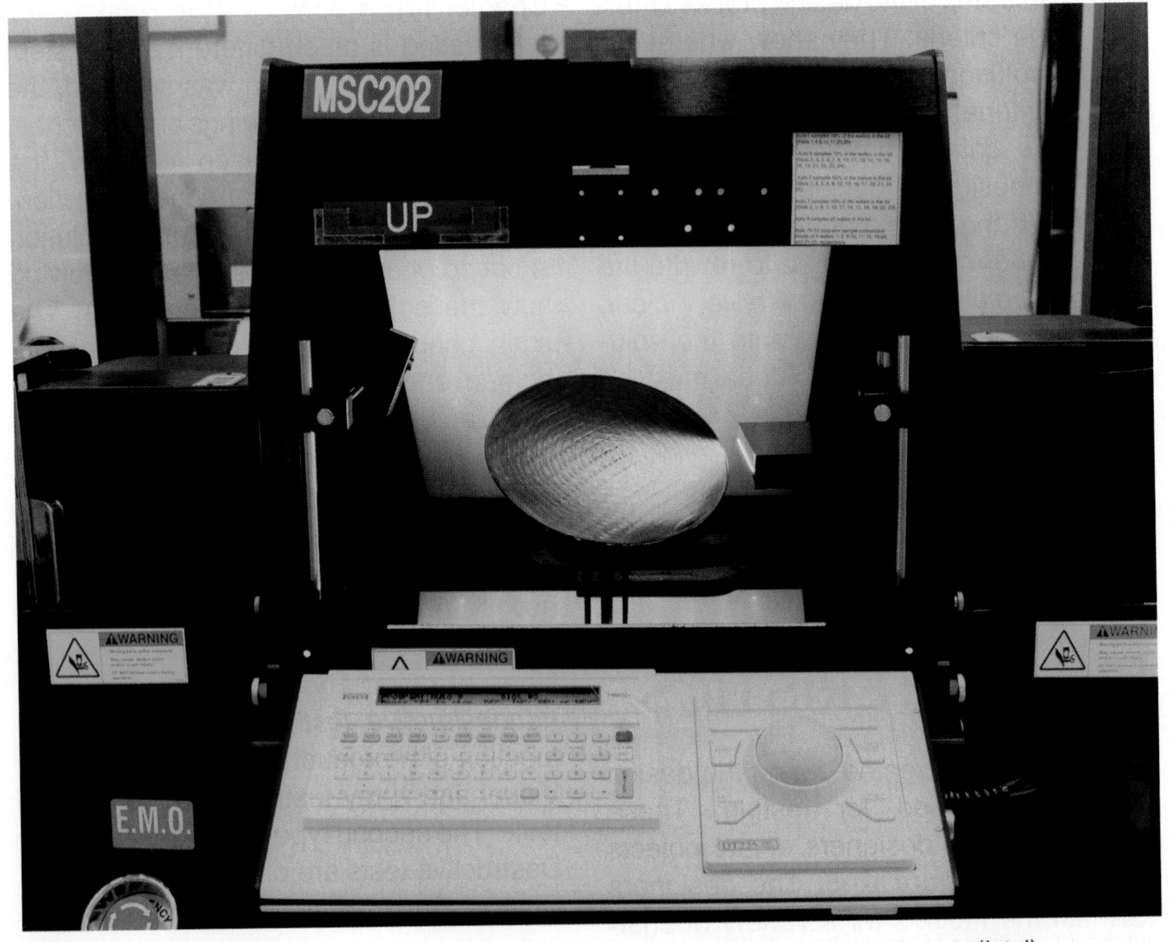

Figure 16-7. Sophisticated tools perform tests to ensure the highest-quality processor. (Intel)

the solution will last. The durability (life expectancy) of a product can be planned. Sometimes, designers plan for the products to wear out after a certain amount of time. For example, a paper plate is designed to be less durable than china. A short life expectancy usually makes products less expensive. Durability is tested using machines that repeat the same motion over and over.

Material properties are characteristics determining how materials function. Materials have many different properties, as discussed in Chapter 5. Designers test the material properties to make sure the materials they selected work correctly. There are many different tests used to test material properties. Designers might apply a sudden load to the design solution to test the tensile strength. See Figure 16-8. Designers might also test how well the product conducts electricity. Sometimes, designers even test to determine how much sound the solution makes.

Testing must be carefully designed to adequately determine the ***reliability*** of the product. Required reliability is defined in the design criteria. This reliability describes the product's performance and durability.

Human Engineering

Human engineering examines the way people interact with the solution. The two main areas in human engineering are physical and environmental. When designers test their solutions, they are concerned with both areas.

The physical component deals with how the user handles the solution. The sizes of both the user and the solution are very important. For example, ladder designers must test their solution to make sure it fits the user. The lengths of the

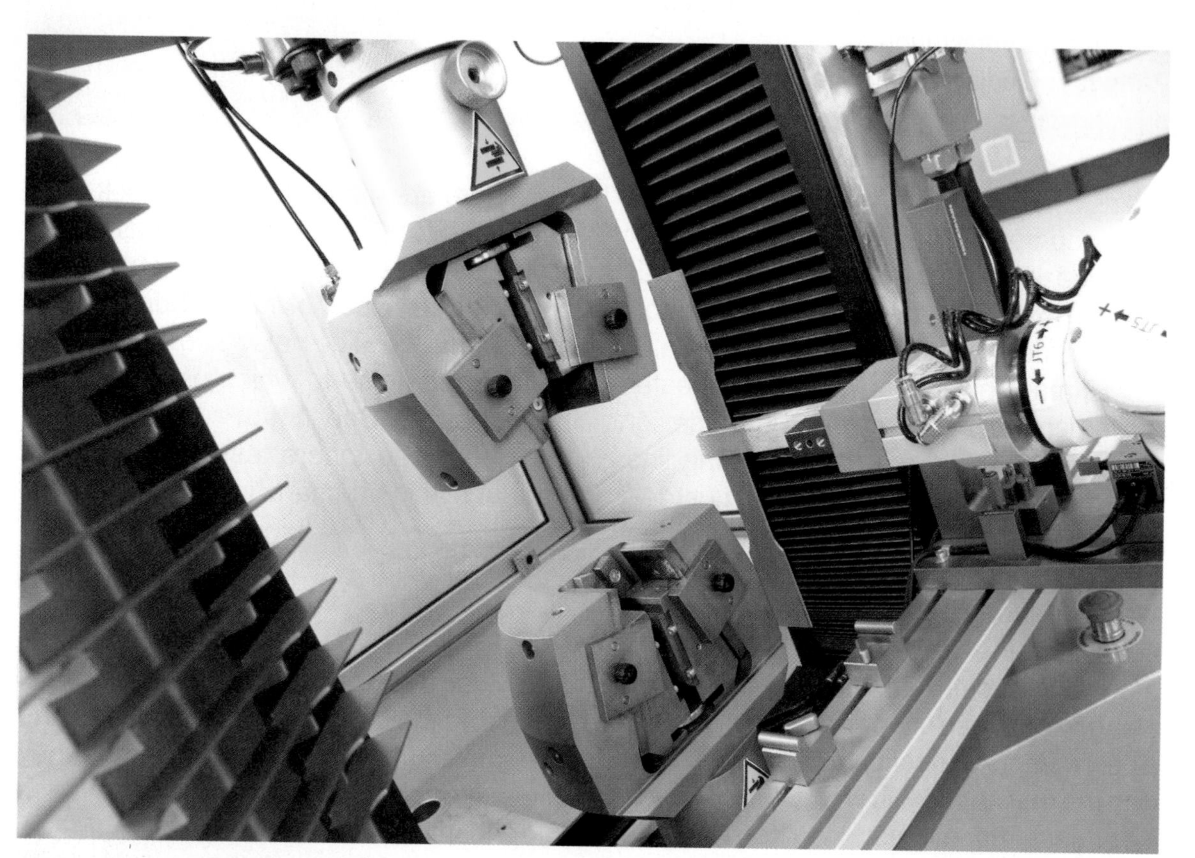

Figure 16-8. Designers use testing devices to determine material properties. This testing machine can be used to test a range of properties including tensile, compression, shear, and bending strength. (Qualitest)

Career Highlight

Quality Control Workers

Manufacturing

Quality Control Workers

The Job: Quality control workers ensure that products being produced are of a high quality. This work often includes inspecting, testing, measuring, and sampling. Most quality control workers work in manufacturing centers. Depending on their specific job, the workers might visually inspect products, physically test products, or even taste or drink product samples. Their job is to find defective products and determine if there is a pattern of defects.

Working Conditions: These workers often work on a factory floor and spend a large amount of time on their feet. Similar to many manufacturing workers, they might work a morning, an afternoon, or an evening shift. They might also work weekends.

Education and Training: Most jobs for quality control workers require a high school diploma and on-the-job training. Some more technical industries, such as medical and pharmaceutical, might require an associate's or bachelor's degree. Certification is also available in some industries.

Career Cluster: Manufacturing

Career Pathway: Quality Assurance

(©iStockphoto.com/leezsnow)

user's arms and legs are important sizes. The steps of the ladder must be the right distance apart to fit most people. See Figure 16-9. The ladder must also be able to be carried to and from the work area.

Human engineering also tests repeated movements. This is done often in the design of systems. Think of a manufacturing line. The engineer designing the line must test all the jobs along the line. It is important that the jobs are safe for the workers.

The second concern in human engineering is the environment. The environment is the heat, light, and noise occurring along with the solution. Designers creating a new computer monitor must be concerned with the environment in which it will be used. They test the monitor in different lights. The glare created in some lights might be harmful to the eyes of the user.

There are many charts and books that help designers test for humans engineering problems. The books show tables and charts of the types of things humans can and cannot do. They also show the levels of light and noise to which people should be exposed. See Figure 16-10. These books help designers test their solutions without placing humans in dangerous situations.

Economics

Most designs are created to solve a need and make a profit. Designers test the economics of a solution to make sure they will make money. They test two variables in economics: cost and market.

The ***cost*** of the solution is divided into two areas: preproduction and production.

Figure 16-9. Ladders must be designed to fit the human user. (Werner Co.)

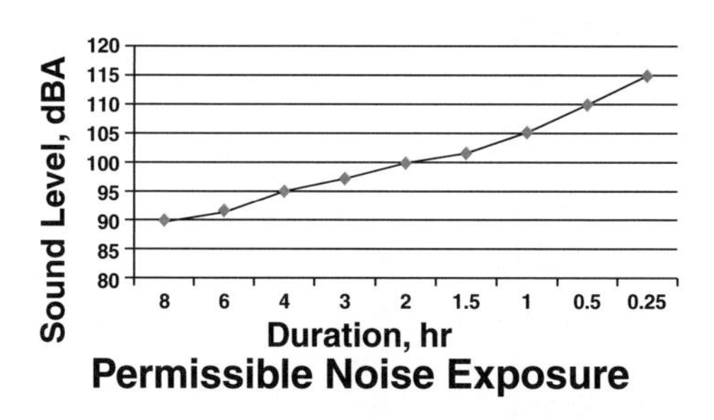

Figure 16-10. This chart shows the levels of sound humans can be exposed to for different amounts of time. (Data from OSHA Standard No. 1910)

The preproduction costs include money spent to design the product. By the testing stage, most of the money has been spent on research and the development of sketches, drawings, and models. More money will be spent to create drawings in the next step.

The production costs include money that will be spent manufacturing the solution. They include everything from the price of materials to the wages of the workers. The costs can be calculated by examining the model. Designers use the model to figure the amount of materials that will be used. They can also figure the amount of time it will take to manufacture the solution.

The second economic factor is the market. The market includes those people buying and selling the product. See Figure 16-11. Consumers are a major part of the market. They are the ones who choose whether or not to buy the product. The designers test the solution by surveying consumers. They again use the model to help the consumers visualize the solution. The second part of the market is the competitors. Competitors are those who make products competing with the solution. Designers might test the products of their competitors to see how their solution compares. They then advertise the features that make their solution

Figure 16-11. Consumers choose to buy the products they like the best.

better than the competition. The designers also study the prices of the competitors' products.

Appearance

The appearance of a product is one of the most important characteristics. This characteristic is also one of the hardest to test. There is no lab test that can give the designers data on how nice a product looks. Appearance is a characteristic people see and have feelings toward. You might have heard the phrase "Beauty is in the eye of the beholder." The same is true for the appearance of a product. Many people can look at the same solution and have different reactions. Some might like the shape. These people might not, however, like the color. Others might think the texture is unpleasant. At the same time, others might like it.

The best way to test appearance is to use the reactions of people. Designers often ask many people about their reactions to a product. The questions might be very general, such as "Do you like the product?" They can also be very specific, such as "On a scale of 1 to 5, how would you rate the size of the display screen?" The answers from the survey are then gathered. The results of the questions are used as data to evaluate the appearance.

Safety

Safety is a major characteristic of a design. This characteristic involves more than you might think. The solution must be safe for consumers to use. This design must be safe for workers to produce. Lastly, the solution must be safe for the environment.

Consumer safety is the most obvious need for safety. Solutions must be tested to ensure they are safe to use. Safety testing can be done by using objects to simulate humans. Automobiles are safety tested with crash-test dummies. See Figure 16-12. The dummies have sensors placed around their bodies to record data. The data tells where the dummy was injured during the test. This allows accurate tests, without putting humans at risk of being hurt. Some products are safety tested using computer simulations. The computer acts as if a human is using the solution and records safety data.

All solutions must be safe for those using them. They should also be safe for those around them. After testing, age ratings are given to some solutions. The age rating identifies the lowest age a person should be to use the product safely. See Figure 16-13. Games and movies with adult content are not rated for children. Designs with small parts are not rated for very young ages because children can swallow the parts.

Figure 16-12. Crash-test dummies allow the safety of automobiles to be tested without hurting humans. (Daimler)

Figure 16-13. This product has an age rating. This rating tells consumers the product is intended for children 18 months and older.

There are several government organizations supervising product safety. The U.S. Consumer Product Safety Commission (USCPSC) watches over everything from household appliances to children's toys. The National Highway Traffic Safety Administration (NHTSA) tests the safety of motor vehicles. The Food and Drug Administration (FDA) oversees the testing of food, medicines, and cosmetics, such as hair spray and makeup.

The organization concerned with how the solution affects the safety of the environment is the Environmental Protection Agency (EPA). Solutions must not harm the environment when used. Testing is done on products to determine their impacts on the environment. Aerosol cans, for example, are products that were found to hurt the environment. The cans released chlorofluorocarbons (CFCs). After testing was done, the CFCs were found to damage the environment. Now, most aerosol cans have no CFCs. In some cases, they have been converted to spray pumps.

Products can also be unsafe for the environment when they are thrown away. Solutions are tested to find how they should be disposed. If some products, such as chemicals and batteries, are poured down the drain or thrown away, they can be very unsafe. These products can contaminate the air and water around them.

Testing Environments

Tests are conducted in three different environments. They can be completed in the laboratory, the field, or virtual space. Each environment has advantages and disadvantages.

The Laboratory

The ***laboratory*** is used when a controlled area is needed. In laboratories, variables, such as temperature, moisture, and amount of light, can be controlled. A ***variable*** is any condition that can be changed. Many variables can affect the outcome of a test. For example, if designers are testing a sound, they do it in a lab where they can control the amount of sound. See Figure 16-14. They make sure the only sounds in the room are coming from the device they are testing. If there are other sounds, these sounds will distort the results of the test.

The labs used in testing are often very specialized. It is common for a lab to be used for only one type of testing. It is too expensive for all designers to have their own labs because the labs are specialized. Some designers use companies that specialize in testing their solutions. This also helps to make sure the testing is accurate.

Figure 16-14. This researcher uses a soundproof laboratory to test the noise the tire produces on the two different surfaces. (Goodyear Tire and Rubber Company)

The Field

The second testing environment is field-testing. ***Field tests*** are conducted in the environments in which the solutions will be used. The solutions are used the same way the consumers will use them. Airplanes are field-tested by flying. See Figure 16-15. Field tests can be conducted in areas that are off-limits to the public. Automobile tests are done on closed-circuit tracks. The car being tested might be the only car on the track. This is done for safety. It would be unsafe to test an automobile at high speeds on roads with other vehicles.

Actual consumers do some field-testing. Prototypes of small appliances, for example, might be given to a group of people to test. These are tested in the users' homes or offices. When the testing is complete, the users report the results of the test to the designers.

Figure 16-15. This experimental aircraft is tested in test flights. (NASA)

Technology Explained

ergonomic environment: Addressing the needs of ergonomics to create a comfortable and efficient environment.

Workers who spend long periods of time doing detailed work with a computer might be susceptible to eyestrain, back discomfort, and hand and wrist problems. Ergonomics is the science of adapting the workstation to fit the needs of the drafter. Applying ergonomic principles results in a comfortable setting and an efficient environment.

There are many types of ergonomic accessories that can improve a computer workstation. In addition, a few things can be done to create a comfortable environment and help prevent injury or strain to the operator's body. The following are specific guidelines to help prevent injuries for various body parts:

- Position the monitor to minimize glare from overhead lights, windows, and other light sources. Reduce light intensity by turning off some lights or closing blinds and shades. You should be able to see images clearly, without glare.
- Position the monitor so it is 18″–30″ from your eyes. This is about an arm's length. To help reduce eyestrain, look away from the monitor every 15–20 minutes and focus on an object at least 20′ away for 1–2 minutes.
- Forearms should be parallel to the floor.
- Periodically stretch your arms, wrists, and shoulders.
- Try using an ergonomic keyboard and mouse. The keyboard keeps the wrists in a normal body position. See Figure 16-B. An ergonomic mouse fits the hands more comfortably than a standard mouse does.
- Adjust the monitor so your head is level, not leaning forward or back. The top of the screen should be near your line of sight.
- Use a comfortable chair that provides good back support. The chair should be adjustable and provide armrests.
- Sit up straight. This maintains good posture and reduces strain. Think about good posture until it becomes common practice.
- Try standing up, stretching, and walking every hour. This also reduces strain.
- Keep your thighs parallel to the ground.
- Rest your feet flat on the floor or use a footrest.
- When taking a break, walk around. This stretches the muscles and promotes circulation through your body.

Figure 16-B. This keyboard was designed to fit ergonomic principles. (Microsoft Corp.)

Virtual Space

The last testing environment is the virtual setting. Virtual testing, or ***computer testing***, is a very popular form of testing. See Figure 16-16. This testing requires a computer model and testing or simulation software. Virtual testing can be less expensive than lab testing and field-testing. This testing can also generate very accurate data. Computer testing is very useful for large and complex solutions that are hard to test in reality. The space shuttle is a solution that has been tested in a computer environment.

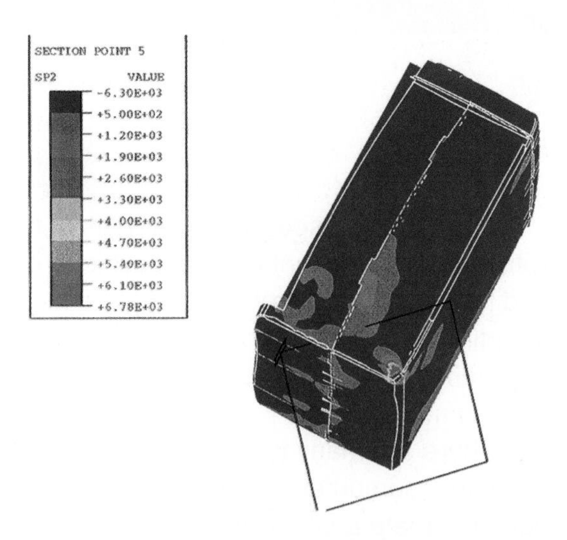

Figure 16-16. The use of computers in testing is very helpful and accurate. (Whirlpool Corporation)

The Testing Procedure

Whether the test is conducted in a lab, in a field, or on a computer, it must follow a certain procedure. Following the procedure helps make the test valid. A valid test is one that is accurate and can be trusted. The testing procedure includes six steps:

1. Determine the purpose.
2. Choose the test.
3. Collect materials and equipment.
4. Conduct the test and gather data.
5. Analyze the data.
6. Evaluate the results.

Determine the Purpose

The purpose of the test is the reason the test is being conducted. This reason should relate back to the criteria and characteristics of the solution. For example, the purpose of the test might be to evaluate the appearance of the solution. Each test should have only one purpose. It would be very hard to conduct a test of both the function and economics of a solution. Conducting more than one test on a solution might be necessary. It might not always be necessary, however, to test all the characteristics of a design.

Choose the Test

Once the purpose is determined, the test must be chosen. There are many different tests that can be conducted on products. Crash tests and vibration tests are common tests done on automobiles. See Figure 16-17. Appearance and clinical tests are typical medication tests. Many products are tested using customer-satisfaction tests.

Figure 16-17. This vibration test is conducted to test the durability of the automobile. (Daimler)

Tests can be conducted using existing tests. The researcher can also create a new test. Testing organizations have collections of tests that have already been developed. The American Society for Testing and Materials (ASTM) is one of the largest organizations that write testing standards. ASTM has over 10,000 standards for materials and tests. Many of the tests ASTM identifies are very specific. One example of an ASTM test is the *Standard Test Method for Measuring Maximum Functional Wet Volume of Utility Vacuum Cleaners*. In other words, the test is used to check how much fluid a utility vacuum cleaner can handle.

ASTM and other testing organizations are excellent sources for finding existing tests. The second option is to create a new test. If the designers create a new test, it must be evaluated to make sure it is valid. This test has to test the right features. The designers must also make sure they control the variables. For example, designers testing the "bounce" in different basketballs must have the same air pressure in each ball.

Collect Materials and Equipment

Whatever test is chosen must be documented. The written procedure for the test should include any necessary information about the test. The test will have sections for materials, procedures, and observations. The materials include the equipment, or items used during testing. Product samples are common testing materials. Testing equipment consists of the devices needed. Bunsen burners, clamps, and wind tunnels are pieces of testing equipment. See Figure 16-18.

Conduct the Test and Gather Data

After the materials and equipment have been collected, the next step is to conduct the test. The test should be conducted by following the written procedure. The procedure is often called the *testing method*. The method must be followed step-by-step, so the test can be repeated with the same results. If the researchers change their procedure, they must make note of it.

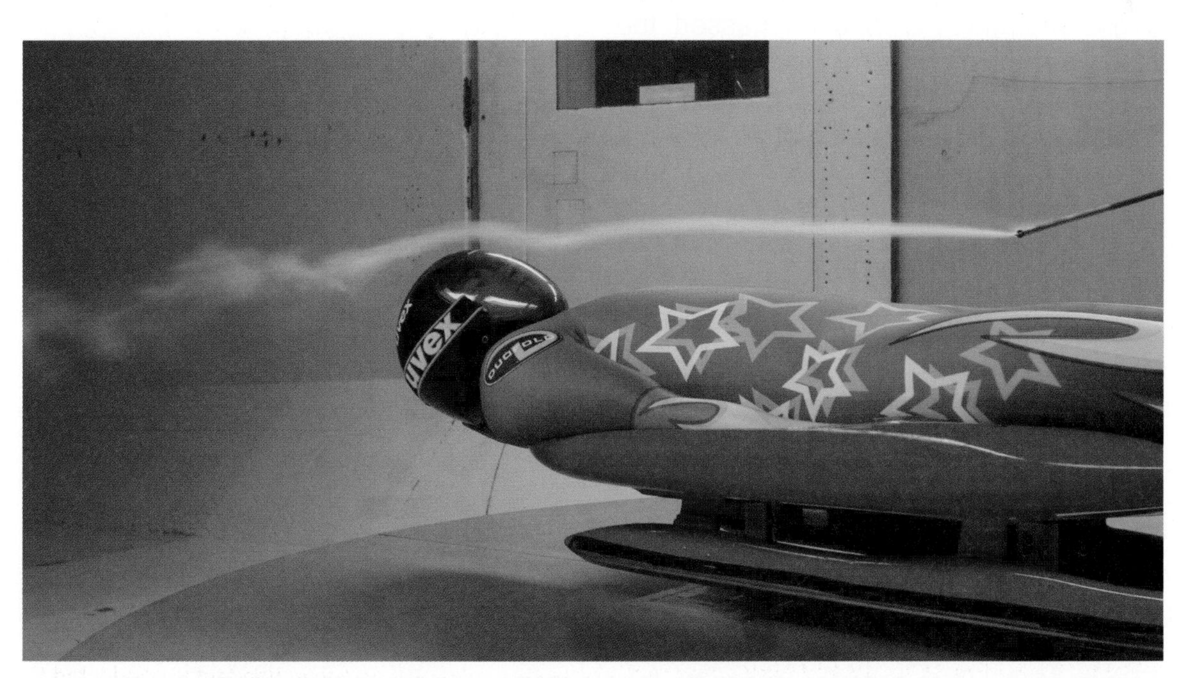

Figure 16-18. Wind tunnels are used for a number of different applications. This is an image of wind tunnel testing of two-time Olympic luge medalist Mark Grimmette. (Courtesy of the San Diego Low Speed Wind Tunnel)

It is important that the data is recorded during the test. The data can come in many different forms. Data can be the amount of times the solution was used before it broke. These facts can be the reactions to the appearance of a solution. The data can even be the force with which a crash-test dummy hit the steering wheel.

Analyze the Data

The designers must analyze the data. This data can be very overwhelming. Data from a single test can cover tens or even hundreds of pages. The designers must break down all the data. They begin by doing any necessary calculations. Sometimes, averages are calculated. Once the calculations are complete, the designers display the data with graphic models (charts, graphs, or tables).

Evaluate the Results

The graphic models are used to understand the results of the tests. The researchers can evaluate the results by reviewing the tables and graphs. They might find that the solution passed the test. The researchers might also find that the solution failed the test. The results of the test help the designers determine if the solution is acceptable.

Assessing the Solution

The final stage in the testing of a solution is assessment. The assessment is the designers' final conclusion on the solution. This conclusion can be made only after the results from all the tests are completed. The designers review the results from tests done on the function, human engineering, economics, appearance, and safety of the solution and create an assessment.

The assessment usually includes a trade-off. ***Trade-offs*** are decision processes the designers must face. They recognize the need for sensible compromises among opposing factors. A trade-off can occur when a solution does not pass all the tests conducted. For example, the test results might show that a solution is safe and functions well. The solution, however, might have some appearance problems. The trade-off comes when the designers have to decide if safety and function are more important than appearance. When trade-offs are made, there is a preference or an exchange for one feature or idea in favor of another.

Once the designers take into account the test results and any trade-offs, they determine a final assessment. The designers have three basic choices. See Figure 16-19. They

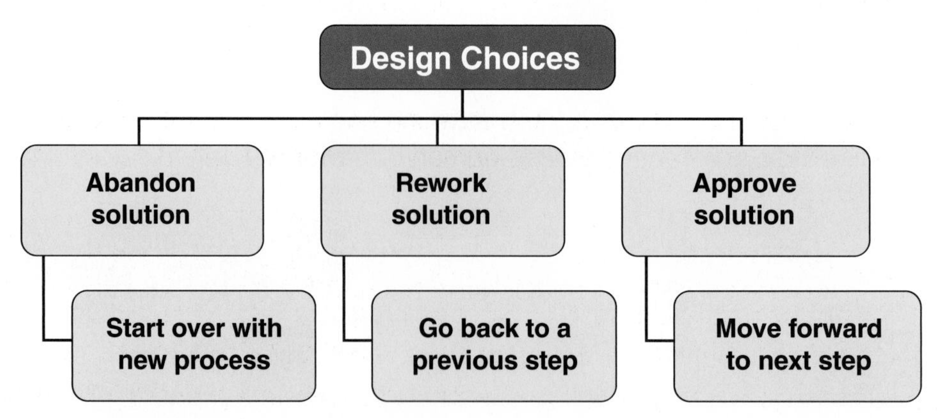

Figure 16-19. Once the testing is complete, the designers must make an assessment. They can abandon, rework, or approve the design.

can choose to completely abandon the solution. This assessment leads the designers back to the first step of the design process. There, they start over from the beginning. The second choice is to rework, or revise, the solution. This is selected when the solution did not do well in one or two tests. This choice might lead designers back to any of the previous steps of the design process. The last assessment that can be made is that the solution meets the expectations. This assessment moves the solution forward to the next step of the design process.

Summary

The testing stage of the design process has two different functions. The first is to test the solution. There are hundreds of different tests that can be conducted on solutions. Five main areas (function, human engineering, economics, appearance, and safety) are tested. Each of these areas can be tested in three different environments: in the laboratory, in the field, and on the computer.

Testing a solution has a procedure. This procedure begins by determining a purpose. A test is then chosen. Materials and equipment are selected. The test is conducted. The data is gathered during the test. This information is analyzed. The results are evaluated. Once the tests have been evaluated, the designers can make an assessment. The assessment of a design determines whether the solution will be discarded, revised, or sent to the next step.

STEM Connections

Science

Conduct a test on a material. Record the data. Report the results.

Curricular Connections

Language Arts

Write a test that can be used to test the function of a solution. Include the purpose, materials, and procedures in the test.

Social Studies

Research how tests are used to study humans and human behavior. Create a display showing how testing products and testing human behavior are both different and similar.

Activities

1. Create a display highlighting the main characteristics tested in product testing. List possible examples of tests that can be used to test each characteristic.
2. Research a major testing or safety organization. Create a presentation explaining the organization. Give examples of the testing this organization conducts.

3. Use the testing procedure to test a solution you have developed. You can research a test or develop your own. Once the testing is completed, make an assessment of your solution.
4. Develop a budget for a product you can make in your classroom lab. Some possible projects include bookends, coatracks, wooden pens, and computer-monitor supports. First, determine the amount of materials needed to make the product. Include a consideration for waste and determine the cost of materials. Determine the time and equipment needed to make the product. Estimate a reasonable hourly wage for labor and a reasonable hourly cost for tools, equipment, and overhead. Develop a marketing plan. Estimate the advertising costs. Calculate the cost per item to produce the product. Suggest an estimated price for the product.

Test Your Knowledge

Do not write in this book. Place your answers to this test on a separate sheet of paper.

1. Why is it necessary to test solutions?
2. A test is a judgment made toward a solution. True or false?
3. When is an assessment made?
4. Design criteria are used in the testing stage. True or false?
5. Select the statement best describing performance:
   A. Performance is how the solution works.
   B. Performance is how well the solution works.
   C. Performance is how long the solution will work.
6. List and summarize the five main characteristics of solutions tested.
7. Name the three testing environments.
8. A valid test is one that can be trusted. True or false?
9. Write the six steps of the testing procedure.
10. A trade-off is a compromise with which designers are often faced. True or false?

Reading Organizer

List the three types of characteristics that were likely tested on an object before it was chosen as a design solution. Give reasons for those tests. Use the examples in the text and the example given below.

Example:

Energy-Saving Lamp

Test	Reason
Performance	Maximum wattage of lamp
Durability	Maximum time lamp was lit

TSA Modular Activity

This activity develops the skills used in TSA's System Control Technology event.

System Control Technology

Activity Overview

In this activity, you will build a computer-controlled mechanical model representing controls for an elevator serving two levels. You will also prepare a report explaining your design and listing directions for operation.

Materials

- A pencil.
- Paper.
- Two touch sensors.
- Two lights.
- Two motors.
- Computer hardware and software control system.

Background Information

- **General.** You will use the sensors, lights, and motors to model an elevator serving two floors. The sensors represent the elevator call button. The motors represent the elevator doors. One represents the first-floor elevator doors. The other represents the second-floor elevator doors. A running motor represents an open elevator door. The lights represent the location of the elevator car. When the first-floor light is on, the elevator car is at that floor. When the second-floor light is on, the elevator is at that floor.

- **Control conditions.** Your control system must adhere to the following guidelines:
  - The elevator car can be at only one floor at any given time.
  - The elevator car must be at the floor before the doors can open.
  - The elevator car cannot leave the floor while the doors are open.
  - When a button is pushed, the elevator car is called to the floor. The doors open and close. The car moves to the other floor. The doors open and close.
  - The car remains at the current floor until called.
  - There must be a two-second delay between when the car arrives at the floor and when the doors open.
  - The doors remain open for five seconds and then close automatically.
  - If the elevator call button is pressed while the doors are open, the doors remain open for five seconds from the time when the button is pushed.
  - The car must wait two seconds after the door closes before going to the other floor.
- **Report.** Your report must include the following:
  - A description of the solution.
  - Instructions for operation.
  - A printout of the control program.

Guidelines

- Create rough schematics of design solutions and control logic.
- Sketch the final design before constructing it.
- After constructing the model and writing the control program, test the model. Before you begin testing, create a list of the various conditions and situations that need to be tested.

Evaluation Criteria

Your project will be evaluated using the following criteria:

- The report.
- The model functionality, dependability, and ingenuity.
- The computer-program logic and functionality.

Automobiles undergo different types of tests. These tests help designers assess a variety of characteristics.

CHAPTER 17

Communicating Solutions

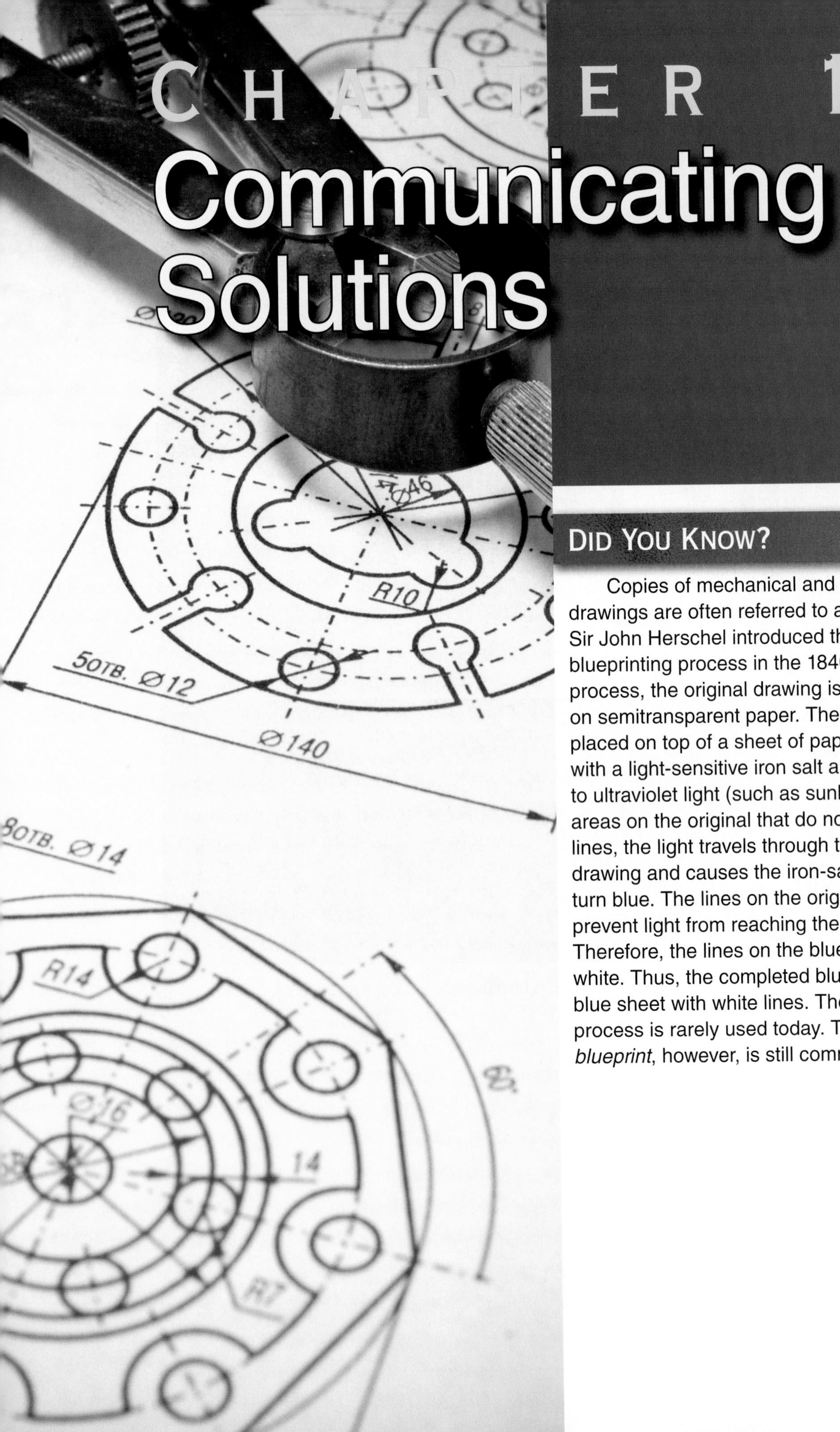

Did You Know?

Copies of mechanical and architectural drawings are often referred to as *blueprints*. Sir John Herschel introduced the blueprinting process in the 1840s. In this process, the original drawing is created on semitransparent paper. The original is placed on top of a sheet of paper covered with a light-sensitive iron salt and exposed to ultraviolet light (such as sunlight). In areas on the original that do not have lines, the light travels through the original drawing and causes the iron-salt coating to turn blue. The lines on the original drawing prevent light from reaching the blueprint. Therefore, the lines on the blueprint remain white. Thus, the completed blueprint is a blue sheet with white lines. The blueprinting process is rarely used today. The term *blueprint*, however, is still common.

Objectives

The information given in this chapter will help you do the following:

- Summarize the three types of documents used to communicate solutions.
- Compare engineering drawings and architectural drawings.
- Summarize the three types of working drawings.
- Explain orthographic drawing and pictorial drawing.
- Give examples of traditional and CAD drawing tools.
- Create a mechanical drawing.
- Summarize the process of receiving approval for a solution.

Key Words

These words are used in this chapter. Do you know what they mean?

alphabet of lines
architectural drawing
assembly drawing
bill of materials
compass
computer-aided design (CAD)
design report
detailed drawing
dimensioning
drawing board
engineering drawing
45° triangle
isometric drawing
multiview drawing
oblique drawing
perspective drawing
plotter
scale
schematic drawing
specification sheet
template
30°-60° triangle
T square
working drawing

Preparing to Read

As you read this chapter, outline the details of the different ways of communicating design solutions. Use the Reading Organizer at the end of the chapter to organize your thoughts.

Have you ever seen or used blueprints? See Figure 17-1. Have you ever used a schematic drawing or seen a drawing showing how a product is put together? If you have, a design has been communicated to you. The objects used to communicate solutions are created in this step of the design process.

The goal of almost every design process is to make a profit. The solution has been designed with this goal in mind. To make a profit, however, most solutions must be mass-produced. The designers do not manufacture the product. They give the solution to others skilled in manufacturing processes. The people who produce the product might not have been part of the design process. Therefore, they might not know very much about the solution. The designers must communicate all aspects of the solution to those who are making the product. See Figure 17-2.

Figure 17-1. Builders use blueprints to communicate solutions. (Habitat for Humanity International)

Figure 17-2. Drawings are used in the manufacturing of products. (©iStockphoto.com/thelinke)

TECHNOLOGY EXPLAINED

parametric sketch: a computer drawing defined by certain relationships.

Many modern CAD software programs use parametric sketches to create solid models. See Figure A. The relationships defining these sketches can be geometric (such as parallel lines), dimensions, or equations. The parametric sketches are used to create solid models. The solid models are then used to generate a number of design documents. Once a solid model is created from a parametric sketch, it can easily be altered. If the design changes, the sketch can be updated. All the parts and drawings associated with the sketch are also updated.

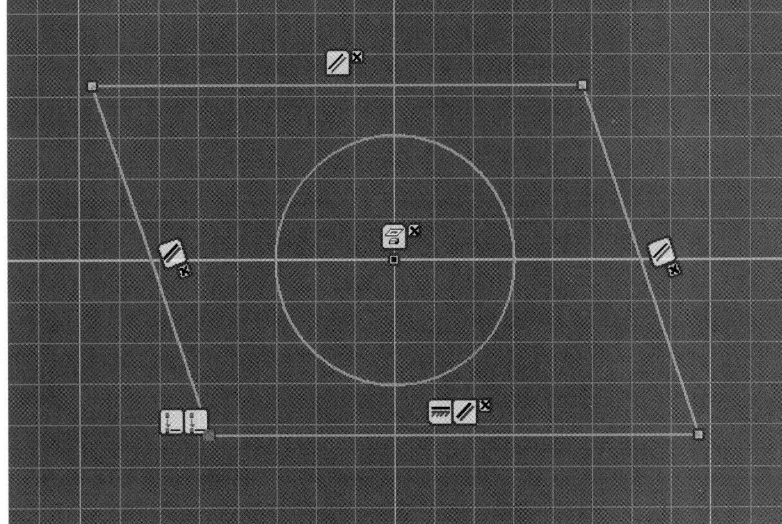

A Paremetric Sketch

Designers who use programs based on parametric sketches complete several steps in the creation of solid models. The steps are similar to the steps used to create pencil sketches. First, the designers rough out the shapes. For example, if designers are creating a wrench, they draw a rectangle with a circle at each end. They then add some features, such as the openings at both ends of the wrench. At this point, the designers begin to add the parameters. Parameters are often called *constraints*. The constraints can be the dimensions, such as the length of the wrench. They can also be geometric constraints, such as that the centers of the circles must be aligned with the center of the wrench.

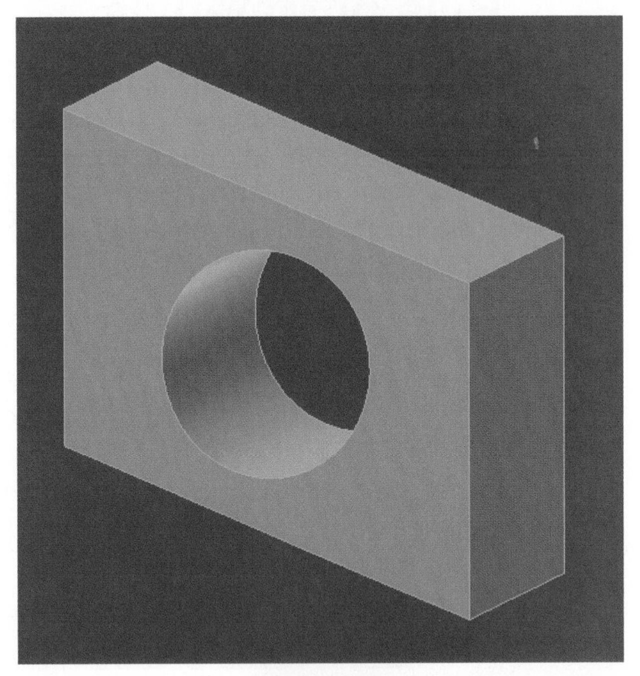

A Solid Model

Once the 2D sketch is constrained, it is given a thickness. This is often done by using a command called *extrude*. Extrusion makes the sketch a 3D model. In the case of the wrench, once it is a model, it can be rounded to fit better in the user's hands. The solid model can easily be changed because it was created with a parametric sketch. If the designers decide to make the wrench longer, they can go back to the sketch and change the dimension constraint. This constraint automatically changes the length of the solid model. The dimension constraint also updates all other drawings using that part. This is an important use of parametric sketches. This use makes altering designs much easier and faster than in traditional sketches.

Figure A. This parametric sketch is constrained so the two horizontal lines and the two angled lines remain parallel. The solid model shows the sketch after it has been given a thickness.

Communication Documents

The best way for designers to communicate a solution is through a set of documents. These documents describe everything the manufacturer must know to produce the solution. Consumers also use design documents when they assemble and service products. There are three major types of design documents:

- Bills of materials.
- Drawings.
- Specification sheets.

Each document has a specific purpose and is needed to produce the solution.

Bills of Materials

A ***bill of materials*** is a document listing and describing the parts needed to build the solution. This document is created in the form of a table. See Figure 17-3. This document lists information about the various parts needed to produce the solution. The columns of the table divide the different types of information. The information given is a part number, the number of items needed, and the name of the part. The size of each part and the material to be used are also listed on the bill of materials. Each row of the table provides information for a different part. Every part used to make the solution is listed on the bill of materials. The main pieces are listed first. The hardware and fasteners, such as hinges, screws, and bolts, are listed last.

Drawings

Drawings are used to communicate the appearance of the solution. They can be used to show the size and shape of the solution or how the pieces of the solution fit together. The drawings are created as mechanical drawings. Mechanical drawing is a type of drawing also known as *drafting*. These drawings are much different from drawings artists create or those designers sketch. Mechanical drawings are very neat and accurate. See Figure 17-4. Mechanical drawing requires specific sets of tools and skills.

Specification Sheets

The last document used to communicate is a ***specification sheet***. Specification sheets are also called *spec sheets*. They are used to describe items that cannot be shown in drawings. There are two main types of spec sheets:

- Material sheets.
- Quality-of-work sheets.

Material specification sheets describe the materials to be used in the product. The bill of materials was used to list the

Part Number	Quantity	Part Name	Size	Material
1	1	Base	3/4" x 4" x 4"	Pine
2	1	Slider	3/4" x 3/4" x 3"	Pine
3	2	Rails	3/4" x 3/4" x 3"	Pine
4	1	Top	1/4" x 2 1/2" x 3"	Hardboard
5	1	Jar	Standard size	Mason jar

Figure 17-3. This bill of materials lists all the parts needed to manufacture a gumball dispenser.

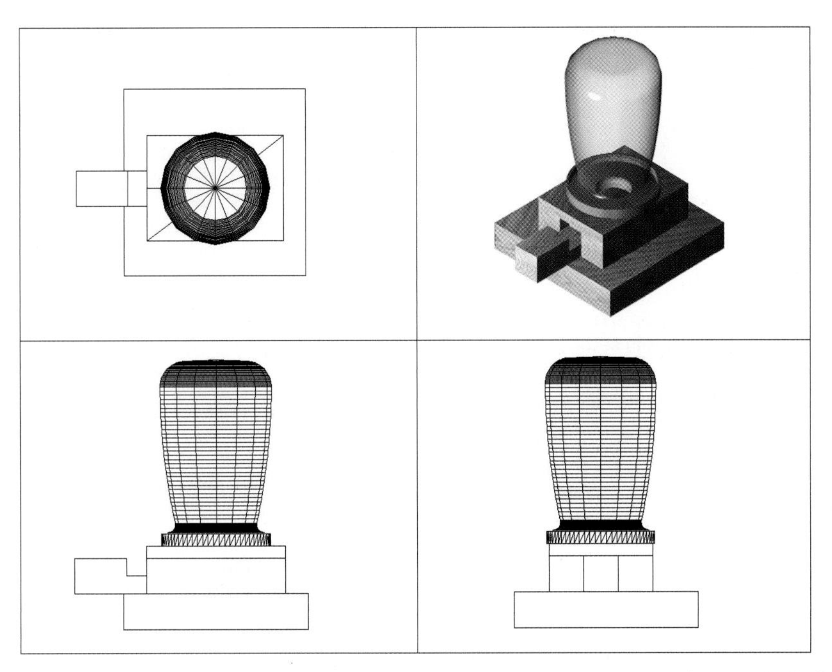

Figure 17-4. This mechanical drawing shows three views and a rendering of a gumball dispenser.

types and sizes of the materials. The spec sheets go into greater detail. They list the exact materials, as well as the quality of materials to be used. For example, lumber can be purchased in several different grades. A spec sheet lists the grade that should be used. Spec sheets are also used to describe the properties of the materials. One type of material spec sheet is an MSDS. See Figure 17-5. An MSDS describes properties such as color, odor, density, and melting point. This sheet also lists how to handle and store the materials. These spec sheets describe any health risks in using the materials.

Quality-of-work specifications explain how well the product will be produced. They list the type of finish on the materials. This can include how well the materials are sanded, painted, or stained. Quality spec sheets also list who is responsible for different stages of the product. This is important in construction. The spec sheet might state that the owner is responsible for laying the carpet and painting the interior.

The bill of materials, drawings, and specification sheets must be exact. These documents are often used when companies make a bid to produce the solution. A bid is a price a company charges to make a product. If a document is wrong, it can cost the designer a lot of money to have it changed once a company has made a bid. The documents are used as part of the contract between designers and manufacturing and construction companies.

Categories of Mechanical Drawings

Mechanical drawings are created to communicate solutions. They can be divided into two major categories:

- Engineering drawings.
- Architectural drawings.

The two categories are used for different types of objects. Tools, machines, toys, and other manufactured products are drawn with ***engineering drawings***. Buildings and structures are drawn with ***architectural drawings***.

Section II - Emergency and First Aid

Primary Route of Entry:
Inhalation
Eyes:
Flush with water for several minutes.
Skin:
Wash with soap and water.
Inhalation:
Remove from exposure.
Ingestion:
Dilute stomach contents with several glasses of milk or water.

Symptoms of Overexposure:
Minimal respiratory tract irritation may occur as with exposure to large amounts of any nontoxic dust.

Medical Conditions Generally Aggravated by Exposure:
None when used as described by product literature.

Additional Information:
None

Section IV - Physical Data

Appearance/Odor:	Black powder / faint odor
Boiling Point:	N.A.
Solubility in Water:	Negligible
Evaporation Rate:	N.A.
Vapor Density (Air=1):	N.A.
Volatile:	N.A. % (Wt.) N.A. % (Vol.)
Softening Range:	43.3°C–60°C (110°F–140°F)
Melting Point:	N.A.
Specific Gravity (H_2O=1):	~1
Vapor Pressure (mm Hg):	N.A.
pH:	N.A.

Section V - Fire and Explosion Data

Flash Point (Method Used):	N.A.
Flammable Limits:	LEL: N.A., UEL: N.A.
NFPA 704:	Health - 0, Fire -1, Reactivity - 0

Figure 17-5. MSDSs explain all the safety hazards associated with materials. Shown are sections from an MSDS for printer toner. (Xerox)

Engineering Drawings

Engineering drawings are used to communicate products to be manufactured. The products can be anything from toasters to space shuttles. See Figure 17-6. Many types of designers and engineers create and use engineering drawings. Aerospace, automobile, electrical, and mechanical engineers all use engineering drawings to create their products. Industrial and product designers also use engineering drawings.

Architectural Drawings

Architectural drawings communicate buildings and structures to be constructed. Floor plans and building details are common architectural drawings. See Figure 17-7. Houses, buildings, bridges, and towers are all drawn with architectural drawings. Land surveyors, architects, and city planners use architectural drawings.

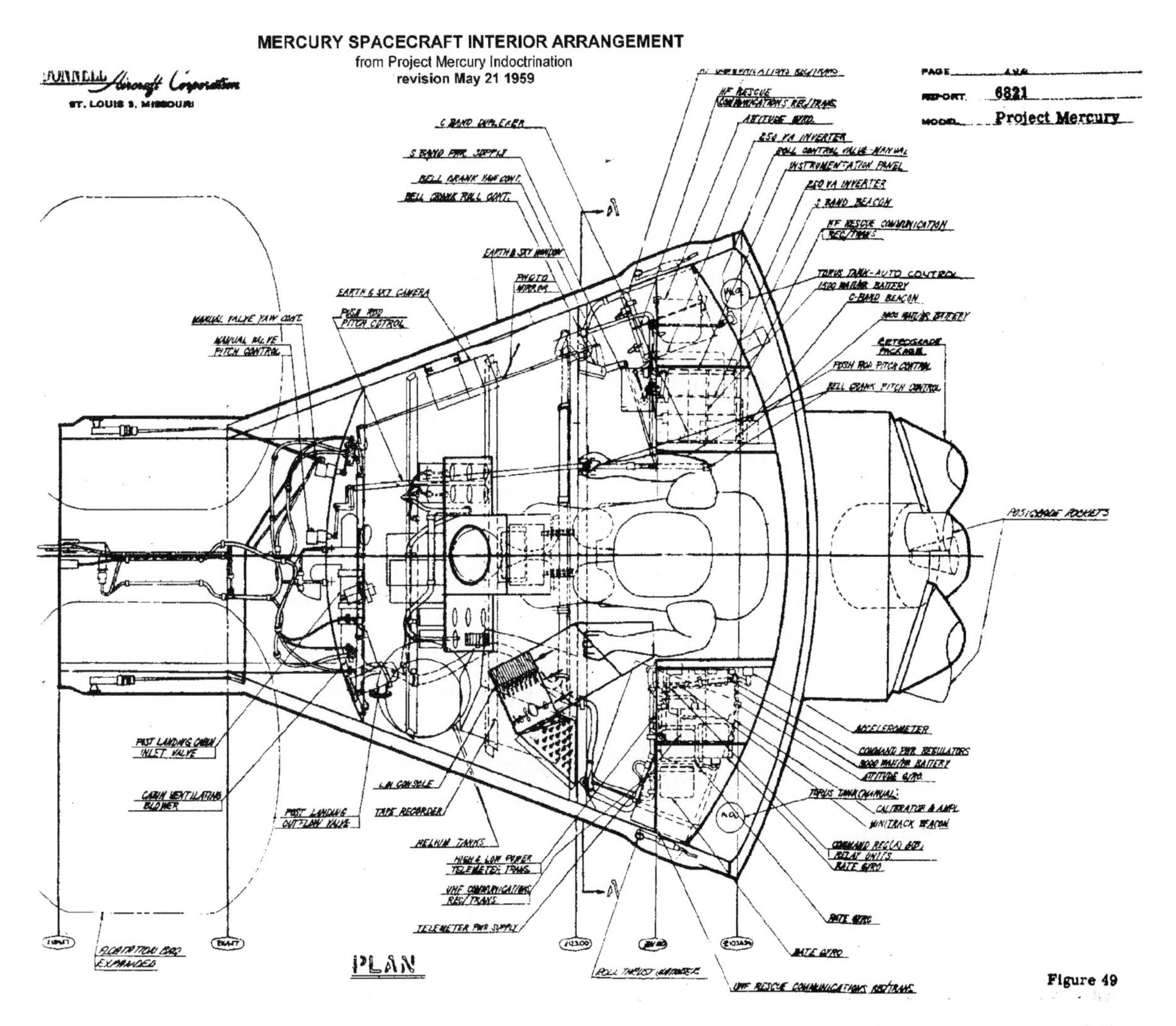

Figure 17-6. Engineering drawings are used in many different industries, including aerospace. (NASA)

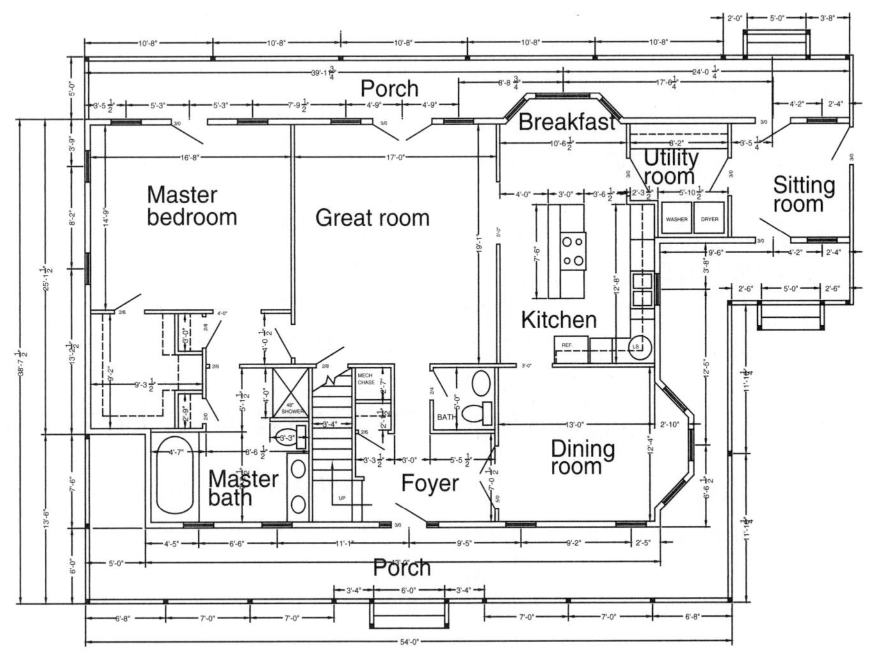

Figure 17-7. Floor plans are examples of architectural drawings.

Types of Working Drawings

Engineering and architectural drawings are drawn as working drawings. ***Working drawings*** are the most complete drawings produced. These drawings display all the information needed to build the solution. They also show how the products are assembled.

Working drawings are drawn to a certain standard. Most designers follow the standards that the American National Standards Institute (ANSI) sets. ANSI has developed a list of standards describing how objects should be drawn and sizes should be shown. Following the ANSI standards ensures that all working drawings can be read and understood. Designers create three types of working drawings:

- Detailed drawings.
- Assembly drawings.
- Schematic drawings.

Detailed Drawings

Detailed drawings are produced for each piece of the solution. See Figure 17-8. The detailed drawings show the exact sizes and shapes of the pieces. The drawings are complete enough that the pieces can be built using the drawings. Detailed drawings use different views to show the object. The designer must choose the views best describing the part.

These drawings describe the size and shape of the piece using dimensions. Dimensions use arrows and numbers to describe size and location features. A floor plan is a detailed drawing including dimensions. A detailed drawing also gives information on the material and finish, using notes. Notes are lines of text describing an object. They give the viewer necessary information about the piece.

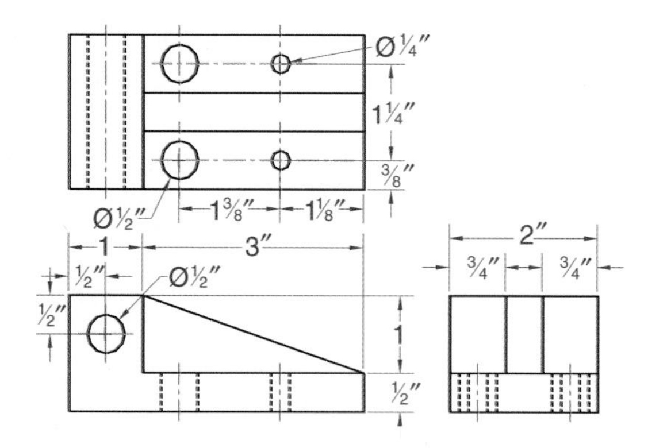

Figure 17-8. Detailed drawings are used to explain the sizes and shapes of individual pieces of the solution.

Assembly Drawings

Assembly drawings show how parts fit together. See Figure 17-9. These drawings can be used to help put the pieces together. Products requiring consumers to assemble them often include assembly drawings. The drawings help visualize how all the parts fit with one another. They are drawn to look three-dimensional with all the pieces pulled apart. Dotted lines are used to show the location of many of the pieces.

These drawings can also be drawn to show an assembled product. They can help show how the device or product functions. The parts are labeled with numbers or letters and arrows, called *leaders*. The numbers or letters match the bill of materials included with the drawing. This helps people identify each part. Assembly drawings do not include dimensions. If the exact sizes of the parts are needed, the detailed drawings are used.

Schematic Drawings

Schematic drawings are used to show systems. A system is a group of objects working together for a common goal. Schematic drawings are used to show how the parts are connected to form

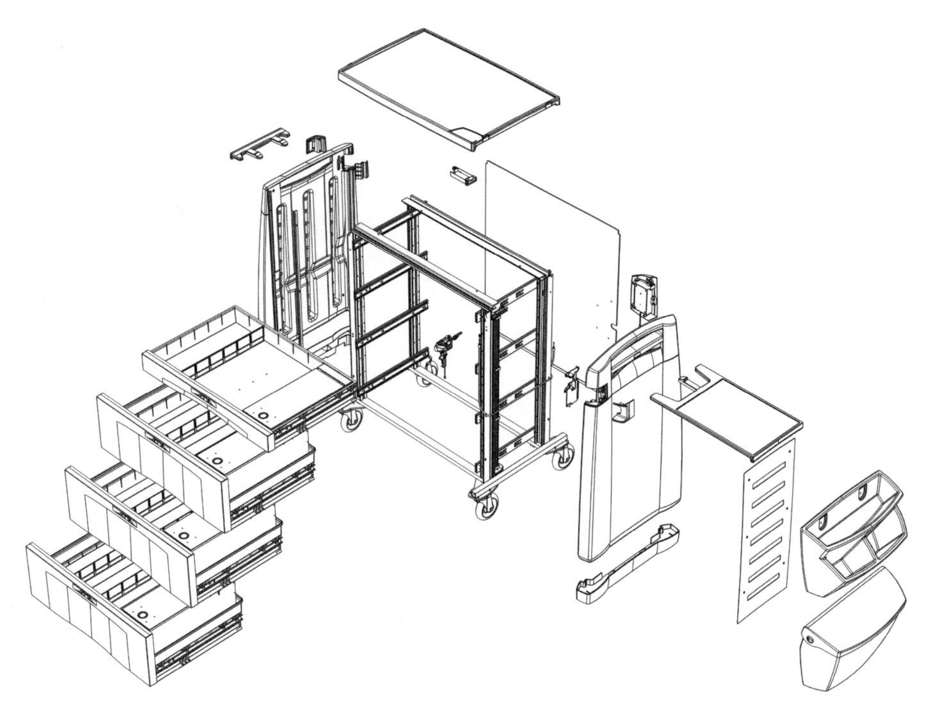

Figure 17-9. Assembly drawings show how products fit together. (Design Central, design firm; Artromick International, client)

a system. See Figure 17-10. These drawings are used for electrical circuits, water pipes, and production lines. Schematic drawings are not drawn to actual scale. They do not communicate size and shape as other working drawings do.

These drawings are used to show the relation of the parts to one another and how the product flows. Electrical drawings show the flow of electricity among parts of the circuit. Piping drawings show how water moves from one location to another. Drawings of production lines communicate how the product moves from one end of the production line to the other.

Schematic drawings use symbols for the components of the system. See Figure 17-11. ANSI sets the symbols so all designers can read the schematic drawings. Lines representing paths are used to connect the symbols. In an electrical drawing, the lines represent wires or copper on a circuit board.

Figure 17-10. A large-scale schematic of the layers of circuits in a computer processor. The final processor is only a few inches in length. (Intel)

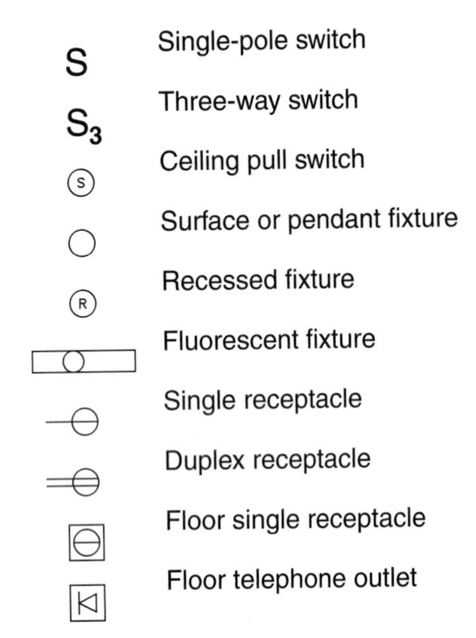

Figure 17-11. Standard symbols are used to make schematics easier to draw and understand.

Drawing Classifications

Most design solutions are physical objects having three dimensions. The objects have widths, heights, and depths. In this step, the designers' job is to decide the best way to display the three dimensions. Designers have two basic ways to communicate solutions:

- Orthographic drawings.
- Pictorial drawings.

Orthographic Drawings

Orthographic drawings are one way of communicating solutions. They show objects in separate two-dimensional views. They are often called ***multiview drawings*** because they use several views to describe the object. When creating multiview drawings, designers must select the best views to describe the object. All objects can be shown in up to six different views. Imagine a typical house for a moment. The house has six different sides. You can easily walk around the house to see the front, the back, and the right and left sides. If you were able to fly over the house, you would see the top. If the house had a crawl space underneath, you would be able to look at the bottom of the house. In an orthographic drawing, each of the six views of the house can be drawn as individual two-dimensional drawings.

Multiview drawings can be very confusing because there can be six different views. To help reduce the confusion, each view has a specific location. The placements of the views are standard. All designers place the views in the same locations. To understand the locations of the views, imagine placing the solution in a glass box and then rotating the box and drawing exactly what you see on each of the six sides. See Figure 17-12. Now imagine the box is hinged on the front, on each side. When the box is opened, each view is in the right place. See Figure 17-13.

Using the glass box to see all six sides is an example of orthographic projection. *Orthographic projection* is the term used to create six views of an object. All six views are not, however, always needed. Some

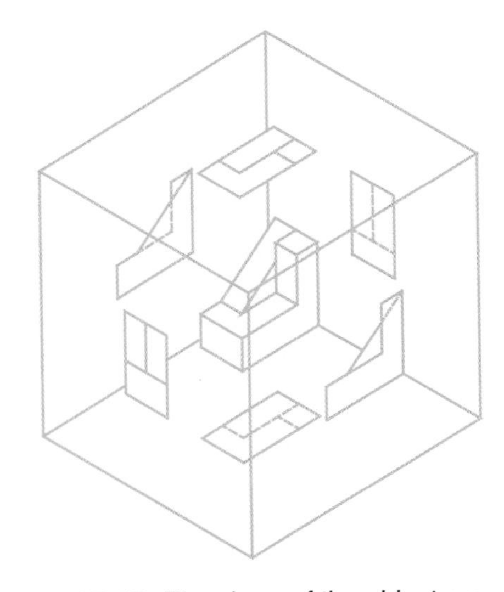

Figure 17-12. The views of the object are projected onto the sides of a box.

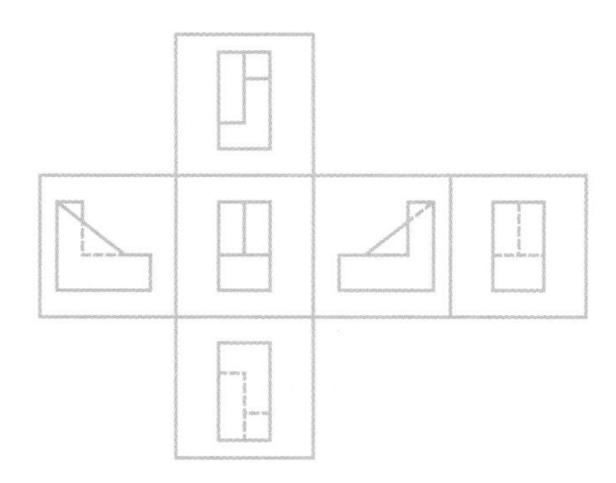

Figure 17-13. When the box is open, the views are in the right place.

objects can be completely described in one or two views. It is actually uncommon to see orthographic drawings with all six views. Designers do not waste the time to create unnecessary views. Drawings with one, two, or three views are the most common and have specific uses:

- **One-view drawings.** These are used to show flat parts and objects. See Figure 17-14. Sheet metal and hardboard pieces are drawn using one view. The top of the object is the view normally shown.
- **Two-view drawings.** These are used when drawing cylindrical objects. Soda cans and baseball bats are drawn with two views. In two-view drawings, the front and side or end are shown.
- **Three-view drawings.** These are used for rectangular objects. They are usually created using the front, top, and right side of the object. Three-view drawings are often used to draw complex parts.

When creating orthographic drawings, the first step for designers is to identify the front of the object. The front is always the side showing the most detail. Sometimes the front of the object is very obvious. For example, the front view of a television set is the side the television screen is on. Other times, however, it is harder to determine which view should be the front view. An

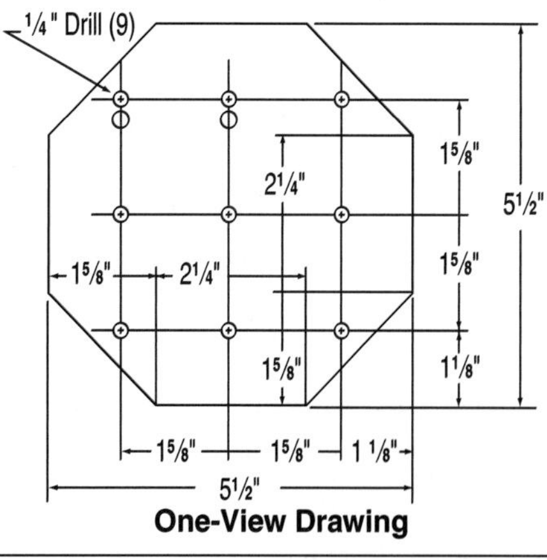

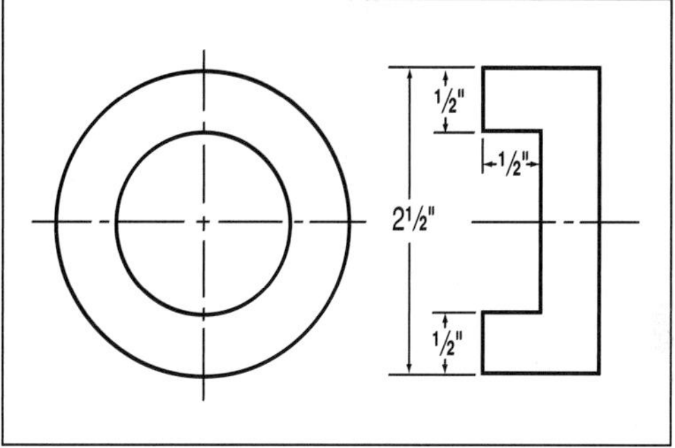

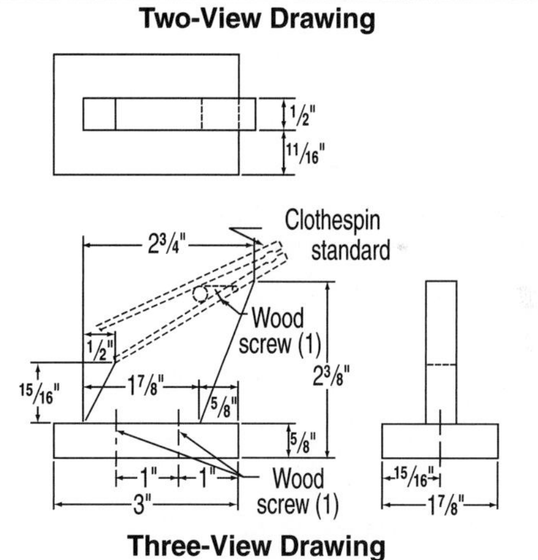

Figure 17-14. Orthographic drawings can be one-, two-, or three-view drawings.

automobile, for example, is drawn with the side of the vehicle as the front of the object. The side is the best front view because it shows the most detail. This view shows the slope and aerodynamics of the car. Once the front of the object is selected, the remaining views are easy. The top view is found by rotating the object down 90°. The right side view can be drawn by turning the object 90° to the left. All the views should be 90° from the view next to them.

Line Types

One challenge with orthographic drawings is that some objects have parts that cannot be seen in all the views. Looking at the front view of a toaster, for example, you are not able to see the slots for the bread. If an object has holes drilled into it, you do not know how deep they go into the object. For this reason, there is a type of line used to show features that cannot be seen. See Figure 17-15. These lines are called *hidden lines*.

A hidden line is just one type of line used in multiview drawings. There are many different types of lines. They are known as the ***alphabet of lines***. See Figure 17-16. The most common lines in the alphabet are the construction line, object line, borderline, hidden line, and centerline. Each line is drawn differently and has a specific purpose:

- **Construction lines.** These are solid lines drawn lightly. They are used to lay out drawings.
- **Object lines.** These are the lines used to show the edges of the object. They are heavy and dark lines.
- **Borderlines.** These are the thickest and darkest lines. They are used to surround the edges of the paper.
- **Hidden lines.** These show the location of objects that cannot be seen. Hidden lines are drawn with short dashes.
- **Centerlines.** These are used to show the centers of any objects with center points, such as arcs, ellipses, and circles. Centerlines are drawn with alternating long and short dashes.

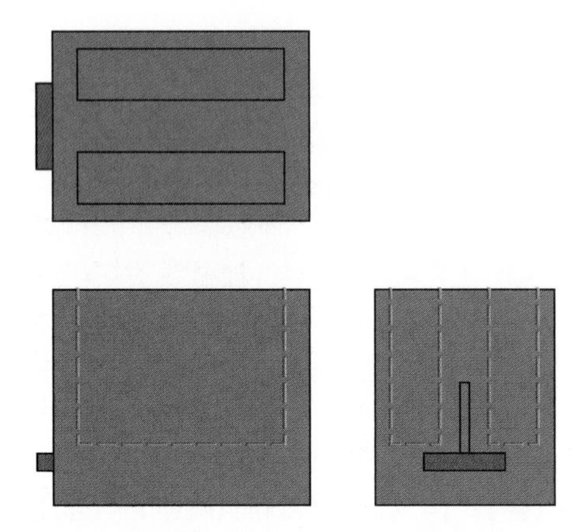

Figure 17-15. Hidden lines are used to show the slots in the toaster.

Dimensions

There are other types of lines used in drawings. ***Dimensioning*** is a process using two types of lines: extension lines and dimension lines. See Figure 17-17. Extension lines are lines drawn from an object to the outside of a view. They show the edge or center of an object. Dimension lines are drawn between two extension lines. These lines have an arrowhead on each end and a measurement in the middle. Together, the extension and dimension lines are used to describe the dimension of an object. The dimension of an object, as described in Chapter 14, is a measurement giving one of three types of information:

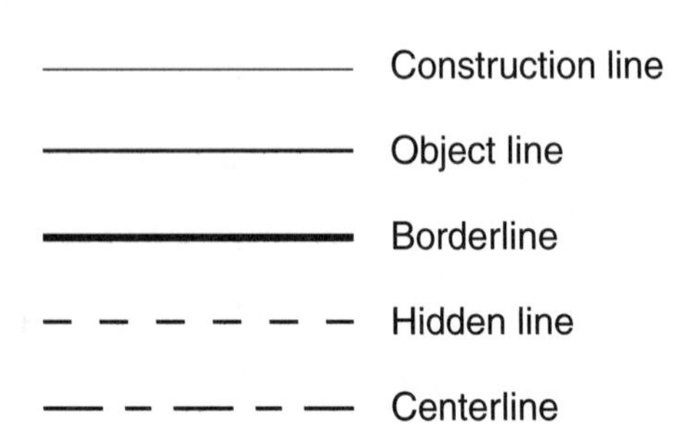

Figure 17-16. The alphabet of lines includes the standard lines used in technical drawing.

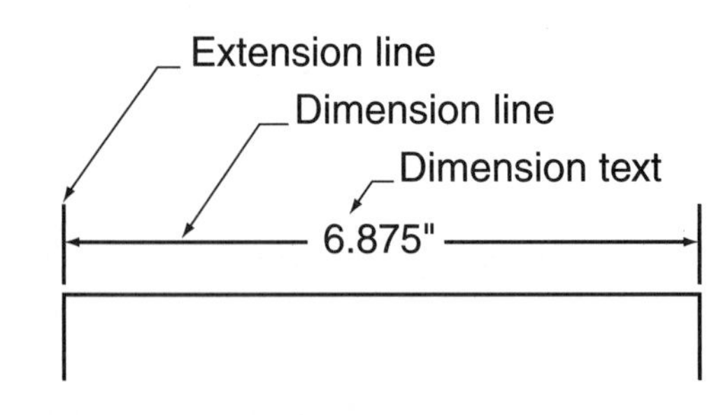

Figure 17-17. Dimensions include extension lines, dimension lines, and dimension text.

- **Size dimensions.** These describe the lengths of objects.
- **Location dimensions.** These show the distance between two features (such as circles or arcs) or between a feature and the edge of the object.
- **Shape dimensions.** These show the angles between different parts of an object.

Dimensions must be displayed clearly so they can be understood. See Figure 17-18. The designers must include all the dimensions needed to produce the object. They should not, however, include more dimensions than are necessary. There are many rules designers use when dimensioning drawings. The following are some rules:

- Dimensions should not be placed inside the object.
- Dimensions should be placed between the views, not along the outside of the drawing.
- Dimensions should be placed on the view best showing the measurement.
- The locations and sizes of all circles and arcs must be shown.

Circles and arcs are dimensioned using leaders. A leader is a line with an arrow on one end and a measurement on the other. The measurement given for a full circle is always the diameter of a circle. The radius measurement is shown for any arc, or incomplete circle. The dimensions used should not distract viewers. They are used to better communicate the sizes and shapes of objects.

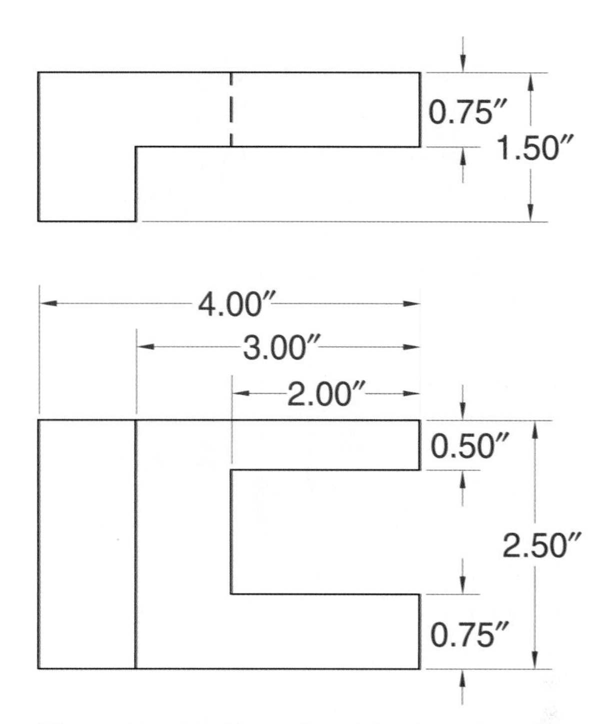

Figure 17-18. Dimensioned drawings should be easy to read and follow the rules of dimensioning.

Pictorial Drawings

Pictorial drawings, as the name suggests, are similar to pictures. See Figure 17-19. They show how objects look to the eye. Pictorial drawings are easier to understand than orthographic drawings. They are often used when the viewer does not understand orthographic drawings. Orthographic drawings can be confusing because they are separated into different views. In pictorial drawings, the views are drawn together. Instead of being a flat drawing of the sides, pictorial drawings are drawn to look three-dimensional. They are easy to understand because they look real.

Figure 17-19. Pictorial drawings appear as the eye sees the objects. (Keith Nelson)

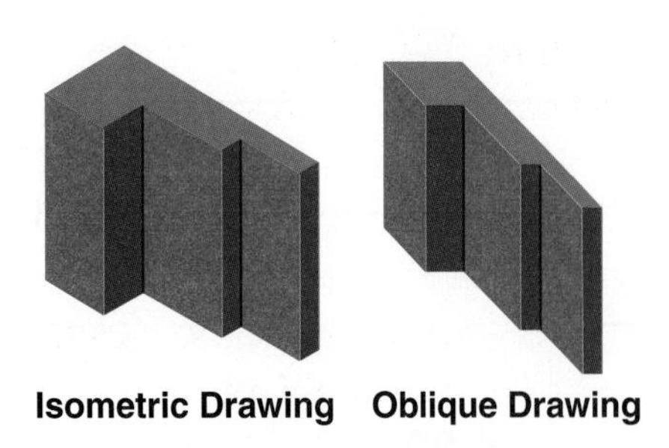

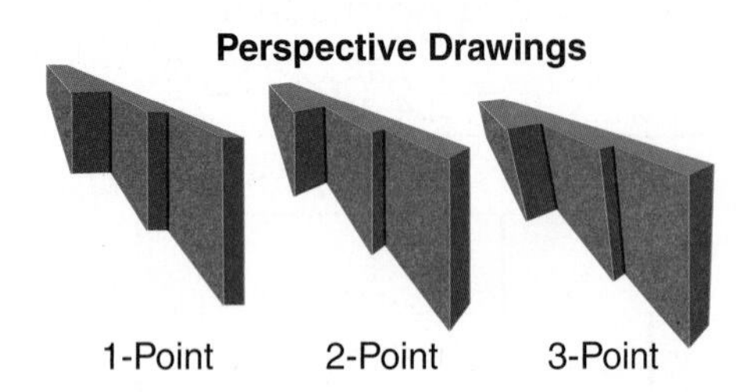

Figure 17-20. Pictorial drawings include isometric, oblique, and perspective drawings.

There are several types of pictorial drawings. See Figure 17-20. These are the three main types:

- Isometric drawings.
- Oblique drawings.
- Perspective drawings.

You might remember these types from the pictorial sketches in Chapter 13. Pictorial drawings, however, are different from sketches. The sketches created in the earlier steps were simply ideas. They were not very detailed. The pictorial drawings in this step are very detailed and accurate. They represent the final solution of the design process.

Isometric Drawings

Isometric drawings are one of the most common types of pictorial drawings. See Figure 17-21. The word *isometric* means "equal measurements." Isometric drawings are created with angles equal to each other. The three lines making the front corner are always 120° from each other. The lines along the bottom of the object are each 30° from horizontal.

All lines in an isometric drawing are drawn to scale. Each line is accurate when it is measured. The length, width, and height of the object are all drawn to scale, or accurate, in isometric drawings. They can be drawn full-scale. Full-scale drawings are used when the object fits on the paper using the actual measurements. These drawings can also be drawn to other scales. Scales are labeled using two numbers. The first number is a length on the paper. The second number is the actual length in real life. For example, a scale of 1/2″ = 1″ means that every 1/2″ on the paper equals 1″ on the object. So, if the object is 3″ long, it is drawn at 1 1/2″. Scales are used when the object is either too big for the paper or too small to be seen.

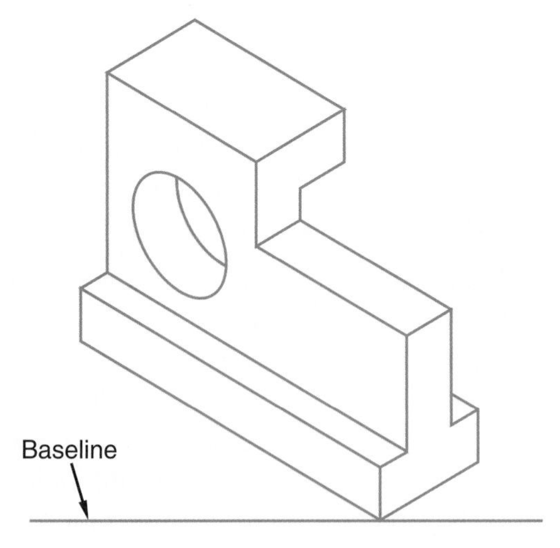

Figure 17-21. Isometric drawings are created so the horizontal lines are 30° from the baseline.

Oblique Drawings

Oblique drawings are not used as often as isometric drawings. They are used to show objects in which one view is the most important, such as kitchen cabinets or television sets. See Figure 17-22. These drawings are created using the front view from an orthographic drawing. The front of an oblique drawing is the only true-size view. Lengths and angles can be measured only on the front view. To give the front view depth, lines angling backward are drawn from the front of the object.

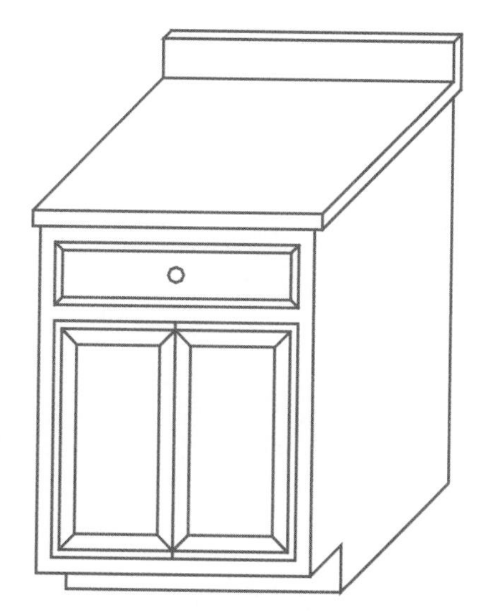

Figure 17-22. Objects with most of the detail in the front view are often drawn with oblique drawings.

The angle of the lines can be drawn at any angle. The most common angles, however, are 30° and 45° angles.

There are three main types of oblique drawings. The only difference among the three types is the depth of the object. When the depth is drawn the true length, the drawing is a cavalier oblique. A drawing showing half of the length is called a *cabinet oblique drawing*. A general oblique drawing is drawn with a depth of anywhere from one-half to full-size. Oblique drawings are mainly used for display drawings. They are not often used for drawings that will be measured because only the front is accurate.

Perspective Drawings

Perspective drawings are the drawings most often used in the presentation of ideas. They are the pictorial drawings most similar to what the eye sees. Perspective drawings are very similar to perspective sketches, discussed in Chapter 13. These drawings, however, are created with drawing tools and are much more accurate than the sketches. They rely on vanishing points and vertical lines. A vanishing point is a spot toward which the lines are pointed.

The number of vanishing points in a perspective drawing determines the drawing's type. A one-point perspective is drawn using only one vanishing point. The one-point perspective drawing resembles the oblique drawing. The front view is flat on the paper. The top and sides of the drawing are at an angle. The difference is that, in the oblique drawing, all the angled lines are parallel, or at the same angle. In the one-point perspective, all the lines are either vertical or angled back to a vanishing point.

A drawing with two vanishing points is called a *two-point perspective*. See Figure 17-23. This drawing looks similar to an isometric drawing. In a two-point perspective, the vanishing points are on opposite sides of the horizon line. Each side of the object in the sketch extends toward one of the vanishing points.

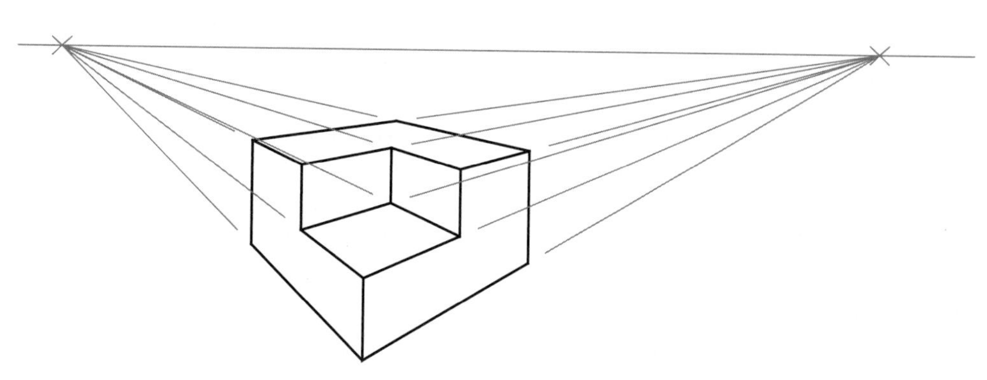

Figure 17-23. A two-point perspective drawing.

Career Highlight

Architectural Drafters

The Job: Architectural drafters prepare the technical drawings and plans construction workers use to build buildings and civil structures. They prepare drawings and specifications showing the architectural and structural features of the structures. The drawings provide construction workers with visual guidelines for the structures. Also, they show the technical details of the structures, provide dimensions, indicate materials, and list procedures. Architectural drafters might specialize in one type of structure, such as residential, commercial, or industrial.

Working Conditions: Most architectural drafters work a standard 40-hour week. Generally, they work in comfortable offices. These offices are furnished to promote and support the drafters' work.

Education and Training: Employers usually want applicants who have completed drafting programs at technical institutes, community colleges, or four-year colleges or universities. They expect applicants to have good drafting and mechanical drawing skills. Employers also expect applicants to have a knowledge of drafting standards and an ability to use CAD and drafting systems.

Career Cluster: Architecture & Construction

Career Pathway: Design/Pre-Construction

A three-point perspective drawing is a drawing that has three vanishing points. Three-point perspective sketches can be used for tall buildings and structures. Three-point perspective drawings are not used as much as one- and two-point perspectives.

Mechanical Drawing Tools

Tools are involved in producing mechanical drawings, as they are involved in almost all uses of technology. They are used to help designers draw straight and accurate lines. These tools are also used to make mechanical drawings easier to create. Without mechanical drawing tools, it is difficult to draw perfect circles or create parallel lines. There are two types of technical-drawing tools:

- Traditional tools.
- CAD tools.

Traditional Tools

You might already be familiar with a few traditional technical-drawing tools. When you think of architects or engineers, you might imagine them using T squares, triangles, and compasses. All these tools are traditional technical-drawing tools. Traditional tools are used when designers and draftspersons create drawings using pencils and paper.

A basic set of traditional tools includes a drawing board, a T square, a 45° triangle, a 30°-60° triangle, a compass, an eraser, templates, and a pencil. See Figure 17-24. The ***drawing board*** provides a smooth surface on which to draw. This board also includes one straight edge for the T square to move along. The ***T square*** is a tool made up of two pieces: the head and the blade. The two pieces are perpendicular to each other. The head rests along the side of the drawing board and keeps the blade horizontal across the paper. The T square is used to draw all horizontal lines in the drawing. This tool is also used as a ledge for other tools, such as triangles.

A standard set of traditional drawing tools includes two triangles. One of the triangles, called the ***45° triangle***, has one 90° corner and two 45° corners. The other triangle is called a ***30°-60° triangle*** and has one 90° corner, one 30° corner, and one 60° corner. The triangles are used with the T square to draw vertical lines. They are also used to draw lines at 30°, 45°, and 60° angles. If both triangles are used together, angles of 15° and 75° can also be drawn.

Figure 17-24. These tools make up a traditional set of drawing tools.

The ***compass*** is the drawing tool used to create circles and arcs. This tool resembles the letter *A*. One leg of the *A* has a small pointed tip. The other holds a pen or pencil. The compass is held at the top and rotated around to draw circles. See Figure 17-25. A tool that is very similar to a compass is a divider. The main difference is that dividers have two pointed tips, instead of one. Dividers are used to transfer measurements without using a scale.

A ***scale*** is a tool used to make measurements. This tool is a specialized type of ruler. Scales have different sides that help designers draw objects larger or smaller than they are in real life. Designers use different types of scales, depending on the industry in which they work. Architects and interior designers use architectural scales. Civil engineers and land surveyors use engineering scales. See Figure 17-26. Industrial designers use mechanical-drafting scales. Each scale has sides that are useful for the type of drawings the designers will create. Architectural scales have sides used in drawings in which 1/8″, 1/4″, 1/2″, and 3/4″ equal 1′. Engineering scales are used for drawings measuring 10′, 20′, 30′, 40′, 50′,

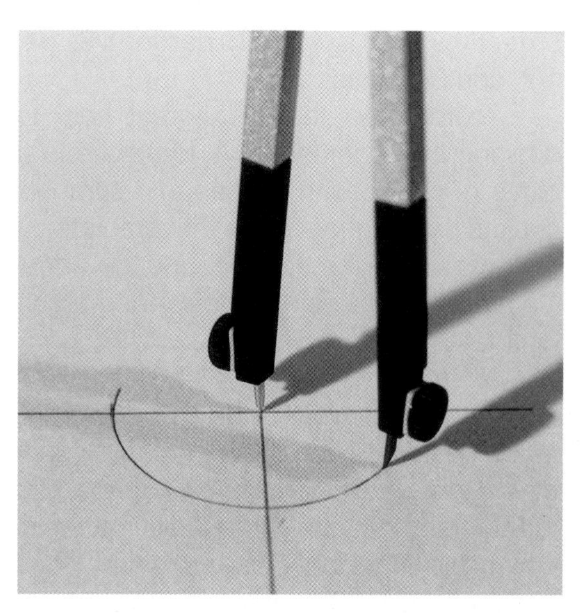

Figure 17-25. Compasses are used to create perfectly round circles.

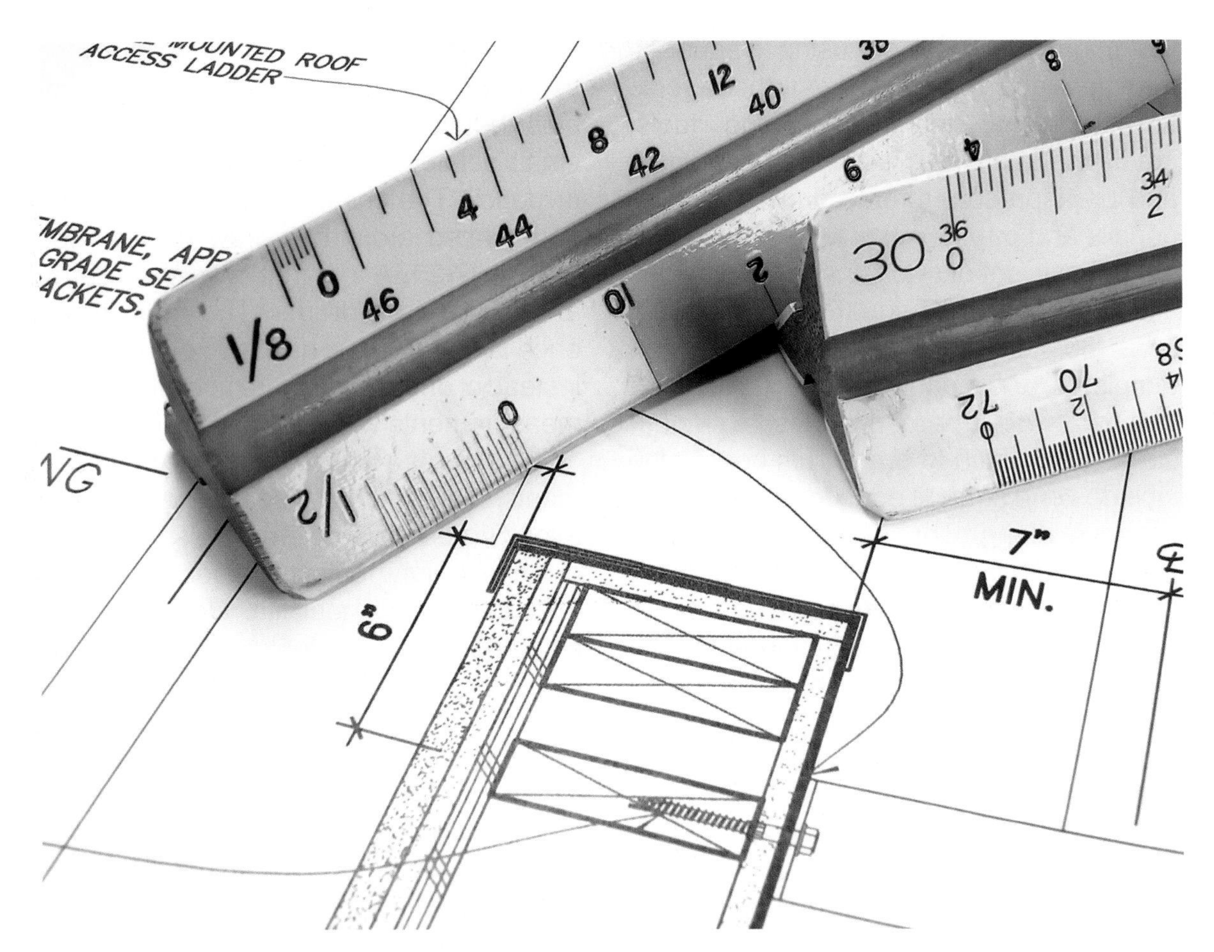

Figure 17-26. Scales help engineers and designers draw large objects on paper. (©iStockphoto.com/cmcderm1)

or 60′ for every 1″. Mechanical-drafting scales have a side for full-scale, divided in sixteenths of an inch; and sides for 1/8, 1/4, 1/2, and 3/4 scales.

Another tool that is a great help to designers is a template. A ***template*** is a piece of plastic with shapes and symbols cut into it. Designers place the template on their drawings and trace around the edge. Common templates are circular and architectural templates. See Figure 17-27.

All the traditional tools are used together to create mechanical drawings. Many designers use them in many industries. Some designers feel it is important to understand how to use traditional tools. Other designers feel these tools are not as important as CAD tools.

Computer-Aided Design (CAD) Tools

Computer-aided design (CAD) utilizes the computer to create technical drawings. See Figure 17-28. Many types of designers use CAD tools that are part of a computer system. A typical CAD system includes a computer, a monitor, a printer or plotter, storage systems, input devices, and CAD software.

The computer used for CAD tools is a bit different from the typical computer used in a household. CAD computers require large amounts of memory and large processors. Some computers used for CAD even have two processors.

Figure 17-27. Templates make drawing standard objects easier. (©iStockphoto.com/sndr)

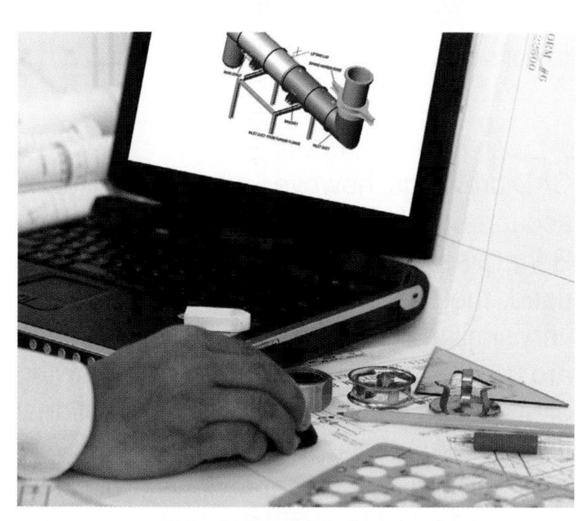

Figure 17-28. Designers often use CAD tools. (Product Development Technologies, PDT)

The computer processor speeds up the time it takes for the computer to perform commands. The memory in a computer allows the computer to handle more functions at once.

Another difference in CAD systems is the way these systems store files. Many CAD drawings are large and take up a lot of disk space. So, designers use tape, CD-ROM, DVD-ROM, or hard backup drives on which they can write information. Using these drives allows the designers to save the drawings they create. This is one advantage of CAD drawings. Designers are able to save the drawings and easily make changes after the drawings are complete.

These designers often use large computer monitors because they must be able to see the drawings they are creating. See Figure 17-29. The bigger monitors allow them to see more of their drawings. The designers are also able to zoom into different areas of the drawings to look more closely. CAD computer monitors must also have a high resolution. The resolution controls how much information can be displayed on the screen at one time.

The tools used in a CAD system often include a type of printer other computer users do not use. A ***plotter*** is a printer able to print on large paper. Many architectural drawings, for example, need to be printed on paper 24″ wide and 36″ long. Plotters allow designers to make prints this big.

The actual computer requirements depend on the CAD software used. There

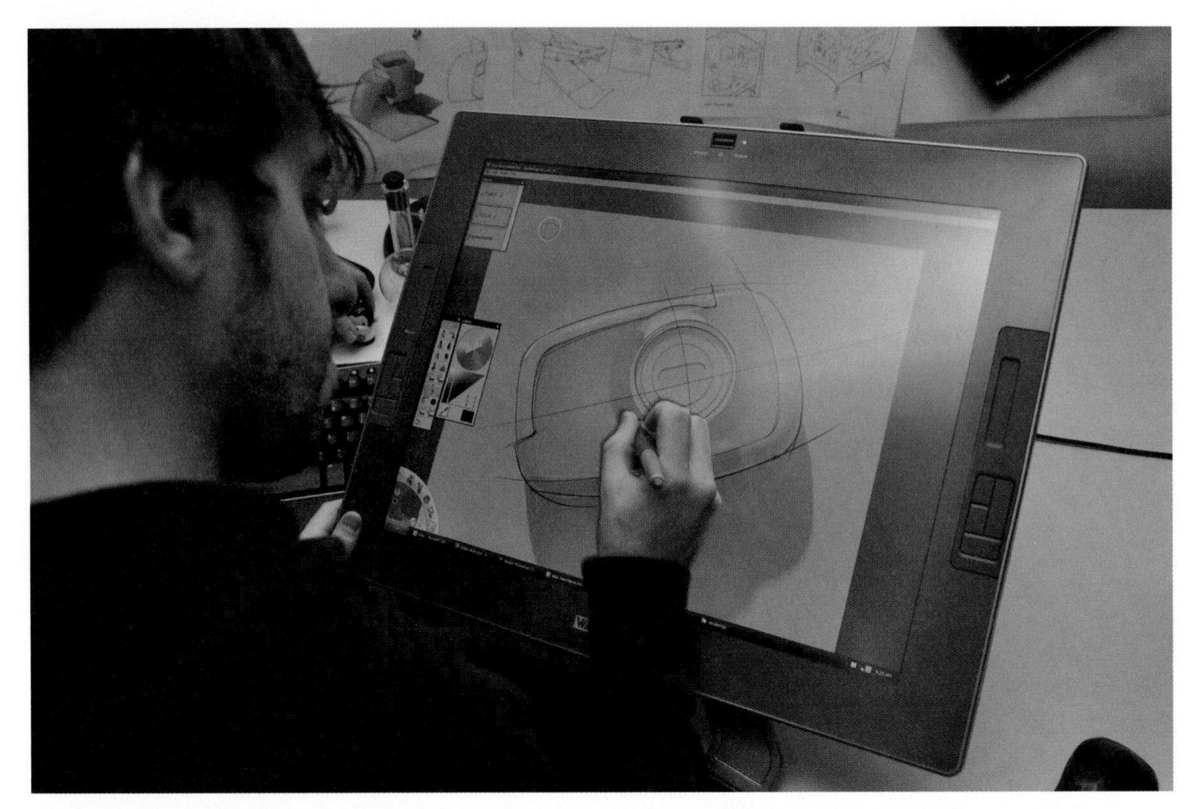

Figure 17-29. Large computer monitors with high resolutions help designers see large drawings. (Design Central)

are many types of CAD software. Software ranges in price from a few dollars to several thousands of dollars. The quality of the software ranges from low quality to outstanding quality. The limitations of the software also vary. Designers purchase the right software for their needs. Some software is created for a typical homeowner to create floor plans and landscape designs. Other software allows designers to create complex parts or even design entire buildings and cities.

Making Mechanical Drawings

Mechanical drawings can be produced with either traditional tools or CAD tools. Each type has advantages and disadvantages. Traditional drawings are often easier to create because the designer does not have to learn a computer-software package. CAD drawings, however, are easier to make accurate and easier to correct or change. Regardless of the type of drawing tools used, the process of creating a drawing is very similar. For example, the following steps are used to create an orthographic drawing of a picture frame using either traditional tools or CAD tools. See Figure 17-30.

1. Decide how many views are needed to show the features of the object.
2. Sketch the views on a sheet of paper.
3. Draw construction lines outlining the front and top views.
4. Add a 45° line. Transfer lines outlining the right side view.
5. Locate the centerlines of any circles or arcs.
6. Add the rest of the lines in all the views.
7. Erase or trim the construction lines.

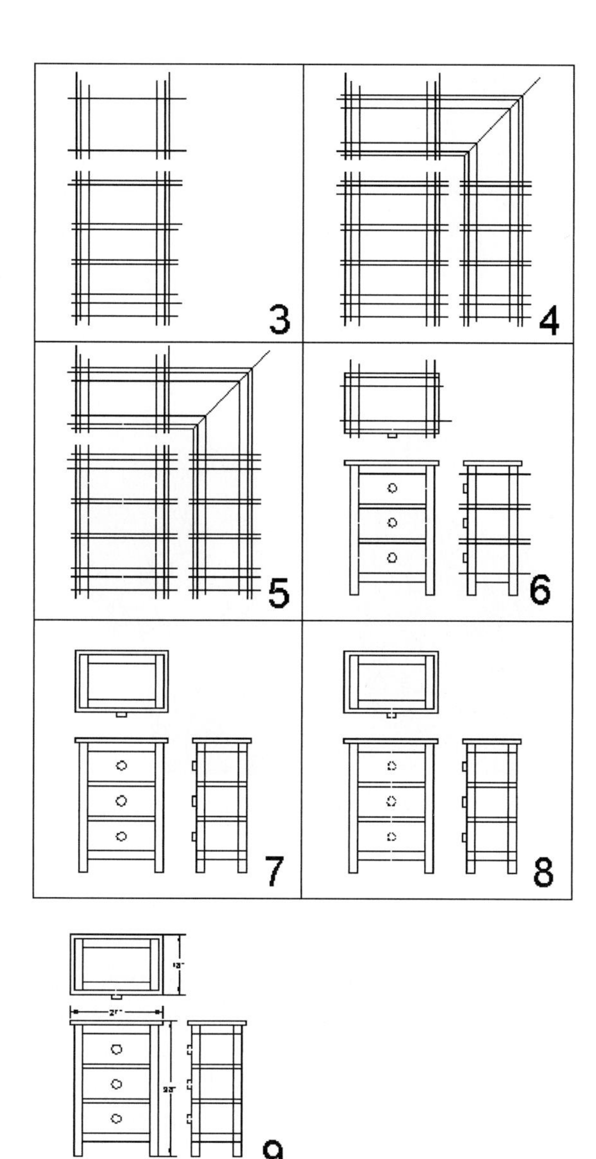

Figure 17-30. These steps are used to create mechanical drawings.

8. Add any hidden lines and additional centerlines needed.
9. Add dimensions.

Documenting and Presenting the Solution

The final stage in communicating solutions is to present the final solution. All products must be approved before they are sent to production. To receive approval, the designers must present the final solution to management. See Figure 17-31. This stage involves creating a design report, or portfolio, and a presentation.

The ***design report***, also known as a *design portfolio* or an *engineering portfolio*, is created to show the entire design process. The portfolio should include documents from each step of the process. This report should include the design brief listing the problem, criteria, and constraints. Next, there should be a review of the research that was conducted. The review should list the types of research used and the findings from the research. The sketches generated and the evaluation grids used to select the best solution should be the next documents in the report. It should be clear that a number of concepts were considered. Also, it should be clear why the final solution was selected. The next documents should show the design as it was developed from a sketch to a final solution. This section includes renderings, models, test results, mechanical drawings, bills of materials, and specification sheets. Lastly, a statement of recommendation should be created that clearly states why the design is being recommended and how it solves the original problem.

This report is distributed to the design management team before a final oral presentation. The oral presentation is the opportunity for the design team to convince the management that the design solution solves the problem. The presentation should highlight the design process. This oral presentation should focus on the problem and how the design team solved the problem. The presentation should follow the same order as the design report. The report and presentation are often the final checkpoints of the design process. If the presentation and reports are approved, the solution goes into production.

Figure 17-31. Designers display their work in presentations to clients or management. (©iStockphoto.com/Yuri_Arcurs)

Summary

Solutions must be communicated to several groups of people. Management must approve the solution before it is manufactured. Manufacturers must understand the solution in order to produce it. Consumers must be able to assemble and use the solution. Designers use three different documents to communicate the solution. They create bills of materials, drawings, and specification sheets, showing and explaining the solution.

The drawings created are called *mechanical drawings*. They are created using either traditional tools or CAD tools. The drawings are very accurate and precise. They can be detailed, assembly, or schematic drawings. Each drawing type has specific uses. Detailed drawings are usually created using orthographic drawings. Orthographic drawings have several views used to describe the object. Assembly drawings are normally pictorial drawings. Pictorial drawings are created to look three-dimensional and are much easier to understand than orthographic drawings. When all the drawings and documents are complete, they are presented to management or the client for final approval.

STEM Connections

Science

Create a drawing that would be used with a science lab report. The drawing should show the equipment and supplies used in the experiment.

Mathematics

Draw a 1″ grid on a simple cartoon image. Using a scale, recreate the cartoon image at three different scales (for example, 1″ = 1/2″, 1″ = 2″, and 1″ = 1/4″).

Curricular Connections

Language Arts

Research a type of material. Write a specification sheet explaining the common uses, material properties, and health risks of the material.

Activities

1. Bring a toy or household object with three or more parts from home. Create a bill of materials for the product.
2. Create a detailed drawing of an object in your classroom (for example, a desk, bookshelf, or chair). Use the correct line types. Add dimensions.
3. Create a set of documents for the solution you have created in the previous chapters. Once the documents are complete, present your solution to the class.

Test Your Knowledge

Do not write in this book. Place your answers to this test on a separate sheet of paper.

1. A bill of materials contains information about the health risks of a material. True or false?
2. A mechanical drawing is an accurate drawing created using drawing tools. True or false?
3. Give three examples of objects that would be communicated by engineering drawings and three that would be communicated by architectural drawings.
4. Name the three types of working drawings.
5. A schematic drawing can be used to show the flow of a product. True or false?
6. Orthographic drawings are also called _____.
7. List four examples of traditional drawing tools.
8. Computers used for CAD are the same as most computers used as household computers. True or false?
9. The steps used to create drawings are very similar when using traditional tools and when using CAD tools. True or false?
10. Explain the process of presenting a solution.

Reading Organizer

On a separate sheet of paper, create a detailed outline based on what you have read about the different ways of communicating design solutions.

Example:

I. Communication documents
   A. Bills of materials
   B. Drawings
   C. Specification sheets

TSA Modular Activity

This activity develops the skills used in TSA's Lights, Camera, Action! event.

Lights, Camera, Action!

Activity Overview

In this activity, you will develop a storyboard, script, production plan, and finished video for one of the following themes:

- A profile of your school.
- A profile of a student or faculty member.
- A profile of a local business or organization.
- A discussion of a community issue.

Materials

- Paper.
- A pencil.
- A three-ring binder.
- A video camera.
- A computer with video-production software.

Background Information

- **Planning the video.** Plan your video as a story, making sure it has a beginning, middle, and ending. List the important points you want to cover, and then use those points to create a simple outline. Think visually. Remember that your story will be told mostly with pictures, instead of words.
- **Storyboard.** To help yourself and the others involved in the project visualize the video, make a storyboard. A storyboard is a series of simple pictures, similar to a comic book, showing each change in what the viewer will see. Make each individual storyboard sketch on a separate page. This allows you to try different combinations to get the most effective sequence for the video. When you have a final sequence, number your pages.

- **Scripting.** Even though the pictures (video) will tell most of the story, the spoken words (audio) will tie the pictures together. Your script will follow the sequence of the storyboard, with a brief description of each numbered shot and the actual words that will be recorded to accompany that shot. Sometimes, audio is not fully scripted.
- **Production planning.** Video is almost never shot in the final program sequence. Therefore, careful planning is needed to make the most effective use of time and resources. For each numbered shot in your storyboard, your production plan should list the location of the shot, any on-camera people involved, any props or special materials needed, and (where relevant) the time of day the shot must be made. For initial planning, note cards with the information for each shot are useful. They can be sequenced to make efficient use of your resources—for example, grouping together all the shots at one location or involving the same people.
- **Shooting the video.** To provide visual interest in the finished video, be sure to vary your types of shots. Don't shoot all close-ups or all wide shots. Suit the shot type to the subject and to the flow of the video as shown on your storyboard. To help identify each shot, make a large card with the scene number written in bold marker. Shoot a few seconds of the card at the beginning of the shot. To provide editing flexibility, shoot a variety of cutaway shots. These are used as transitions between other shots. For example, if you are showing a school assembly program, include a few close-ups of members of the audience listening to the speaker. When shooting interviews, you normally concentrate on the person being interviewed. When the interview is finished, shoot some close-up shots of your reporters asking questions and some of them just looking interested and nodding.
- **Camera and microphones.** If you are shooting in a location where you depend on battery power for the camera, be sure to have one or more fully charged backup batteries available. When lighting conditions change (such as moving from indoor to outdoor settings), be sure to check the camera's white balance. Adjust, if necessary, to avoid adding a color cast to your video. When shooting general scenes, such as a cafeteria or sporting event, record the live audio. This audio can later be used in postproduction. For recorded narration or interviews, avoid using the camera's built-in microphone, if possible. Much better audio quality results from using a microphone (wired or wireless) placed close to the subject being recorded.
- **Postproduction.** When editing your scenes, use various visual effects, such as dissolves, fades, and zooms. Music, sound effects, and additional voice-over narration can be added during postproduction. Use cutaway shots to avoid disturbing jump cuts in the video. These cuts occur, for example, when an interview must be edited to eliminate unwanted material. If the camera is focused on the speaker, changes in expression or head position between the adjoining shots make the cut obvious. By inserting a second or so of video showing the interviewer looking interested, the change will not be noticed.

Guidelines

The final video must be two-and-a-half to four minutes in length. The following items must be included in your final printed report:

- A cover page.
- A table of contents.
- A storyboard.
- A script.
- A description of editing techniques.

Evaluation Criteria

Your project will be evaluated using the following criteria:

- The storyboard, script, and production plan.
- The correlation of the storyboard to the video.
- The technical quality of the video.

Drawings are one way of communicating design solutions. The tools shown in this image are necessary for designers to accurately communicate solutions this way.

CHAPTER 18

Improving Solutions

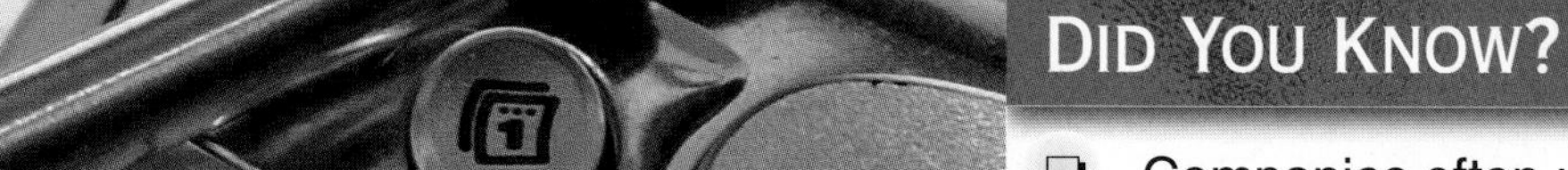

Did You Know?

- Companies often update and improve their logos. A logo is a symbol of a corporation's identity. Logos can be traced back to ancient Greece. There, craftspeople marked their work with their own symbol. Today, logos are used as the main symbols of companies.
- Logos are designed so they are easy to recognize and remember. They should also be pleasing to look at and in good taste. It would hurt a company's image if the company's logo was disrespectful and unsightly. Most importantly, the style of the logo must be up-to-date. Think of several major companies. Have they updated their logos in your lifetime? Are their new logos more up-to-date?
- A logo must be able to be reproduced easily. For this reason, most logos use only one, two, or three colors. The easier the logos are to reproduce, the less expensive it is to print them on a number of different items.

Objectives

The information given in this chapter will help you do the following:

- Explain the importance of improving solutions.
- Summarize the four main reasons for improvement.
- Summarize the four major sources of information about products.
- Explain the process of improving solutions.

Key Words

These words are used in this chapter. Do you know what they mean?

brand name
building code
competition
competitor
consumer
internal source
patch
profit
recall
regulation
standardization

Preparing to Read

As you read this chapter, choose objects whose designs have been improved over time. Correlate those improvements to the four categories of reasons for improvements. Use the Reading Organizer at the end of the chapter to organize your thoughts.

Have you ever played a game you wish had more options, had trouble finding information on a Web site, or taken a toy back to the store because you did not like it? If so, designers want to know about your dissatisfaction. They are concerned about how well people like their products. Designers work hard to improve their designs, even after the designs have been produced.

In this step of the design process, designers search for different ways to improve their products. It might seem to you that, once designers reach this stage, they are done with the design. The designers have gone through many steps to get to this stage. They have done everything from identifying a problem all the way through solving it. This included conducting research and tests and creating sketches, models, and drawings. The solution has probably even been manufactured and sold by this point. See **Figure 18-1.** You might think the designers are finished with the product by now.

Designers, however, are never really done with a solution. Design is a process that does not stop. Designers can always go back and make changes to a design. The design process is similar to a ladder. See **Figure 18-2.** At this step of the ladder, designers can choose to step backward to a previous step. When the designers went through the process the first time, they created a solution. This solution was the solution that seemed to be the best at the time. The designers used research, surveys, and tests to determine the solution they designed. As time goes on, however, many things change. The best

Figure 18-1. A finished image of a mobile assistant.

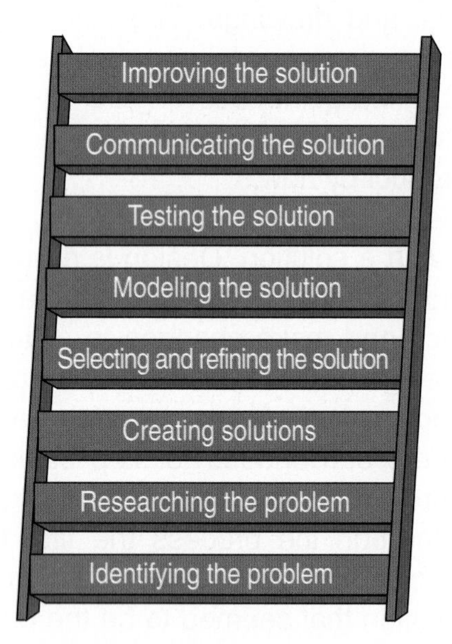

Figure 18-2. The steps of the design process can be seen as a ladder.

design at one moment might not be the best solution at a later time. For example, imagine an automobile designed in 1945. See Figure 18-3. Is this automobile the best solution in today's world? The antique car does not have air-conditioning, power steering, a CD player, or an air bag. This car is not desirable and would not sell very well. The design is obsolete, or out-of-date. Some designs take a long time before they become obsolete. Other designs are obsolete in a very short amount of time. Consider a computer. A computer designed several months ago might already be outdated. In a matter of a few months, there are new computers that are faster and better.

Reasons for Improvements

Keeping a product from becoming outdated is a concern for designers. If their product is out-of-date, the company will suffer. Designers must improve their solutions to keep them current. The concern to improve products can be grouped into four categories:

- Profit.
- Competition.
- Function.
- Safety.

Figure 18-3. This antique auto does not have many of the features found in cars today. (Ford Motor Company)

Profit

Profit is a major factor in product improvement. This factor is the amount of money left over after the company's bills are paid. Designers work to increase their profit in two different ways. They can raise the price of the product. This increases the income from the solution and is the easiest way to increase the profit. People, however, might choose not to pay the extra money for the product. If you went to the store to buy a game costing five dollars more than it did the day before, would you pay for it? You would probably find another product to buy with your money. If all people stopped buying the product, the company would lose money, instead of making more. See Figure 18-4. If the company raises the price, it should also improve the function of the product. Improving the product helps people feel that the extra money is worth spending.

The second option designers have to increase profit is to lower their expenses. If it costs less to produce and sell the product, the designers will have more profit. There are several places designers can try to lower their costs. One place is in manufacturing. See Figure 18-5. Designers can hire a manufacturing engineer to examine how the product is produced. They might find that the manufacturer has waste materials that can be used to make more products or that they have more workers than they need. Unfortunately, this can cause companies to lay off workers. The designers might also be able to use less expensive materials to produce the products.

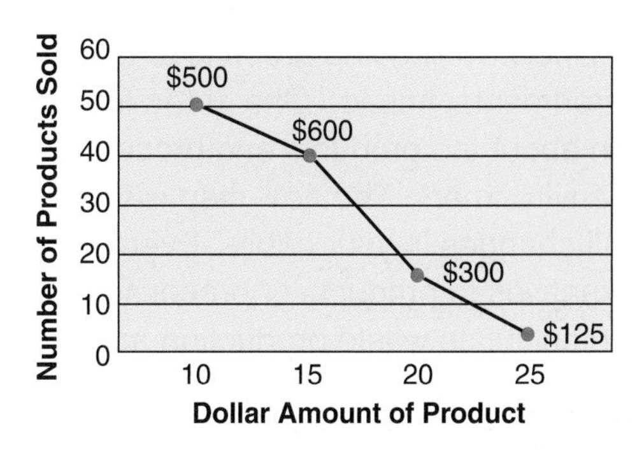

Figure 18-4. This graph shows how many products will sell at each price and how much money they will generate.

Figure 18-5. Manufacturing is a major part of a company's costs.

If they can lower the cost of materials, they can increase profit.

Competition

The second reason for improvement is competition. ***Competition*** is the term used when two or more companies sell the same types of products. Designers improve their products to keep ahead of their ***competitors***. Competition is an important part of business. Businesses compete to gain customers. Competition is great for ***consumers*** because it helps keep prices down. This part of business also keeps the quality of products at a relatively high level. Designers want consumers to choose their products over the products of another company. To ensure this happens, they make their products more appealing. They might begin by improving the design of a product's package. The package is the first thing that catches the consumer's eye. See Figure 18-6. Many consumers are willing to buy the first product they

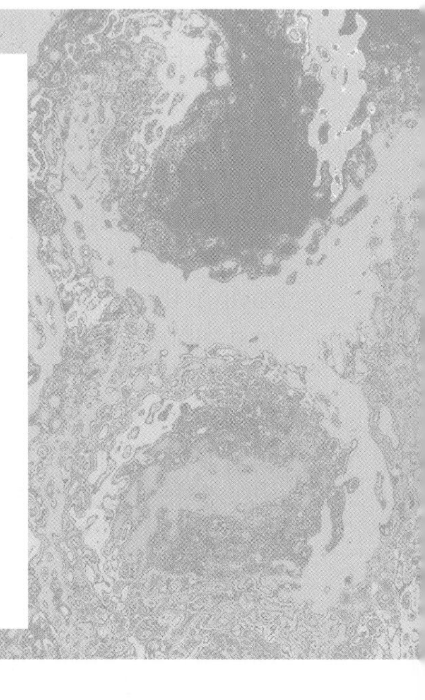

Think Green

Reduction

One way of going green is to take steps to reduce your environmental impact. One of the first steps toward reduction is to learn about the products and processes that have harmful effects on the environment. The next step is to combat those effects by making small changes in your habits. Even just being more mindful of green alternatives to products or processes can help. Larger steps would be to calculate waste production and carbon dioxide emissions. The attempts individuals and companies have made toward reduction have been found to have a great impact on the environment by impacting it less.

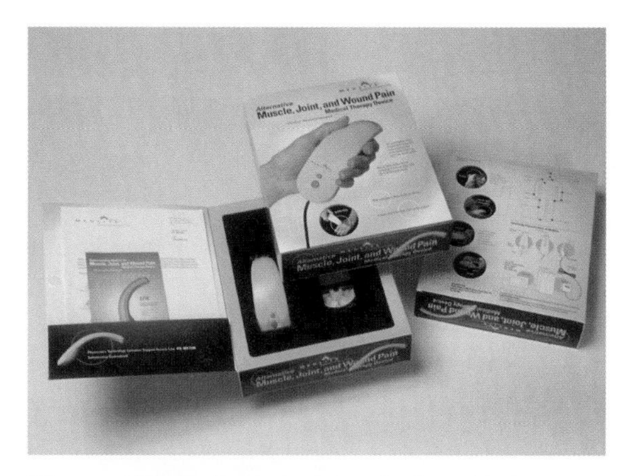

Figure 18-6. Packages are designed to be attractive and grab your attention. (Design Central, design firm; Medlife, client)

see. Some consumers, however, need more than a nice package. Designers also improve the function of the solution. They include more or better features than the competitors' products have.

Designers also improve products to build a brand name. A ***brand name*** is the title of a line of products. If a consumer likes one product with a certain brand name, they are more likely to buy another product with the same name. For example, Frito-Lay® makes Ruffles® potato chips. Someone who likes these chips would be more likely to buy another type of chip Frito-Lay makes than chips from another company.

Function

Function is the third reason to improve a solution. This reason was mentioned as a part of profit and competition. Function is also, however, a major reason for improvement on its own. Throughout the design process, sketches and drawings were analyzed. Prototypes and mock-ups were tested and evaluated. It is still possible, however, that the solution does not function as well as it can. There might be certain uses that were not tested. The product might fail when it is used in those manners. All products are subject to failure. Once these failures are discovered, the designers must improve the solution to eliminate the problems.

When a product has problems or issues needing to be addressed, they are handled quickly. This is very common in software design. When software designers discover a problem, they improve the solution and create a patch. A ***patch*** is a small piece of software that fixes a known problem. The piece of software can be installed on top of the current software. When larger changes are made to software, designers create new versions. New versions of software are created after all the problems from the original software have been identified. They also have changes in the functions or features of the program. If the changes are major, the version is given a new whole number (such as version 6.0). See Figure 18-7. If the changes are small, it is given a decimal number (such as version 3.2).

Safety

Safety is another major reason for improving solutions. The safety of the solution was tested during the design. Some safety issues, however, are found after the product has been produced. When a major safety problem is found, a product recall is issued. A ***recall*** is a request that those who purchased a product bring it back. In return, the consumers receive a new product. In other cases, their original product is altered to make it safe. Recalls are common in products such as food and children's products. If it is found that food might be contaminated, the company recalls it. Companies want to ensure all their products are safe. Many companies rely on product registration to issue recalls. They send a notice to the registered owners of the product.

When safety concerns that can lead to death are found, mandatory recalls

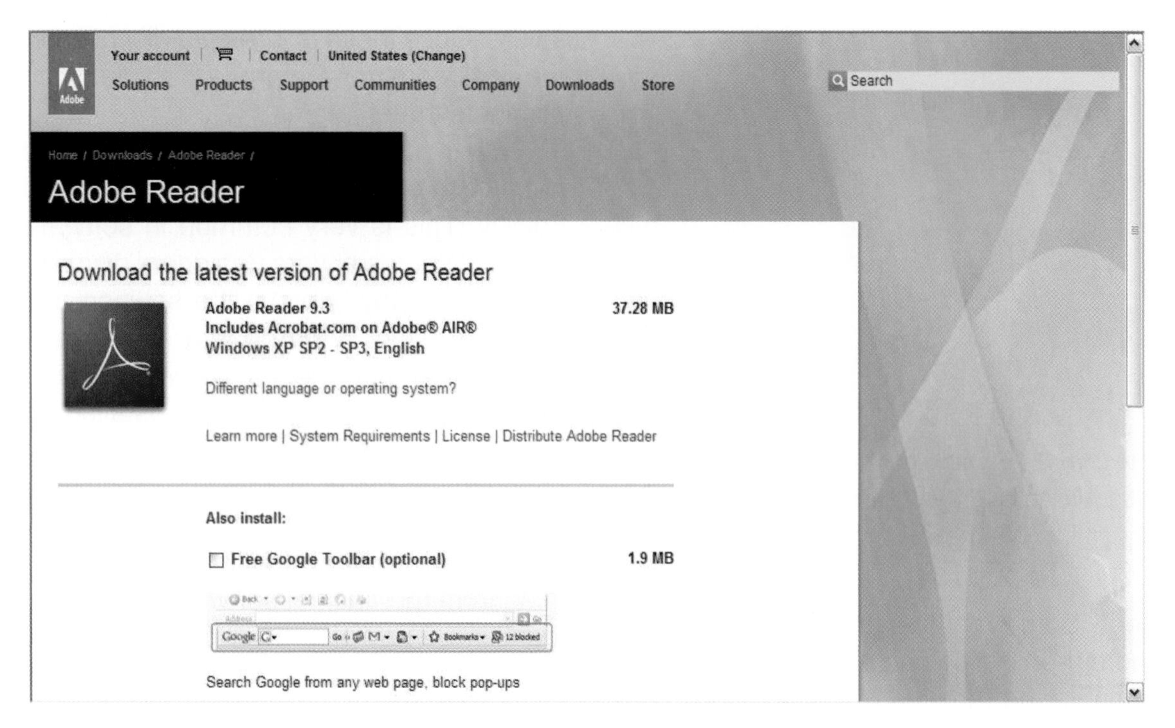

Figure 18-7. New software is often given a new version number.

are issued. A mandatory recall means, if you do not return the product, you are using it at your own risk. Not all safety issues require mandatory recalls. Some issues are very minor and are handled using voluntary recalls. A consumer with a product that has been voluntarily recalled has the option of returning it for a new and improved product. Consumers are not risking serious injury or death, however, if they choose not to return the product.

Sources of Improvement

Designers rely on a number of different sources of information about their products. These sources help them determine the types of improvements needed. They include consumers, competitors, government organizations, and internal sources.

Consumers

Consumers might be the best source of information for improving solutions. See Figure 18-8. Consumers are the users of products. After using a product, they know what they like and do not like about it. Designers receive feedback from consumers in several ways.

One source of feedback is when customers return items. Designers and companies collect a list of reasons products are returned. They might find that many people are returning the product for the same reason. The designer uses this information to redesign and improve the solution.

Another way they receive information is through complaints. Companies and designers take complaints seriously. Companies want their customers to be satisfied and to enjoy their products. When customers are unhappy with a product, the company tries to improve the product. With the use of the Internet, it is much easier for

Figure 18-8. The users of a product can be great sources of information.

customers to reach companies and voice their opinions. Most companies have Web sites including the phone numbers, street address, and e-mail addresses that can be used to contact them.

Consumers should know it is also important to provide positive feedback. See Figure 18-9. Through returns and complaints, designers receive negative feedback. Designers also need information, however, on the positive features of the solution. This information helps the designers change only the bad parts of the design.

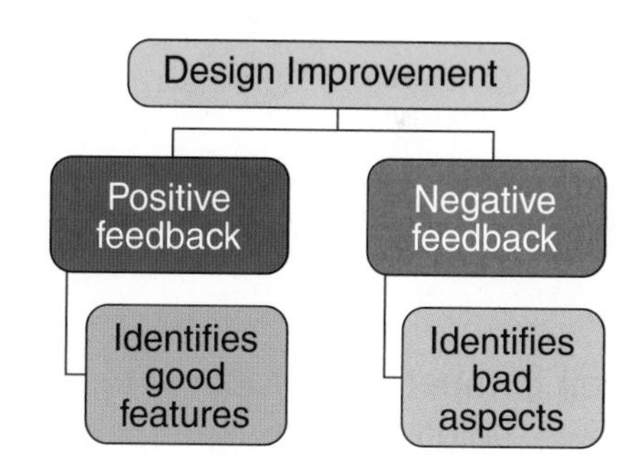

Figure 18-9. Both positive feedback and negative feedback are helpful to designers.

One way designers identify the good and bad features of a solution is through product samples. Companies often distribute free samples of their products. Comment cards are handed out along with the products. Consumers are asked to try the product and fill out the comment cards. The comment cards are written so the company can get as much feedback as possible. The company can then make improvements without taking away good features.

Competitors

Competitors are an excellent source of information. They are the companies selling the same types of products. Companies

Career Highlight

Building Inspectors

The Job: Building inspectors examine buildings and other structures to make sure they meet local and state codes and regulations. Building codes list the specifications that must be followed when buildings are constructed. Buildings are inspected during several phases of construction.

Working Conditions: Inspectors often work 40-hour workweeks. They spend most of their time either on job sites or reviewing construction documents. Their work often requires them to climb ladders and crawl into crawl spaces to check the construction of buildings.

Education and Training: The requirements of a building inspector differ depending on the state. Most, however, do not require more than a high school diploma. Building inspectors do often have job experience in the skilled trades. Most inspecting jobs do require certification. Some require the passing of an examination.

Career Cluster: Architecture & Construction

Career Pathway: Construction

(©iStockphoto.com/lisafx)

use their competitors' products to improve their own solutions. Designers compare their own products to competing products. They must be careful, however, that they do not create the exact same products.

Most products are either copyrighted or patented. Copyrights and patents are legal rights making it illegal to copy or reproduce a product. The process of receiving a copyright or patent usually takes place after a product has been designed and tested. This is why some design projects are kept secret during their developmental stages. Many solutions are uncovered at a public event. Automobiles are one example of a product revealed at large events. See Figure 18-10.

Government Organizations

Government organizations are a source of improvement different from the others. When designers use consumers and competitors as sources of information, they choose whether or not to make the changes. Government organizations, however, do not give the designers an option. Many government organizations

Figure 18-10. Auto shows are used to uncover new products and vehicles. (Ford Motor Company)

make regulations. ***Regulations*** are rules and laws designers must follow.

The importance of safety in the workplace cannot be overemphasized. Some aspects of safety include working safely with hazardous materials and wastes, working safely with machinery and tools, working safely with electricity, preventing fires, and understanding how to extinguish any fires that might occur. There are many codes, laws, standards, and regulations related to safety and technology systems. Often, such standards are adopted into laws. These laws are designed to protect workers and the general public from injuries and other health risks. In addition to safety standards specified by law, most companies have safety standards developed for their specific systems and processes. Many companies have safety training sessions to ensure that employees are familiar with the hazards and safety regulations in their workplace.

Building codes are common sets of government regulations. They are regulations that local governments set. These codes inform builders and designers of the rules they must follow in designing and building structures. See Figure 18-11.

Figure 18-11. Builders must check local building codes when constructing new buildings. (Habitat for Humanity International)

Architects and other designers must follow the building codes as they design buildings. They must also follow the codes when they improve or renovate buildings. When builders renovate old homes, they must update everything to meet the new building codes. For example, if you are renovating an older home, you might need to replace parts of the electrical wiring because the current wiring does not meet today's building codes.

All designers must deal with government regulations of some sort. Chemists design medicines. Some designers, such as chemists, have many regulations set by the FDA. The U.S. Department of Agriculture (USDA) sets the regulations for those who raise animals and crops. The Federal Communications Commission (FCC) regulates television and radio programs.

Technology is powerful. If it is not controlled or used properly, it has the potential to cause great harm. These regulations exist to protect society and the environment. Thus, it is imperative that they be followed. In many cases, not adhering to such regulations puts you and others in danger. Not adhering to these regulations also can result in legal consequences. Before engaging in any technological activity, be sure to research the regulations governing it and follow them precisely.

Internal Sources

Internal sources provide information on improvements from within the design company. The sources can be members of the design team or people from other departments, such as marketing or manufacturing. They try to improve products to help their company. The need to lower manufacturing costs can come from internal sources. Lowering manufacturing costs can have many benefits for the company.

Standardization is another improvement coming from internal sources. This improvement is a process used to make standard parts. Standard parts can be used for more than one product. See Figure 18-12. For example, a company making television sets can design the same remote control to work on all their TVs. It is less expensive to design and manufacture one remote control than many different models.

Figure 18-12. Batteries come in standard sizes so they fit many different products. (Duracell)

Making Improvements

Making improvements on a design is a continuous process. This process uses the same design process that was used to create the product the first time. Once the designers choose to improve their product, they must choose where they are going to begin. The designers take steps back down the design process. Remember, we described the design process as a ladder. Each step of the process is a rung on the ladder. The designer can choose to revisit and redo any of the previous steps.

If the problem is a small issue, the designers might need to go back only a step or two. For example, imagine designers created a stapler. The problem with the product is that the staples do not fit inside of it. The designers might decide to go back to the step of communicating designs. There, they can change the size of the stapler on the working drawings. The designers submit the new drawings to the manufacturing team. The improved staplers are then produced. The problem is solved.

Unfortunately, it is not always this easy. There are times when the designers must go all the way back to the first step of the design process. If the product does not fit the users' needs or function correctly, the designers might have to begin with a new design brief. Other times, the designers are able to go back to a step in the middle of the process. Whichever step the designers go to is the point where they start the process over. The designers must continue through the rest of the process. They cannot skip any of the steps just because they completed them when they designed the old product.

SUMMARY

Design is a never-ending process. Solutions can always be improved. A design that was a good solution at one time will eventually become outdated. Designers have several reasons to improve their solutions. These reasons range from profit and market share to function and safety. Designers receive information about the need for product redesign from various sources. Customers and competitors are two sources of information. Other sources are government agencies and the company itself.

Whatever aspects the designers improve, they must go back and redo the steps of the design process. Some improvements require the designers to go back to the very beginning and redefine the problem. Other problems with the design can be solved by going back only a step or two. Innovative designers often check their designs to see if improvements are needed. It is important that designs are improved. This is the only way our technology and knowledge will advance.

STEM CONNECTIONS

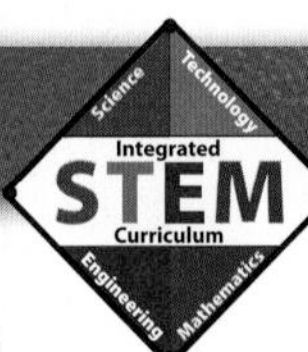

Science

Research the USDA and FDA regulations controlling the medicines and foods we use and eat. Choose two regulations from each agency. Present the information to your class.

CURRICULAR CONNECTIONS

Language Arts

Write a letter to a company regarding one of the company's products. In the letter, provide positive and negative feedback.

ACTIVITIES

1. Choose a modern product and the original version of it. Research the improvements that have been made to keep the product up-to-date.
2. Gather several household products or toys. Develop a poster board showing different ways you can improve the products.
3. Interview a designer about the importance of improving designs.

Test Your Knowledge

Do not write in this book. Place your answers to this test on a separate sheet of paper.

1. Once a design reaches the improvement stage, it is complete. True or false?
2. Give three examples of products that have become obsolete.
3. _____ is the amount of money left after the bills are paid.
4. Define *competition*.
5. Recalls are rare in food and children's products. True or false?
6. Returns and complaints provide _____ feedback.
7. Competitors are great sources of information. True or false?
8. Standard parts are pieces that can be used to make _____ product.
9. Improving solutions is a process. True or false?
10. Why might a designer have to start over at the beginning of the design process?

Reading Organizer

On a separate sheet of paper, list objects whose designs have been improved. Give reasons for those improvements. Use the four categories of reasons for improvements, and if possible, give details in your reasons.

Improved Object	**Reason for improvement**
Example: Brakes of an automobile	Safety: The existing design of the brakes was thought to be potentially hazardous

CHAPTER 18 ACTIVITY A

COMMUNICATION DESIGN

THE CHALLENGE

Design and produce a 30-second commercial advertising the technology-education course you are currently taking.

INTRODUCTION

Every day, new products, systems, and structures are designed. Some products are even designed to sell other products. Television commercials are an example of this type of design. Designers creating commercials use a design process including the following steps:

1. Identifying a problem.
2. Researching the problem.
3. Creating solutions.
4. Selecting and refining the solution.
5. Modeling the solution.
6. Testing the solution.
7. Communicating the solution.
8. Improving the solution.

This activity gives you an opportunity to complete the same steps, while designing a television commercial.

EQUIPMENT AND SUPPLIES

- A pencil.
- Paper.
- Video cameras.
- Recording media.
- Activity sheets from the Student Activity Manual.
- Costumes.
- Props.

Procedure

Your teacher will divide you into design teams of three or four students. Each group will complete the following steps. Pay close attention to each step because some are to be done as a group. Others are to be done individually.

Identifying a Problem

1. Review the design challenge.
2. As a group, create a design brief for this challenge. The design brief should include the problem statement, criteria, and constraints.

Researching the Problem

3. As a group, create a survey that will help you gather information about different parts of the design. Ask questions to determine answers to the following questions:
   - What information should be included in the video? (What projects or activities are important parts of the course?)
   - Should there be teacher, student, or administration interviews?
   - What types of music would make good background music for the video?
4. Distribute the survey to your classmates.
5. Collect the surveys. Record the answers.
6. Make a list of the most popular responses.

Creating Solutions

7. Using the survey responses, each member of the group should create a storyboard for the video. The storyboard should include images of the different scenes. The images can be rough sketches. Also include the music or sound effects that will be used. If you have several ideas for the video, you can create more than one storyboard.

Selecting and Refining the Solution

8. Gather as a group. Review each of the storyboards.
9. Group members should explain their own storyboards.
10. Record your reactions (both positive and negative) to each of the storyboards.
11. When all storyboards have been presented, use your notes to help you select the one best fitting the design brief.
12. Make any changes necessary to the selected storyboard.
13. Have one student redraw the final solution.
14. The rest of the group members should begin writing a script.

Modeling the Solution

15. Gather the equipment needed to record your video.
16. Line up all the actors you need.
17. Record your video.

Testing the Solution

18. As a group, watch the video you have recorded.
19. Make notes on things you like and dislike about the video.
20. Share your notes with the rest of the group.
21. Look back at your design brief. Answer the question "Does the video solve the problem and meet the criteria and constraints?" If so, you are ready to share the video with the class. If not, you might need to go back and redesign the parts that do not solve the problem.

Communicate the Solution

22. Once the video passes your test, share it with your teacher and classmates.
23. View the rest of the videos your classmates made.

Improving the Solution

24. After viewing, write down suggestions that could have helped to improve your own solution and the solutions of your classmates.

Challenging Your Learning

After viewing the commercials the other groups in your class made, write a short essay describing the positive aspects of the other videos. Describe changes that you could have made to make your video better.

SECTION 4 TECHNOLOGICAL CONTEXTS

In Chapter 2, you learned about systems. Now, you are about to study a number of different systems. Each one is a complete, distinct system that is dependent on others. Some of them use many of the same resources. They have, however, different processes and outputs. The outputs can be new products, information, energy, or services for people. This section deals with the following seven systems and their processes:

- **Agriculture.** This system plants, grows, and harvests food crops and fibers used for clothing.
- **Construction.** This system transforms manufactured products and materials into structures we need for living, learning, and working.
- **Energy conversion.** This system changes the forms of energy to make them more useful.
- **Communication.** This system transforms information into organized designs, meaningful instructions, and purposeful skills.
- **Manufacturing.** This system turns raw material into automobiles, clothing, toothpaste, and thousands of other items we use every day.
- **Medicine.** This system promotes good health and treats illnesses and physical conditions.
- **Transportation.** This system uses energy from chemicals, sunlight, and wind for power. This power is used to move people and materials.

Technology Headline:

Personal Rapid Transit Systems

In the late 1960s and early 1970s, engineers developed the first working personal rapid transit (PRT) system in Morgantown, West Virginia. Several decades later, the promising possibilities of this technology have yet to be realized. A cross between an automobile and railcar, a PRT system would make city travel cheaper while allowing passengers to easily reach their desired destinations.

PRT systems resemble a solitary railcar, detached and operating as an individual unit. Unlike a subway or high-speed rail, the PRT system does not operate on a set schedule. Rather, a passenger calls a car to the terminal where they are waiting and then enters their specific destination into the fully automated PRT system. Essentially, PRT systems offer the convenience of a taxi without having to navigate congested streets or highways. As a bonus, trips on a PRT system would be less expensive than a taxi ride, and passengers traveling in small groups would enjoy relative privacy.

Lower fares on a personal rapid transit system are due in part to construction costs that are estimated to be considerably lower than a traditional subway or elevated railcar system. Personal rapid transit cars are small—most models are designed to hold approximately six people—and therefore, require a scaled-down infrastructure, which would cut costs. This would reduce fares and provide an inexpensive alternative to automobiles that run on gasoline—an increasingly expensive fuel. PRT systems are powered by electricity with an onboard battery used in case of electric failure.

PRT systems would act on a supply-and-demand basis. Individuals will arrive at stations and input their destinations. If an individual is the only passenger to select a specific location, within a certain amount of time—typically no more than five minutes—a car will arrive to shuttle this individual alone. However, if a group of people should happen to choose the same destination, a car will arrive immediately to shuttle them. The number of people required to automatically trigger a car without the required wait time would vary between PRT systems and depend on the system's maximum capacity.

The Morgantown system operates as a true PRT system only some of the time. During peak hours, cars run on a schedule, shuttling people between stops that are known to be popular. When traffic is lighter, the PRT system allows passengers to once more select their destinations and be shuttled on their own timetables.

PRT systems offer an attractive alternative to high taxi fares, rush hour gridlock, and packed subway cars that do not afford much privacy. The technology to implement this system is available; hopefully, these railcars will be incorporated into our cities someday. They might even make a morning commute faster and more convenient—perhaps it will even be enjoyable!

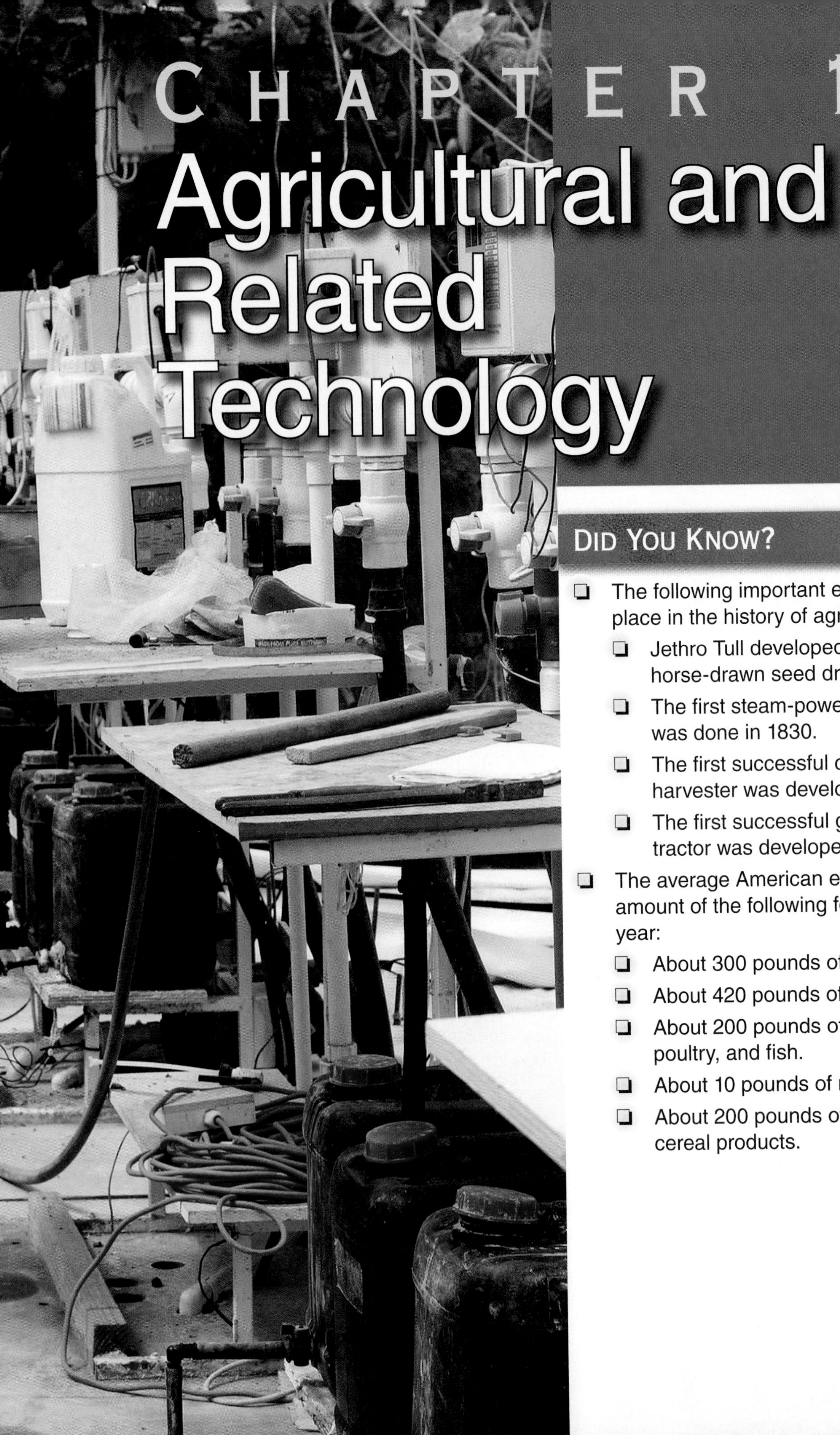

CHAPTER 19

Agricultural and Related Technology

DID YOU KNOW?

- The following important events took place in the history of agriculture:
  - Jethro Tull developed the first horse-drawn seed drill in 1701.
  - The first steam-powered plowing was done in 1830.
  - The first successful combine harvester was developed in 1836.
  - The first successful gasoline tractor was developed in 1892.
- The average American eats this amount of the following foods each year:
  - About 300 pounds of fruit.
  - About 420 pounds of vegetables.
  - About 200 pounds of meat, poultry, and fish.
  - About 10 pounds of nuts.
  - About 200 pounds of flour and cereal products.

Objectives

The information given in the chapter will help you do the following:

- Explain what agriculture is.
- Compare the two major types of agriculture.
- Give examples of the six major groups of crops grown on farms.
- Summarize the seven main types of machines used in growing and harvesting crops.
- Recall the five main types of livestock farms.
- Explain how technology is used in aquaculture.
- Explain how biotechnology can be used in agriculture.
- Summarize the five methods of food preservation used today.
- Explain the concept of an artificial ecosystem.
- Give examples of ways quality control is used in agriculture and biotechnology.

Key Words

These words are used in this chapter. Do you know what they mean?

agricultural technology
agriculture
animal husbandry
animal science
aquaculture
artificial ecosystem
baler
biotechnology
combine
crop
cultivator
disc
drip irrigation
forestry
gene-splicing
genetic engineering
grain drill
harrow
harvest
hydroponics
irrigation
pest control
pivot sprinkler
plant
plant science
plow
sprinkler irrigation
swather
tillage
tractor

Preparing to Read

Look carefully for the main ideas as you read this chapter. Look for the details that support each of the main ideas. Use the Reading Organizer at the end of the chapter to organize the main and supporting points.

Early nomads are thought to have become the world's first farmers. They ***planted*** seeds of certain grasses for food. These grasses produced a new ***crop*** we now call *grain*. The nomads built villages as they waited for the seeds to grow and ripen. All this took place about 10,000 years ago. The building of villages happened in several places, including areas that are now Jordan, Iraq, and Turkey. The villagers tended the crops with crude hoes and bone sickles (cutters). This was some of the first technology people developed.

Today, modern farming uses both science and technology. Science is used in many agricultural activities, including cross-pollinating crops and crossbreeding livestock. These types of science are called ***plant science*** and ***animal science***. Likewise, science is used to describe the seasons. Scientific knowledge of weather helps guide planting and harvesting. Technology, however, has caused massive changes in farming. See Figure 19-1. New and modern agricultural machines and equipment allow fewer people to grow more food. Also, technological advances have helped people preserve and store food for later use. These and other advancements are part of ***agricultural technology***.

What is agriculture? ***Agriculture*** is using science and technology in planting, growing, and harvesting crops and raising livestock. This practice includes using materials, information, and machines to produce the food and natural fibers needed to maintain life.

Main Types of Agriculture

Agriculture takes place on the farms and ranches of the world. See Figure 19-2. Individuals using small plots of land called *gardens* also practice agriculture.

This practice has two main branches. Crop production grows plants for human food, animal feed, and natural fibers to meet daily needs. This production produces ingredients for medicines and industrial processes. Crop production provides plants for landscaping needs. This production produces trees for ornamental uses and wood-product needs.

Animal husbandry involves breeding, feeding, and training animals. These animals are used for food and fiber for humans. In some cases, they are used to do physical work. Many animals are raised for hobbies, such as horseback riding, and for dog and horse racing.

Agricultural Crops

The crops raised today have evolved from specific regions of the world. Corn, beans, sweet potatoes, white potatoes, tomatoes, tobacco, peanuts, and sunflowers came from North and South America. China and central Asia gave us peas, sugarcane, lettuce, onions, and soybeans. Rice, citrus fruits, and bananas

Figure 19-1. Technology has greatly changed farming. This pea combine harvests green peas.

Figure 19-2. Crops and livestock are grown on farms and ranches. (U.S. Department of Agriculture)

came from Asia. The Middle East, southern Europe, and North Africa gave us the common grains. This region also was the home of sugar beets, alfalfa, and most grasses. Crops can be divided into several major groups:

- **Grains.** These are members of the grass family grown for their edible seeds. See Figure 19-3. This group includes wheat, rice, corn (maize), barley, oats, rye, and sorghum. Grains are the main food energy source for about 75% of the world's population. All the grains can be used in processed human food. Also, they can be used to produce oils, such as soybean, corn, and canola oils. Grains are widely used for animal feed.

Figure 19-3. Grains are a major food crop. (Deere and Company)

- **Vegetables.** Plants grown for their edible leaves, stems, roots, and seeds are called *vegetables*. See Figure 19-4. These plants provide important vitamins and minerals for the daily diet. Vegetables include root crops, such as beets, carrots, radishes, and potatoes. They also include leaf crops, such as lettuce, spinach, and celery. Other vegetables provide food from their fruit and seeds. This group includes sweet corn, peas, beans, melons, squash, and tomatoes. Vegetables such as cabbage, cauliflower, and broccoli are called *cole plants*. These plants are widely grown on commercial farms called *truck farms*.
- **Fruits.** Other plants cultivated for their edible parts are called *fruits* and *berries*. See Figure 19-5. They include apples, peaches, pears, plums, and cherries, which are grown in temperate climates. Citrus fruits (oranges, lemons, limes, grapefruits, and tangerines), olives, and figs are grown in warmer climates. Tropical fruits include bananas, dates, and pineapples. Smaller fruits and berries include grapes, strawberries, blackberries, blueberries, raspberries, and cranberries. Fruits are grown on farms called *orchards* or *berry farms*.
- **Nuts.** Plants grown for their hard-shelled seeds are called *nuts*. They include walnuts, pecans, almonds, filberts (hazelnuts), coconuts, and peanuts. All, except peanuts, are grown on trees in orchards. Peanuts grow underground on plants grown on farms.
- **Forage crops.** Plants grown for animal food are classified as forage plants. These plants include the hay crops, such as alfalfa and clover. Also, grasses used for pasture and hay are included in this group.
- **Nonfood.** Plants are grown for uses other than food. These plants include tobacco, cotton, and rubber. They also include nursery stock grown for landscape use and Christmas trees.

Figure 19-4. These watermelons are a vegetable crop. (U.S. Department of Agriculture)

Figure 19-5. These apples are almost ready to be harvested.

Technology in Agriculture

Crops, as all living things, have a life cycle. They are born when seeds germinate. Crops grow and reach maturity. They can be ***harvested*** or allowed to die. Farming takes advantage of this cycle through four processes. These include planting, growing, harvesting, and in some cases, storing. See Figure 19-6. The edible parts of many crops are processed into food. For example, wheat can be processed into flour. Flour can be further processed, with other ingredients, into bread.

At one time, growing crops was labor-intensive. It took many people to grow the food and fiber people needed. Today, technological advancements allow a few people to grow food for many people in a relatively short period of time. Farming has become equipment intensive. New tools and machinery have been designed to make work easier and more productive. Fewer people are involved with producing food now. More people are needed, however, for processing, packaging, and distributing it.

Growing crops involves a number of technological devices and systems. These specialized pieces of equipment are used to improve the production of food, fiber, fuel, and other useful products. These devices can be divided into the following classifications:

- Power (pulling) equipment.
- Tillage equipment.
- Planting equipment.
- Pest-control equipment.
- Irrigation equipment.
- Harvesting equipment.
- Storage equipment.

Power, or Pulling, Equipment

People tamed and trained animals to pull loads in the Stone Age. By 3500 BC, oxen were used to plow fields. Until the twentieth century, animals provided the majority of the pulling power needed in farming. See Figure 19-7. During the 1900s, the ***tractor*** replaced animal power on most farms. In the early 1900s, the modern all-purpose tractor was developed. By the 1950s, there were more than 3.5 million of these tractors in use. Today, the farm tractor can be found in all parts of the world. These devices provide the power to pull all types of farm equipment. There are two basic types of tractors: wheel tractors and track machines. See Figure 19-8. Both of these types of machines have the following features:

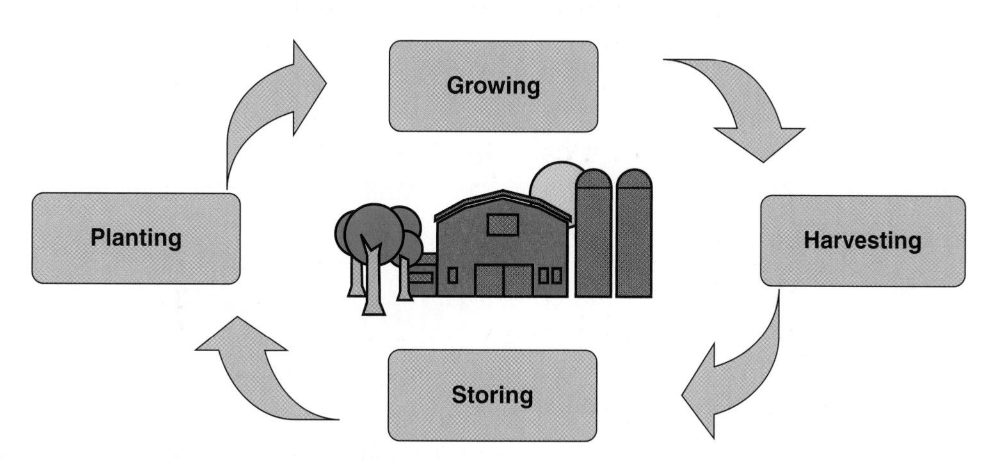

Figure 19-6. The processes in growing crops.

Figure 19-7. Up until the mid-1900s, horses, mules, and oxen provided the pulling power needed on farms. (U.S. Department of Agriculture)

A Wheel Tractor

A Track Machine

Figure 19-8. Wheel and track tractors are used on modern farms. Can you see the major parts of the tractors described in the text? (Deere and Company)

- A power source (engine).
- A way to transmit the power (transmission and drive train) for pulling a load.
- A method of controlling speed and direction.
- Traction devices (wheels or tracks).
- An operator's area (seat, cab, and controls).
- A hitch onto which equipment can be fastened.

Tillage Equipment

Tillage equipment is designed to break and pulverize the soil. This equipment develops a seedbed for the seeds and plants. The cornerstone of tillage is the moldboard ***plow***. This plow can be traced back to tree branches and antlers used to prepare the soil. In the 1800s, iron and steel plows were developed. They, similar to modern plows, had a blade-shaped plowshare that cut, lifted, and turned over the soil. See Figure 19-9.

Discs (or disc plows) are also used to prepare the seedbed. Sometimes these machines are used after the plow. At other times, the disc is used instead of a plow to start preparing a field for planting. A disc, as its name implies, is a series of curved discs on a shaft. See Figure 19-10. When the machine is pulled through the ground, it slices and crumbles the earth.

Generally, a ***harrow*** is used after plowing and discing are completed. A

Figure 19-9. A plow is used to cut, lift, and turn the soil. (Case IH)

Figure 19-10. A disc cuts and pulverizes the soil. Note the close-up of the discs on the right. (Deere and Company)

harrow is a frame with teeth. These teeth can be spikes or spring shaped. The harrow is dragged over the ground to give the soil tilth. Tilth means fine and crumbly soil.

Today, farmers are reducing the amount of ***tillage*** they do. This is especially true on soil that can erode easily. This technique is called *minimum tillage*. Minimum tillage uses small amounts of work to prepare the soil. Often, a set of chisels on a frame is used to open the soil. In some cases, no-tillage systems are used. In these cases, crops are planted without working the soil from the previous crop. Special planters slice open the soil and plant the seeds.

Planting Equipment

Once the soil is prepared, fertilizer must be applied. Fertilizer is a liquid, a powder, or pellets containing important chemicals. This substance primarily delivers nitrogen, phosphorus, and potassium to the soil. Other nutrients are also in many fertilizers. Fertilizer can be applied before, during, or after planting seeds. This substance can be applied with special equipment or along with a seed planter. Often, dry fertilizer is scattered (broadcast) before planting. Liquid and gaseous (anhydrous-ammonia) fertilizer is applied by injecting it into the soil. A machine with a series of knives is pulled over the ground. The liquid or gaseous fertilizer is injected into the trench the knives create.

The seeds must be planted to start the crop cycle. Over most of history, this was done by hand. In the early 1700s, however, a new machine was developed. This machine is called the *seed drill*, or ***grain drill***. See Figure 19-11. As it is pulled along, it opens a shallow trench. A seed is then dropped. The trench is closed, covering the seed.

Other planting machines have been developed for potatoes and corn. See Figure 19-12. Specialized machines are used to plant vegetable plants, such as tomatoes and cabbage.

Figure 19-11. A grain drill in use. (Deere and Company)

Pest-Control Equipment

In nature, not all plants that sprout live to maturity. Diseases and insects kill some of the plants. Neighboring plants crowd out others. Farm crops face the same dangers. A number of machines have been developed to help control these pests.

Figure 19-12. This farmer is planting corn with a six-row corn planter. Fertilizer is applied from the front hoppers. Corn seed is in the back hoppers. (Deere and Company)

Cultivators are used to remove weeds and open the soil for water. These machines are a series of hoe-shaped blades pulled through the ground. See Figure 19-13. The blades break the crust and allow rain and irrigation water to enter the soil. They also cut off and pull out weeds. (A weed is any out-of-place plant.)

Sprays can be applied to control weeds and insects. Those controlling weeds are called *herbicides*. Insect sprays are called *pesticides*. Ground equipment or airplanes can apply both of these materials. The equipment has a tank, a pump, and spray nozzles on a boom. As the plane or ground applicator crosses the field, a mist of ***pest control*** is applied. See Figure 19-14.

Figure 19-13. This farmer is cultivating a row crop (corn). Notice the tank in the front of the tractor. Fertilizer and an herbicide are being applied as the plants are being cultivated. (U.S. Department of Agriculture)

Irrigation Equipment

In parts of the world, rainfall is sufficient to raise crops. Many places, however, are too dry for successful farming. In some of these areas, ***irrigation*** is used. This is artificial watering to maintain plant growth.

Irrigation systems can be traced back to 2100 BC. The Egyptians developed systems using the water from the Nile to irrigate crops. Their systems, similar to all irrigation systems, contained these elements:

- A reliable source of water.
- Canals, ditches, or channels to move the water.
- A way to control and distribute the water.

Figure 19-14. A field is being sprayed from the air with chemicals protecting it from weeds and insects.

The source of water is usually a lake, a river, or an underground source (aquifer). A dam at its outlet often controls water in a lake. This kind of barrier backs up river water to form a reservoir. See Figure 19-15. A well and pump obtain underground water. A series of canals and pipes moves the water from the source to farm fields. One of three basic methods is used to apply the water to the land.

Flood irrigation is used on level fields where there is a lot of water available. A sheet of water advances from a ditch across the field. Lateral ditches and pipes with holes along their lengths supply the water to one side of the field. Gravity causes the water to flow across the field. Other ditches or pipes might carry off excess water. In row crops, furrows (small ditches) between rows of plants are used to move the water from one end of the field to the other. See Figure 19-16. Pipes or tubes are used to control the water entering each furrow.

Figure 19-15. This reservoir behind Grand Coulee Dam provides water for power generation and irrigation.

Sprinkler irrigation is used to better control the amount of water used and the area watered. These systems involve a water source, a pump, main (distribution) lines, lateral (sprinkler) lines, valves, and sprinkler heads. The pump forces water into the main distribution lines. The water flows through them to lateral pipes, which have sprinkler heads attached at set intervals. Valves between the main and lateral lines can shut off or control the water flow.

Figure 19-16. These strawberry plants are being flood irrigated. (U.S. Department of Agriculture)

The water in the lateral lines enters the sprinkler heads, which spray water onto the land.

Sprinkler systems can have a number of straight sprinkler lines. Each line can apply water to a long, narrow band across the field. The water is allowed to run for a set time to irrigate the bands on each side of the sprinkler lines. These lines are then moved by hand or rolled to the next position. Here, they apply water to the next bands.

A large number of straight lines can be used to cover the entire field. When the lines are turned on, the entire field is irrigated at once. This eliminates the need to move individual lines. These lines are called *solid-set sprinklers*.

Other sprinkler systems are called ***pivot sprinklers***. See Figure 19-17. These systems use one long line that is attached at one end to a water source. The line pivots around the point of attachment on large wheels powered by electric motors. This long line is constantly moving very slowly in a circle. Sprinkler or mist heads apply the water as the line pivots.

Drip irrigation is the third type of irrigation. The system uses main lines to bring water near the plants. Individual tubes or emitters bring water from the main lines to each plant. This system ensures each plant is properly watered. Drip irrigation reduces the amount of water lost to evaporation.

Harvesting Equipment

Once a crop reaches maturity, it must be harvested. Each type of crop has its own special harvesting equipment. ***Combines*** are used to harvest grains. See Figure 19-18. A combine is a combination of two early farm machines: the header (cuts heads from the grain) and thresher (removes grain from chaff). The combine cuts off the tops of the plants containing the grain. The heads and straw move into the machine. A cylinder causes the grain to break away from the heads. Blasts of air and screens separate the grain from straw, chaff, and weed seeds. The grain is moved into storage hoppers on the machine. The unwanted materials are conveyed out the back of the machine and dropped onto the ground.

Figure 19-17. The pivot sprinkler system is irrigating crops. (U.S. Department of Agriculture)

Figure 19-18. This combine is harvesting grain on a hillside. The header (cutting area) is at the same angle as the hillside. The machine itself, however, is kept level. This is necessary to make the grain separation efficient. (Deere and Company)

Other crops use different harvesting machines. A combine or a special corn-picking machine can harvest corn. Mechanical pickers are used to harvest almost all cotton grown in the United States. Vegetables can be harvested by special-purpose machines or by hand. See Figure 19-19. Special machines dig and collect onions and potatoes. See Figure 19-20. Fruits are usually picked by hand and placed in boxes. Special machines can, however, be used. They shake the trees, causing the fruit to fall into raised catching frames. Nuts are also harvested in this manner.

A series of machines harvest hay. A mover might cut the plants and let them fall on the ground. After the hay has dried for a day or more, a rake is used to gather it into windrows (bands of hay). In other cases, a windrower, or ***swather***, might be used. This machine cuts and windrows the hay in one pass over the field. See Figure 19-21. After the hay has dried, it is usually baled. A hay ***baler*** picks up a windrow and conveys it into a baling chamber. There, the hay is compressed into a cube. Wire or twine is tied around the cube to maintain its shape. The

Figure 19-19. This special-purpose machine harvests green beans. The machine is unloading a batch into a truck. The truck will move the beans to a freezing plant.

Figure 19-20. This harvesting machine is picking up onions that were machine dug and placed on top of the ground to dry. Workers of this machine separate clods and rocks from the onions.

Figure 19-21. The swather is cutting and windrowing hay. (Deere and Company)

Figure 19-22. A bale of hay leaving a baler. (Deere and Company)

finished bale is ejected out the back of the machine. See Figure 19-22. Special balers have been produced to make round bales.

Storage Equipment

Many crops are stored before they are sent to processing plants. Grain is stored in silos or buildings at grain elevators. See Figure 19-23. Hay is stored in hay barns, which are roofs attached to long poles. These buildings generally do not have enclosed sides or ends. Many vegetables and fruits are stored in climate-controlled (cold storage) buildings. The crops are transported to processing plants throughout the country and world, as demand requires.

Figure 19-23. These trucks are bringing grain to an elevator for storage.

Raising Livestock

Farmers and ranchers raise large numbers of livestock. This livestock includes cattle, sheep, goats, horses, swine (pigs), and poultry (chickens and turkeys). These animals are primarily raised to provide meat, milk, or materials for clothing.

In historical times, many farms raised a few of each of these animals. Today, most livestock are raised on single-purpose farms. These farms include the following:

- **Cattle ranches.** They raise beef cattle for meat and hides.
- **Dairies.** They raise dairy cattle primarily for milk.
- **Swine farms.** They raise hogs for meat and hides.
- **Horse farms.** They raise horses for pleasure riding and racing.
- **Poultry farms.** They raise turkeys for meat and chickens for meat and eggs.

A number of different technologies are involved in livestock raising. Specialized equipment and practices are used in the care of animals. Many livestock operations require buildings to house animals and process feed. See Figure 19-24. These buildings are erected using construction technology. Livestock production also includes machines that grind and mix feed for the animals and distribute the feed to the animals. Feed troughs or bunkers are required so the animals can eat grain and hay. See Figure 19-25. Water must be provided using manufactured pumps and tanks. Finally, machines and equipment are used to dispose of the animal waste (such as manure). See Figure 19-26. Often, this animal waste is used to fertilize crops or for compost used by plant nurseries and home gardeners.

Special Types of Agriculture

There are several unique types of agriculture practiced in North America. The three main types are hydroponics, aquaculture, and forestry:

- ***Hydroponics.*** This consists of growing plants in nutrient solutions without soil. Hydroponic systems supply nutrients in liquid solutions. The plants grow in a porous material.

Figure 19-24. These chickens are contained in a large house on a poultry farm.

Figure 19-25. Look at the cattle feedlot. There is a feed mill in the background. Fences contain the animals within a set area. Feed bunkers can be seen in the foreground.

Figure 19-26. This hog farm contains pens in buildings and a waste-containment lagoon (upper left of photo).

Figure 19-27. These workers are using equipment to harvest catfish at a fish farm.

- ***Aquaculture.*** This is growing and harvesting fish, shellfish, and aquatic plants in controlled conditions. Aquaculture uses ponds, instead of soil, to grow its crop. See Figure 19-27.
- ***Forestry.*** This includes growing trees for commercial uses, such as lumber and timber products, paper and pulp, and chips and fibers.

Technology Explained

landscape plant: a plant grown in a controlled environment. These plants provide flowers and vegetables for home gardeners and landscape contractors.

Below is a picture story of how technology is used to produce these plants:

1. Seeding trays are selected.

2. Planting mix (soil) is automatically placed in the trays.

3. Seeds are automatically placed in each cell of the trays.

4. The soil and seeds are sprayed with water.

5. The moist trays are placed in controlled environments until the seeds sprout.

6. The trays with new plants are placed in greenhouses.

7. The plants are transplanted into trays with larger cells or into plastic retail packs.

8. Throughout the growing process, the plants receive automatic watering.

9. Light and humidity are controlled to produce an ideal growing environment.

10. The plants are put into greenhouses when they are ready for sale.

11. The finished plants are loaded onto carts and made ready for delivery to retailers.

12. The plants are transplanted into home gardens.

Agriculture and Biotechnology

A specific technology that has greatly impacted agriculture is ***biotechnology***. The term *biotechnology* is fairly new. The practice, however, can be traced into distant history. Evidence suggests that the Babylonians used biotechnology to brew beer as early as 6000 BC. The Egyptians used biotechnology to produce bread as far back as 4000 BC. Both of these activities are directly related to agriculture.

Modern biotechnology can be traced back to at least World War I. Scientists used an additive to change the output of a yeast-fermentation process. The result was glycerol, instead of ethanol. The glycerol was a basic input to explosives manufacturing.

During World War II, the next stride in biotechnology took place. This involved the production of antibiotics (antibodies). These drugs are also the products of fermentation processes.

What is biotechnology? Specifically, biotechnology deals with using biological agents and principles in processes to produce commercial goods or services. The biological agents are generally microorganisms (very small living things), enzymes (a special group of proteins), or animal and plant cells. They are used as catalysts in the selected process. The word *catalyst* means they are used to cause a reaction. The catalyst does not, however, enter into the reaction itself.

Agricultural biotechnology is a type of biotechnology. This biotechnology consists of techniques used to create, improve, or modify plants, animals, and microorganisms. Agricultural biotechnology can be used to improve many different activities impacting agriculture.

Biotechnology can be used to combat diseases. This was the first major use of biotechnology. For example, insulin used to treat people who have diabetes can be produced less expensively using biotechnology. Also, enzymes reducing blood clots can be produced using biotechnology. Golden rice providing infants in developing countries with beta-carotene to fight blindness is a result of biotechnology. Biotechnology can be used to promote human health. The nutritional values of foods can be improved using biotechnological techniques.

Another use of biotechnology is fighting animal diseases. See Figure 19-28. Biotechnology was used in developing a vaccine for shipping fever. This disease is a major factor in feedlot deaths. Biotechnology was also used in producing a vaccine protecting wild animals against rabies.

This science is a major factor in increasing crop yields. Biotechnology has helped produce more food on the same number of acres. This factor has allowed farmers to feed more people using the same effort. For example, biotechnology was used to produce soybeans resistant to certain herbicides (weed sprays). Also, it was used to develop a cotton plant resistant to major pests.

Biotechnology can be used to supplement the common techniques of selective breeding and pollinating. This specific application is often called ***genetic engineering***. Genetic engineering enables people to move genes in ways they could not before. This engineering allows people to develop plants and animals with desirable traits.

Gene-splicing is based on a major discovery called *recombinant DNA*. The structure of DNA is a double helix (spiral) structure. This arrangement consists of a jigsawlike fit of biochemicals. The two strands have biochemical bonds between them.

The DNA molecule can be considered a set of plans for living organisms. This molecule carries the genetic code determining the traits of living organisms.

Figure 19-28. Biotechnology can be used to fight diseases in animals.

Scientists can use enzymes to cleanly cut the DNA chain at any point. The enzyme selected determines where the chain is cut. Two desirable parts can then be spliced back together. This produces an organism with a new set of traits. The process is often called *gene-splicing*.

Gene-splicing allows scientists to engineer plants having specific characteristics. See Figure 19-29. For example, resistance to specific diseases can be engineered into the plant. This can reduce the need for pesticides to control insect damage to crops.

This activity has received many headlines in newspapers and magazines. Gene-splicing is controversial. Some people think it will make life better. Others think we should not change the genetic structures of living things.

Food Production

People must eat to live. We eat many different kinds of food, prepared in a variety of ways. Few crops produced through agricultural technology are sold in their natural states. Most foods have been processed or preserved in some way by the time we eat them. Food processing is one of the most important of all agricultural-technology processes.

Food production is primarily the job of farmers. Many other people, however, handle the food the farmers produce before it reaches the consumers. After being harvested, most food products go to processing plants. The food products must then be transported to consumers. The transportation chain they follow includes truck, airfreight, and train personnel; food brokers; distributors; food wholesalers; and finally, food retailers, such as your local grocery store. See Figure 19-30.

Figure 19-29. Biotechnology can be used to develop crops yielding more food and resisting diseases and pests.

Figure 19-30. Foods must go through a lengthy transportation chain before you can buy them in your grocery store.

Food Processing

The conversion of harvested crops into food that is ready to eat is called *processing*. This is not just a single process, however. Processing includes many different processes, which vary depending on the food being processed. At the processing plants, skilled workers perform many jobs, including sorting, washing, peeling, slicing, roasting, grinding, canning, flash freezing, boxing, cooking, adding preservatives, and packaging.

Early humans ate their food exactly as they found it. They ate fruits, nuts, leaves, and roots as they gathered them. These humans even ate the fish and game they hunted without cooking it. These people began cooking, with the discovery of fire, about 1 million years ago. This is the first way food was processed.

Cooking makes food softer and easier to digest. The heat kills bacteria that might cause the food to rot or make you sick.

Most modern cooking appliances use heat from electricity or gas flames. Microwave ovens use radio waves to heat the food.

Some foods are sold ready to eat. Other foods come partly prepared. You must cook these types of foods, such as frozen pizzas, TV dinners, and cake mixes, before you eat them.

Grains are processed in mills. They can be ground, sifted, steamed, shredded, or toasted. These grains are then made into products such as flour, bread, and cereal products. See **Figure 19-31.** Fruits and vegetables must be peeled before being processed. Sometimes fruits are crushed to make juice. Milk is processed into whole milk, skim milk, condensed milk, cheese, butter, and cottage cheese.

Some processed foods contain additives. Additives are chemicals that improve the food or keep it from spoiling. Preservatives, flavorings, and dyes are all additives.

Figure 19-31. Grains can be processed into many popular food items, including bagels.

Food Preservation

Most food spoils after a while. Bacteria, mold, and insects feed on it. Food preservation is one of the oldest technologies humans used because food is so important to survival. Some of these technologies have been used for thousands of years. Over time, people began to understand why these methods work. They learned how to create the conditions preventing the growth of microorganisms that spoil food. These are some of the techniques used today for preserving food:

- Refrigeration and freezing.
- Canning.
- Dehydration.
- Chemical preservation.
- Irradiation.

Other methods of food preservation include freeze-drying, pasteurization, fermentation, carbonation, smoking, and cheese making.

Refrigeration and Freezing

Refrigeration and freezing are the most popular forms of food preservation used today. See **Figure 19-32.** Long ago, people used to bury food in snow or ice to keep it fresh. This is because cold temperatures cause bacteria to stop growing. Refrigeration slows the growth of bacteria so food stays fresh longer—usually a week or two—instead of spoiling in just a few hours. Freezing stops bacterial growth all together. This form of preservation keeps food fresher than refrigeration does. These methods are so popular because they have very little, if any, effect on the taste and texture of most foods.

Bacteria stop growing below 14°F. They do not, however, die. To destroy all bacteria, food is steamed before freezing. In food preserved in this way, once the food is thawed, bacteria can grow again.

Vegetables, fish, and poultry are

Figure 19-32. These quick- (flash-) frozen peas are ready for packaging. (©iStockphoto.com/Fiolika)

Figure 19-33. This canning process is used to preserve fish. (©iStockphoto.com/Anutik)

frozen by dipping the packaged foods into tanks of freezing saltwater. The same process is used to freeze canned juices. A spray of liquid nitrogen with a temperature of –320°F freezes more expensive foods, such as shrimp. Meat is frozen by traveling through a tunnel, as fans blow air at a temperature of –40°F on it. This process is called *blast freezing*.

Canning

Canning uses a cooking process to preserve food in glass jars or metal cans. See Figure 19-33. Canning uses heat to remove oxygen from the container, kill microorganisms in the food, and destroy enzymes that can spoil the food. During the canning process, the can or jar is filled with food. The air is pumped out to form a vacuum. The container is sealed. The food is heated and then cooled to prevent it from becoming overcooked. This heating inactivates enzymes that can change the food's color, flavor, or texture. Canning is used to preserve many foods, including fruits, vegetables, jams and jellies, soups, and juices.

Dehydration

Dehydration is the process by which foods are dried to preserve them. During this process, most water is removed from the food. This increases the concentrations of salt and sugar. These high concentrations kill any bacteria. Dried foods kept in airtight containers can last a relatively long time.

Early humans dried meat and fruit in the sunshine. Today, many foods, such as powdered milk, soup, potatoes, dried fruits and vegetables, beef jerky, pasta, instant rice, and orange juice, can be dehydrated. See Figure 19-34. You might eat them as they are sold, or you might need to add water. Drying often completely changes the taste and texture of foods. Many new foods created by dehydration have, however, proven to be just as popular as the original forms.

Chemical Preservation

People have used salt, sugar, and vinegar to preserve foods, especially meat, for thousands of years. When salt or sugar is used, the process is called *curing*. See Figure 19-35. The salt or sugar dissolves in the water in the food and kills the bacteria. Today, this process is most often used to create foods such as "country ham," dried beef, corned beef, and pastrami. When vinegar is used, the process is called *pickling*. This process has been used to preserve meat, fruits, and vegetables. Today, however, it is used almost exclusively for making pickles, or pickled cucumbers.

Figure 19-34. Dried fruits are commonly eaten as snacks and included in breakfast cereals and bars. (©iStockphoto.com/dkgilbey)

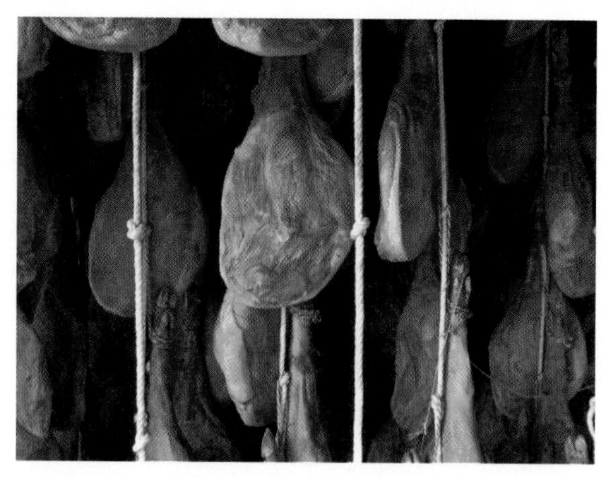

Figure 19-35. The production of a dry-cured, Spanish ham called *jamón serrano*. (©iStockphoto.com/Miguelito)

THINK GREEN

Sustainable Agriculture

You have learned about the various aspects of agriculture in this chapter. *Sustainable agriculture* takes all those aspects and works toward developing more environmentally friendly ways of continuing to use these resources. Sustainable agriculture doesn't use factory farming, which takes a more industrial approach, and can pollute the soil, water, and air with various wastes. Sustainable agriculture uses natural fertilizers and crop rotation, as well as other measures, to raise livestock and grow crops responsibly in order to leave little to no impact on the environment. Local farmers' markets are an example of sustainable agriculture.

New chemicals, such as nitrites, benzoates, and sulfites, are used today as preservatives. They either inhibit the growth of bacteria or kill them. These chemicals can be found on the ingredient lists of many different foods. For example, sulfur dioxide preserves fruits used to make jam, fruit juice, and dried fruits. All these chemicals can be harmful if used in large amounts. The FDA limits the amounts of these chemicals allowed in foods.

Irradiation

Irradiation is the process by which X-ray radiation is used to kill bacteria in food. This process kills bacteria without significantly changing the food. Irradiation can occur after foods have been packaged, which is a big advantage. If a food is sealed in plastic and then irradiated, the food becomes sterile and can be stored on a shelf without refrigeration for a long time. The FDA has approved the irradiation of chicken and beef. The use of this technique can prevent many forms of food poisoning. Large amounts of radiation must be used, however. Many people are not sure whether or not it is safe. People generally do not like the term *radiation*, so this process is still not very common in the United States.

Artificial Ecosystems

Artificial ecosystems are human-made complexes reproducing some facets of the natural environment. They can be used to study agricultural processes and systems as these processes and systems would be useful to biological ecosystems. Some examples of artificial ecosystems are terrariums and the hydroponics stations discussed earlier. These ecosystems function as part of a larger closed system supporting living organisms.

A terrarium is used to nurture plant or animal life in an enclosed environment. This ecosystem acts as a complete habitat using all the systems of life, such as food, water, shelter, and space. Terrariums can be used for decoration and enjoyment. They can also be used, however, to study the ways in which the elements of an ecosystem depend on one another. Some greenhouses can be considered large-scale terrariums. They can be used to grow plants and animals in areas differing from the natural habitats of these plants and animals. See Figure 19-36.

Figure 19-36. Artificial ecosystems, such as this greenhouse, allow plants to grow in controlled environments. (Agricultural Research Service, U.S. Department of Agriculture)

A hydroponics station is used to grow plants in a mineral nutrient solution, instead of soil. Similar to terrariums, these stations are controlled environments supplying the light, humidity, food, and water the plants need for growth. They are an alternative to traditional agriculture for farmers in areas with poor soil or where plants are grown year-round in confined areas. See Figure 19-37. Using hydroponics stations,

Figure 19-37. This worker is picking strawberries grown using hydroponics. (Ken Hammond)

CAREER HIGHLIGHT

Agricultural Managers

Agriculture, Food & Natural Resources

The Job: Agricultural managers manage the daily activities of farms, ranches, nurseries, and other agricultural establishments. Their activities deal with running the farm, rather than performing actual production activities. Agricultural managers hire and supervise workers, who perform the production tasks. These managers determine what crops to plant, oversee production activities, hire workers, assign duties to workers, and oversee maintenance activities.

Working Conditions: They spend time in the office and in the field, supervising workers. Work for livestock-farm managers is continuous throughout the year. Crop-farm managers usually work long hours during the planting and harvesting seasons. During the rest of the year, they have a reduced work schedule as they plan for next season's activities.

Education and Training: Growing up on a family farm was once the major training program for farmers and farm managers. The increasingly complex scientific and business aspects of farming, however, require advanced education from community colleges or universities. Many younger agricultural managers hold a bachelor's degree in business, with a concentration in agriculture. Additionally, they often need farm- or ranch-work experience.

Career Cluster: Agriculture, Food & Natural Resources

Career Pathway: Agribusiness Systems

farmers can grow vegetables and other plants in the middle of a desert.

Managing an artificial ecosystem entails collecting facts to plan, organize, and control processes, products, and systems. Operating such a system necessitates absolute control and cultivation. Temperature, nutrients, light, water, air, circulation, waste recycling, and monitoring of insects all require management in order for the system to function well.

QUALITY CONTROL IN AGRICULTURE AND BIOTECHNOLOGY

Quality control is critical in agricultural technology and biotechnology. Organizations such as the U.S. FDA regulate technology in the agricultural and biorelated industries. The U.S. FDA regulates food to ensure public health. Agricultural processes must adhere to FDA guidelines. Food products must be FDA approved before they can be sold. Food products and medicines must be tested extensively before they can be sold to consumers to ensure public health.

There are many areas of quality control. In livestock farming, for example, the feed given to the animals must meet quality standards. The animals must be carefully checked to ensure that they have no harmful diseases. The preparation of meat for consumption must occur in a sanitary environment. In crop raising, crops are inspected for diseases or other signs of distress. Fertilizers must be approved for use on crops. Fruits and vegetables are inspected after they are harvested.

Food-processing systems also have quality controls in place. Since these systems prepare food for human consumption, a breakdown of the quality control system can have health risks for the consumer. Therefore, the systems must be carefully designed to ensure that the food products do not become contaminated or spoiled. Food product is inspected as part of the system. All materials the food product contacts must be kept clean and sanitary.

Summary

Agriculture involves growing plants and animals on farms and ranches. This practice includes crop production and animal husbandry. Agriculture involves the use of many types of machines. This practice uses pulling, tilling, planting, pest-control, irrigation, harvesting, and storage equipment. Agriculture also uses structures and equipment to feed and care for cattle, hogs, horses, and poultry. A special kind of agriculture is aquaculture. Aquaculture involves raising fish and plants in controlled water environments.

Often, agriculture employs biotechnology. Biotechnology applies biological organisms to production processes. We use biotechnology to produce new strains of plants and drugs.

Technology is also used in food production. The food agriculture produces must go through many processes before reaching consumers. Technology is used to make sure the food stays fresh.

Natural agricultural environments can be recreated artificially. They can be used for enjoyment or educational purposes. Terrariums and hydroponics stations are the most common types of these artificial ecosystems.

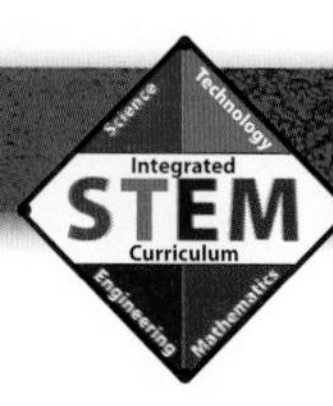

STEM Connections

Science

Research how cross-pollination and crossbreeding are used to improve animal and plant species.

Mathematics

Measure the school plot and convert the measurement to hectares and acres.

Science

Research DNA and its application to genetic engineering.

Mathematics

Research the units of measurement used for agricultural crops. Develop a poster or another display presenting and contrasting these units of measurement.

Curricular Connections

Social Studies

Develop a timeline for the invention of agricultural equipment. Select one machine and write a report on its inventor, operation, and importance.

Activities

1. Select a food you eat. Describe how its basic ingredient was planted, grown, and harvested.
2. Go to a local farm-implement dealer or locate an appropriate site on the Internet. Gather information on how a specific piece of farm equipment operates.
3. Build a diorama (model) of a farm. Show the technology used on it.

Test Your Knowledge

Do not write in this book. Place your answers to this test on a separate sheet of paper.

1. What is agriculture, and why is it important?
2. What are the two main branches of agriculture?
3. Members of the grass family that have edible seeds are called _____.
4. Grass and hay crops grown for animal feed are called _____.
5. The four processes in growing crops using technology are _____, _____, _____, and _____.
6. Name and describe three types of farms on which livestock is raised.
7. Growing and harvesting fish in controlled conditions is called _____.
8. Using biological agents to produce goods is called _____.
9. Producing a new organism by cutting and joining genes is called _____.

10. Label the following types of food preservation:
    A. The most common forms of food preservation today. _____
    B. Used to preserve pasta, instant rice, beef jerky, and powdered soup. _____
    C. Used to make pickles and corned beef. _____
    D. Can be used to preserve meat. People are concerned, however, about its safeness. _____
11. Summarize the purpose of artificial ecosystems.
12. Why is it extremely important to have effective quality control in food production and preservation?

Reading Organizer

Draw a bubble diagram for each main idea in the chapter. Make each of the main ideas the central bubble, while using details in smaller bubbles to surround the main points. An example from this chapter is shown.

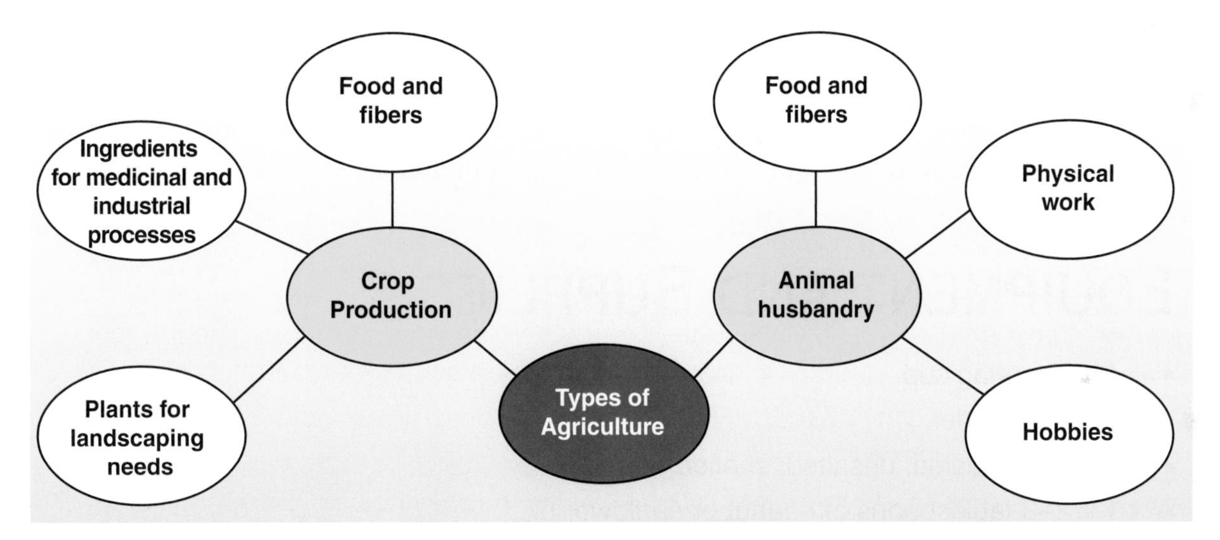

CHAPTER 19 ACTIVITY A

AGRICULTURE AND BIORELATED TECHNOLOGY

INTRODUCTION

Many agricultural products are processed into foods people eat. One of these foods everyone knows about is peanut butter. Peanuts have been around for centuries. They have been found in mummy tombs in Peru. Peanut butter, however, is a product of recent history. A famous physician, Dr. John Kellogg, developed it in 1890. He also invented cornflakes.

Dr. Kellogg developed peanut butter as a protein substitute for patients with no teeth. Abrose Straub developed the first machine to make peanut butter in 1903. Later, the famous agricultural scientist, Dr. George Washington Carver, developed an improved version of peanut butter. In 1922, the commercial production of peanut butter was developed. This process kept the oil from separating in the peanut butter. Today, more than half the American peanut crop is processed into peanut butter.

EQUIPMENT AND SUPPLIES

- A measuring cup.
- A food blender.
- 1 cup of roasted, unsalted, shelled peanuts.
- 1 1/2–3 tablespoons of peanut or safflower oil.
- 1/2 teaspoon of salt.

PROCEDURE

1. Place the peanuts and salt into a food blender.
2. Blend the peanuts and salt in the food blender. Add small amounts of oil to make the peanut butter smooth.
3. Store in an airtight container in the refrigerator. Use within two weeks.

   This makes about 1 1/2 cups of peanut butter.

CHALLENGING YOUR LEARNING

Use the Internet or other research media to discover other products that can be made from peanuts. Describe how one of these products is made.

The food shown in this image were farmed using the four main processes of agricultural technology: planting, growing, harvesting, and storing.

CHAPTER 20

Construction Technology

Did You Know?

- The pyramids were built to protect the tombs of the early pharaohs. These pharaohs were the ancient kings of Egypt. There were about 100 pyramids built near present-day Cairo.
- The Appian Way was the first major Roman road. This road was built in 312 BC. The Romans built roads throughout their empire. The empire stretched from Scotland to the Sahara Desert and from Spain to Syria.
- Eleanor Coade developed a new building material in 1722. This material was called *Coade stone*. Coade stone was a type of concrete molded to look similar to stone.
- William Jenney developed a new method of construction. This method was used to build the Home Insurance Building in Chicago. The new method of construction gave birth to a new type of building—the skyscraper.

- Rural settlements can be called *villages* or *hamlets*. Urban settlements are often called *towns*, *cities*, or *metropolises*. Villages and hamlets usually have a few houses, one or two churches, and some stores. Cities have many houses, stores, churches, factories, and schools. They also have hospitals; water, power, and transportation systems; banks; post offices; and recreational centers.

Objectives

The information given in this chapter will help you do the following:

- Explain how early humans developed shelters.
- Summarize the five main types of buildings.
- Give examples of civil structures and discuss their uses.
- Explain the steps in the construction process.

Key Words

These words are used in this chapter. Do you know what they mean?

apartment building
boundary
civil structure
commercial building
condominium
construction
contractor
flooring
foundation
industrial building
infrastructure
insulation
joist
mechanical system
public building
rafter
religious building
residential building
roof
specification
stud
superstructure
survey
utility

Preparing to Read

As you read this chapter, outline the details of the different types of constructions and the methods used in construction technology. Use the Reading Organizer at the end of the chapter to organize your thoughts.

Throughout history, people have sought places to live. Our early ancestors lived in and under natural features, such as caves and trees. As they developed farming techniques, people started living in villages. These villages contained houses for a number of families. They also had buildings in which to store grains and other foods. As civilization progressed, people built more structures. Their villages started to become small cities. See **Figure 20-1.** There were simple homes for most of the people. The cities also had fancy houses, however, for leaders and rich people. These cities had buildings for businesses and government activities. In or around the cities, people built temples to their gods. See **Figure 20-2.** People built roads to travel on and bridges with which to cross rivers. They also built dams to contain water supplies and aqueducts to move water from one place to another. There are remains of many of these ancient structures throughout the world. You can see the pyramids in Egypt, the Colosseum in Rome, and Mayan temples in Central and South America.

Figure 20-1. The remains of Pompeii. Pompeii was a city built in what is now Italy. In 79 AD, an eruption on Mount Vesuvius destroyed it.

Figure 20-2. The remains of a temple in Herculaneum, an ancient city in Italy.

The kind of structures people built reflected the people's lifestyles. People who moved in search of food needed simple, portable structures. The early plains Indians used tents for their villages. Today, nomads in Central Asia still use this type of structure. People who farmed or lived in an area with ample food supplies built more permanent structures. See **Figure 20-3.** These people developed wood or stone structures.

Today, people continue to build structures. Many of us live in neighborhoods made up of things people have erected. We call these buildings, roads, and monuments *constructed works*. They are structures erected on the sites where they are to be used. The act of building these structures is called ***construction***. The design and construction of structures for service and convenience have developed from the advancement of methods for measurement, controlling systems, and the knowledge of spatial relationships.

Construction is a series of carefully planned events. This area of technology uses materials, work, processes, and equipment to build a structure on a site. The structures people design and build meet many different needs. These include buildings and civil structures. See **Figure 20-4.**

Figure 20-3. Examples of early buildings. On the left is a Sioux dwelling of the western plains. (Smithsonian Institution) On the right is a cliff dwelling that ancient Pueblo Indians built. These Indians lived in Arizona, New Mexico, Utah, and southern Colorado.

Figure 20-4. Construction technology is used to erect buildings and civil structures.

Buildings

Some of the earliest constructed works were buildings. Buildings are structures providing protection and safety for humans and their possessions. They serve a variety of purposes. Today, there are at least five different types of buildings:

- Residential.
- Commercial.
- Industrial.
- Public.
- Religious.

Residential Buildings

Shelter is one of our basic needs. This type of establishment protects us from the weather and outside threats. Homes make us safer and more comfortable. They contain the basic spaces people need. Most

residential buildings (homes) have living, food-preparation, entertainment, sleeping, storage, and sanitary (bathroom and laundry) spaces. They might have attached or detached automobile storage. Several different types of homes are built. See Figure 20-5. Single-family dwellings stand generally alone on a plot of land. These buildings are designed for one family to occupy. Duplexes are buildings containing separate spaces for two families. These spaces are often side by side in the same building. Condominiums and apartment buildings have several living units joined together. In ***condominiums***, a separate family owns each living unit. A company or one person generally owns an ***apartment building***. The apartment units are rented to the people living in them. Apartment buildings and many condominiums are multistoried (have more than one floor or level) buildings. A single, large building can have many separate dwelling units.

Commercial Buildings

Each of us does some business in a building. See Figure 20-6. We buy groceries or clothing in a building. Each of us receives medical or dental care in

A Single-Family Dwelling

Condominiums

Figure 20-5. People live in different kinds of structures. These include single-family dwellings, apartments, and condominiums.

Figure 20-6. People take care of daily business and shopping in commercial buildings.

a building. We might buy an airline ticket at a travel agent's office or have our car repaired in a garage. As we travel, we might stay in a motel or hotel. All these activities are part of commerce and take place in ***commercial buildings***. These buildings can be professional offices, shopping centers, supermarkets, lodging establishments, or repair facilities. The businesses occupying these buildings provide us with services or sell us goods.

Industrial Structures

All societies produce goods and services they need. They make products in factories. Societies generate electricity in power plants. They develop news and entertainment programs in radio and television studios. These activities are called *industrial activities* or *productive activities*. They require special structures we call ***industrial buildings***. See Figure 20-7. These structures provide workspace and shelter for people, materials, and equipment. They generally contain office space, production or service areas, storage areas, and worker-support areas (such as restrooms, locker areas, and cafeterias).

Figure 20-7. Industries use buildings in their business activities. (Alaska Airlines)

Public Buildings

People use their tax money to erect special-purpose buildings. These buildings are designed to meet public needs. They provide areas for administration, police and security, fire protection, health care, education, and other government functions. Government agencies, such as cities and towns, school districts, states, and the federal government, build them. Most ***public buildings*** are paid for with tax money. These buildings can be schools, government office buildings, courthouses, jails, and monuments.

Religious Buildings

People erect buildings in which they can practice their religious activities. See Figure 20-8. ***Religious buildings***

Figure 20-8. Churches, synagogues, and mosques are religious buildings. This is the National Cathedral in Washington, DC. (©iStockphoto.com/lillisphotography)

Career Highlight

Painters

Architecture & Construction

The Job: Painters apply coatings, such as paint, stain, and varnish, to buildings and other structures. They prepare the surfaces to be covered, apply a primer or sealer, and apply the finish coat.

Working Conditions: Most painters work 40 hours a week or less. They must stand for long periods, do a considerable amount of climbing and bending, and work with their arms raised overhead.

Education and Training: Painters can learn their trade through on-the-job training or a two- to four-year apprenticeship program. An apprenticeship program combines on-the-job training with classroom instruction.

Career Cluster: Architecture & Construction
Career Pathway: Construction

are called a number of different names, including *churches*, *cathedrals*, *synagogues*, and *mosques*. These buildings are used for worship, fellowship, education, and other activities associated with religion. In many cases, community-service activities are also provided in the buildings. These activities might include counseling, collecting food and clothing for the needy, providing shelter for the homeless, and conducting community-interest programs (such as music and speakers).

Civil Structures

Civil structures are constructed works supporting the public interest or other technological activities. They are built for the convenience of all the people who live in an area or a nation. Civil structures include constructed works that are not buildings. Examples of civil structures are streets and roads, bridges, tunnels, railroad lines, airport runways, sewers, pipelines, dams, ponds and reservoirs, communication towers, sports fields, monuments, and water towers. See **Figure 20-9.**

Bridges and overpasses take people and vehicles over rivers, canals, and other obstacles. Overpasses cross obstacles, such as other roadways or railroad tracks. Airport runways provide landing strips for airliners and other airplanes. Pipelines move petroleum from wells to refineries.

Sewers carry waste away from homes and other buildings. Water towers and reservoirs store water for drinking, firefighting, and industrial and commercial activities. Pipes and cables supply water, fuel, and electricity. Communication towers broadcast radio, television, and telephone signals. Power transmission lines move electricity from generators to homes, factories, and businesses.

The Foundation

The Superstructure

Figure 20-9. Earth is moved and compacted to make the foundation for a road. The asphalt superstructure is then laid down. (Natchez Trace Parkway, National Park System)

Monuments are structures built to honor people, ideas, and events. See Figure 20-10. Almost everyone can recognize a picture of the Statue of Liberty in New York. This statue was a gift from the French people to the United States. Likewise, most people can recognize the Washington Monument and the Lincoln Memorial in Washington DC.

Constructing a Structure

Construction is a technological activity. This activity requires a series of actions that have to be done in the right order. These actions are part of a technical process called the *construction process*.

Figure 20-10. The Jefferson Memorial in Washington, DC. This memorial is a type of public structure. (©iStockphoto.com/compassandcamera)

This process generally follows these eight steps:

1. Preparing to build.
2. Preparing the site.
3. Setting foundations.
4. Building the framework.
5. Enclosing the structure.
6. Installing utilities.
7. Finishing the interior.
8. Finishing the site.

Preparing to Build

A structure starts with an identified need. A family needs a new home. A community needs a new courthouse. A business needs a new store. A doctor needs a new office. The list can go on and on.

The need must be changed into a design for the building. The selection of designs for structures is based on many factors. Builders need to consider style, convenience, cost, climate, and function. They also must pay attention to building laws and codes. These laws and codes are typically part of the city or county regulations for construction. Once all these factors have been considered, a suitable design can be created.

The construction industry is subject to many regulations. Zoning laws control the types of structures that can be constructed in each section of a community. Building codes identify the processes and resources that can be used for each aspect of construction. Typical laws dictate that construction is performed in compliance with the Americans with Disabilities Act (ADA), the National Electrical Code (NEC), and OSHA regulations. Government officials make periodic inspections.

The design is described with a set of plans. See Figure 20-11. The new structure is shown on architectural drawings and specifications. The drawings show the shape and size of the proposed structure and the arrangement of spaces within the structure. They also show the location of features, such as windows and doors, and how foundations, floors, walls, and roofs are to be constructed. Electrical, heating, and plumbing systems are shown. The plans often include additional information sheets called ***specifications***. These information sheets are descriptions telling

Figure 20-11. A model of a house on top of architect drawings for the structure. (©iStockphoto.com/Franck-Boston)

people how the work must be done. They also explain what quality of materials to use. The plans and specifications tell the builder what the structure will look like and how to build it.

Once the plans are developed, the future owner must obtain the money to build the project. Public projects get funds from taxes. Government officials might take the money from existing tax revenue. In some cases, they might have to levy a new tax to pay for the project. Individuals pay for private projects. Some people might already have the money in the bank. Many people, however, have to borrow the funds from a bank to build the structure.

With the money secured, someone must locate and obtain a building site. A dam or a bridge is built on a river. There are few choices of location. A house, a store, an office, or a factory is different. There are many places where these structures can be built. It is important to choose the right place. A house should be in a residential area. See Figure 20-12. The house should also be close to things the owner needs. Being near shopping areas, schools, parks, and jobs might be important. In contrast, an industrial building needs to be away from residential areas. This building needs to be located near transportation, communication, water, sewage, and power systems. These elements are called the ***infrastructure***.

When land is bought, the new owner must know where the property begins and ends. The defining lines are known as ***boundaries***. They are established by doing a survey. Also, soil tests are often needed. Builders need to know how well soil will support a structure.

Also, a ***contractor*** or several contractors must be selected. On many building projects, a general contractor is hired to oversee all the work. A general contractor manages materials, other contractors, and workers.

Contractors are responsible for hiring workers. They also get materials and equipment needed to construct the structure.

Figure 20-12. Homes should be built in residential neighborhoods.

Think Green
Green Architecture

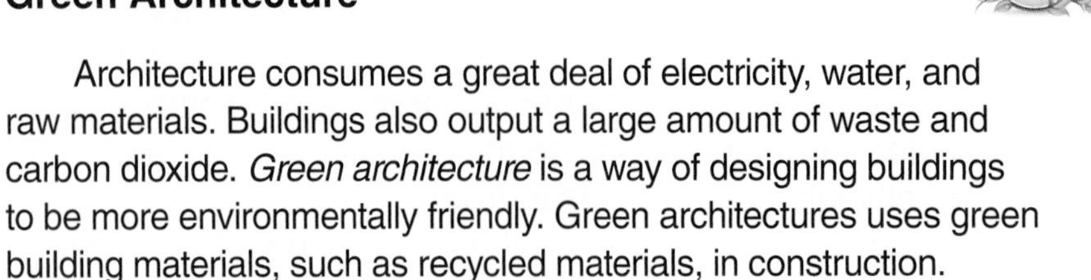

Architecture consumes a great deal of electricity, water, and raw materials. Buildings also output a large amount of waste and carbon dioxide. *Green architecture* is a way of designing buildings to be more environmentally friendly. Green architectures uses green building materials, such as recycled materials, in construction. Buildings designed with green architecture may use energy-efficient or renewable-energy sources. Water use is designed to be more efficient. With green architecture, even indoor air and water quality may improve.

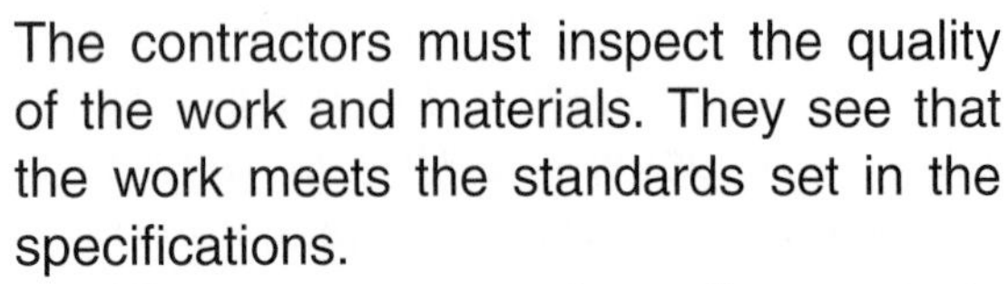

The contractors must inspect the quality of the work and materials. They see that the work meets the standards set in the specifications.

There are several quality controls in construction systems. Structures are designed based on building-code requirements, which generally focus on the safety aspects of the structure. The building specifications list quality requirements for materials. Standard construction materials, such as steel bolts and plywood panels, are rated before arriving at the job site. Materials produced on-site, such as concrete, must be tested in the field. Throughout the construction process, inspectors review the work to ensure that the contractor is meeting the building-code requirements. See **Figure 20-13.**

Figure 20-13. Throughout construction, building inspectors ensure that the structure meets building and safety codes. (©iStockphoto.com/jhorrocks)

The last step before building can start is getting a permit. Most communities control what is built in their area. Factories should not go up in a residential neighborhood.

Officials of a city or county review the plans. They want to know that the building meets local building codes and zoning ordinances. When the plans are approved, a building permit is issued. This document allows the contractor to start work.

Preparing the Site

Few building sites are ready for a new structure. There might be old buildings that must be torn down. Brush and trees might need to be removed. Rocks and debris must be hauled away. The ground might need to be leveled. See Figure 20-14. Often the topsoil is scraped away and stored for later use. Sometimes, temporary buildings and service roads are built to aid the construction of permanent structures. All these activities help prepare the site for the new structure.

Once the site is prepared, a structure can be located on it. The structure must be kept a certain distance from the property of others. Local restrictions dictate these distances. Work crews must ***survey*** the site. See Figure 20-15. These crews measure distances from the boundary lines of the property. Stakes are driven to show where to place the structure's foundation.

Setting Foundations

A constructed structure has two major parts. See Figure 20-16. There is the *substructure* below the ground. This substructure is often called the ***foundation***. The foundation connects the structure to the earth. Also, it spreads out the weight of the structure so the structure does not sink into the ground. The other part of the structure is the ***superstructure***, which is usually seen above the ground. The superstructure is the part of the structure that is used. This part is the reason the structure is built.

Structures rest on foundations. A foundation can be constructed of such materials as packed earth, gravel, wood,

Figure 20-14. This hillside lot is being prepared as a building site.

Figure 20-15. This surveyor is surveying a building site. (U.S. Department of Agriculture)

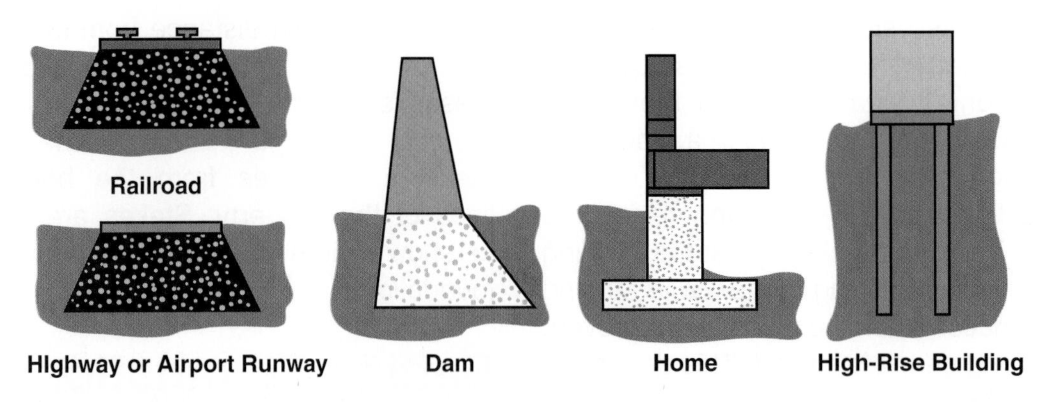

Figure 20-16. Samples of superstructures and substructures.

concrete, or steel. Wood or steel poles, called *pilings*, can be sunk into the ground. These poles support the structure when the soil is not firm.

There are several different types of foundations used for buildings. The type of structure determines the type of foundation needed. Three common types are spread, raft, and pile foundations. See Figure 20-17.

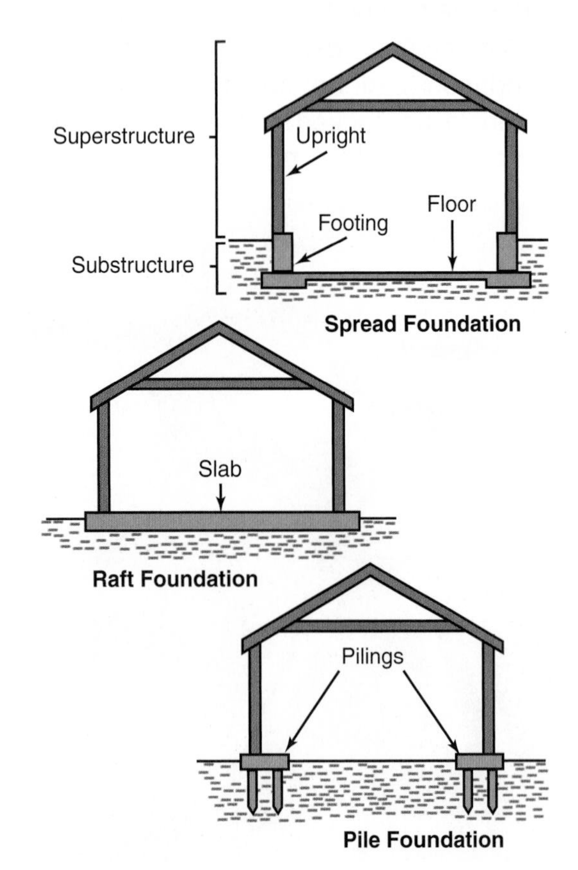

Figure 20-17. Some kinds of substructures and superstructures for houses.

Foundations start with excavating. This process prepares the ground for the foundation. Holes or trenches (ditches) are dug in the ground. It is important to place the foundation on solid soil or rock. Also, the foundation footings must be below the frost line. The frost line is the deepest level at which the ground freezes during the winter. Otherwise, freezing and thawing of the earth damages the foundation and the superstructure.

One of the most popular materials for foundations is concrete. This material is always used for raft and pile foundations. Concrete is a mixture of cement, gravel, and water. Freshly mixed concrete does not hold its shape. Therefore, temporary structures are needed to aid in the construction of these foundations. The concrete must be poured into forms until it sets. See Figure 20-18. After the concrete becomes solid, the forms can be removed. If a footing is needed for the foundation, the footing is poured first. The foundation form is then erected. The foundation is poured.

Building the Framework

A framework, or superstructure, is the part of the structure built on top of the foundation. The superstructure can be made of steel, concrete, adobe (straw and mud

Figure 20-18. A set of forms with the concrete curing between them.

blocks), wood, or other material. There are several types of frameworks. The most common is called a *framed superstructure.* See Figure 20-19. Other types include solid frameworks (for example, a dam) and air-supported frameworks (for example, a sports dome).

Let us explore common frame construction, as it is used in a house. The superstructure of the house is built on top of the foundation. The framework has three main parts:

- Floor.
- Wall.
- Roof.

The superstructure must be securely fastened to the foundation. Bolts or straps are embedded in the top of the foundation wall. Wood pieces are attached to the bolts or straps. These parts make up the sill plate. See Figure 20-20.

Floor Framing

The floor frame is fastened on top of the sill plate. ***Joists*** extend from one wall to the other. See Figure 20-21. The joists are placed on an edge and nailed to a header. They must carry the weight of the floor, furniture, appliances, and people that will be on them.

Sometimes, the foundation walls are far apart. The joists cannot reach from wall to wall. Also, the span might be too great for them to carry the weight that will be in the house. An extra support is then placed between the two walls. This steel or wood support is called a *girder.*

Wall Framing

In most houses, walls are made of wood framing. Recently, steel framing is being used in some homes. Wall frames

A Bridge's Framework

A Building's Framework

Figure 20-19. Can you see the frameworks, or superstructures, of the bridge and the office building?

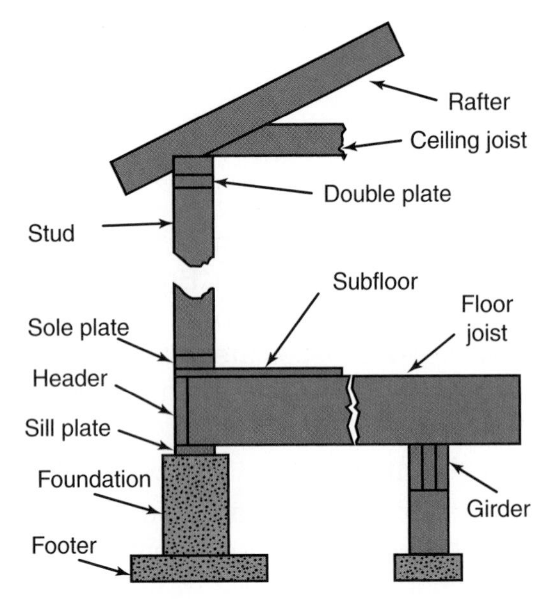

Figure 20-20. The parts of the framework of a frame-built house.

Figure 20-22. A framed wall for a new home. Can you see the plates, studs, and headers?

are designed to be light and strong. They include plates, studs, and headers. See Figure 20-22.

Plates are the horizontal parts at the top and bottom of the wall. They hold the ***studs***, which are the vertical parts. Studs are 2 × 4 or 2 × 6 boards or metal shapes spaced 16″ or 24″ apart. The spaces between the studs are later filled with insulation. Headers are beams placed above door and window openings. They support the weight over these openings.

Roof Framing

Buildings can have flat or pitched ***roofs***. A pitched roof is attached to the walls. This roof rises several feet to the center of the house. The high part of the roof is called the *ridge*. The low ends are called the *eaves*.

The roof-framing members are called ***rafters***. They are generally spaced 16″ to 24″ apart. At the ridge, the rafters are fastened to a ridge board. In some cases, trusses are used for the roof framing. These are manufactured

Figure 20-21. An example of floor framing.

Technology Explained

tower crane: a self-raising crane used to lift materials on high-rise building projects.

Building high-rise structures offers a number of unique challenges. One challenge involves lifting the structural steel members that will be part of the building into place. This is often done with tower cranes, such as those shown in **Figure A.**

Figure 20-A. This tower crane is used to lift building materials into place.

A tower crane has several parts: the tower sections, main jib (cross arm), slewing (turning) gear, and cab. A truck-mounted crane lifts the base unit of the tower crane into place. The height of the crane needs to be increased as the building is constructed. The crane rises under its own power. As shown in **Figure B,** several steps are involved.

When additional height is needed, a device called a *climbing frame* is used. The frame is positioned beneath the cab. Hydraulic cylinders lift the jib and cab unit above the tower. As shown in the middle drawing of **Figure B,** the crane then lifts another section and swings it into the opening in the climbing frame. The new section is bolted in place to produce a stable tower. This procedure is repeated as many times as necessary. The climbing frame is generally removed when it is no longer needed.

The tower crane uses a trolley, cables, and a hoist (a power-driven drum) to lift loads. The trolley is a frame with pulleys that can be moved along the main jib. A cable extends from the cab out to the trolley and down to the load. This rope is wound around the hoist drum to lift the load. When the load is at the right height, the main jib turns. The trolley moves along the jib to position the load over the structure. The load is then lowered into position. When the project is completed, the tower crane is disassembled and moved to the next construction site.

A typical tower crane can have a maximum unsupported height of about 265′ (80 m). This crane can be much taller, if it is tied to the building as the building rises around the crane. This type of crane can have a maximum reach of about 230′ (70 m) and maximum lifting power of about 20 tons (18 metric tons).

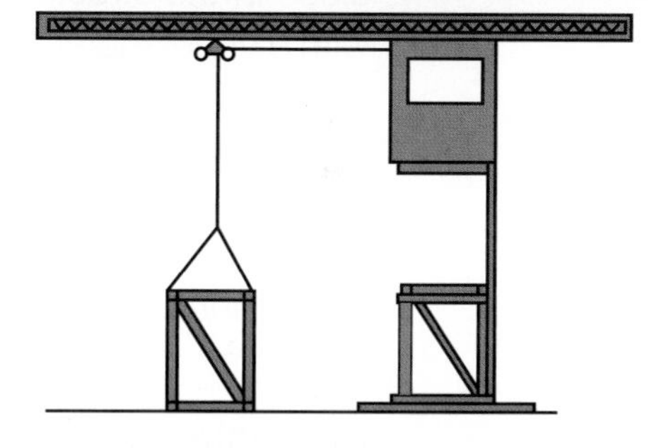

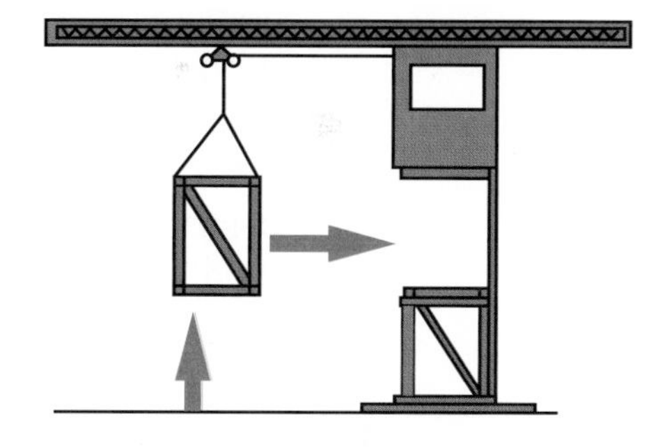

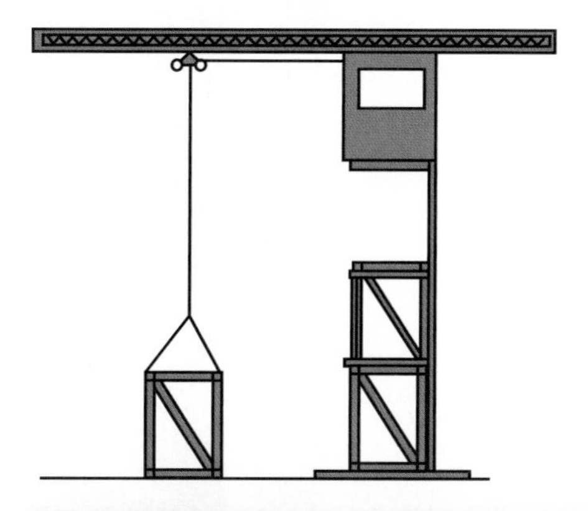

Figure 20-B. How a tower crane increases its height. The climbing frame lifts the cab and jib. A new section is put in place.

elements that include the ceiling joists and rafters in one unit.

Enclosing the Structure

Buildings are designed to serve many purposes:

- Keep out the weather (rain, snow, wind, heat, and cold).
- Protect people and goods.
- Give privacy.

To accomplish these tasks, the building must be enclosed. Builders perform many tasks to close in the structure. They cover the frame with a sheet material, such as plywood, flake board, oriented strand board (OSB), or insulating board. The material closes the structure and makes the frame more rigid. When this material is applied to outside walls, it is called *sheathing*. Sheathing used on roofs is known as *decking*.

Enclosing Wall Materials

After the sheathing is applied, windows and outside doors are installed. See Figure 20-23. The windows and doors enclose important openings giving people sunshine and giving air access to the structure.

Sheathing encloses and strengthens the frame. This is usually 4′ × 8′ sheet material, such as OSB, plywood, or a similar material. The sheathing, however, is not very attractive. To improve the beauty of the building, attractive weatherproof materials cover the sheathing. There are many different types of exterior building materials. The material might be brick, stone, wood siding, aluminum, vinyl (plastic), brick veneer, glass, wood panels, rocks, logs, or shingles.

Enclosing the Roof

Roofs are more likely to leak than walls are. Special waterproof materials must be used to keep the building dry. Shingles are normally used on all pitched roofs.

A shingled roof is installed by attaching many small pieces of waterproof material. This material must withstand many kinds of weather conditions. As each piece is installed, it overlaps the one beneath it. See Figure 20-24. Shingles are made of many different materials, such as wood, slate, ceramic, aluminum, fiberglass, or asphalt.

Installing Utilities

Buildings usually contain a variety of subsystems, including electrical, water, waste-disposal, climate-control, communication, and structural subsystems. Most of these subsystems are referred to as ***utilities***. Utilities are services coming into

Figure 20-23. This worker is installing vinyl windows in a wall of a house.

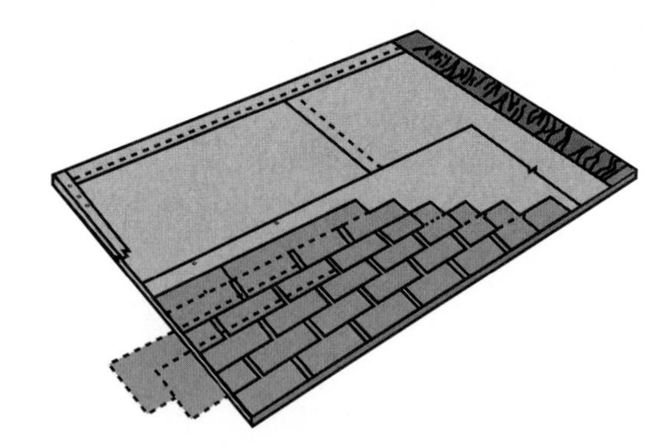

Figure 20-24. Shingles are used on many homes. Notice how they overlap to provide a waterproof surface. (Asphalt Roofing Manufacturers Association)

a building. They are supplied from pipes and cables. Most buildings have electrical wiring. They also have a plumbing system to bring in water and carry away wastes. In many homes, an air-conditioning system supplies cool air. Telephone and television cables support entertainment and communication systems. Sometimes, pipes bring in natural gas for heating. All these systems are hooked up to outside utility lines. The community or private companies supply these lines.

In a building, these systems are usually called the ***mechanical systems***. They must be installed before the inside of the house is finished. Installing mechanical systems is called *roughing in*. Usually, you cannot see the systems. Mechanical systems are placed inside the walls and under floors. If you go into a basement or under a house, you might see the pipes, ducts, and electrical wiring.

The steps of roughing in are usually carried out in a special order. Ductwork or pipes for the heating system are installed first. They carry warm or cool air from a furnace or an air conditioner. These pipes go to each room of the house. They are fitted between floor joists and wall studs.

Plumbing systems are installed next. See Figure 20-25. This system of pipes carries liquids and gases. Plumbing in a

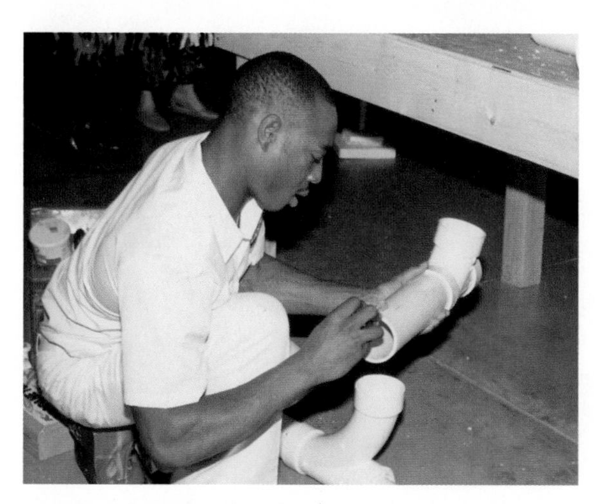

Figure 20-25. Plumbing-system pipes supply pure water and carry away wastewater.

building is needed to accomplish several purposes:

- Provide fresh, pure water.
- Remove wastewater.
- Carry fuel to furnaces, water heaters, and stoves.

Similar to the ductwork, plumbing is installed in the open spaces between the wall studs and floor joists.

Electrical systems distribute electrical power. They also are used for communication. Electric power runs appliances and is used for lighting. Telephones and intercom systems use electrical impulses to carry speech. Some homes have home-theater systems requiring wiring to various speakers. Since electrical wiring is small, it is installed last.

Finishing the Interior

Once the utilities are roughed in, the interior can be finished. The walls and ceilings can be insulated. Also, they can be covered with protective and decorating materials. Finally, appliances, cabinets, and fixtures can be installed.

Insulating

Outside walls and ceilings of the house must be insulated. ***Insulation*** is a material that resists heat passage. This material keeps heat inside during cold weather. In hot weather, it keeps the heat outside. Insulation is very light, soft material made in several forms. Loose insulation can be blown into ceilings. Long strips of insulation, called *batts*, can be placed between studs and joists. See Figure 20-26.

Enclosing Interior Walls and Ceilings

Once the insulation is in, workers can enclose the interior. First, walls and ceilings are covered to provide a smooth surface. This is generally done with drywall or wallboard. See Figure 20-27. Drywall is manufactured by putting a layer of gypsum (a chalky substance) between two layers

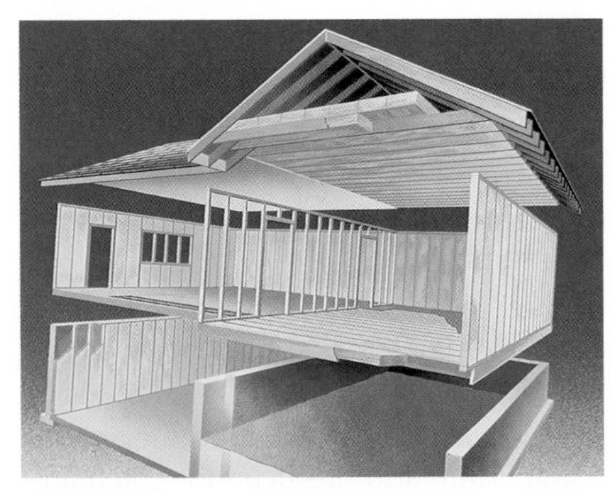

Figure 20-26. Insulation is made from materials that do not conduct heat. Placing it in walls and ceilings prevents heated air from passing in or out of a building. (Owens Corning)

of heavy paper. Drywallers attach large sheets of drywall with glue or nails. Seams and nail heads are concealed with a special tape and filler.

Finishing the Inside

Finishing includes jobs making the interior of the building attractive. These tasks generally follow a certain order:

1. Painting and decorating.
2. Installing finished flooring.
3. Installing window, door, and baseboard trim.
4. Installing electrical and plumbing fixtures and accessories.
5. Cleaning up.

Painting and decorating are usually done first. They are done to beautify the interior. Paint protects wood and drywall surfaces. Wallpaper, wood paneling, and ceramic tile are other choices for wall coverings.

Finished ***flooring*** is usually installed after the painting and decorating are done. Installing it at this point keeps the floors from being damaged during the painting process. Flooring materials are designed to wear well and look attractive. Many

Figure 20-27. These workers are attaching drywall to the interior walls and ceilings. (U.S. Department of Agriculture)

Figure 20-28. This ceramic tile is a good entry floor covering.

different materials make good finished flooring. Wood, carpeting, linoleum, and ceramic tile are among the most commonly used. See Figure 20-28. These materials are fastened to the subfloor, or underlayment, with special nails or adhesives (glues).

The next steps involve trimming the structure and installing accessories. Trim consists of the decorative wood or plastic strips covering joints. Joints appear where floors, walls, and ceilings meet. They are also where window and door frames meet walls.

Next, room doors are carefully hung on their hinges. Closet doors are installed. Edges are dressed (planed) so they fit openings.

Cabinets are then installed in kitchens, bathrooms, and other rooms. These cabinets provide storage space and support countertops. During this installation, the countertops and kitchen appliances are installed.

Finally, hardware and accessories are installed. Hardware includes doorknobs, latches, catches, and brackets. Other accessories include closet shelving and towel bars. Plumbing faucets and electrical fixtures are among the final items installed.

Throughout all these interior activities, the workers clean up the various work areas. They also appropriately dispose of scrap and waste. This reduces the chances of damage to the structure and injury to the workers.

Finishing the Site

Construction is not complete until the building site is finished. During construction, the site becomes cluttered. Scraps of building materials are everywhere. There might be piles of dirt. Several things must still be done:

1. Clearing the site.
2. Leveling and grading the ground.
3. Creating walks and drives.
4. Adding landscaping.

Clearing the Site

Clearing of the site might need to be done before any other finishing steps are taken. Some of the dirt might not be needed. This dirt must be hauled away. Rocks, trash, and scraps of building materials must be removed. Temporary buildings and fences are taken down.

Leveling and Grading the Ground

Earth might have to be moved. Holes might have been dug for foundations. Some of the earth is pushed back to fill in around the foundation. This is called *backfilling.* The earth is shaped around the structure. Soil might be moved from one spot and placed in another, so it is more pleasing. Topsoil might be returned to areas that will have plants and a lawn.

Creating Walks and Drives

Walks and drives give users access to a building. Drives must have a heavy base of gravel. The surface might be finer gravel, concrete, or asphalt. Sidewalks can be constructed of concrete, natural stone, wood, or masonry units. See Figure 20-29.

Figure 20-29. This natural stone makes for a decorative walkway.

Adding Landscaping

Landscaping is a way of making the site more attractive. This modification includes planting trees, shrubs, and flowers. Landscaping also often includes planting grass or putting down sod. Ground cover, such as bark or rock, is sometimes used where grass is not wanted. This cover keeps soil from washing away and covers unattractive soil.

SUMMARY

Construction is building a structure at the spot where the structure will be used. Our ancestors knew little about putting up buildings. They lived in caves or built crude shelters that did not last. In time, people began to settle in one place and erect permanent buildings.

Structures of today are built for several purposes. Residences and commercial buildings shelter people where the people live and work. Industrial buildings house companies making products and providing important services. Public buildings provide space for tax-supported activities. Religious buildings provide places for worship and fellowship. Civil structures help move people and materials or provide some other benefit to the public. Construction processes include acquiring land; preparing the site; putting in foundations; and building the frame, or superstructure. They also include installing mechanical systems and finishing the structure and site.

STEM Connections

Science

List the major materials used in construction. Investigate their properties. Examples of properties include strength, hardness, and color.

Mathematics

Measure the rooms in a home. Develop a chart comparing the sizes of these rooms. Use square footage.

Science

Design and conduct a test on the wear resistance of several different flooring materials.

Curricular Connections

Social Studies

Research the types of housing people developed in both of the following:

- One region or country at different points in history.
- Different cultures or countries at the same point in history.

Social Studies

Describe how the following affect the type of housing a society develops:

- Climate.
- Natural materials available.
- Lifestyle.

Activities

1. On your way to school, look for construction projects. Make a list of these projects. Try to separate the projects into one of the following groups: residential buildings, commercial buildings, public buildings, and religious buildings.
2. Locate a map of your city. Mark where residential, industrial, and commercial areas are located. Explain why you think they are located where they are.
3. List the reasons your school is located where it is.

Test Your Knowledge

Do not write in this book. Place your answers to this test on a separate sheet of paper.

1. A building, road, or monument is called a(n) _____.
2. Match the types of buildings with the right descriptions. You will use some answers more than once.

   _____ Can be used as a grocery or clothing store.
   _____ Contains spaces for living, food preparation, and sleeping.
   _____ Can be called a *church*, *synagogue*, or *mosque*.
   _____ Can be used as a jail or firehouse.
   _____ Can be apartments.
   _____ Can be used as factories.
   _____ Used by businesses providing services or selling goods.

   A. Residential.
   B. Commercial.
   C. Industrial.
   D. Public.
   E. Religious.

3. Paraphrase the definition of *civil structure*.
4. Building a structure begins with recognizing a(n) _____.
5. The part of the structure connected to the earth is called the _____.
6. The frame of the house includes what three major parts?
7. Explain the three main purposes buildings serve.
8. Inside a building, the heating ducts, electrical wiring, and plumbing pipes are known as the _____.
9. Materials used to keep heat in or out of a building are called _____.
10. Planting the lawn and shrubs around a home is called _____.

Reading Organizer

On a separate sheet of paper, create a detailed outline based on what you've read about construction technology.

Example:

I. Constructing a structure
   A. Building the framework
      1. Floor framing

CHAPTER 20 ACTIVITY A

CONSTRUCTION TECHNOLOGY

INTRODUCTION

There are structures everywhere you look. Some of these structures are houses, schools, hospitals, stores, and factories. Around them are streets and roads, parking lots, and power lines. Underground, there are storm sewers, water mains, and gas lines. Elsewhere, there are airports, railroad lines, dams, and bridges. These are all part of our constructed world. They are the results of people using construction technology. In this activity, teams in your class are going to work together to construct a structure. They will use construction technology to build a model of a storage shed.

EQUIPMENT AND SUPPLIES

- Scale lumber.
- Building materials:
  - 1/4″ × 5/8″ × 16″ pine (scale 2 × 4 × 8).
  - 1/4″ × 1″ × 16″ pine (scale 2 × 6 × 8).
  - 1/8″ × 5/8″ × 16″ pine (scale 1 × 4 × 8).
  - 8″ × 16″ six-ply poster board (scale 1/4″ × 4′ × 8′ plywood).
  - 8″ × 16″ matboard (scale 1/2″ × 4′ × 8′ plywood).
- Rules.
- Squares.
- Miter boxes.
- Backsaws.
- Coping saws.
- Utility knives.
- Hammers.
- 5/8″ × 18 brads.
- Adhesives.

PROCEDURE

Your teacher will divide the class into groups to construct the various parts of the shed. The main responsibilities of each group are as follows:

- Group 1—Front wall.
- Group 2—Left sidewall.
- Group 3—Right sidewall.
- Group 4—Rear wall.
- Group 5—Floor.
- Group 6—Rafters.

Each group should complete the following steps:

1. Carefully study the plans provided:
    - Figure 20A-1—Pictorial view.
    - Figure 20A-2—Floor plan.
    - Figure 20A-3—Floor-joist plan.
    - Figure 20A-4—Front elevation.
    - Figure 20A-5—Left-side elevation.
    - Figure 20A-6—Right-side elevation.
    - Figure 20A-7—Rear elevation.

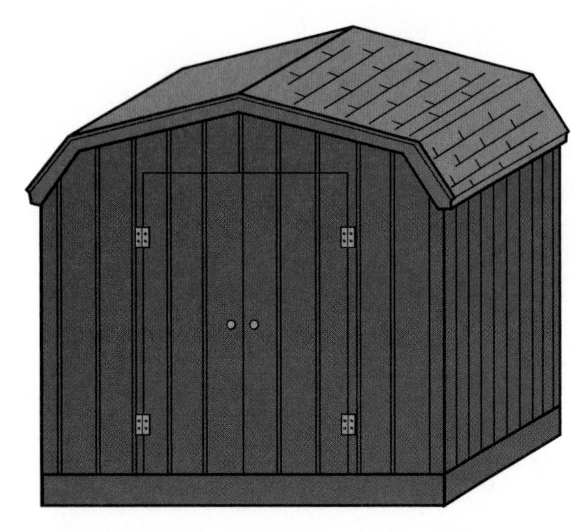

Figure 20A-1. A pictorial drawing of the storage shed.

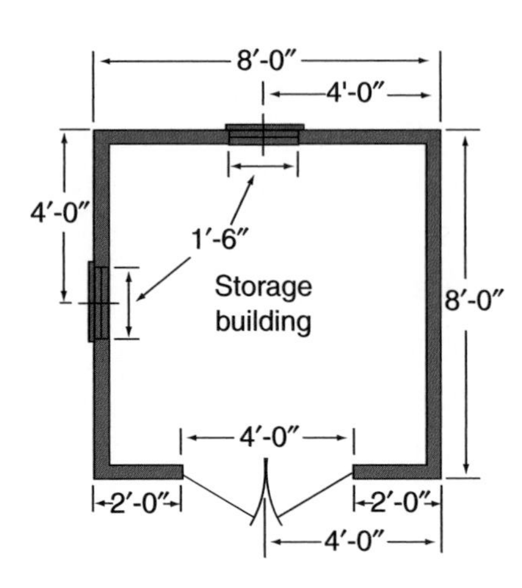

Figure 20A-2. The floor plan.

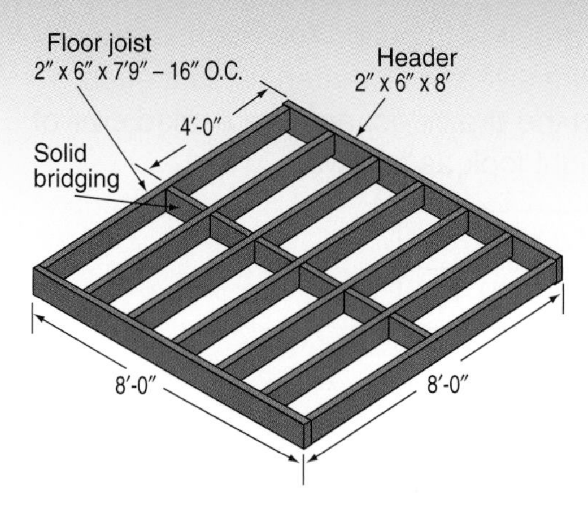

Figure 20A-3. The floor-joist plan.

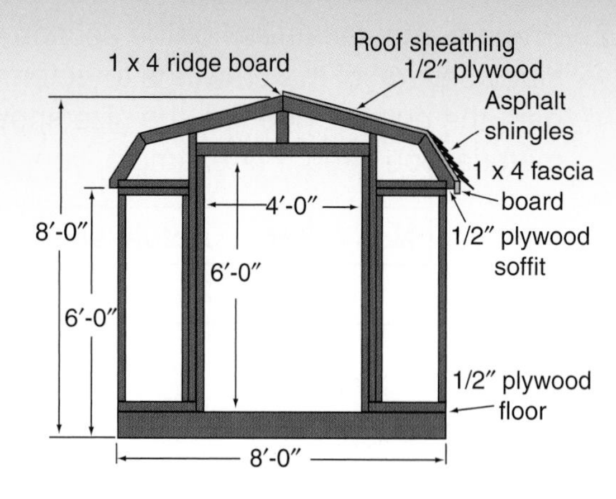

Figure 20A-4. The front elevation.

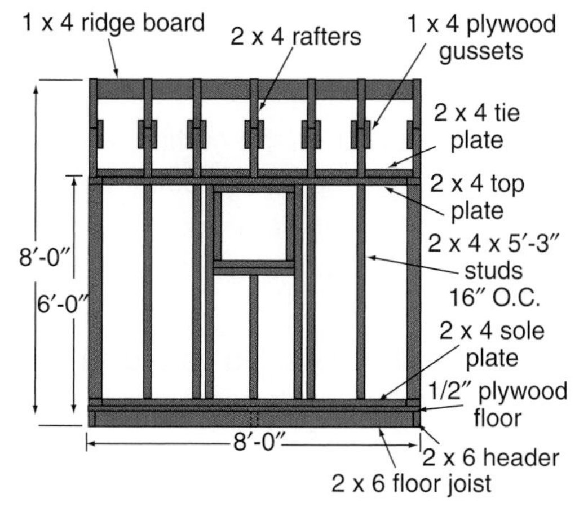

Figure 20A-5. The left-side elevation.

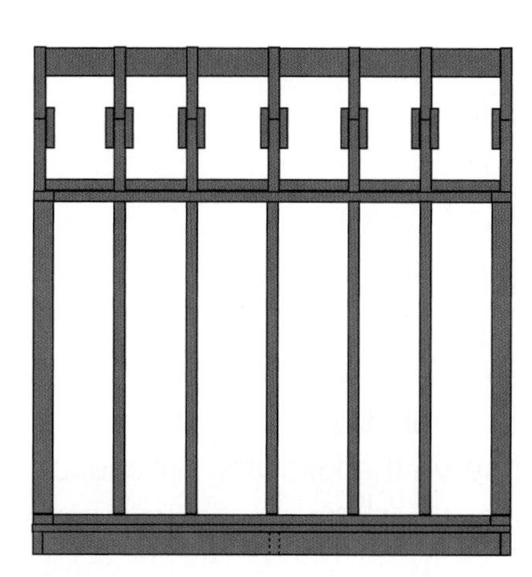

Figure 20A-6. The right-side elevation.

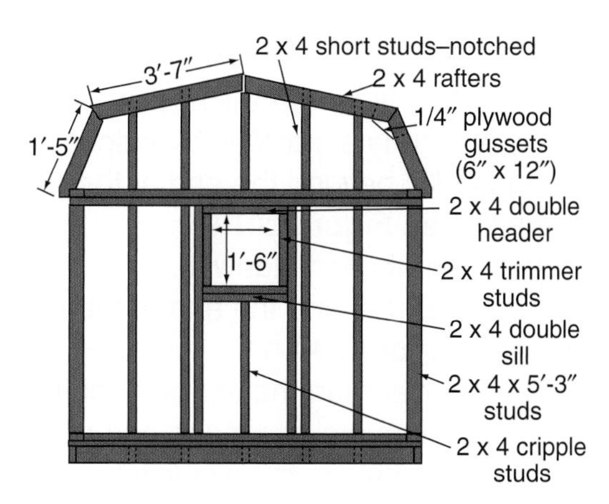

Figure 20A-7. The rear elevation.

2. Prepare a materials list for the section your group is to build. You might need to study several of the drawings to determine the sizes of all materials. List each part, the number needed, the size shown on the drawings, and the scaled size of material you need. For example, your list might look as follows:

No.	Part	Actual size	Scaled size
8	studs	2 × 4 × 6′	1/4″ × 5/8″ × 12″

NOTE

The actual length is divided by six to determine its scaled length.

3. Get the materials your group needs to make the shed section.
4. Cut all materials to their correct sizes.

SAFETY

Be careful when using hand tools with cutting edges. Never touch the cutting edge with your hand. Cut away from any part of the body. Carry sharp-edged and pointed tools turned downward and away from the body. Never carry sharp tools in your pockets. Store tools not in use. Always check with your instructor for safety instructions with any tool.

5. Have your teacher check the materials.
6. Assemble the assigned shed section according to the drawings. Be careful to observe the following precautions:
   - Group 1—Front wall. See **Figure 20A-8.**
     a. The tie plate is 5/8″ shorter than the top plate on both sides of the door. This lets the side-section tie plate overlap and connect the sides to the front.
     b. There are cripple (short) studs under the door header.
   - Group 2—Left sidewall. See **Figure 20A-9.**
     a. There are cripple studs between the window header and sill.
     b. The drawing appears to call for three studs on each side of the window. Actually, there are two studs separated by a space.
     c. The tie plate extends 5/8″ beyond the end of the top plate.
   - Group 3—Right sidewall. See **Figure 20A-10.**
     a. This wall is different from the left sidewall.
     b. The tie plate extends 5/8″ beyond the end of the top plate.
   - Group 4—Rear wall. See **Figure 20A-11.**
     a. The tie plate is 5/8″ shorter than the top plate on both sides of the door. This allows the side-section tie plate to overlap and connect the sides to the front.
     b. There are cripple studs between the window header and sill.
     c. The drawing appears to call for three studs on each side of the window. Actually, there are two studs separated by a space.
   - Group 5—Floor. See **Figure 20A-3.**
     a. The floor joists are fabricated first.
     b. The floor-joist assembly is then covered with 1/2″ plywood (matboard). See **Figure 20A-7.**

Figure 20A-8. The front-wall assembly drawing.

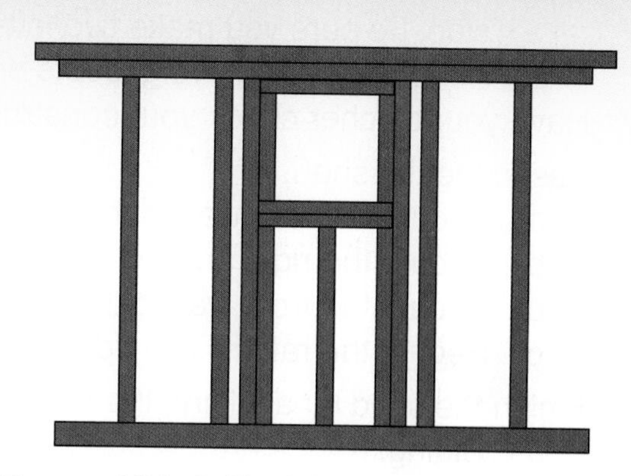

Figure 20A-9. The left-sidewall assembly drawing.

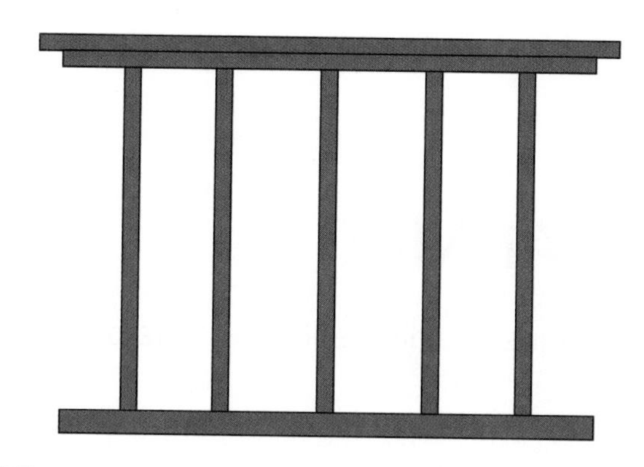

Figure 20A-10. The right-sidewall assembly drawing.

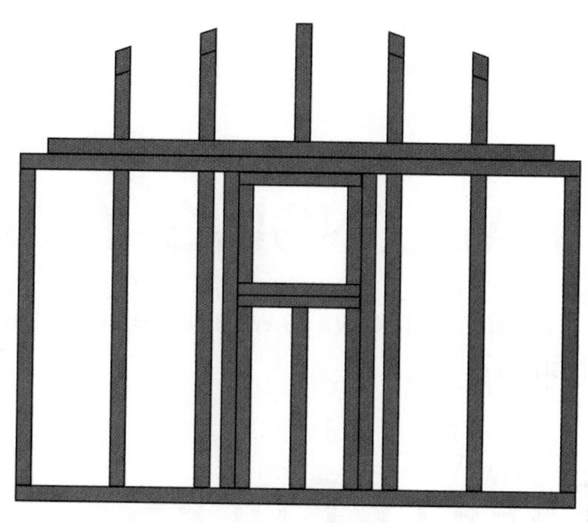

Figure 20A-11. The rear-wall assembly drawing.

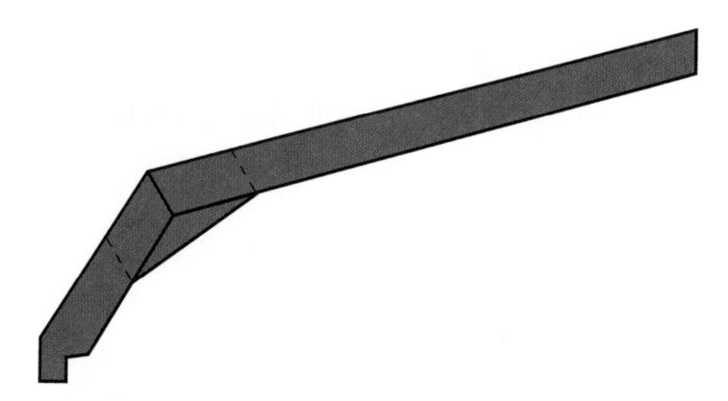

Figure 20A-12. The rafter assembly drawing.

- Group 6—Rafters. See Figure 20A-12.
  a. Four sets of rafters have 1/4″ plywood (poster board) gussets on only one side. Be sure you make two left-side and two right-side rafters.
  b. All other rafters have gussets on both sides of the joist.

7. Have your teacher check your constructed sections, or parts.
8. Assemble the shed.
   a. Secure the rear wall to the floor.
   b. Secure the right and left walls to the floor.
   c. Secure the front wall to the floor.
   d. Secure the rafters and ridge board in proper position.
9. Finish the shed by applying the following:
   a. Siding.
   b. Roof sheathing.
   c. Shingles (abrasive paper).
10. Fabricate and install a door. (optional)
11. Install a window—a plastic square. (optional)
12. Paint the shed. (optional)

Challenging Your Learning

In what three ways would the construction techniques be different if you were building a house?

Safety Rules

- Keep fingers 6″ away from saw blades.
- Do not attempt to make miter cuts or crosscuts freehand. Use the miter box.
- Extreme care is important when using the utility knife. Replace dull blades on utility knives.

TSA Modular Activity

This activity develops the skills used in TSA's Engineering Structure event.

Engineering Structure

Activity Overview

In this activity, you will create a balsa-wood bridge and perform destructive testing to determine its failure weight (the load at which the bridge breaks).

Materials

- Grid paper.
- Cutting devices.
- 20′ of 1/8″ × 1/8″ balsa wood.
- A 12″ × 18″ or larger pinboard.
- Glue.

Background Information

- **General.** There are several types of bridges: beam, truss, cantilever, suspension, and cable stayed. The length of the span and available materials generally determine the type of bridge used in a particular situation.
- **Gussets.** Gussets are plates connected to members at joints to add strength. These plates are normally used in steel construction. The structural steel members are welded or bolted to the gusset. When designing your bridge, include a gusset at each joint, if possible.
- **Wood properties.** Due to its molecular structure, wood can normally carry a larger load in tension than it can in compression. Also, a shorter member can carry a greater compressive load than a longer member can.

Guidelines

- You must create a scale sketch of the bridge before building.
- Pieces of balsa wood can be glued together along lengthwise surfaces. No more than two pieces of balsa can be glued together. You cannot use an excessive amount of glue.
- The bridge length should be between 8″ and 14″. The width should be between 2 3/4″ and 4″.
- The bridge design must take into account the loading device. Your teacher will provide specific guidelines for the bridge length and width and the required details for attachment of the loading device.
- Your bridge will be weighed before a load is applied.

Evaluation Criteria

Your project will be evaluated using the following criteria:

- Accuracy of the sketch, compared to the completed bridge.
- Conformance to guidelines.
- Efficiency (failure weight ÷ bridge weight).

There are different types of buildings and structures. You may see residential buildings near public or commercial buildings, with civil structures nearby.

Chapter 21
Energy-Conversion Technology

Did You Know?

- One-fourth of the world's population consumes three-fourths of the world's energy production. This group of energy consumers lives in the industrialized countries.
- The Grand Coulee Dam on the Columbia River in Washington is 5223′ long and 550′ high. This dam was made from 11,975,000 yds^3 of concrete. The Grand Coulee Dam has 33 generators that can produce 6809 megawatts of electrical power.
- On the average, the yearly cost of operating a clothes dryer is over $47. A personal computer and monitor cost over $24 a year to operate.
- Coal, oil, and natural gas provide more than 85% of the energy used in the United States. Alternative (such as solar, wind, and geothermal) energy sources provide less than 5%.

Objectives

The information given in this chapter will help you do the following:

- Recall what energy conversion is.
- Explain how an electric generator works.
- Summarize several types of heat engines.
- Explain how a battery works.
- Explain how a fuel cell works.
- Summarize the common types of solar converters.
- Explain how wind and water turbines are used to generate electricity.
- Summarize the major ways energy and power can be transmitted.
- Explain what would happen if there was no quality control for energy systems.
- Summarize the importance of energy conservation.

Key Words

These words are used in this chapter. Do you know what they mean?

active solar system
anode
cathode
chemical converter
conductor
conservation
electricity
electromagnetic induction
energy converter
external combustion engine
fluid converter
four-cycle engine
fuel cell
gas turbine
heat engine
internal combustion engine
jet engine
mechanical converter
passive solar system
rocket engine
solar cell
solar converter
step-down transformer
step-up transformer
thermal converter
transformer
transmission
Trombe wall
turbine
water turbine

Preparing to Read

Make a list of the types of energy converters described in this chapter. As you read, list examples of each type of converter. Use the Reading Organizer at the end of the chapter to organize your thoughts.

As you learned in Chapter 6, energy is the ability to do work. Using energy is essential for meeting basic human needs. Energy helps us have a better standard of living. Using energy contributes to people living longer and healthier lives.

People have learned to use energy over thousands of years. This started with human beings learning to make fire. As history progressed, farmers used animals as a source of energy to do work. Later, people used mechanical energy to power machines. They harnessed wind power and waterpower for their workshops and mills. The Industrial Revolution added new ways to power devices. Coal and steam power laid the foundation for new ways to do work. See **Figure 21-1.** Today, society uses more recent developments in power generation. These developments include the internal combustion engine and large-scale electricity generation.

Figure 21-1. An early traction steam engine and threshing machine working in a field at a wheat festival. (©iStockphoto.com/tilo)

Conversion and Converters

Throughout history, people have depended on energy. They have used biological (human and animal), chemical, mechanical, solar, and hydraulic sources. People developed ways to convert this energy from one form to another. They have used it to do many different types of work.

Energy converts easily from one form to another. This allows us to make energy more usable. Power systems are used to propel and provide force to other technological products and systems. For example, when a person rides a bicycle, chemical energy stored in the body is changed into mechanical energy (pedals moving a chain, which moves the back wheel). An automobile engine converts heat energy into mechanical energy so the automobile can be moved. A streetcar converts electrical energy into mechanical energy (motion). See Figure 21-2.

People have developed many types of converters. See Figure 21-3. An ***energy converter*** is a device that changes one type of energy into a different energy form. (Refer back to Chapter 6 to review the types of energy.) There are five main types of converters:

- ***Mechanical converters.*** These convert kinetic energy to another form of energy.
- ***Thermal converters.*** These convert heat energy to another form.
- ***Chemical converters.*** These convert energy in the molecular structure of substances to another energy form.
- ***Solar converters.*** These convert energy from the Sun to another energy form.
- ***Fluid converters.*** These convert moving fluids, such as air and water, to another form of energy.

Figure 21-2. A streetcar converts electrical energy into mechanical energy so it can move. (©iStockphoto.com/Joe_Potato)

	Chemical	Electrical	Heat	Light	Mechanical	Acoustical (Sound)
Chemical	Food Plants	Battery Fuel cells	Fire Food Hot water boiler Steam boiler	Lamp Gas lantern Candle Firefly	Gas engine Human muscle Animal muscle	Smoke alarm Exploding matter
Electrical	Battery Electrolyte	Diode Transistor Transformer	Electric blanket Hair dryer Toaster	Lightbulb TV screen Lighting	Electric motor Relay (type of magnet)	Horn Loudspeaker Thunder
Heat	Distilling Vaporizing Gasifying	Thermocouple Thermopile	Heat exchanger Heat pump Solar panel	Fire Lightbulb	External and internal combustion engines Turbine	Explosion Flame tube
Light	Camera film Plant growth through sunlight	Photovoltaic (solar) cell Photoelectric cell	Heat lamp Laser	Laser Reflector	Photoelectric door opener	Sound track of movie Videodisc
Mechanical	Gunpowder	Alternator Generator	Brake Friction	Flint Spark	Flywheel Pendulum Water stored in tower	Wind instrument Voice
Acoustical (Sound)	Hearing	Hearing Microphone Telephone	Sound absorption	Color organ	Ultrasonic cleaner	Megaphone

Figure 21-3. Energy can be changed from one form to another. The chart shows you what is used to make each kind of change. For example, to change electrical energy into heat, you can use a hair dryer.

Energy can be converted several times before it is used. Consider generating ***electricity***. We change the heat in steam to mechanical energy in the motion of a generator. This motion is changed into electrical energy. If the electricity is used to light a room, the lightbulb converts the electricity into light and heat energy. Even animals and plants can be considered energy converters. They take in food and change it into either chemical energy or mechanical energy. Let us look at some examples of common energy converters.

Mechanical Converters

Nearly one-fourth of all energy used in the United States and Canada is converted into electricity. This energy is then used in different ways. Some energy provides light for homes and offices. See Figure 21-4. Some energy is used to power electric motors. Another portion is used for heat in buildings and industrial processes. Still more is used to run various electrical and electronic devices.

A machine called an *electric generator*, or *alternator*, provides most electrical energy. This generator converts mechanical energy (machine motion) into electrical energy. A ***turbine***, or an engine, usually turns the generator at high speeds.

The operation of an electric generator is based on an important scientific principle Michael Faraday discovered in 1831. He found that, if a ***conductor*** is moved through a magnetic field, an electrical current is set up (induced) in the conductor. This process is called ***electromagnetic induction***.

The wire is part of a path along which electricity can flow. This path is called a *circuit*. Imagine that you can see inside a very simple electric generator. This generator is a loop of wire turning in the space between the poles of a magnet. See Figure 21-5. As the loop spins, it moves through the magnetic field. This makes the

Figure 21-4. Electricity is used in many ways. In this photograph, you can see electricity used to light the office area.

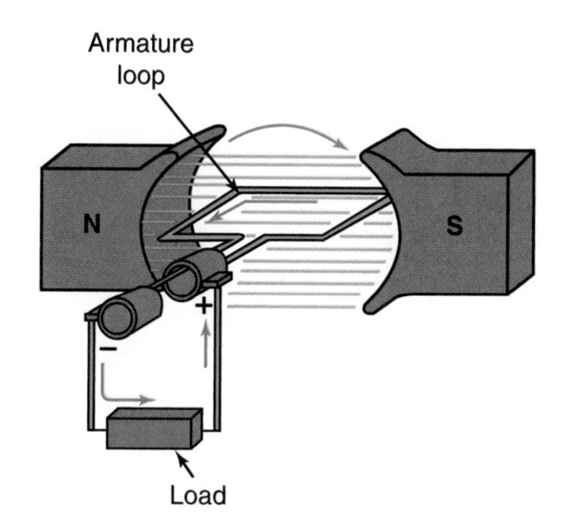

Figure 21-5. A simple generator consists of a magnet and a loop of wire. The wire loop rotates through the magnetic field.

Figure 21-6. An internal combustion engine is used to power this race car.

electricity move in one direction through the wire. As the loop continues to rotate, it becomes parallel to the magnetic field. The flow stops for a moment because wires are not cutting through any lines of magnetic force. As the loop continues, current starts up again. Now it moves in the opposite direction through the wire loop. This type of electricity is called *alternating current.* Electricity in North America changes direction 120 times a second. This electricity is called *60-Hz alternating current.* In other areas, electricity might be 50 Hz.

In reality, an electric power station is simply a large generator. Homes, businesses, and factories are parts of the circuit, as are streetlights, emergency sirens, and other devices. They make up what is called the *electric power system.*

Thermal Converters

Heat engines are thermal converters. An engine is a machine that converts energy into motion. See Figure 21-6. Oil, gasoline, steam, or electricity usually supplies the energy. The output is almost always rotary motion of a shaft.

Engines can be classified in a number of ways:

- The use for which the engine is designed: automobile, locomotive, or aircraft.
- The form of energy used: steam, compressed air, or gasoline.
- The type of cycle: Otto (in ordinary gasoline engines) or diesel.
- The type of motion produced: reciprocating or rotary.
- The place where chemical energy is converted into heat energy: internal or external combustion.
- The cylinder position: v, in-line, or radial.
- The number of piston strokes needed to complete a cycle: two or four.
- The cooling method used: air or water cooled.

Engines are sometimes called *motors.* The term *motor,* however, is generally used to describe devices that transform electrical energy into mechanical energy.

Internal Combustion Engines

In an ***internal combustion engine***, the fuel is first drawn into the engine. The fuel is then ignited and burned. The heated gases expand so rapidly that they appear to explode. This provides the power to create the desired motion.

TECHNOLOGY EXPLAINED

turbine: a device that transforms energy from a flowing stream of fluid into rotating mechanical energy.

Humans have harnessed the energy of nature to do work for centuries. Early efforts used animal power. Later, the forces of the wind and flowing water were harnessed. The first efforts in this arena involved the windmill and waterwheel. See **Figure A.** Both of these devices are effective. They, however, provide limited amounts of power.

Figure 21-A. This wind turbine was derived from an early windmill.

The device that has served best in capturing the energy of flowing fluids is the turbine. Turbines can be used to convert energy from wind, steam, or running water into rotating motion. One common use of turbines is in hydroelectric power plants. These turbines work at about 90% efficiency. This means they convert 90% of the energy from the rushing water into rotating mechanical energy.

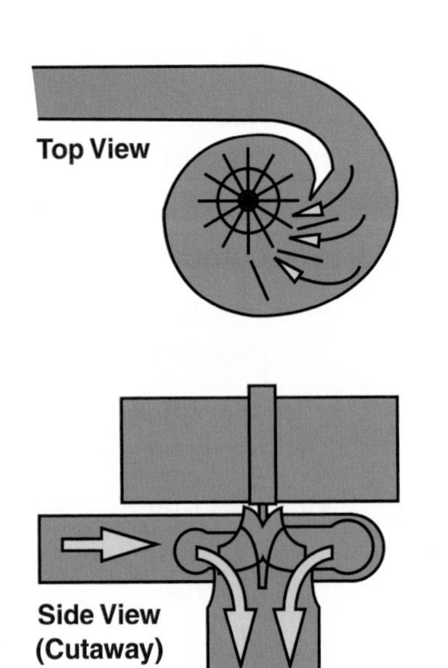

Figure 21-B. Two views of the Francis water turbine.

The first water turbine was developed in France in 1827. Benoit Fourneyron invented it. A modern water turbine is the Francis turbine. See **Figure B.** In this device, water is piped into the turbine. The water swirls through the area between the casing and blade unit. The casing of the turbine includes a series of vanes that catch the rushing water and direct it onto the blades. These vanes are set so the force of the water is tangent to the rotational plane of the blades. This angle captures the maximum amount of energy from the water. The turbine shaft is usually connected directly to a generator.

Figure 21-C. A cutaway model showing the inside of a steam turbine.

Another common type of turbine is the steam turbine. See **Figure C.** The steam turbine uses a series of blade units similar to those in a jet engine. The hot steam enters at one end of the turbine. As the steam moves through the unit, it forces the blades to turn. The steam cools and expands as it travels across the blade units. To account for this, each succeeding set of blades is slightly larger in diameter. The steam finally condenses into water at the back of the turbine and is returned to the boiler for reheating. Steam turbines are used in coal-fired, natural gas, and nuclear power plants. They also power many oceangoing ships.

All engines operate in much the same way. A sealed chamber takes in a mixture of fuel and air. This mixture burns there. See Figure 21-7. There are four steps, or strokes, for a typical piston engine:

1. Taking in a fuel-air mixture (intake stroke).
2. Compressing the mixture (compression stroke).
3. Burning the fuel (power stroke).
4. Exhausting the gas (exhaust stroke).

This design is known as a ***four-cycle engine***. Another type is the two-cycle engine. Unlike the four-cycle engine, it burns a fuel charge every time the piston is at the top of the cylinder. The intake and compression occur in one cycle (upward motion). The power and exhaust are in the other cycle (downward motion).

Jet engines, gas turbines, and rocket engines are jet propulsion devices. They burn a mixture of fuel and air. Jet propulsion devices are used primarily in high-speed aircraft, missiles, and spacecraft. They are also used in portable electric-generating stations and experimental cars.

Jet engines

In a ***jet engine***, the compressor draws in air. See Figure 21-8. The compressor is a cone-shaped cylinder with rows of small fan blades. As the air is forced through the compression stage, the air's pressure rises significantly. In some engines, the air pressure rises to 30 times the normal pressure. This high-pressure air enters the combustion area. Here, fuel injectors inject a steady stream of kerosene, jet fuel, propane, or natural gas. A "flame holder" ignites the fuel. The hot gases enter the turbine section. One set of turbines drives the compressor. Another set of turbines drives the output shaft. The final set of turbines spins freely without any connection to the rest of the engine.

Gas turbines

If you go to an airport, you see the commercial jets that carry many people. Huge turbofan engines power most commercial jets. See Figure 21-9. These engines are a type of engine called ***gas turbines***. They create power from high-velocity gases leaving the engine.

Rocket engines

When they think about engines, most people think about rotation. This rotation is used to drive other devices. ***Rocket engines***, however, are different. They are reaction engines based on a principle of physics. This principle suggests that, for every action, there is an equal and opposite reaction.

An inflated balloon can show the principle of a reaction engine. See Figure 21-10. In a closed balloon, the air pressure is equal

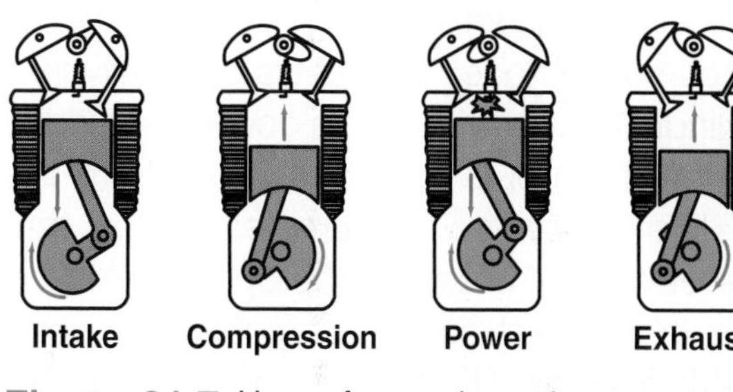

Figure 21-7. How a four-cycle engine operates. As the piston moves downward, it pulls fuel into the combustion chamber. (This chamber is the space above the piston.) The piston squeezes fuel charges into a small space. The fuel charges burn. Hot gases force the piston downward. This is the power that does work. Burned gases are pushed out of the combustion chamber.

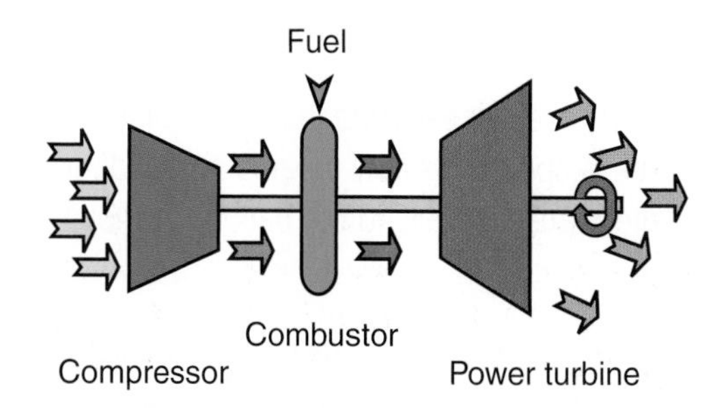

Figure 21-8. Gas turbine engines have three parts: a compressor to compress the incoming air; a combustion area burning fuel and producing high-pressure, high-velocity gas; and a power turbine that uses the energy from the gas flowing from the combustion chamber.

Figure 21-9. Jet engines power most commercial aircraft. (©iStockphoto.com/Deejpilot)

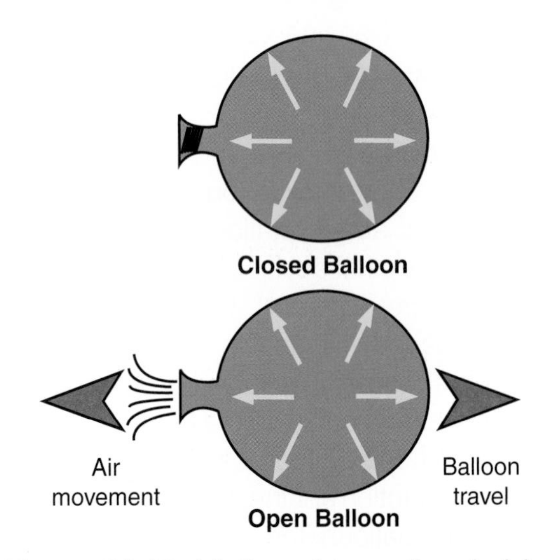

Figure 21-10. A balloon shows us the principle of reaction engines. Air inside the balloon pushes on all parts of the balloon equally. The balloon does not move. Air escapes from the balloon when the neck is open. This flow creates an equal and opposite reaction, causing the balloon to move.

in all directions. The balloon remains in place. If the air is allowed to quickly escape from the balloon, motion is created. The internal air is thrown out the nozzle. This creates an equal and opposite reaction. The balloon moves in the opposite direction.

Similar to a balloon, a rocket engine is throwing mass. This engine burns solid or liquid fuel to create hot gases. These gases leave the rocket-engine nozzle at a high speed in one direction. This causes the rocket to move in the opposite direction. Unlike a jet engine, a rocket carries its own oxygen. This allows the engine to operate in deep space, where no oxygen exists.

There are two major types of rocket engines: solid fuel and liquid fuel. Solid-fuel rocket engines were invented hundreds of years ago in China. The principle of a solid-fuel rocket is simple. See Figure 21-11. This rocket uses a fuel that burns very quickly, but does not explode. The fuel is generally in a cylindrical form with a hole down the middle. When the fuel is ignited, it burns from the middle outward. This creates gases that travel out the nozzle and propel the rocket. Fireworks use solid-fuel rockets to send them into the air. Also, model rockets are lifted with solid-fuel rockets.

Robert Goddard developed the first liquid-fuel rocket engine in 1926. The operation of a liquid-fuel rocket is fairly simple. See Figure 21-12. This rocket uses a fuel and an oxidizer. For example, Goddard used gasoline and liquid oxygen. The fuel

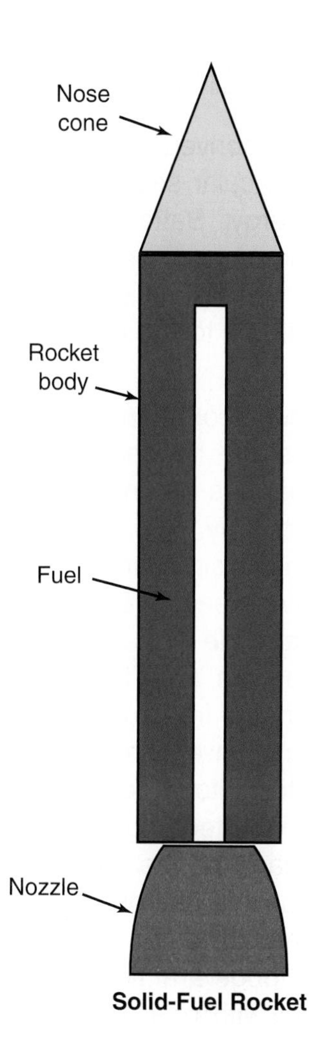

Figure 21-11. A cross-sectional view of a solid-fuel rocket.

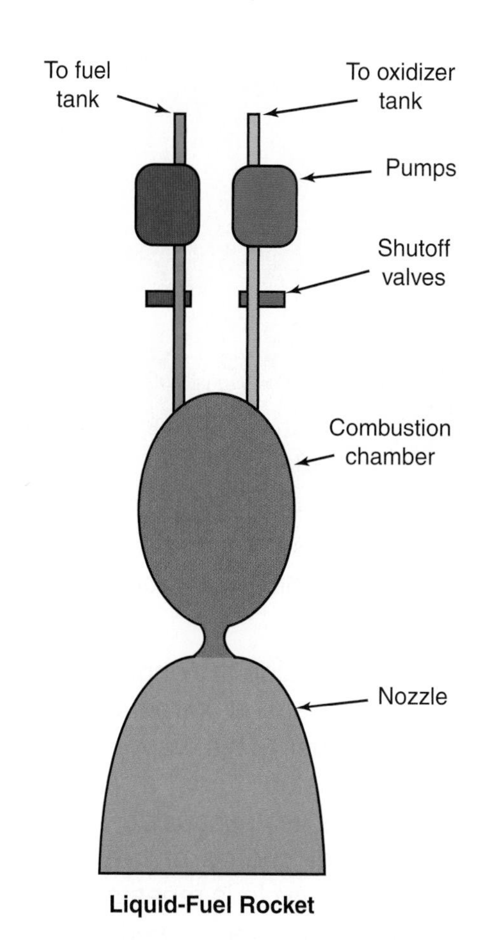

Figure 21-12. A cross-sectional view of a liquid-fuel rocket.

and oxidizer are pumped into a combustion chamber. They are burned in the chamber to create a high-pressure, high-velocity stream of gases. These gases flow through a nozzle accelerating them further. They leave the nozzle at speeds of up to 10,000 mph. This stream of gases propels the rocket. Most commercial and many military rockets use liquid-fuel systems. See **Figure 21-13.**

Figure 21-13. A liquid-fuel rocket launch. (©iStockphoto.com/cornishman)

External Combustion Engines

An ***external combustion engine*** burns fuel outside of the engine chambers. Only the expanding fluid enters the engine. The steam engine is an example of an external combustion engine. See **Figure 21-14.** Coal, wood, or some other

Figure 21-14. This stationary steam engine was used in logging operations in the early 1900s.

fuel is burned outside the engine to create steam. The steam is introduced into a cylinder chamber. This vapor moves a piston. The piston turns a crankshaft. Cooled steam is exhausted from the engine. Another charge of live steam is introduced. The cycle is repeated.

Steam engines are used very little today. Some historical railroads and collectors maintain steam engines. See Figure 21-15.

Figure 21-15. This historic railroad in Colorado still uses a steam engine.

Chemical Converters

Chemical converters change the energy found in molecular structures into another form of energy. Batteries and fuel cells are the most common chemical-energy converters. We will look closer at each of these converters to see how they work.

Batteries

The most common chemical-energy converter is the household battery. This battery changes chemical energy into electrical energy. All batteries have a cathode (the positive area) and an anode (the negative area). See Figure 21-16.

The ***cathode*** is made from a mixture of chemicals. In an alkaline battery, the cathode is made from manganese dioxide, graphite, and an electrolyte. This mixture is compacted into a hollow cylinder. The cylinder is inserted into a steel container. The steel can and the compacted chemicals become the cathode. They make up the positive charge of the battery.

The cathode and anode cannot come into contact with one another. Therefore,

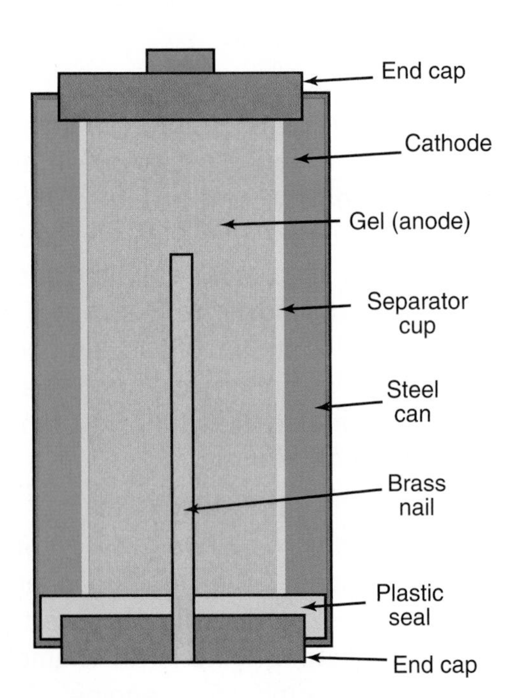

Figure 21-16. A cross-sectional view of a simple battery.

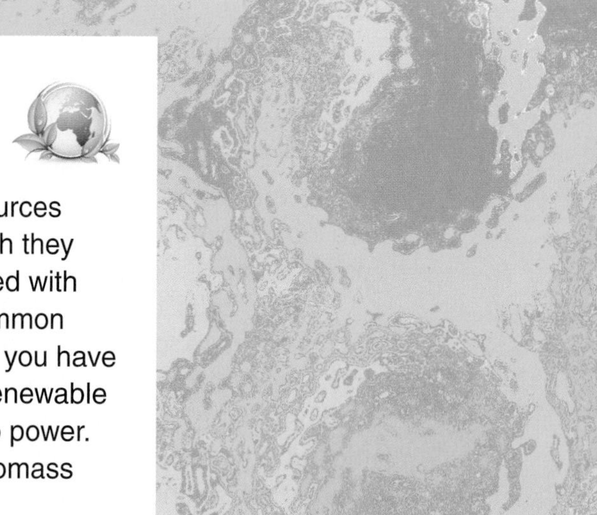

THINK GREEN
Renewable Energy

Fossil fuels and other sorts of exhaustible energy resources are eventually going to run out. Because of the rate at which they are decreasing, renewable energy sources are being studied with the hope of reducing the use of exhaustible resources. Common renewable energy sources include the Sun and the wind. If you have seen solar panels and wind turbines, you have seen how renewable energy may be harnessed. The energy is then converted to power. Other sources of renewable energy include hydropower, biomass energy, and geothermal energy.

a separator is placed next to the cathode. The ***anode*** is made from zinc powder and several other materials. This side is produced as a gel. This gel is inserted into the steel can against the separator. Finally, a brass nail (acting as the current collector), a plastic seal, and a metal end cap are attached. The brass nail collects the current the battery produces.

Fuel Cells

Another chemical-energy converter is the ***fuel cell***. This converter looks similar to an electrical storage battery. See Figure 21-17. The fuel cell produces electricity from a fuel and oxygen. This converter does not, however, burn the fuel.

A fuel cell has a container to hold an electrolyte, such as phosphoric acid. Two carbon plates are placed into the electrolyte. These are the terminals that allow electricity to flow through them.

To start the cell, oxygen and hydrogen (or another fuel) are fed into it. The fuel loses electrons to one of the carbon plates, causing it to become negatively charged. Meanwhile, oxygen is fed to the other plate, or terminal. The oxygen collects electrons from the second plate, causing the plate to have a positive charge.

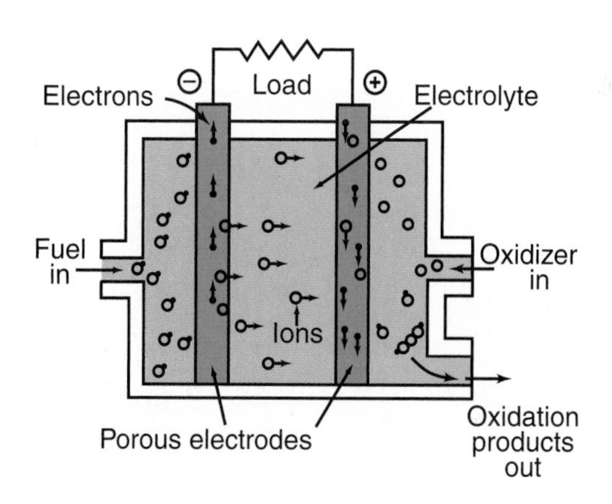

Figure 21-17. The fuel cell makes electricity directly from fuel.

The fuel cell can now provide electric current. This current becomes active when a load, such as a light, is connected into the circuit. Free electrons on the negative plate travel through the conducting wire. They flow through the lightbulb and back into the fuel cell through the other plate. In the electrolyte, ions of the hydrogen combine with oxygen ions to form water.

Solar Converters

Many electrical devices never need batteries. Some devices use solar converters instead. These converters change energy from the Sun into another form of energy.

Solar Cells

There are calculators that never use batteries. Some emergency road signs or call boxes can be far away from electrical lines. Satellites have unique power systems. There are buoys in harbors and lakes that are not powered by batteries. ***Solar cells*** power these devices. See Figure 21-18.

Solar cells used for these and other devices are photovoltaic cells. The cells get their name from two terms: *photo-* means "light," and *voltaic* means "electricity." Therefore, the term *photovoltaic* means "obtaining electricity from light."

Photovoltaic cells are made from a special material called a *semiconductor.* See Figure 21-19. The most common material for these cells is silicon. When light strikes the silicon semiconductor, this material absorbs it. The light energy causes electrons in the semiconductor to become loose. The electrons can flow freely within the material. This flow of electrons is called *current.* If metal contacts are placed on the top and bottom of the cell, current can be drawn off for use.

Figure 21-18. A photovoltaic cell in use.

Solar Heating Systems

Many homes and some businesses use solar systems for heat. These systems are another useful type of solar converter. Homes and businesses might use one of two types of systems:

- ***Passive solar systems*** use no moving parts to capture and use solar energy.
- ***Active solar systems*** use moving parts to capture and use solar energy.

Passive solar systems

There are three major types of passive solar-energy systems: direct gain, indirect gain, and isolated. Direct gain is the simplest passive solar-energy system. See Figure 21-20. Sunlight enters the house

Figure 21-19. A close-up view of a solar cell.

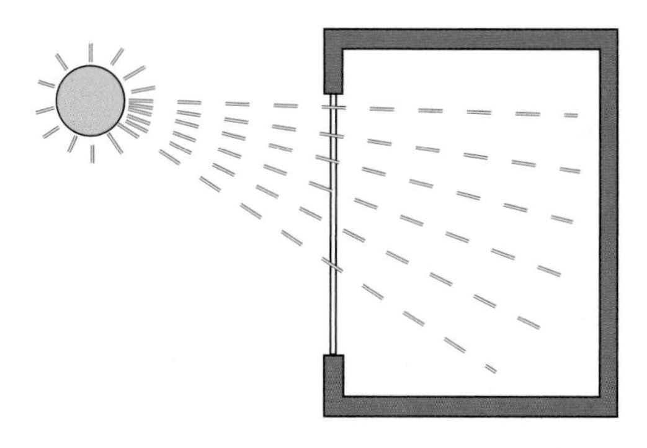

Figure 21-20. A direct-gain solar heating system.

through collectors. These collectors are usually windows facing south. The sunlight then strikes masonry floors or walls. These materials absorb and store the solar heat. The surfaces of the masonry are generally a dark color because dark colors absorb more heat than light colors do. At night, the heat stored in the masonry heats the room.

An indirect-gain passive solar system has thermal storage between the windows and living spaces. See Figure 21-21. This storage is usually called a ***Trombe wall***. The Trombe wall is an 8″- to 16″-thick masonry wall with a layer of glass mounted about 1″ in front of the wall. The wall's dark colored outside surface absorbs solar heat. This stored heat radiates into the living space.

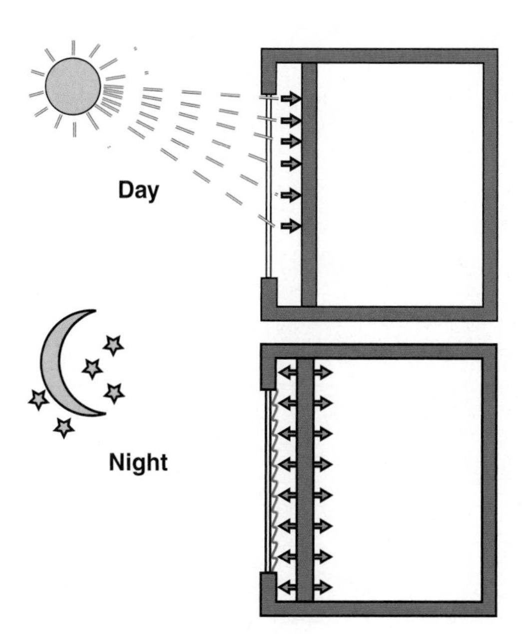

Figure 21-21. An indirect-gain solar heating system. This system gathers solar energy during the day and releases the energy at night.

A sun space, or an isolated solar system, is often called a *solar room*, or *solarium*. The simplest sun-space design uses vertical windows with no overhead glazing (windows). Sun spaces experience high heat gain and high heat loss through their windows. Many sun spaces are separated from the home with doors or windows because of this fact. The heat generated in the sunroom can be moved to other parts of the building through vents or ducting.

Active solar systems

Active solar systems can be used to provide hot water and heating for homes. These systems are either hot air or water systems. In the hot air system, solar energy is used to heat air. See Figure 21-22. The heating system operates similar to any forced-air heating system. Fans and ducts circulate the air. Some of the air is used to heat the home during the day. Additional hot air is used to heat a rock storage area. During the night and on cloudy days, the hot rocks are a heat source. Air is blown over the rocks to heat the home. In most of these systems, an auxiliary furnace provides heat during long cold or cloudy periods.

Hot water systems allow the solar energy to heat water. The hot water is circulated through the house. This water enters heat exchangers, allowing it to heat air in the various rooms. The hot water can also be used to heat tap water for household use.

Active solar systems can also be used to make electricity. A common way to generate electricity is called *concentrated solar-power technologies*. These systems use reflective materials, such as mirrors, to concentrate the Sun's energy. See Figure 21-23. The energy is focused on a

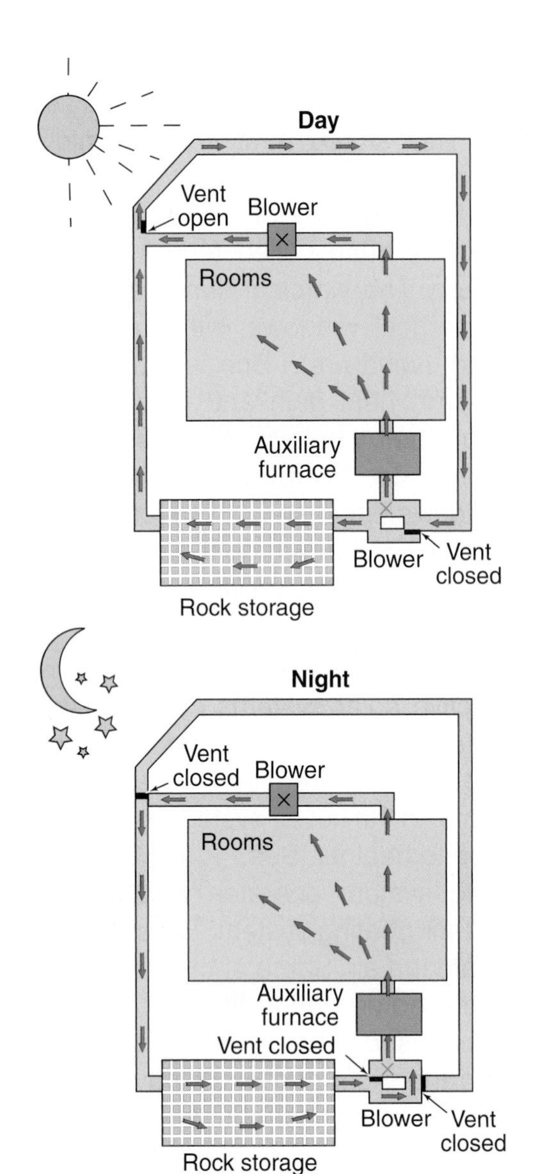

Figure 21-22. An active solar, hot air system.

Figure 21-23. The tower in the center of this photo contains many mirrors. These mirrors concentrate solar energy on the collectors on the ground.

specific point. The intense rays are used to turn water into steam. The steam is used to drive turbines, generating electricity.

Fluid Converters

Fluid converters are yet another way to change energy into a usable form. They convert energy from moving fluids into other forms of energy. Wind and water turbines are the most common fluid converters.

Wind Turbines

Wind is a free, but somewhat undependable, energy source. When it blows, it provides a great deal of energy. On a calm day, however, there is little or no wind to capture.

The energy from wind is generally changed into mechanical or electrical energy. A sail changes wind energy into mechanical energy. The energy is used to move the boat through water. Likewise, farmers have used wind energy to do work for many years. They have used windmills to pump water for household and livestock use. See **Figure 21-24.** In Holland, windmills are used to pump water from low-lying areas. See **Figure 21-25.**

Figure 21-24. This windmill is used to pump water for cattle in Nebraska.

Figure 21-25. A replica of a Dutch windmill.

Today, the wind is becoming a more common way to make electricity. Wind energy is available throughout the world. In the United States, the Atlantic coast, the Pacific coast, the Texas Gulf coast, the Great Lakes, and portions of Alaska and Hawaii have winds strong and sustained enough to produce wind-powered electricity. Also, exposed crests of long ridges and the summits of many mountains are good wind-generation sites. Wind is a renewable energy source that can replace more traditional fossil-fueled power plants. These plants produce large quantities of emissions that contribute to air pollution.

Wind is used to spin the blades of a wind turbine. To be effective, wind speeds must be above 12–14 mph. Speeds below this do not turn the turbines fast enough to generate electricity. The blades turn a shaft attached to a gear transmission box. The transmission turns a high-speed shaft. This shaft turns an electric generator. Commonly used turbines produce about 50–300 kilowatts of electricity each. A kilowatt is 1000 watts. Therefore, a 50-kilowatt (50,000-watt) wind turbine can produce enough electricity to light 500 100-watt lightbulbs.

Large wind turbine installations (wind farms) are being built in many parts of the country. See Figure 21-26. Today, wind power generates about 100 gigawatts (billion

Figure 21-26. Part of a large wind farm. (Olivier Tétard)

watts) of electric power. This amounts to over 1% of the world's generated electric power. This number is growing each year, however, as more wind farms come on-line.

Water Turbines

Water turbines are widely used in hydroelectric power plants. They use the energy of moving water to make electricity. Possible sources of water are fast-flowing rivers and reservoirs behind dams. The water flows from the river or reservoir through a pipe called a *penstock*. From there, it flows into an intake shaft that brings the water into contact with the turbine. The turbine catches the water. The water's force causes the turbine to turn. The turbine spins a shaft connected to a generator to produce electricity. See Figure 21-27.

Energy and Power Transmission

Energy is not always used where it is produced. Many times, it must be moved to where it is needed. Moving it to where it performs work is called ***transmission***. Three major types of transmission are mechanical, fluid, and electrical.

Mechanical Transmission

Mechanical transmission uses physical objects to move power and energy from

Turbines

The Blades of a Turbine

Figure 21-27. Views of the turbines at the Hoover Dam generating station and the blades of a water turbine. (©iStockphoto.com/StephanHoerold; ©iStockphoto.com/cristinaciochina)

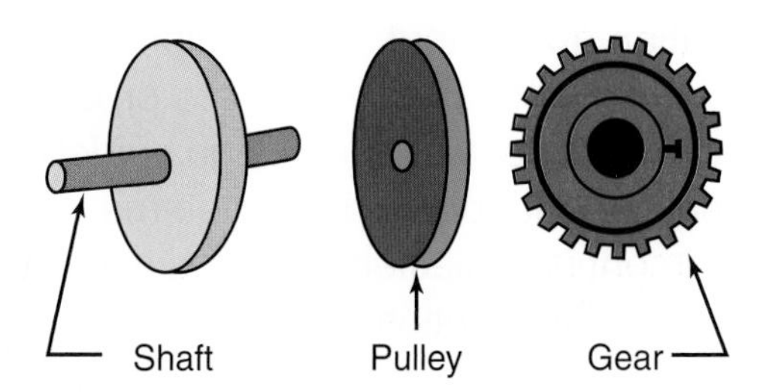

Figure 21-28. Some examples of mechanical means of power transmission.

one place to another. This can be done in a number of ways. The three simplest techniques are shafts, belts and pulleys, and gears. See Figure 21-28.

Shafts are rods that can be used to transmit power. An example is a camshaft in an automobile. The camshaft transmits rotary power along its length. Also, the cams change the rotary power to reciprocating (up-and-down) motion.

Belts are another way of transmitting mechanical power. The system includes at least two pulleys. One pulley is on some source of power. The other one is on the load. The belt transmits the power from one pulley to the other. A factory machine that an electric motor operates might use a belt to transmit the mechanical power of the electric motor.

Gears transmit mechanical power with teeth meshing with other gears. Sprockets are special types of gears that are toothed wheels. A chain transmitting power links two sprocket wheels. A bicycle is a good example of this kind of transmission. See Figure 21-29.

Figure 21-29. Bicycle sprockets and chains move power from the pedals to the rear wheel.

Fluid Transmission

If a power source exerts force on a fluid (liquid or air), the fluid transmits the power to where it is needed. The fluids must be contained so the force exerted on them travels to where the force is needed. Usually, the fluids are held inside pipes and cylinders. A common example of fluids transmitting energy is the brake system of an automobile. See Figure 21-30. The driver's foot pressing down on the brake pedal provides the energy. A piston attached to the brake pedal presses on fluid. The pressure forces the fluid through lines (tubing) to each wheel. Other pistons are attached to the brake shoes.

Electrical Transmission

Electricity is the movement of electrons through materials called *conductors*. The movement of electrons is easy to transmit. Lines carry electricity from the power plant to wherever it is to be used. The lines are conductors made of either copper or aluminum.

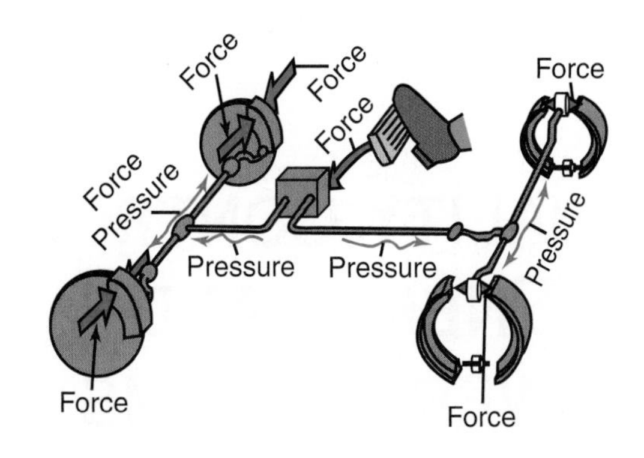

Figure 21-30. Automobile brakes are an example of a system transmitting energy through fluids. (EIS Div., Parker-Hannifin)

Figure 21-31. Electric transformers increase or decrease live voltage. (©iStockphoto.com/jlsohio)

Electric current leaving the generating station is fed into transformers. These ***transformers*** are devices increasing or decreasing the force (voltage) of the electric current. **Figure 21-31.**

A transformer increasing the voltage is a ***step-up transformer***. These transformers boost the voltage from around 13,800 volts to as high as 700,000 volts. The current is then carried to the point of use. There, ***step-down transformers*** reduce the voltage.

Quality Control in Energy and Power Systems

Quality control in electrical and power systems focuses on monitoring system operation and output. The equipment and machines within the system are produced using the inspection and design guidelines typical of manufacturing systems. This equipment is subjected to periodic maintenance and inspection. Most electrical and power systems have a built-in monitoring system. When the operation or output is beyond specified limits, the system might signal an alert to an operator. The operator can then make required adjustments. In some cases, the system might shut down automatically.

Using and Conserving Energy

Energy and power technology has many benefits. This technology makes life better. Energy and power technology promotes global transportation and communication.

Business and industry rely on energy and power technologies to provide electricity so machines, lights, computers, and equipment can operate. Mechanical power transmission is critical in the manufacturing and automotive industries. Nearly all communication systems and devices need electrical power, sometimes provided by batteries, to operate. Businesses' dependence on electricity is such that the operations of nearly all businesses are greatly reduced if the business loses electrical power.

Our technological society requires a great deal of energy. The rate at which energy is being used in the world is increasing. Natural resources might be depleted in the future before other energy resources are available to replace them.

We can, however, do our part to slow down the rate at which energy is being used. Much of the energy used in our environment is not used efficiently. ***Conservation*** is the act of making better use of energy. Some efforts are already being made to save energy:

- Manufacturers are building more energy-efficient products. For example, automobiles are being made with more fuel-efficient engines. Furnaces are more energy efficient. Home appliances use less energy.
- Buildings are better insulated. Higher insulation standards are being used in new construction. This means thicker insulation in ceilings and walls. Triple glazing (three panes of glass) is available in windows.
- People are beginning to recycle more materials. It takes large amounts of energy to make a ton of aluminum from bauxite ore. Converting scrap aluminum requires much less energy.

People have personal roles in conserving energy. We have formed habits that waste energy. Consider how we use electricity. How many times have you left radios and lights on when leaving a room?

Our standard of living is the highest in the world. We have many benefits that are the results of modern technology. Also, we have a responsibility to do what we can to conserve natural resources for future generations. What are some of the things you and your family can do to conserve? Is there anything that can be done in your school to use less energy? What can governments do about promoting energy conservation?

Summary

We use energy to do all kinds of work. In many cases, energy is transformed from one form to a more usable form. Energy is converted and used through an energy system. Such a system has an energy source, a converter, a means of transmission, controls, and a load. Much of our available energy is wasted. Throughout the energy-conversion process, concern for energy conservation must be shown.

STEM Connections

Mathematics

Make a graph showing the energy consumption of major home appliances.

Science

Explain the scientific principles of distillation used in petroleum refining. Describe the scientific principles used in electricity generation and use (for example, motors and lightbulbs).

Mathematics

Prepare a display explaining the measurement systems used for various energy uses (such as volts, amperes, joules, and kilowatts).

Curricular Connections

Social Studies

Compare the energy use of various countries in the world. Contrast the energy sources used. Map the locations of major energy resources in the country and the world.

Activities

1. Visit an electrical power plant to learn how it operates. Write a report on what you see. Use a computer to prepare the report, if you can.
2. Build a model of a windmill or waterwheel. Explain how it works.
3. Make a list of devices you see in your community using energy in any form. Place them in the order of their importance to the well-being of the community.
4. Prepare a list of the energy-using devices you use each day. Describe how you can conserve energy while using them.

Test Your Knowledge

Do not write in this book. Place your answers to this test on a separate sheet of paper.

1. Energy is the ability to do _____.
2. A(n) _____ converts mechanical energy into electrical energy.
3. The two types of rocket engines are _____ and _____.
4. Internal combustion and steam engines are _____ engines.
5. Summarize how batteries work.
6. Explain how fuel cells work.
7. Another name for a solar cell is a(n) _____.
8. The two types of solar heating systems are _____ and _____.
9. Direct gain and indirect gain are examples of _____ solar heating systems.
10. Match the types of converters with the right descriptions and give an example of each:
    - _____ Converts liquids into another form of energy.
    - _____ Converts the Sun's energy into another form.
    - _____ Converts heat energy into another form.
    - _____ Converts kinetic energy into another form.
    - _____ Converts energy in the molecular structure of a material into another form.

    A. Mechanical converter.
    B. Thermal converter.
    C. Chemical converter.
    D. Solar converter.
    E. Fluid converter.
11. _____ are used to change wind energy into electricity.
12. _____ change the energy in flowing water into electricity.
13. Three major means used to transmit power are _____, _____, and _____.
14. Transformers used to increase electrical voltage are called _____.
15. Why is conserving energy important?

Reading Organizer

On a separate sheet of paper, list the different types of energy converters. Give examples of each type.

Type of Converter	Examples of Converter
Example: Mechanical converter	Electric generator

CHAPTER 21 ACTIVITY A

ENERGY-CONVERSION TECHNOLOGY

INTRODUCTION

Energy is defined as "the ability to do work." Much of this work involves converting energy from one form to another. Water is stored behind a dam (potential energy). When the water is allowed to fall through pipes to a turbine (kinetic energy), the energy of falling water turns the turbine (mechanical energy). The turbine converts the motion into electrical energy. The electrical energy is transmitted over power lines. These lines have resistance. The resistance produces heat (thermal energy). At the destination, the electrical energy is converted to light, heat, or mechanical energy. In this activity, you will explore one type of energy conversion—light energy into heat energy.

EQUIPMENT AND SUPPLIES

- A heat-lamp test stand.
- Large (42-oz.) juice cans or #10 cans (from the school cafeteria) that have been painted various colors (black, red, blue, green, yellow, brown, and white). These cans should each have a hole in the center of the bottom. A rubber grommet with a hole the diameter of the thermometer should be in each hole.
- A thermometer reading up to 200°F.

PROCEDURE

1. Obtain a heat-lamp test stand.
2. Obtain a painted juice or vegetable can.
3. Record the distances and times that will be used in the experiment. Your teacher will establish these parameters.
4. Set up the experiment as demonstrated. See **Figure 21A-1.**
5. Take the reading at distance #1 for each time.
6. Allow the can to cool.
7. Repeat steps 5 and 6 for distances #2 and #3.
8. Report your results to the class.
9. Record the results from other groups on the Laboratory Sheet.
10. Complete a Laboratory Summary Sheet.

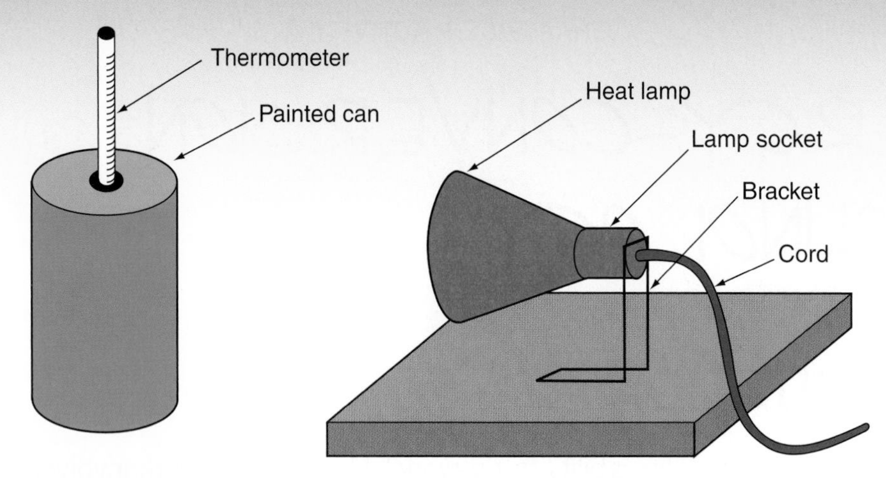

Figure 21A-1. The experiment setup.

Challenging Your Learning

Use the results of your research to answer the following questions:

1. Is a black or white car best for the Desert Southwest?
2. Would a black roof or white roof absorb the most radiant energy?
3. Why would a person buy a red car, even though it absorbs more radiant energy than a white one?

TSA Modular Activity

This activity develops the skills used in TSA's Cyberspace Pursuit event.

Activity Overview

In this activity, you will create a Web site composed of four components:

- An overview of your school's technology-education program.
- General information about your school.
- Historical information about your school.
- A page of links to related or interesting Web sites.

Materials

- Paper.
- A pencil.
- A computer with Internet access and Web page–development software.

Background Information

- **Design.** Careful planning is critical when developing a Web site. Create a sketch for each Web page. List the elements to be included on each page. Include lines showing the links between Web pages. Consider how a user will navigate within the Web site to be sure you have the necessary links.
- **Navigation and functionality.** The visual appearance of your Web pages is important. The appearance is not as important, however, as easy navigation and functionality. Design your Web pages so links can be easily located. Select easy-to-read type for links.

- **Type.** Too many fonts and type sizes can make your Web pages unattractive. Vary the size and font based on the function of the text. Titles are meant to draw attention. Body type should blend in to the overall design and use a typeface, size, spacing, and justification comfortable for reading. If you use an unusual type font, a person viewing your Web page might not have that font on their computer. In this situation, another font is substituted. This would cause your Web page to have an unintended (and, most likely, less attractive) appearance. Use common type fonts, such as Arial, Times New Roman, Tahoma, or Courier. If you want to use an unusual font, create the text as an image file and insert it into the Web page.
- **Images.** Images can add to the visual appeal of your Web pages. Using too many images, however, can clutter a page and cause it to load slowly. Use a compressed-image file format (such as JPEG) for faster loading. Also, use the lowest acceptable resolution.

Guidelines

- Review at least five Web sites, evaluating the design of the sites, in terms of attractiveness and usability.
- Your Web site must have a home page containing separate links to each of the four components.
- There is no minimum or maximum number of pages for the individual components.
- The home page and each of the four components must contain both text and graphics.
- All pages must include a link to the home page.
- Use pencil and paper to prepare a rough sketch for each page and an organization chart showing how the pages are linked.
- After you have developed the rough sketches, create the Web pages.
- Test the completed design to make sure all links work properly.

Evaluation Criteria

Your Web site will be evaluated using the following criteria:

- Web-page design.
- Originality.
- The content of the Web pages.
- Functionality.

Active solar systems can be used to collect solar energy to provide heat water and homes. Solar systems may also be able to provide electricity for homes.

CHAPTER 22

Information and Communication Technology

Did You Know?

- The Bible was the first book printed with Johannes Gutenberg's moveable type. This book was, therefore, the first printed book.
- The first machine with a keyboard and levers was the tachygrapher. This machine was invented in 1823 and is the ancestor of the typewriter.
- The first pictures made from a camera were called *daguerreotypes*. They were named after the inventor Louis-Jacques-Mandé Daguerre.
- The first camera with a lens was invented in 1840.
- The first American president to appear on television was Franklin Delano Roosevelt. The black-and-white image was broadcast from the 1939 New York World's Fair.
- Charles Babbage invented an "analytical engine" in 1835. This engine did calculations and stored data. The analytical engine is an ancestor of the modern computer.

Objectives

The information given in this chapter will help you do the following:

- Explain the concept of communication.
- Compare communication and communication technology.
- Summarize several ways in which people communicate.
- Compare graphic communication and wave communication.
- Give examples of the four types of communication systems.
- Outline and explain the communication process.
- Summarize the four steps in the process for producing a message.
- Compare computer hardware and software.
- Explain the role of quality control in producing communication media.

Key Words

These words are used in this chapter. Do you know what they mean?

audience assessment
audio recording
broadcast message
carrier
central processing unit (cpu)
communication
communication satellite
communication technology
compact disc read-only memory (CD-ROM)
comprehensive
computer
decode
digital videodisc (DVD)
digitize
downlink
electromagnetic radiation
encode
film message
flexography
format
graphic communication
hardware
hypertext-markup language (HTML)
input device
interference
layout
medium
memory
monitor
operating system
output device
photographic system
printing system
published message
receive
receiver
rough
storyboard
telecommunication system
transmit
transmitter
uplink
video recording
wave communication
World Wide Web (WWW)

Preparing to Read

As you read this chapter, outline the details of communication technology. Use the Reading Organizer at the end of the chapter to organize your thoughts.

The modern world is vastly different from the past because of technology. Probably the most important technologies that have changed our life are communication and information-processing technologies. These technologies started with the development of tools to support speech. The written word was developed and started with hieroglyphics. The Egyptians developed hieroglyphics. A number of developments followed this, including moveable type, printing presses, telephones, radios, televisions, computers, and cellular telephones.

Today, humans communicate constantly. They talk to one another. People read what other humans write. They listen to voices on radios. Humans view and listen to television programs. They are in constant communication through electronic media. In fact, humans cannot avoid communication. Anytime people send or receive information, they are communicating and processing information. See Figure 22-1. Think about how many times you receive and send messages in one day.

Communication and Information

People often use the terms *information* and *communication* together. These terms are related. Each, however, means something different. Let's review two words you learned in Chapter 7: *data* and *information*. Data includes individual facts, statistics (numerical data), and ideas. These facts and ideas are not sorted or arranged in any manner. Information is data that has been sorted and arranged. This sorted data consists of organized facts and opinions people ***receive*** during daily life. Changing data into information is called *data processing*, or *information processing*. This processing involves gathering, organizing, and reporting data so it is useful to people. Changing data into information is often done using information technology.

Data and information are of little use, however, unless they are shared. This is what communication does. ***Communication*** is the act of exchanging ideas, information, and opinions. This

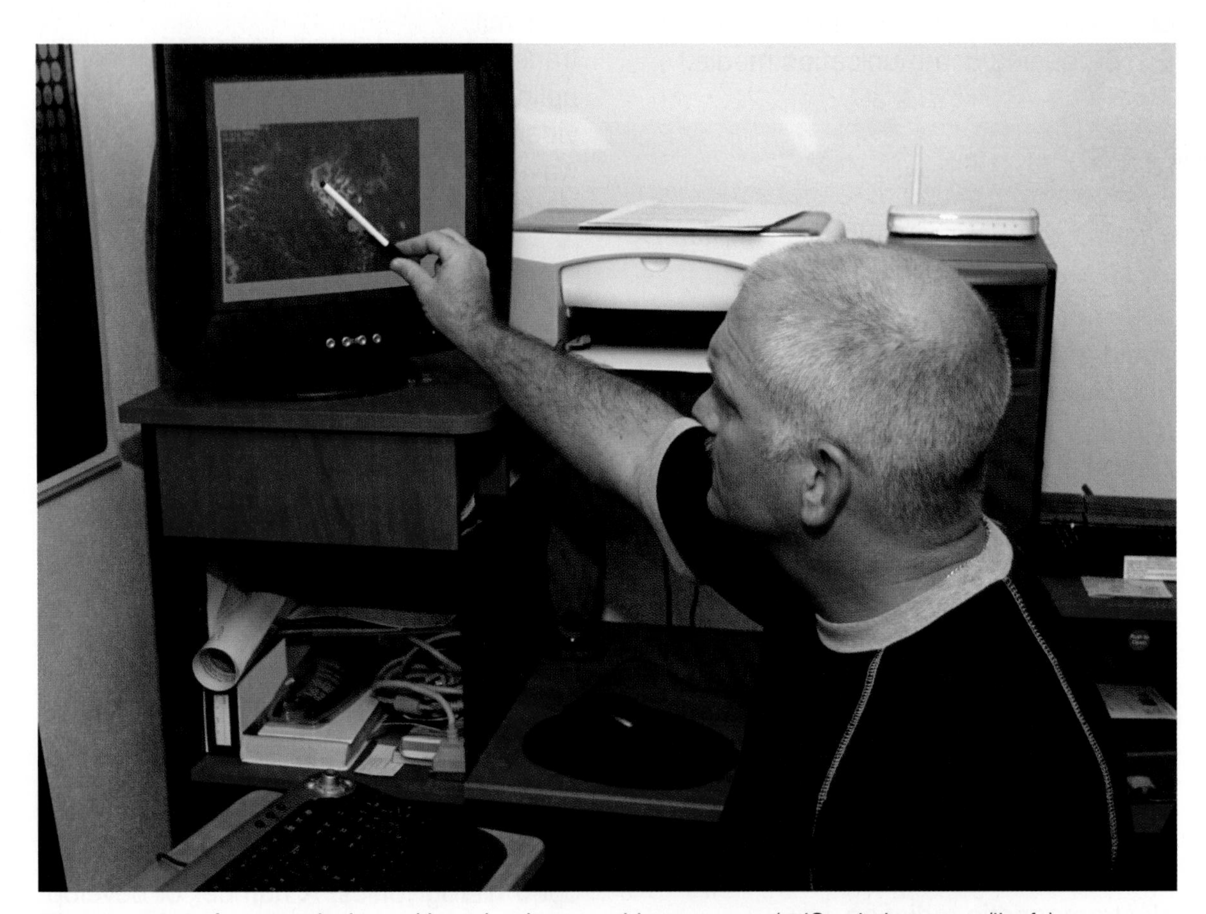

Figure 22-1. A meteorologist tracking a hurricane on his computer. (©iStockphoto.com/lisafx)

act involves a message someone wants people or machines to receive. When we extend communication by using technical equipment, it is called ***communication technology***.

How We Communicate

People must process information before they can communicate. You might know something. The knowledge is in your mind. This fact belongs to you. No one has access to it. There are a number of ways you can share your knowledge. See Figure 22-2. You can use your ability to speak and tell someone what you know. Also, you can write a paper containing the information you have. A person can read the paper to learn what you know. You can draw a picture communicating the information. The picture shows a person what you know. You can take a photograph communicating what you know, use a symbol, or develop a sign representing the information. The use of drawings, symbols, and measurements promotes clear communication by offering a general language with which to convey ideas. Technical systems use specific symbols and terms.

So far, we have discussed people directly communicating with other people. This is face-to-face communication. People show pictures and symbols. They can also use formal language. These are common types of communication. They, however, use little technology. There are no technical means used. Machines and equipment are not involved in the communication. A communication system is not present.

Communication Technology

Communication technology uses equipment and systems to send and receive information. See Figure 22-3. This technology communicates information using either graphic systems or wave systems.

Figure 22-3. This sound-mixing board is used in sound-recording studios.

Figure 22-2. People can communicate with pictures, symbols, writing, or speech. (©iStockphoto.com/technotr)

Communication

Graphic

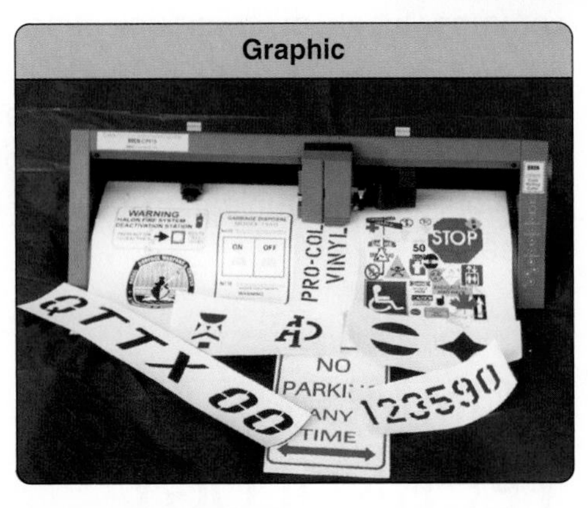

Electronic

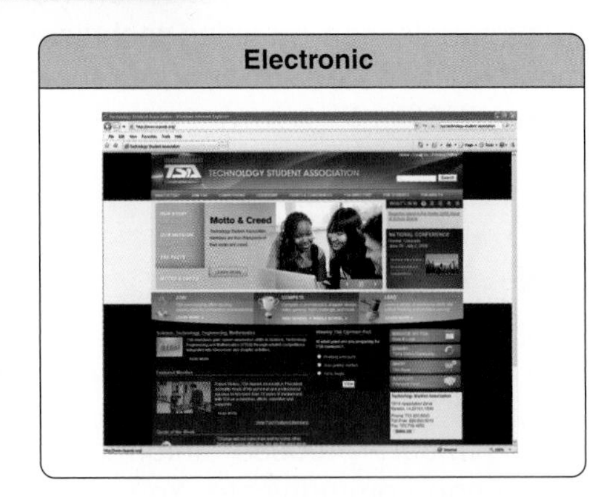

Figure 22-4. The two major communication systems are graphic and electronic systems. (Bren Instruments Inc., TSA)

See Figure 22-4. The term *graphic* comes from a word meaning "to draw or write, as on paper." The word *wave* refers to radio waves, a kind of energy.

Graphic Communication

Information can be communicated using drawings, pictures, graphs, photographs, or words on flat surfaces. See Figure 22-5. These types of messages are called ***graphic communication***. In graphic communication, paper, film, or another flat surface carries the message. Graphic-communication systems produce printed graphic and photographic ***media***. They communicate through drawings, printed words, and pictures. The systems also use photographic prints and transparencies (slides and motion pictures). Graphic messages are visible at all points as they move from the sender to the ***receiver***.

Figure 22-5. A newspaper is an example of a graphic-communication medium.

Wave (Electronic) Communication

Wave-communication systems depend on an energy source called *electromagnetic radiation*. ***Electromagnetic radiation*** is energy moving through space in waves. This radiation travels at the speed of light—186,000 miles per second. See Figure 22-6.

Sound waves are much slower than electromagnetic waves. They travel at different speeds in different media. In air, they travel at about 750 mph at sea level. They travel about four times faster in water.

People can use light, sound, or electrical waves to send information. The information is coded at the source and then ***transmitted*** (sent) to the receiver. There, the code must be changed back to information. This type of communication is often called *electronic communication*.

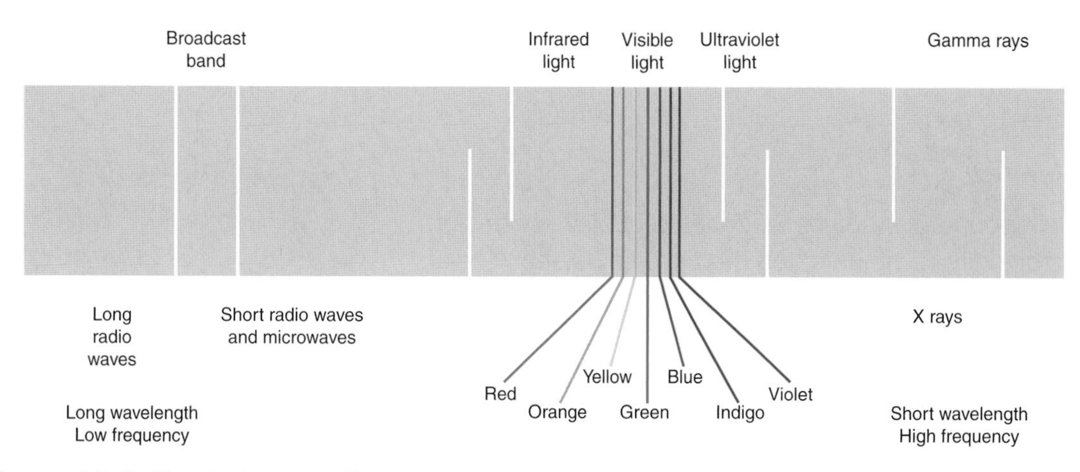

Figure 22-6. The electromagnetic spectrum.

See **Figure 22-7.** These systems often use electronic equipment to code and transmit the information.

For example, suppose your favorite sports figures are interviewed. They speak into microphones. Speech is changed into pulses of electromagnetic energy. This energy is transmitted from the radio station's broadcast tower. Your radio's antenna picks up some of these pulses of energy. Various parts of the radio process the energy so the radio speaker can change

Cameras Capture Signals

Equipment Processes Signals

Signals Are Sent

Signals Are Received

Figure 22-7. A typical communication system. Cameras capture pictures and sound. Complex equipment processes the signals. The signals are then sent through the air. The viewer gets the message on a receiver (TV set). (©iStockphoto.com/asterix0597, Westinghouse Electric Corp.)

Career Highlight

Desktop Publishers

The Job: Using computer software, desktop publishers design printed and electronically communicated media. They select formats and then combine text, photographs, drawings, and other graphic elements to produce publication-ready material. These publishers might create graphics, convert photographs and drawings into digital images, manipulate digital images, and design page layouts for a wide variety of materials, such as books, calendars, magazines, newspapers, and packages.

Working Conditions: Desktop publishers usually work eight-hour days in office areas. They are subject to stress. These publishers are also subject to the pressures of short deadlines and tight work schedules.

Education and Training: Most people working in desktop publishing qualify for their jobs by taking classes. The also might complete programs at vocational schools or colleges. Associate's or bachelor's degree programs in graphic arts, graphic communications, or graphic design can provide an avenue for entry into desktop publishing careers.

Career Cluster: Arts, Audio/Video Technology & Communications

Career Pathway: Printing Technology

the electrical pulses back into audible sound. You hear the voices. Between the microphones and the speaker, however, no one can hear or see the messages.

Graphic Communication versus Electronic Communication

Electronic communication has many advantages over graphic communication. A book can hold only so much information. Readable type can be only so small. A book can be only so thick and still be useful. ***Wave communication***, however, carries vast amounts of information in almost no space. A space satellite communicates a library of information in a matter of seconds. Fiber-optic systems are said to carry encyclopedias of information per second. A CD can hold several books of information on a single disc. ***Digital videodiscs (DVDs)*** and Blu-ray discs can hold an entire movie on a disc, with room to spare.

You cannot, however, take wave media everywhere with you. A signal from a local radio station can reach only so far.

A cell phone works only when a tower is nearby. Books and magazines, however, are portable. They can also be available at any time or place. Books and magazines capture ideas and visual images. Electronic devices may use wireless networks to obtain different types of media. However, you may not always be able to connect to a wireless network to receive this media.

Therefore, we need different types of communication systems for different uses. Wave communication is quick and inexpensive. Current events reach us almost as they happen. Electronic communication has made our world smaller. This communication has caused us to live in what many people call a *global village*. Print communication media is slower. This media, however, can be used almost everywhere. Print communication media can be easily selected, used, and stored. This media lasts a lifetime.

With the advent of the Internet, graphic communication is delivered through electronic means. This merging of systems allows people to quickly receive vast quantities of text and photographic information (graphic communication). Also, sound information and video information that radio and television (electronic communication) previously delivered are available.

Types of Communication Systems

Communication technology is used for four distinct types of communication:

- People-to-people communication.
- People-to-machine communication.
- Machine-to-people communication.
- Machine-to-machine communication.

See **Figure 22-8.** Each of these information and communication systems affects our daily lives. We come into contact with each type as we interact with our environment. People create information and communication technology systems to gather facts, manipulate data, and

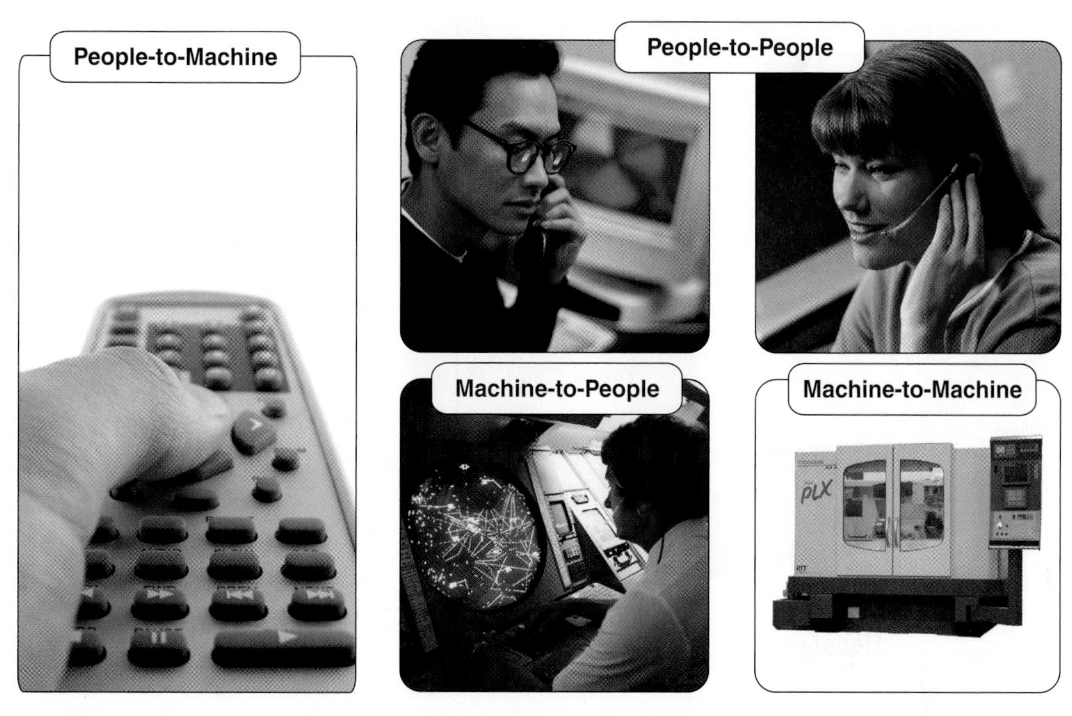

Figure 22-8. The types of communication. (Hirdinge, Inc.)

communicate information more effectively. Information is transmitted and received using various systems.

All information and communication systems have similar components. See Figure 22-9. They all have a device to develop and send the message. This device is often called a ***transmitter***. All systems must have a channel to carry the message. This channel is called the ***carrier***. The carrier can be the airways, a wire, or optical fibers. Finally, each system has a device to gather and process the message. This device is generally called a *receiver*. In addition to these components, many systems have means to store and retrieve information. Music can be stored on CDs. Movies can be stored on DVDs. Data can be stored on computer hard drives.

People-to-People Communication

Everyone communicates with other people in one way or another. Humans have developed complex technological systems to improve their communication. Machines and devices have been produced to help us communicate better and easier.

Figure 22-9. A simple communication system.

Technology Explained

bar code reader: a system used to read numerical codes on packages and tags.

Business and industry strive to be as efficient as possible. This allows companies to produce and sell products at competitive prices. One laborsaving device is the bar code reader. See **Figure A.** Most people have seen a bar code reader at the supermarket or discount-store checkout. The reader is connected to a cash register (called a *point-of-sale terminal*) and one or more computers. See **Figure B.**

The bar code system uses binary numbers. Binary numbers represent numbers and letters with a series of ones and zeros. Bar codes use white spaces to represent zeros and black bars to represent ones. To see an example of a bar code, look on the back cover of this book.

A laser beam reads the bars. The beam hits a spinning mirror. The mirror spreads the beam out so the beam can read the bar codes. The laser beam strikes the package as the package is moved past the reader. The black bars absorb the laser light. White spaces reflect back into the reader and onto a detector. The detector receives the reflected light and changes the light into pulses of electricity. These pulses are amplified and decoded.

The bar code information is fed to the in-store computer. This computer has the description and price for each product in the store in its memory. Price information is sent to the point-of-sale terminal. There, it is printed on the receipt and shown to the customer on a digital display. The in-store computer keeps track of the sales information it receives from the scanner. A record of this information can be sent to the store's central office. The central office can use this information to track sales and order new inventory.

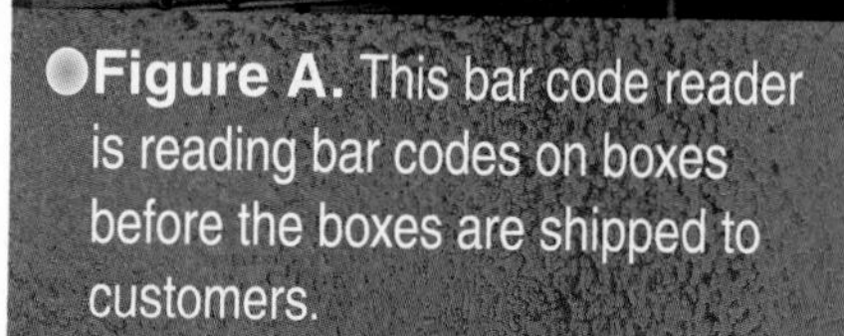

Figure A. This bar code reader is reading bar codes on boxes before the boxes are shipped to customers.

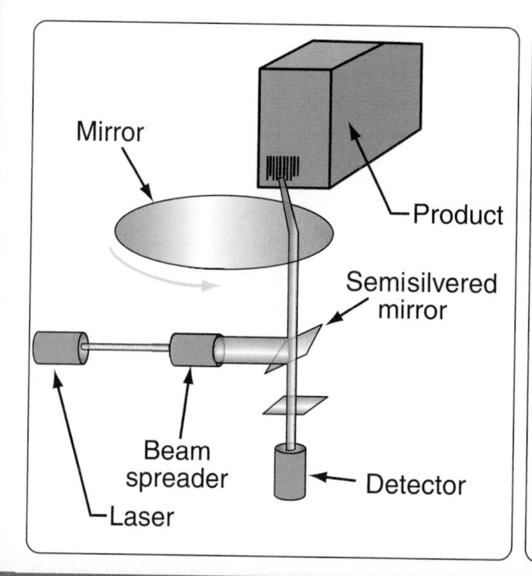

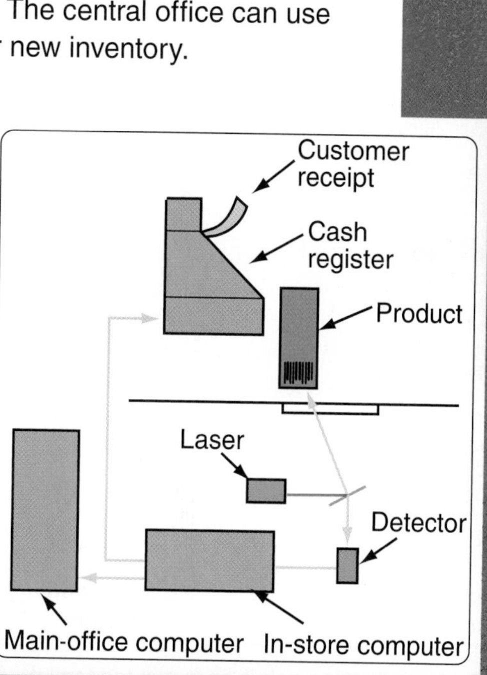

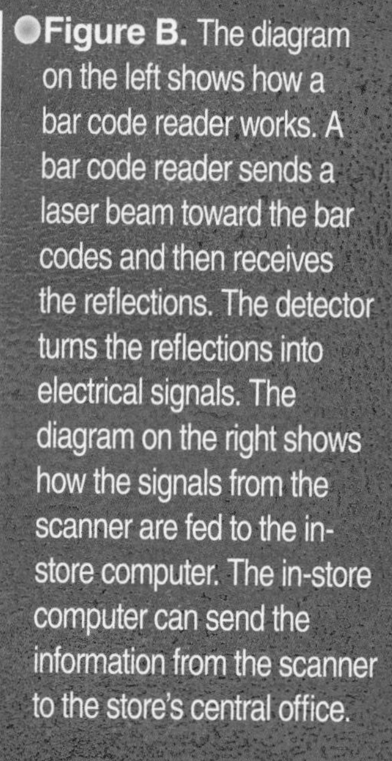

Figure B. The diagram on the left shows how a bar code reader works. A bar code reader sends a laser beam toward the bar codes and then receives the reflections. The detector turns the reflections into electrical signals. The diagram on the right shows how the signals from the scanner are fed to the in-store computer. The in-store computer can send the information from the scanner to the store's central office.

People-to-people communication includes a number of basic systems:

- Telecommunication systems.
- Audio and video recording systems.
- Computer (Internet) systems.
- Printing systems.
- Photographic systems.
- Drafting systems.

Each of these systems is in wide use today. All of them have their places and serve specific purposes.

Telecommunication Systems

Communication systems exchanging information over a distance are called ***telecommunication systems***. There are two major types of telecommunication systems. These are individual telecommunication systems and mass telecommunication systems.

Individual telecommunication systems have been designed to let one person communicate with another individual. These were the earliest telecommunication technologies. Two systems of this type were developed in the 1800s.

The first was the work of Samuel Morse in the 1830s. This system is the telegraph. The telegraph depends on an electrical circuit (a complete pathway through which electricity flows) to carry the message. (The circuit is the carrier.) An operator uses a special code consisting of a series of dots (short electrical pulses) and dashes (long pulses) representing letters and numbers. See Figure 22-10. The operator produces the pulses by pressing a key (the transmitter). At the other end of the circuit, a sounder (the receiver) is activated. The sounder changes the pulses into clicks that can be read as dots or dashes.

The second personal telecommunication system was the telephone. Alexander Graham Bell invented this device in 1876. The telephone uses an electrical circuit to connect the two communicators' phones with a central office. The handset contains a transmitter that changes sound waves into electrical pulses. Also, it contains a receiver that changes the electrical pulses back into sound.

Often, the two telephones are connected to different central offices. The message is then transmitted to the central office on wires. From there, the signal moves between the two offices. This can be done on wires, by microwaves

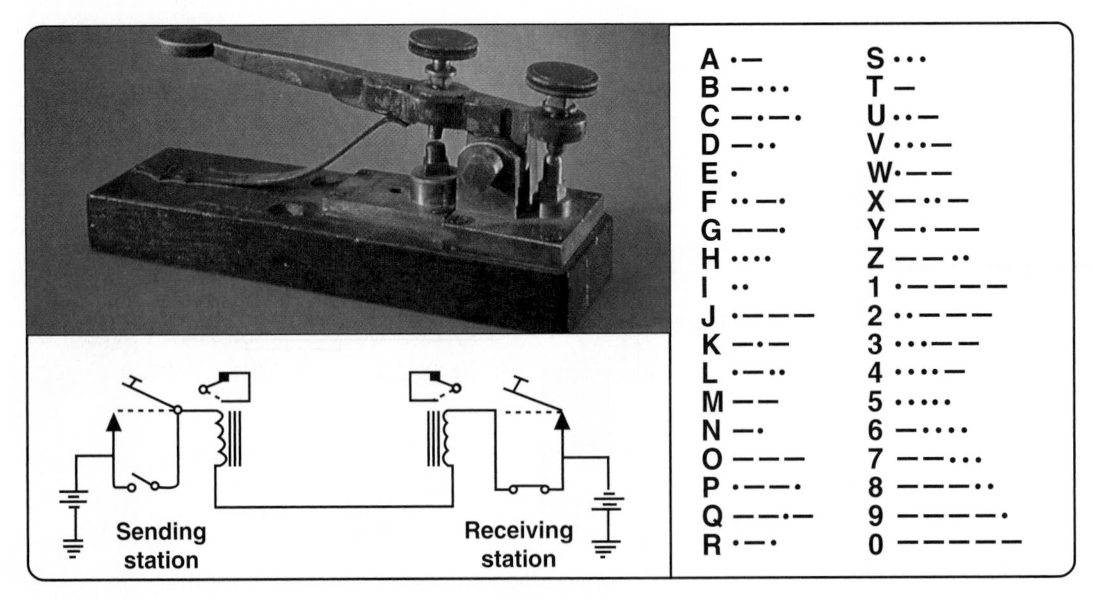

Figure 22-10. An early telegraph is shown in the top left. A schematic of a closed loop is in the bottom left.

(high frequency waves) between towers or satellites, or by glass fiber cables (fiber optics). See Figure 22-11. Today's cellular telephone operates in a similar manner. The main difference is that this telephone is really a low-power radio transmitter. The cellular telephone directly broadcasts its signal to a relay tower. The tower connects the signal to a Mobile Telephone Switching Office (MTSO). The MTSO connects the signal to normal telephone landlines. From there, the signal travels as in a normal telephone call.

Two mass telecommunication methods have been designed to reach large groups of people. These are radio and television. See Figure 22-12.

They operate very much alike. These systems use microphones and cameras to change sound and light into electrical pulses. Transmitters process these pulses into high frequency waves. These waves are carried to a broadcast tower. They are radiated (projected) over a large area, using the atmosphere as the carrier. A receiving antenna captures these waves. The receiver separates a signal from other waves. Speakers and television picture tubes change the signal back into sound and light. Today, television signals are also broadcast by satellite (dish) systems and over cables. These systems replace local broadcast towers. Signals are carried over cables connected to each home or beamed from a satellite to a personal dish receiver. Satellite radio systems use satellites to broadcast sound to receivers.

Figure 22-11. A communication relay tower.

Radio and television broadcast systems use two types of signals. Radio and most television stations use the very high frequency (VHF) bands (groups of signal frequencies). Some television stations use the ultrahigh frequency (UHF) bands.

The advent of space exploration ushered in a new phase of telecommunications. This allowed the use of ***communication satellites***. These satellites are broadcasting stations in outer space. They can receive messages from one place on Earth and then relay the messages back to Earth at a distant place. See Figure 22-13.

A narrow beam of microwaves, carefully aimed to hit the satellite, carries the message. The signals are beamed back to Earth at many different angles and to more than one Earth station. The satellite must always be in the same position above the Earth because of the need to aim the narrow beam. This station must complete one orbit every 24 hours. The satellite travels at the same speed as Earth's rotation. For this reason, its orbit is called *geosynchronous* (at the same time as Earth). If the satellite were to travel slower or faster, the microwave link (contact) would be broken.

The Control Room

TV Broadcasting

Figure 22-12. The control room is the nerve center of a television-broadcasting activity. (Viacom International, Inc.; ©iStockphoto.com/DeshaCAM)

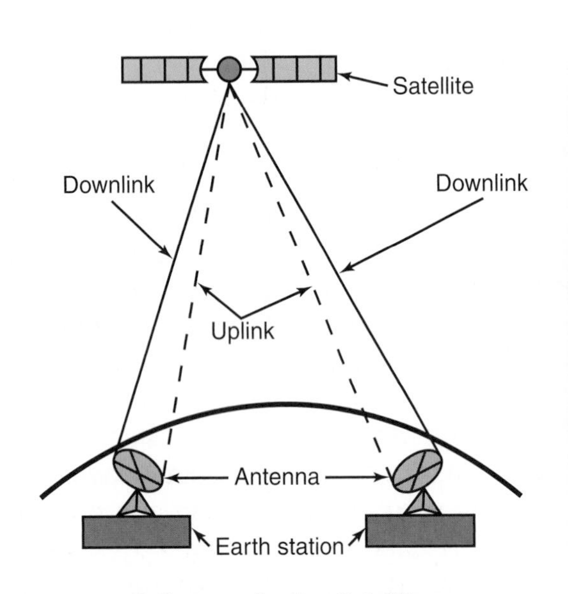

A Communication Satellite

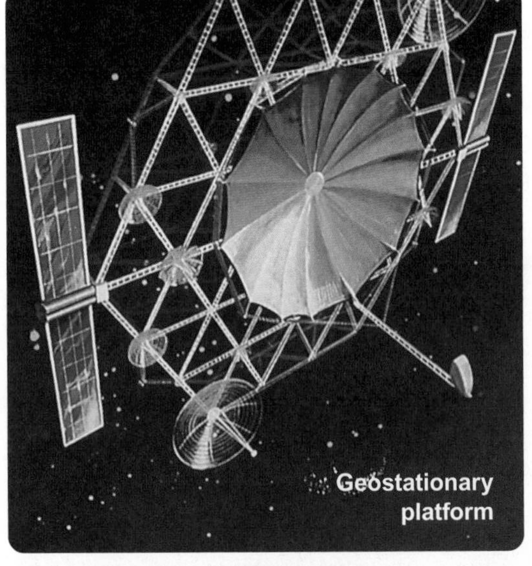

A Communication Platform

Figure 22-13. Communication satellites high in space can receive and send signals from Earth. On the right is a space-communication platform of the future. A powerful antenna will be able to link up with small Earth stations. You might be able to use it for your own personal communication link! (NASA)

Communication satellites are placed about 22,300 miles above Earth. There, they can transmit to over 40% of Earth's surface. (A message beamed to a satellite from Boston or Montreal can be received in Germany.)

Earth stations transmit the microwave signals through bowl-like antennae. The upward signal is called the ***uplink***. The signal relayed back is called the ***downlink***. Signals can also be relayed from one satellite to another.

Audio and Video Recording Systems

Audio recordings and ***video recordings*** are extensions of radio and television communications. They record the same type of information radios and televisions broadcast. These recordings can be seen as warehouses for audio and video information.

These communication systems can be traced to Thomas Edison's invention of the phonograph. This 1877 invention allowed people new freedom. No longer

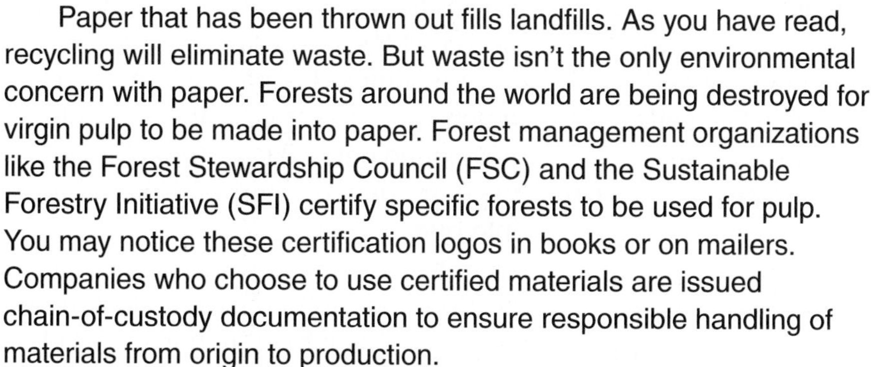
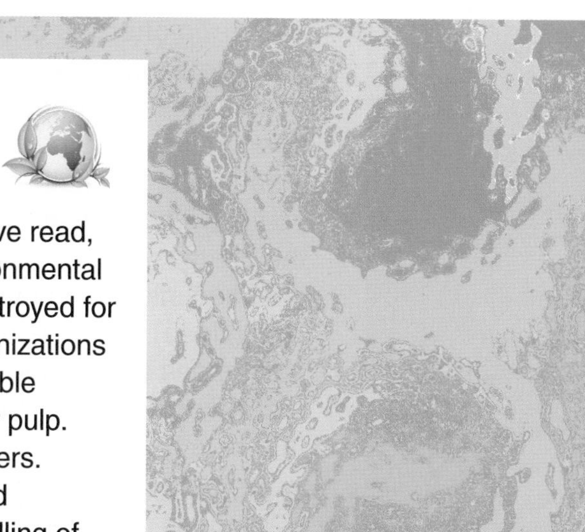

Think Green

Green Paper

Paper that has been thrown out fills landfills. As you have read, recycling will eliminate waste. But waste isn't the only environmental concern with paper. Forests around the world are being destroyed for virgin pulp to be made into paper. Forest management organizations like the Forest Stewardship Council (FSC) and the Sustainable Forestry Initiative (SFI) certify specific forests to be used for pulp. You may notice these certification logos in books or on mailers. Companies who choose to use certified materials are issued chain-of-custody documentation to ensure responsible handling of materials from origin to production.

did they have to listen to a speaker or a concert when someone else scheduled it. They could select the time and place they wanted to listen to the speech or music. See Figure 22-14. Audio and video recordings use one of three technologies:

- **Grooves in discs.** The first audio records used this recording technology. Records were produced with wavy grooves on their surface. A stylus (needle) vibrated as it followed the grooves on a moving record. This vibration produces a small electrical signal. This signal is amplified (made stronger) and changed into sound waves. Few people use this type of recording today.

Figure 22-14. Sound and pictures have been recorded using grooved discs, tapes, and optical recorded discs.

- **Magnetic charges on a tape.** To produce a magnetic recording, coated plastic ribbons are fed through a recording unit. The unit produces a pattern of magnetic charges in the coating. The playback unit reads the charges. The charges are then changed into sound (audio) and light (video) messages. Cassette audiotapes and videocassette recorder (VCR) tapes are produced in this way.
- **Digital codes on a disc.** This is the technology used for audio, video, digital video recorders (DVRs), and ***compact discs read-only memory (CD-ROMs)***. Sound and light waves to be recorded are ***digitized*** (numbered). Each frequency is assigned a specific number. The number is recorded on the disc on microscopic pits and flats. The playback unit reads the code with a laser beam. The digital readout is converted back to very accurate pictures or sound.

Each of these systems has advantages and disadvantages. Records were manufactured inexpensively, scratched easily, and lost quality quickly. Tape recordings were small and unbreakable and could have new material recorded over a used tape. They lost quality over time and sometimes jammed in the player. CDs, CD-ROMs, DVRs, and DVDs produce high-quality reproduction and maintain their quality over time. They are, however, fairly expensive.

Computer (Internet) Systems

Computers are the main tools used for networking information and communication technologies. A development that has become a key communication medium is the Internet. This system allows people to communicate through their computers. The U.S. Department of Defense started the Internet in 1966. Designed to allow researchers to use their computers to communicate, the system divides messages into packets (small parts). These packets each have an address for a destination. They do not, however, have to travel the same route to their destination. When all the packets arrive, the message is reassembled. The system showed so much promise that other people also developed networks. In 1982, the networks were merged, or internetworked (Internet), into one.

The amount of information available on the Internet is huge. Until recently, however, this information was hard to find and use. An innovation that has helped solve the problem is the ***World Wide Web (WWW)***. Documents accessible on the Web are formatted in a language called ***hypertext-markup language (HTML)***. This language allows users to select words, known as *hot links*, to jump from one document to another. For example, the word *airplane* might be highlighted in a document. If a person clicks on the word, additional documents about airplanes are made available.

A second advancement was Mosaic. This system was the first set of programs called *browsers*. Browsers are computer software programs that allow you to use keywords, such as *airplane*, to search for related documents on the Internet. People with a computer containing the proper software and Internet access can have information from around the world immediately at their fingertips. See Figure 22-15.

Printing Systems

Printing systems are the next type of people-to-people communication. They include all systems used to produce letters and pictures on nontreated paper. See Figure 22-16. The printing revolution started with Johannes Gutenberg about 1450. Until that time, scribes produced most books by copying manuscripts one at a time. This process was slow and open to

Figure 22-15. Computers are now a part of global communication with the Internet.

Figure 22-16. High-speed printing presses produce thousands of newspapers each day. (The Chicago Tribune Co.)

many errors. Gutenberg developed moveable type (individual letters on pieces of lead). This development allowed the printer to arrange the type for an entire page. A number of copies of the page could then be produced. This printing from raised letters of type is called *letterpress printing*, or *relief printing*. See Figure 22-17. Other more efficient processes are rapidly replacing it. One similar process is called ***flexography***. This process uses raised type, similar to letterpress printing. Flexography, however, uses a synthetic rubberlike sheet, very similar to a rubber stamp. This sheet is used to print the desired message.

Another more efficient printing process is called *offset lithography*. This process uses a negative of the page to be printed. The negative is placed on an offset plate. The plate has a photographic coating that bright light can expose. The negative allows light to expose the plate where the letters are to be. When the plate is developed, the type to be printed is on the plate.

Letterpress Printing

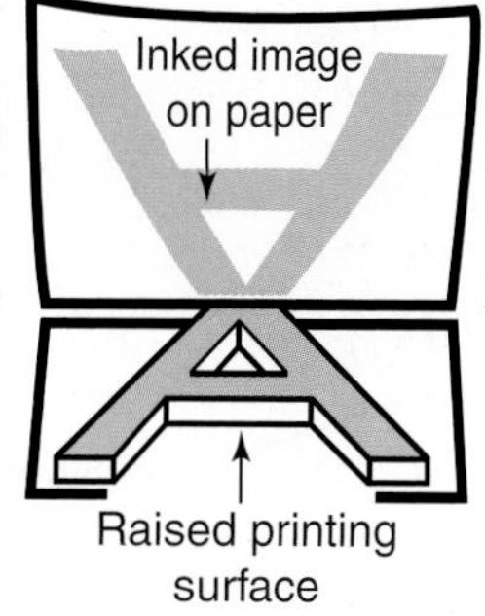

Images on Paper

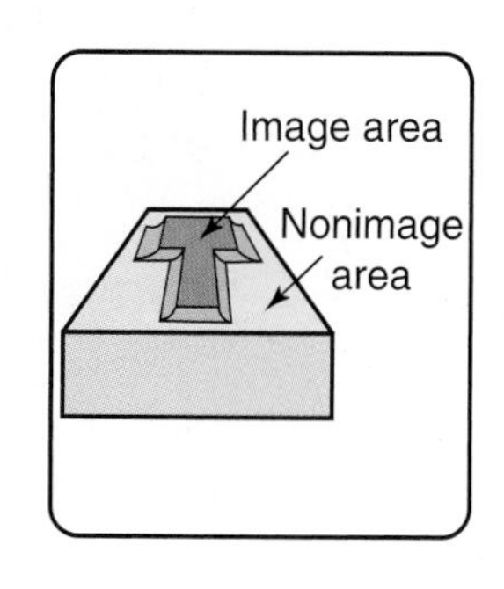

Raised Surface

Figure 22-17. Letterpress printing produces an image on paper. A raised surface makes the image. (Graphic Arts Technical Foundation)

The plate is placed on a special press. This press coats the plate with a liquid (fountain solution). This liquid does not stick where the letters are on the plate. Ink is then applied. The ink does not stick where the fountain solution is. Therefore, ink is on only the type portion of the plate. The plate then presses against a blanket. The blanket picks up the ink. This sheet transfers the ink image to the paper. See Figure 22-18.

Other printing processes include intaglio, screen printing, and electrostatic reproduction. Intaglio prints from a recessed image. This process is often called *etching* and is used for high-quality printing. Stamps and paper money are printed by this process.

Screen printing forces ink through a fabric screen to produce an image. See Figure 22-19. Fabrics, posters, and T-shirts are screen printed.

Electrostatic reproduction is rapidly becoming a major printing process. The office photocopier is an example of an electrostatic machine. High-speed machines of

Figure 22-18. The offset-printing process produces an image on paper from a flat printing surface. (Graphic Arts Technical Foundation)

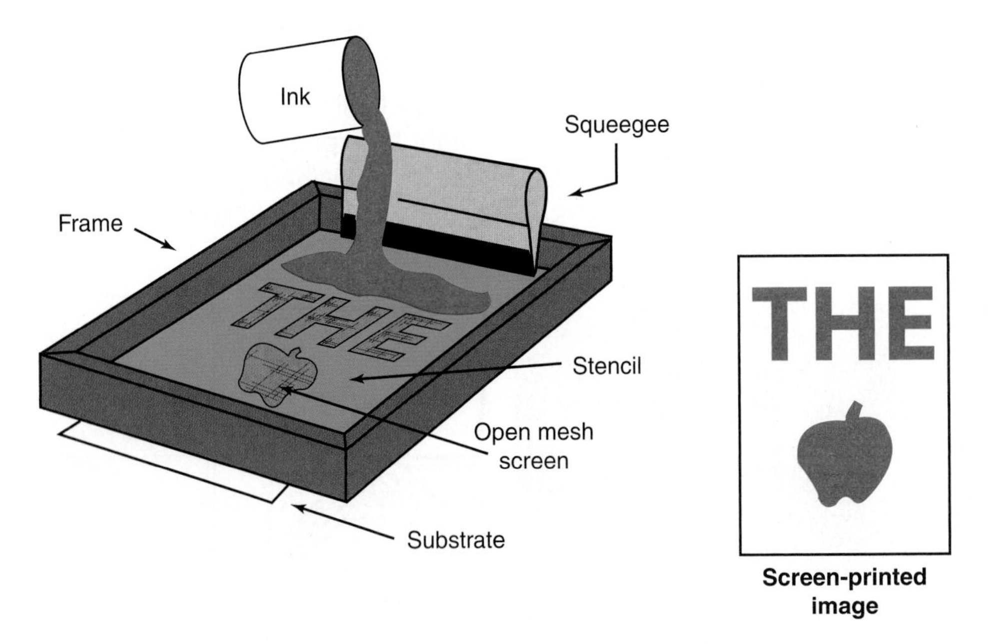

Figure 22-19. The screen-printing process produces an image by forcing ink through openings in a screen carrier. (Graphic Arts Technical Foundation)

this type can print several thousand copies an hour, rather inexpensively.

Photographic Systems

Directly related to printing are ***photographic systems***. These systems use photographs to convey messages. People on an airliner review emergency procedures with a series of pictures. The message was carefully designed and produced. The pictures tell a story.

This use of photography is very much different from personal snapshots. Personal photographs are used to capture moments of time. They generally are not planned to communicate a message. Therefore, most personal photography is not done for communication purposes. This photography is done to record a bit of history or an important occasion.

Originally, photography was a film-based system. The image was captured with a camera on a chemically coated sheet or strip of film. Subjecting the film to a series of chemicals developed the film. The result was a negative (reverse) or positive (actual) film image. See Figure 22-20. The positive image, such as that on photographic slides or movie films, can be projected onto a screen for viewing. The negative image can be used to make photographic prints or in printing processes.

In recent years, photography is rapidly becoming a digital process. A digital camera captures images as electronic signals (ones and zeros) stored on memory devices. See Figure 22-21. The codes can be downloaded, manipulated with computer programs, and printed with color ink-jet or laser printers.

Drafting Systems

The last type of person-to-person communication is drafting. This communication technique uses lines and symbols to communicate designs. Drafting includes engineering drawings used to show parts and products to be manufactured and architectural drawings for buildings. See Figure 22-22. Using sketching and drawing techniques to communicate designs is covered in Section 3 of this book.

Figure 22-21. The digital camera is a recent development that captures images as electronic code. (©iStockphoto.com/jsemeniuk)

Positive Image

Negative Image

Figure 22-20. The image on the left is a positive image. The one on the right is a negative image.

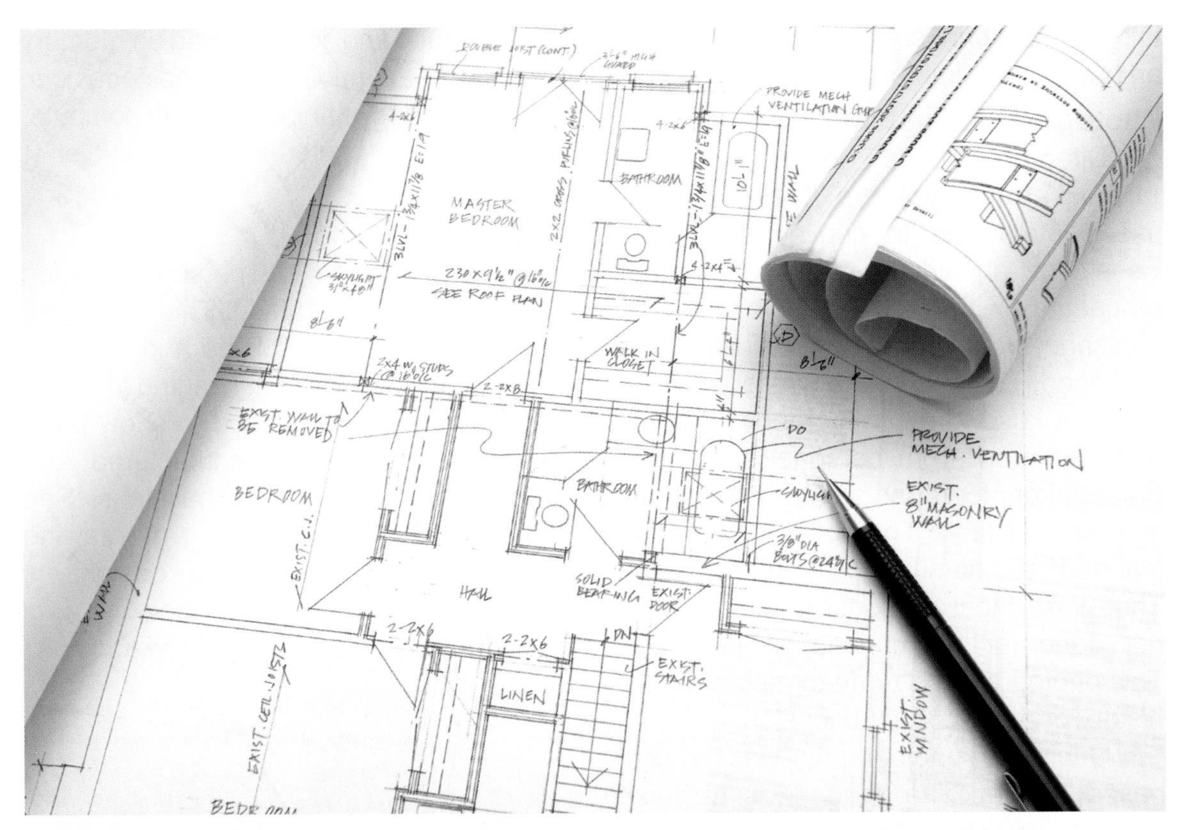

Figure 22-22. An architectural drawing for a new house. (©iStockphoto.com/Branislav)

People-to-Machine Communication

People communicate to machines daily. We set controls "telling" machines how to operate. See Figure 22-23. We set the thermostat to communicate the room temperature we want to a heating and cooling system. To communicate the speed at which we want to travel, we set a speed-control system in a car.

Also, we write computer programs to tell computers what to do. We can tell computers to print letters, draw lines, calculate costs, or perform hundreds of other acts. To ensure the clock signals a time to go to bed or begin a new day, we set time on digital clocks and timers. The timer tells us when food is done in a microwave or an oven. Symbols, or icons, are used on many computers, elevators, and telephones to stand for ideas and to communicate what should be done when the symbol is pressed or used.

Figure 22-23. This worker is communicating to a machine using a control panel. (Ford Motor Company)

Machine-to-People Communication

Directly related to people-to-machine communication is machine-to-people communication. The machine we communicate to, through switches and dials, responds. This device presents us with meter readings, flashing lights, or alarms.

Pilots have many machine-to-people communication systems on the aircraft flight deck. Lights tell them if the engines are running properly. Alarms sound as the plane approaches stall speed. Video screens display the information the radar system gathers. See Figure 22-24.

In our automobiles, gauges and lights are also used to communicate. The fuel gauge tells the driver when to buy more gasoline. The oil light warns the operator of low oil pressure. A blue light tells the driver the headlights are on high beam. A flashing light tells the driver the turn signals are operating.

Machine-to-Machine Communication

The most recent communication systems have machines providing information to machines. CAD systems help people design parts. See Figure 22-25. These systems do drafting. They can direct machines to produce parts.

Computer-aided manufacturing (CAM) uses computers to directly control machine operations. See Figure 22-26. The computer can direct the machine to run at specific speeds. This device can set material feed rates. The computer can cause the machine to change cutting tools. All this occurs without human action. Even more complex systems totally merge the design and manufacturing activities. The complex systems are called *computer-integrated manufacturing (CIM)*. These are but a few examples of machines communicating to other machines.

Figure 22-24. The instruments on this flight deck of a jet airliner are examples of machine-to-people communication. (©iStockphoto.com/A330Pilot)

A CAD System

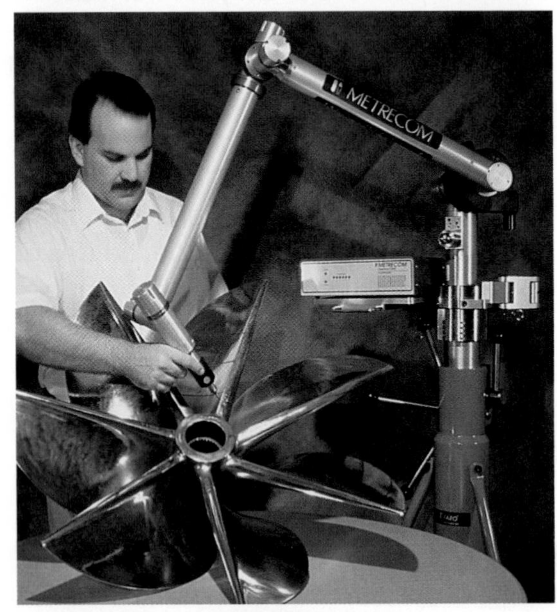

An Industrial Part

Figure 22-25. A CAD system was used to design this industrial part. (Faro Technologies, Inc.)

Figure 22-26. This robotic production line is an example of CAM. (AMP, Inc.)

The Communication Process

You have now read about many communication systems. These systems are made up of a source, an encoder, a transmitter, a receiver, a decoder, and a destination. A communication system is similar to other systems in that it includes inputs, processes, outputs, and sometimes feedback. Each of these systems follows a basic communication process. This process has four major steps. See Figure 22-27. These steps move the information from the sender to the receiver:

1. Encoding.
2. Transmitting.
3. Receiving.
4. Decoding.

At any point along the process, the information can be stored. Later, it can be retrieved from storage.

Encoding

Communication involves exchanging information between a sender and a receiver. The sender must first decide what the receiver wants to know. See Figure 22-28. The message must then

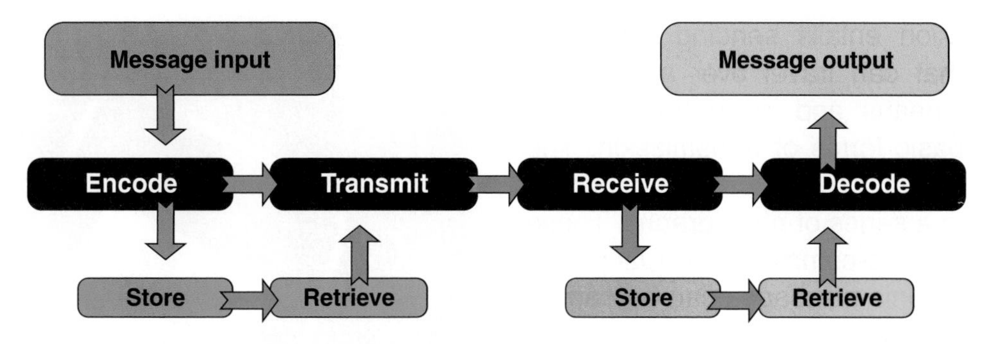

Figure 22-27. The communication process.

Figure 22-28. This designer is developing the message and layout for an advertisement.

be designed to attract the attention of the receiver. Finally, the message must be communicated in a form the receiver will understand. All these tasks are parts of encoding. To ***encode*** means "to change the form of a message."

Messages can be encoded in a number of ways. See Figure 22-29. Information can be encoded using symbols and graphics. Symbols can be used to convey the message. A red, octagonal street sign means "Stop!" Two people shaking hands means agreement. A skull and crossbones means poison. All these are symbols carrying meaning.

Other messages can be written. The information can be communicated using language. This book is communicating a message in this manner. Still other messages are encoded for electronic communication. Pictures and sound are changed into electrical impulses with cameras and microphones.

Figure 22-29. Highway signs use shapes, words, and symbols to communicate.

Transmitting

Once the message is encoded, it must be delivered to the receiver. Remember, we are talking about communication technology. Therefore, a technical means (machine or equipment) must be used. Switching circuits allow signals to be sent back and forth in the communication process. A network is a system connected by communication lines to transfer information from one device to another.

We have already presented the basic ways of transmitting the message.

Transmission entails sending signals in a form that can travel over a distance. Graphic means and wave transmission are the basic forms of transmission. The message can be printed on paper or carried on a series of photographs. These are the graphic-communication technologies. Electronic messages can be carried on sound, light, or radio waves. See Figure 22-30. We can use radio or television transmitters to send our information. The light a laser produces can also carry our telephone conversations. A public-address system can use sound waves to broadcast our ideas.

Figure 22-31. The computer is receiving messages from the operator at the keyboard.

Receiving

The transmitted message must be received. See Figure 22-31. The message must arrive at a desired location and be available to the receiver. This communication the transmitter sends is often in a special form. The message can be a series of electrical pulses on a wire. The communication can be radio waves varying in strength (AM radio) or frequency (FM radio). The message can be in digital code.

Often, receivers are electronic devices that change the transmitted code back into a form people can understand. They can change radio waves into electrical pulses. Receivers also can change digital code into printed words.

Figure 22-30. This television set is used in the first stage of transmitting a message using electronic media.

Decoding

The final act in the communication process is putting meaning to the message, or ***decoding***. Information must be decoded in order to be understood by the reader. Decoding is the opposite of encoding, with data being changed back to symbols and graphics. The people receiving the message must take action. They must read the printed word, listen to the broadcast, or watch the television program. This is not, however, enough. The people must place meaning on the message.

For example, you might hear someone shout, "Fore!" First, you must decide if the word is *fore* or *four.* You then must put it in context (compare it with the situation). If you are on a golf course, you should become alert. "Fore!" means someone is hitting a golf ball. If you are cooking at a fast-food outlet, you might take other action. The manager might have told you to make four hamburgers.

Career Highlight

Computer Systems Analysts

Information Technology

The Job: Computer systems analysts apply computer-technology knowledge and skills to solve computer problems for a company. They help the company use its computer equipment and personnel efficiently. Computer systems analysts can plan for and develop computer systems or develop new ways to use the present systems. They work with both computer hardware and software to develop effective systems, such as business, accounting, inventory, scientific, or engineering applications.

Working Conditions: Most computer systems analysts work a normal 40-hour week in offices or laboratories. Some analysts work at remote sites and telecommute (work at home using a computer connected to the employer's network).

Education and Training: A bachelor's degree in computer science, information science, or Management Information Systems (MIS) is required for many computer systems analyst jobs. Some employers want analysts with a master of business administration (MBA) degree, with a concentration in information systems.

Career Cluster: Information Technology

Career Pathway: Programming and Software Development

A message is effective only if it is received. Time can be wasted designing, producing, and sending messages that are never received. How many tornado, flood, and hurricane warnings are transmitted, but ignored? Highway speed limit signs are often viewed with indifference. Health warnings on cigarette packages are sometimes disregarded.

Storage

At any point in the communication process, the message can be stored. The message can be recorded on video or audiotape or stored on CDs, DVDs, CD-ROMs, or computer discs. The printed message in the form of pages, magazines, or books can be placed in a warehouse.

Retrieval

Later, the message is taken out of storage. The message reenters the communication process. Most television programs are taped (storage) for later broadcast (retrieval). Books are stored in libraries. The reader must retrieve and decode the messages.

Interference

Throughout each step of the communication process, interference can develop. ***Interference*** is anything that makes the message less clear. For the end user, interference includes things such as static on a radio broadcast, smudges on printed pages, missing elements of a Web page, and ghosts on television programs. People who operate communication systems work hard to reduce interference to a minimum.

Communication Messages

Formal communication messages are carefully planned and produced. See Figure 22-32. The production of these messages involves four major steps:

1. Designing the message.
2. Preparing to produce the message.
3. Producing the message.
4. Delivering the message.

These steps apply to all specific activities: publishing, filmmaking, and broadcasting. You will see there are some activities each group does alike. Other activities belong only to a specific industry.

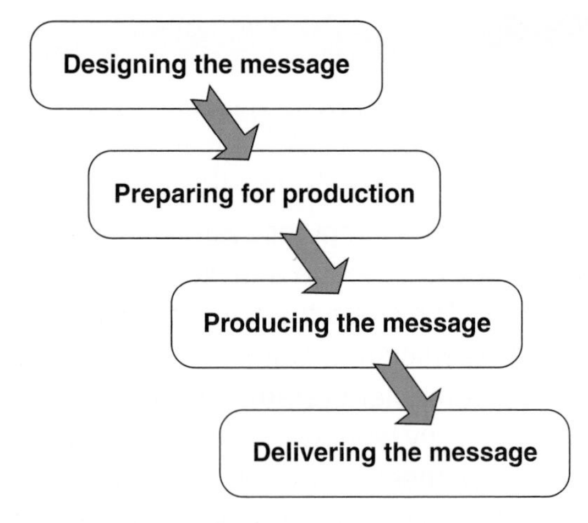

Figure 22-32. The steps in designing a communication message.

Published Messages

Publishing includes all activities producing a ***published message***. This includes newspapers, magazines, books, greeting cards, and flyers. Publishing is part of a larger activity called *printing*. In addition to publishing, printing includes the production of forms, stationery, and other nonmessage materials.

Designing Published Messages

Mass-communication messages are designed to reach large audiences. They are planned to inform people, entertain people, or cause people to take action. There are, however, many audiences for published materials. Therefore, the designers must gather information about a specific audience they want to reach. In particular, the following questions must be answered:

- Who is the audience? (For example, is it young people, sports fans, business leaders, or senior citizens?)
- What gets the audience's attention? (For example, is it words, pictures, or comic presentations?)

- What does the audience value? (For example, is it status, financial security, fun, freedom, or power?)

Factors such as the intended purpose and nature of the message also influence the message's design. All these factors should be taken into account when the message is created and transmitted to a specific audience. Gathering this information is called ***audience assessment***. The information provides the base for designing the message. These factors give the designer guidance and direction to complete several specific tasks.

The first design task involves selecting a ***format***. This is planning for the physical size and shape of the message carrier. The carrier can be an 8 1/2″ × 11″ flyer. The message carrier can also be a large billboard on a busy highway. A newspaper can be a tabloid (smaller size) or regular format. A magazine can be any size. The design can cover an entire page or part of a page.

Next, the designer gathers information. The message might be designed to sell products (an advertisement). See Figure 22-33. Information about the product's features and operation are then needed. This can be matched with audience information to determine what to say. The designer then writes copy for the message. This copy is the message the media is to carry. The message might say, "You need this product," or "Your life will change if you use this product."

Newspaper reporters also gather information. They interview people associated with a story. These reporters then select and combine information to produce interesting copy. The publication might be about a movie star or sports figure. Biographical and current information about the person must then be gathered.

Finally, the designer selects illustrations to support the copy. These are drawings and photographs attracting attention and adding interest to the message. See

Figure 22-33. This designer is developing a layout for an advertisement. (©iStockphoto.com/nicholas_)

Figure 22-34. This photograph of an archeological dig can be used in a story about discovering our past.

Figure 22-34. These illustrations might help the news story convey information. They might show a famous figure at work and play.

Preparing to Produce Published Messages

The format, copy, and illustrations must be brought together. They must be placed in a pleasing arrangement. In designing advertisements, designers first sketch out their ideas. ***Roughs*** (initial sketches) are prepared to show some basic ideas. See Figure 22-35.

The better ideas are converted into refined sketches. These sketches show more detail and thought. Finally, ***comprehensives*** are prepared. These are more complete sketches for the message. They describe the arrangement of the elements, including type, illustration, and white space. Type size and style are called out (written on the sketch).

Likewise, newspapers have layouts designed for their audiences. *USA Today* and other newspapers use colorful formats with many photographs, charts, and graphs. The *Wall Street Journal* caters to its audience with a large amount of text.

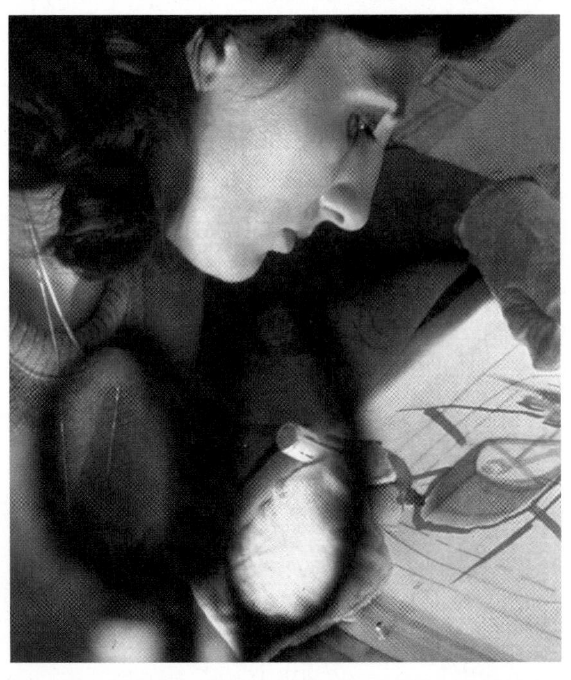

Figure 22-35. This artist is preparing roughs for an advertisement. (American Petroleum Institute)

Producing Published Messages

The message is now ready to be produced. The advertisement, magazine, or newspaper is ready to be printed. The first major task is to schedule the production through the plant. Resources must be allocated to each production task.

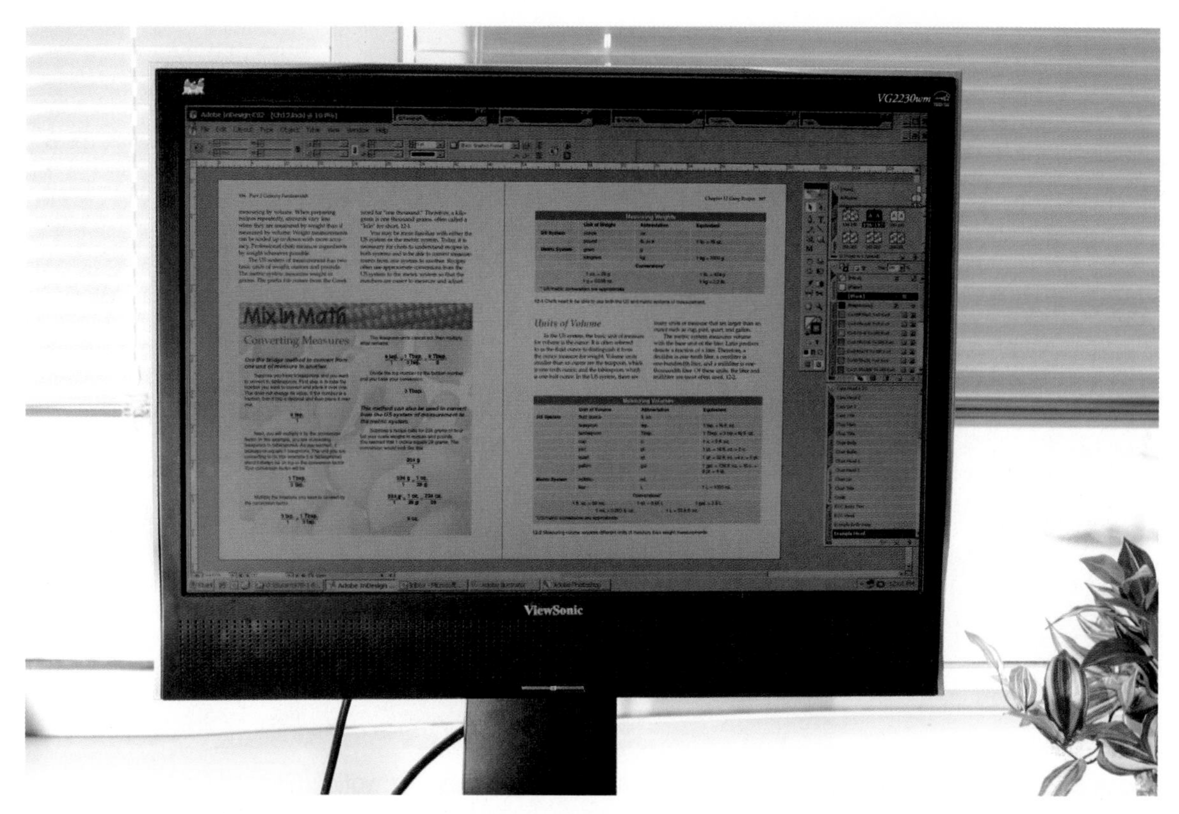

Figure 22-36. Components of a published message are inserted into a layout for a printed product.

The comprehensive is used to set type, size illustrations, and prepare photographs. These components are typically inserted into a ***layout*** in a software program. See Figure 22-36.

The layout is used to make printing plates or silk screens for the production process. These items are called *image carriers*. They carry the image from the press to the paper.

Once the large carriers are produced, the communication product must be printed. See Figure 22-37. Ink must be applied to the substrate (such as paper, plastic, or foil). As described earlier, this can be done by one of five basic processes:

- **Letterpress (or relief) printing.** This is printing from a raised surface.
- **Offset lithography.** This is printing from a flat surface.
- **Intaglio (or gravure) printing.** This is printing from a recessed surface.
- **Screen printing.** This is printing by forcing ink through openings in a screen.
- **Xerography.** This is printing using electrostatic means.

The printed materials are then finished. This step involves one or more processes such as folding the sheets, collating (gathering) the sheets in their proper order, and binding the sheets into books or magazines. Other finishing processes include drilling holes for notebook paper; perforating sheets so they tear out easily; applying adhesive on labels; and laminating sheets with a protective, plastic coating.

Delivering Published Messages

Delivering a printed message uses some standard distribution methods. Subscribers might receive the message in newspapers and magazines. Books are available in stores and libraries. Billboards, posters in store windows, and mailed flyers are still other distribution techniques.

The task is to select a distribution method that reaches the identified audience. Using *Sports Illustrated* magazine to reach large numbers of older women is unwise. A

Figure 22-37. The printed word brings us information from around the world. (Gannett Co.)

message in *Seventeen* magazine, however, reaches many young women.

Film Messages

Film messages are photographs and transparencies. Transparencies include movies, slides, and filmstrips. These messages are designed to present information, change attitudes, or entertain.

Designing Film Messages

Similar to published works, film messages are based on the results of audience assessments. Moviemakers have a good idea about what teenagers go to see. These movies are greatly different from what grandparents feel are good movies.

The first design decision concerns format. Will a motion picture or a series of photographs present the message? How long should the presentation be? Should it be color or black-and-white? Answering these questions establishes the format.

Next, research must be done. Information about the subject must be developed. Educational films require much research. A good example of this type of film is the *National Geographic* video series of programs.

Entertainment films often take less research. Sometimes a film is based on a novel. By writing the novel, the author will have already done most of the research.

The third step is script writing. Most films use one or more actors to portray a story or present information. The actions and words of the actors must be described in writing. This writing is called the *script*. See Figure 22-38. The script carefully describes all events in the film.

Stage sets are designed and built to support these actions. See Figure 22-39. Also, many times, on-site filming locations are selected and used. The sets and locations provide the realism for the message.

Preparing to Produce Film Messages

The production of film messages must be scheduled. The efficient use of resources must be planned. The cast and production crew must then be hired.

Various Voices off: *H'ya Buck!...Howdy, Buck!...How's things in Bisbee, Buck? Have a good trip?*

Meanwhile the SHOTGUN GUARD, who has guarded the treasure box from Bisbee, jumps down to the sidewalk.

SHOTGUN GUARD: *So long, Buck!*

Men begin unhitching the horses. BUCK acknowledges the cheery greetings as the WELLS FARGO AGENT in Tonto pushes his way through the crowd.

WELLS FARGO AGENT: *Howdy, Buck. Got that payroll for the mining company?*

Buck kicks the box which is under his seat.

BUCK: *She's right here in this box.*

The WELLS FARGO AGENT climbs up to the top of the coach, calling to a colleague as he does so.

WELLS FARGO AGENT: *Give us a hand with this box, Jim.*

BUCK: *Jim, I'll pay you that $2.50 when I get through.*

JIM: *Okay.*

The two agents get the box down and carry it off between them—BUCK looks over his shoulder to the other side of the coach.

BUCK: *Now you kids, get away from them wheels!*

He starts to get down and calls out to the men who are leading the horses away.

BUCK: *Well...sir, we ran into a little snow up there, quite bad, so you fellers better prepare for a good frost.* He jumps down and disappears round the side of the coach. The Tonto Hotel is seen on the other side of the road.

Medium shot of the stagecoach as BUCK comes round to open the coach door.

Figure 22-38. A page from the script for *Stagecoach*, a John Ford and Dudley Nichols film. Lorimer Publishing, in London, printed the script.

Directors must stage the various scenes. They decide how each scene is shot. Placements of cameras and lights are considered. Also, directors determine how to use extras (people who add interest to the scene). The camera, lighting, wardrobe, and set crews work closely with directors.

Figure 22-39. This outdoor set can be used in making movies and television shows. Notice the walkways at the top of the "buildings."

The cast learns lines and movements from the script. They then come together to rehearse. Changes are made in the script and staging, until the director, cast, and crew are happy with the production.

Producing Film Messages

After the final rehearsal, filming can start. The actors complete each scene they rehearsed. (The scenes are seldom shot in the order outlined in the script.) Wise use of resources (such as stage time, natural light, and location availability) dictates the proper order in which to shoot the scenes.

The crew and cast might shoot each scene a number of times to develop several different effects. Later, the film is cut and spliced to combine the various scenes. See **Figure 22-40.** These scenes are edited into a final product.

Delivering Film Messages

Most movies are first distributed to theaters. Special companies schedule the films and collect royalties for the film producers. Other films are made for television. They are distributed to the television stations or cable networks. After they

Figure 22-40. This underwater scene will be spliced to become part of an adventure movie. (©iStockphoto.com/dsabo)

are shown on these outlets, many films are converted into DVDs. These discs are distributed through rental businesses or for purchase in stores.

Broadcast Messages

Broadcast messages are communications that radio stations, television stations, and cable television systems carry. These include entertainment, information, news, sports events, and advertisements. Regular programming follows the basic steps outlined for films. Scripts are prepared. Crews and actors are hired. The production is rehearsed and filmed. The product is edited into its final form.

The major difference between films and broadcast messages is in the way these messages reach viewers. Film producers expect people to pay to see their programs in a theater. Broadcast programming sells advertising time to enable "free" delivery of the product. The word *free* is in quotes because the delivery is not free. We pay for radio and television programming every time we purchase an advertised product. Likewise, cable and satellite television channels exist through advertising and payments the cable and satellite companies make. When we sign up for cable or satellite services, we subscribe to certain channels. These channels are sold in packages of several channels per subscription tier. In a similar manner, we pay for part of the production cost of magazines and newspapers with our subscriptions. We pay the rest of the costs, however, through the advertisers.

Broadcast advertisements follow the design steps used for print advertising described earlier. The only difference is in the layout. Radio advertisement uses a script to lay out the ad. Television uses a ***storyboard***. See Figure 22-41. The storyboard shows each shot for the advertisement. Also, the script is included.

Producing broadcast advertising follows the film model. The director and actors rehearse the advertisement. Filming and editing then take place. Finally, time is purchased to present the advertisement on the broadcast station.

Most 60-second advertisements are shot a number of times (sometimes into the hundreds) to get just the right effect. The ads are sponsors' ways to convince people to buy their products. A great deal of money is spent to produce and air the short messages. Therefore, sponsors want their messages to be nearly perfect.

Figure 22-41. An example of a storyboard for a television commercial. (Clorox Co.)

Information Technologies

Many people call the times we live in the *Information Age*. This is because of the vast amount of information available. Today, we have access to more information than at any other time in history. The challenge is to access and use this information.

The primary technology allowing us to deal with large amounts of information is computer technology. The computer is a processing machine. See **Figure 22-42.** This machine uses electronic parts and circuits to process information.

Computer systems have two basic elements: hardware and software. The ***hardware*** is the equipment used. This

Figure 22-42. A computer receives, processes, and stores information.

equipment includes the computer, printers, scanners, and data-storage devices. Software includes the instructions causing the computer to do specific tasks.

Computer Hardware

These systems include a number of different types of hardware. See Figure 22-43. There is the computer itself. Many computers are general-purpose information-processing devices. We often call them *personal computers (PCs)*. These computers take information from a person, another device, or a network and process the information. The processed information is shown on a monitor, stored on a device, or sent to another location on the network.

In addition to PCs, there are mainframe computers that do the same tasks on networks. They process vast quantities of information for businesses, banks, and industrial companies. Also, there are specialized computers that do only one task. For example, a global positioning system (GPS) unit can handle signals only from GPS satellites. A computer in a washing machine can control only that machine. Video game systems, such as the Xbox® video game system, the Wii® video game system, and the PlayStation® system, are specialized computers for playing games. See Figure 22-44. A computer has four basic parts:

- **An *input device*.** These devices allow for inputting information into the computer. They include keyboards, scanners, and network connectors. Input devices take the information and change it into electrical signals the computer can understand.
- **A *central processing unit (cpu)*.** This is the brain processing the information. The cpu follows the program (set of instructions) fed into it. This unit oversees everything a computer does.

Figure 22-43. Today, it is important to be able to use computers as tools. Computers are used to communicate, process data, and control machines.

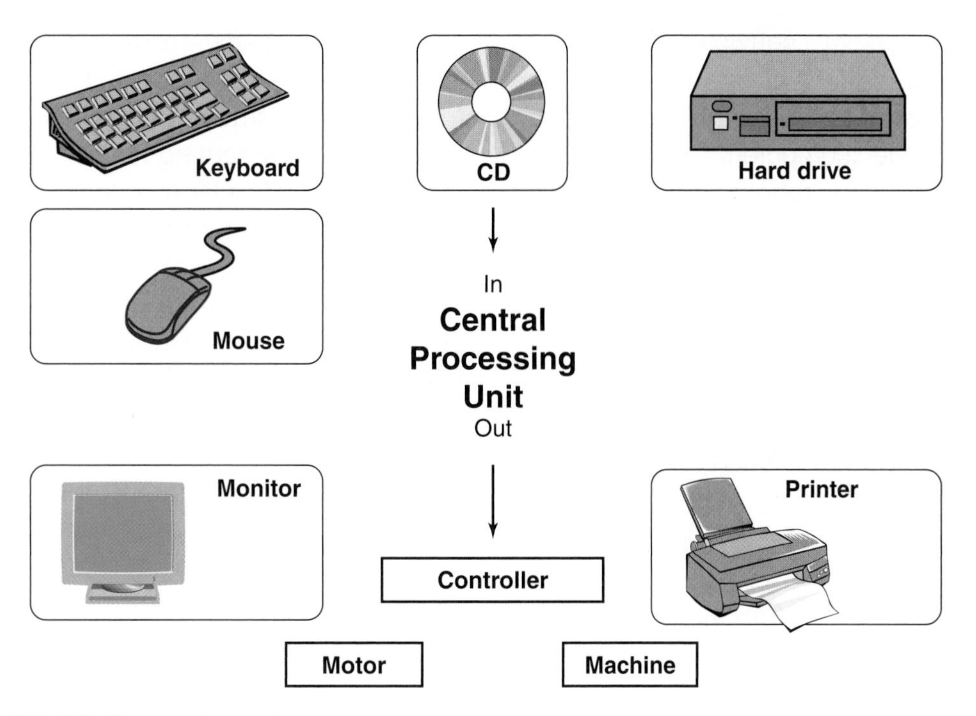

Figure 22-44. A computer system.

- A ***memory.*** This is the place where the information is stored. A memory also stores the program of instructions. There are several specific types of memory in a computer. Two types are more important than the others:
  - **Random-access memory (RAM).** This type of memory temporarily stores information with which the computer is currently working.
  - **Read-only memory (ROM).** This is a permanent type of memory storage. ROM stores important data that does not change.
- ***Output devices.*** They receive and act on information the computer provides. These devices include printers, monitors, and modems. A ***monitor*** is the primary device for displaying information from the computer. A modem is the standard method of connecting to the Internet. Other output devices are removable storage devices. They are used to store data. These storage devices include disc, USB flash, CD-ROM, and DVD-ROM drives. Still other output devices include speakers, machines connected to the computer, and motors.

Computer Software

Computers need instructions to operate. For some computers, the instructions are simple and built into the system. For example, the computer controlling a microwave has one set of tasks to perform. This computer uses simple input and output methods. A keypad is used to input instructions. A liquid crystal display (LCD) shows the output.

Operating Systems

Many computers, however, perform a variety of tasks. They need varying instructions and control. Programs, or software, provide these instructions. All desktop computers have a basic ***operating system*** overseeing all operations. Common operating systems are the Windows® operating system, Linux® software, and Mac OS®

operating system software. There are hundreds of other operating systems available for special-purpose applications, such as mainframe computers, manufacturing applications, and robotics. An operating system does two things in a computer system:

- The operating system manages the computer system. This system controls the hardware and software of the computer system. The operating system manages such things as the processor, memory, and disc space.
- The operating system provides a consistent way for application software to deal with the hardware.

Application Software

The instructions for the computer are called *programs* and *subroutines*. Another name for all the programs and subroutines is *software*. The software tells the computer what to do and how to do it. The computer follows your instructions exactly. In doing so, it does something useful. This can include drawing a line, manipulating text in a document, balancing the colors in a photograph, locating information on the Internet, or balancing a checkbook. Many people are familiar with software such as the Word® program, Excel® spreadsheet software, Photoshop® software, Illustrator® software, the PowerPoint® presentation graphics program, TurboTax® software, Netscape Navigator® software, and AutoCAD® software. These are but a few of the thousands of general and specific applications available to individuals and businesses. Without this software, a computer cannot do complex tasks or solve problems. See **Figure 22-45.**

Figure 22-45. Computer software allows this person to prepare, edit, and print a document. (Courtesy of Dell Inc.)

Quality Control in Communication Systems

Quality control in communication systems ensures the intended message reaches the receiver. In graphic communication, the production of the message follows acceptable standards for text and image preparation. Messages are often proofed prior to being sent to the receiver. Printing systems are monitored to ensure that the duplicates match the original message. Electronic communication uses industry-accepted standards and protocols to ensure that appropriate devices can transmit and receive messages.

Summary

Communication and information technologies are part of every person's life. We use graphic and electronic communication media daily. This media allows us to send or receive information. Information technology allows us to access and use vast quantities of data. Communication and information technologies inform us or cause us to take action. Without communication and information technologies, we would know little about the world around us.

STEM Connections

Integrated STEM Curriculum

Science

Research a device that has helped change the way we communicate or process information. Describe the scientific principles it uses.

Mathematics

Research and describe the measurement system printers use. Compare this system to standard and metric measurement systems. Prepare a display explaining it and its origins.

Curricular Connections

All Subjects

Read a magazine or newspaper article about communication, information, or computer technology. Highlight examples of the uses of communication and information technologies. List parallels in the ways you use these technologies.

Social Studies

Research the development of one communication device. Prepare a report or display on the device and its history.

Activities

1. Interview someone in the communication or information-processing field. Find out what this person does on the job. Prepare a report (with photographs, if you have access to a camera). As part of the report, indicate what you would like about such a job and what you think you would not like.
2. As a group activity, plan and produce a school newspaper or broadcast of the day's activities at the school. Assign teams to the following steps:
   A. Determine what media (print, radio, or telecast) will be used.
   B. Report the news (collect news items by interview).
   C. Edit the news items.
   D. Perform other tasks, such as typing, printing, and broadcasting.
3. Computers can be used to manage production systems, as well as to explore solutions to problems. Use the Internet to search for information on how computers work.

Test Your Knowledge

Do not write in this book. Place your answers to this test on a separate sheet of paper.

1. What is information?
2. What is communication?
3. Give five examples of ways to communicate.
4. What makes communication a technology?
5. Indicate which of the following are examples of communication technology. There can be more than one answer.
   A. Thinking about what you will do today.
   B. Talking to a friend.
   C. Listening to music on a CD player.
   D. Tapping a friend on the shoulder.
   E. Tasting your dessert.
   F. Videotaping a school play.
   G. Drawing a design for a poster.
6. List the six types of people-to-people communication.
7. Indicate the type of communication system used in each situation described:
   _____ Listening to a newscast on the radio.
   _____ Clock chimes sounding the hour.
   _____ Setting the alarm on a clock radio.
   _____ A thermostat controlling the operation of a furnace.
   A. People-to-people communication.
   B. People-to-machine communication.
   C. Machine-to-people communication.
   D. Machine-to-machine communication.
8. List the four major steps in the communication process and summarize each of them.

9. What are the four steps in producing communication messages?
10. The two basic elements of computer systems are _____ and _____.

Reading Organizer

On a separate sheet of paper, create a detailed outline based on what you've read about communication technology.

Example:

I. Communication technology
   A. Graphic communication
      1. Printed graphic and photographic media
      2. Photographic prints and transparencies

CHAPTER 22 ACTIVITY A

INFORMATION AND COMMUNICATION TECHNOLOGY

INTRODUCTION

We communicate daily. Some of our communication is verbal (speech). People talk to other people. Some communication is visual. We write messages for others to read.

Often, we use a technological device to help us deliver our message. We might use printing presses or cameras to produce two-dimensional media, printed pages, or photographs. Other times, we use electrical signals or electromagnetic waves. We use the radio, television, telephone, teletypewriter, or telegraph. These devices move a message from the source to the receiver. This activity lets you build a telegraph system. You will then use this technological system to convey a message.

EQUIPMENT AND SUPPLIES

- Two 3/4″ × 3 1/2″ × 5″ pine boards.
- Two 5/8″ × 5 1/2″ × 28-gauge sheet steel.
- Six 1/2″ brads.
- A 6d box nail.
- Twelve 3/8″ × No. 6 pan-head sheet metal screws.
- Two mini buzzers—Radio Shack No. 273-055 or equivalent.
- Four quick wire disconnects—Radio Shack No. 274-0315 or equivalent.
- A 9-volt battery.
- A holder.
- 25′ of 3-conductor wire or 75′ of single-conductor wire.

PROCEDURE

Your teacher will divide you into teams of four students. Each team will be split up into two groups:

- Group A will build telegraph station #1.
- Group B will be responsible for building telegraph station #2.

How to Build

Each member of every group should complete the following steps:

1. Carefully study the plans for the telegraph set. (Group A—See Figure 22A-1. Group B—See Figure 22A-2.)
2. Select a piece of lumber for the base of the set.

3. Lay out the locations of the key, buzzer, and wire disconnects.
4. Select a piece of 28-gauge sheet steel to make the telegraph key.
5. Lay out the pattern on the metal. See Figure 22A-3.
6. Cut the metal to size.
7. Punch the two screw holes.
8. Bend the metal to form the key. See Figure 22A-4.
9. Drive the 6d box nail into the correct position. Allow 1″ of it to remain above the surface of the wood.
10. Attach the telegraph key, buzzer, and wire disconnects.
11. Wire the telegraph station according to the original drawing.

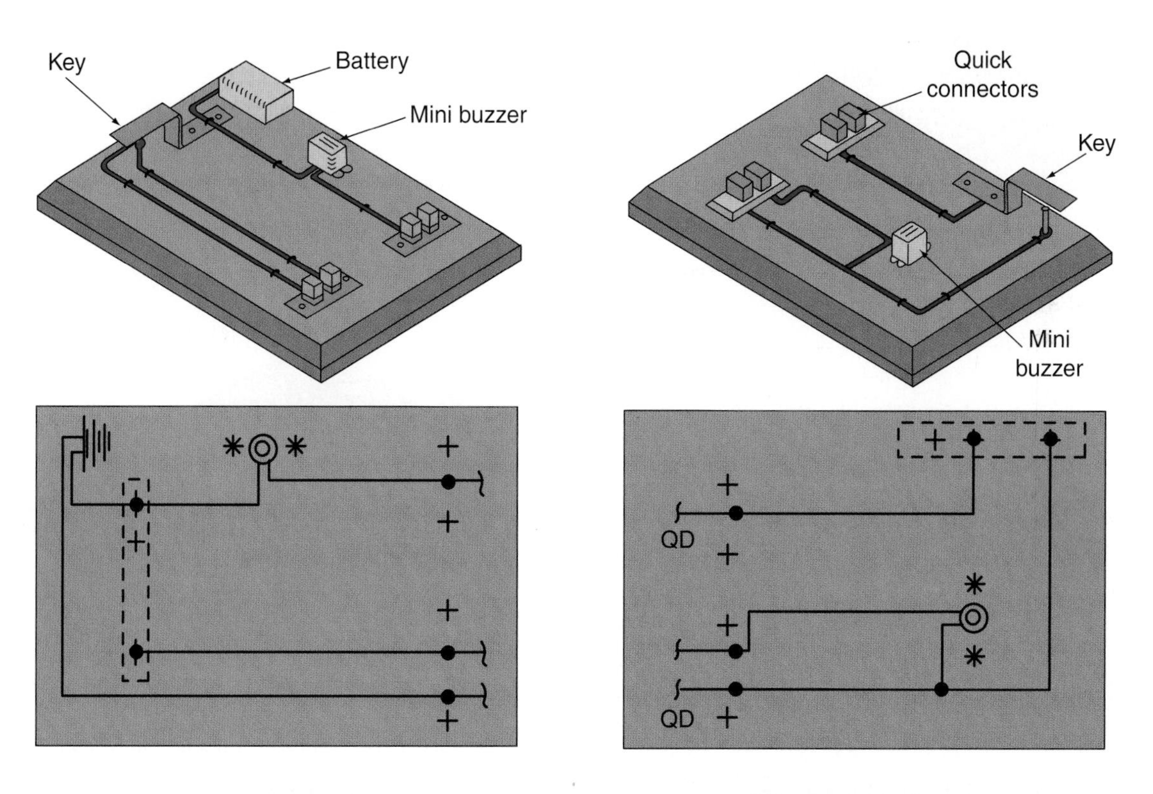

Figure 22A-1. Group A will construct this station. This station is different from the one Group B is building.

Figure 22A-2. Group B will construct this station. This station is different from the one Group A is building.

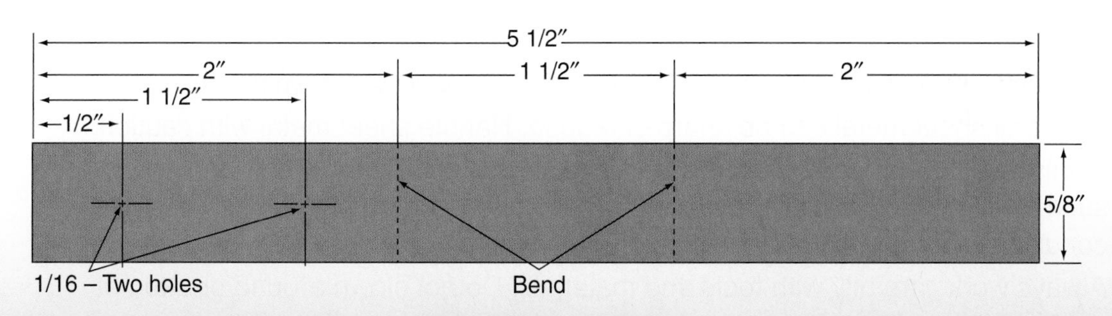

Figure 22A-3. The pattern for the telegraph key.

Sending the Message

1. Wire telegraph station #1 to telegraph station #2. See Figure 22A-5.
2. Complete an assignment using the telegraph system to communicate messages between Group A and Group B.

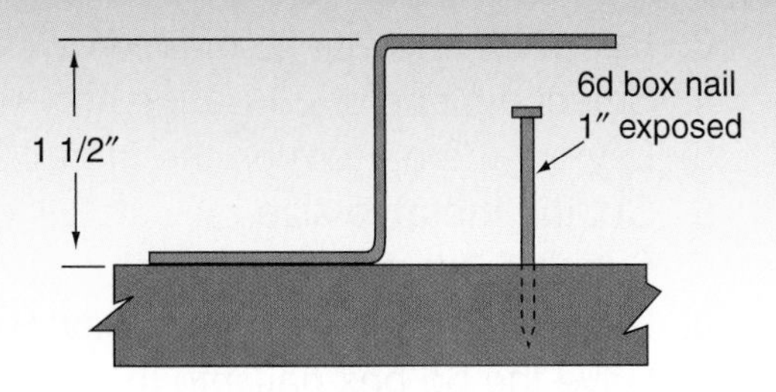

Figure 22A-4. The telegraph key should be bent to this shape.

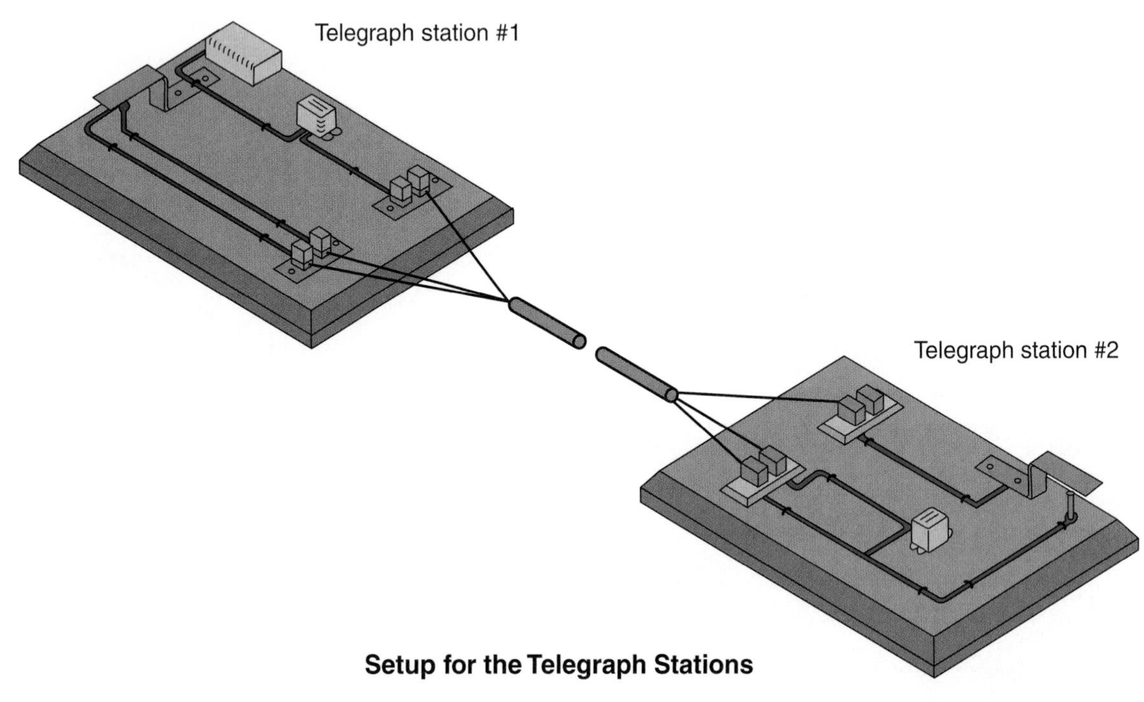

Figure 22A-5. Wire the two telegraph stations together as shown.

Challenging Your Learning

Name two devices that have replaced the telegraph. Why are these better ways to communicate?

Safety Rules

- Be careful when using tin snips. Keep your free hand well away from the cutting edges.
- Edges of sheet metal can be sharp or jagged. Handle sheet metal with caution.
- In this activity, you will use only a 9-volt power supply. There is no danger from such low voltage. Still, it is possible to get an uncomfortable shock under some conditions. Do not touch bare wires or terminals.
- Always work carefully with tools and materials. Do not clown around or attempt practical jokes. Clowning around can cause injury to yourself or others.
- If you are uncertain about what to do, ask your instructor. Don't guess!
- Do not attempt to connect the telegraph stations to any power source other than the 9-volt battery.

TSA Modular Activity

This activity develops the skills used in TSA's Graphic Design event.

Activity Overview

In this activity, you will create a one-color graphic design appropriate for the cover of this *Technology: Design and Applications* textbook.

Materials

- Paper.
- A pencil.
- A computer with graphic design software and clip art.
- A printer.
- 8 1/2″ × 11″ × 1/4″ white foam-core board.
- Vellum overlay.

Background Information

- **General.** Consider the principles of design as you develop your project. Locate and size elements to achieve balance and proportion. Use contrast to emphasize key elements. Select elements to provide rhythm.
- **Elements.** The cover can include graphic elements, such as photographic images, clip art, or original design elements. When selecting illustrations, consider the content of the illustration (is it appropriate for the cover?) and its final printed size (will it be printed so small that details are lost?).
- **Design issues.** Consider the follow items as you develop your design:
  - The title should be a prominent feature and should be the most readable text.
  - The subtitle should also be easy to read.
  - The cover should be attractive.
  - The cover should have a theme related to the theme of the textbook.

GUIDELINES

- Identify good design examples from a selection of other textbook covers.
- Your textbook cover must include the title, subtitle, and authors' names.
- Develop sketches and a rough layout using paper and pencil.
- After you have developed the rough layout, create the design using the graphic design software.
- The project is produced as a black-and-white, camera-ready image for one-color duplication. Submit the final design on white foam-core board with a vellum overlay for protection.

EVALUATION CRITERIA

Your project will be evaluated using the following criteria:

- Design elements.
- Attractiveness.
- Relation of the cover theme to the textbook content.
- Camera readiness and the appropriateness of the design for duplication.

The way we communicate is constantly changing because of newer technologies. Thanks to wireless connections and portable hardware, we can now communicate in several ways no matter where we are.

CHAPTER 23
Manufacturing Technology

Did You Know?

- Manufacturing had its start in the New Stone Age, with grinding corn, baking clay, spinning yarn, and weaving textiles.
- Manufacturing in early civilizations concentrated on commonly used products, such as pottery, oils, cosmetics, and wine.
- The first mass-produced pencils were made in Nuremberg, Germany in 1662.
- Eli Whitney proposed the idea of interchangeable parts in 1798. This idea made it possible to produce goods quickly because standard parts were used.
- Henry Ford and his colleagues first introduced a conveyor belt to an assembly line to make magnetos (a type of generator) in 1913. He then introduced the idea for the manufacturing of automobile bodies and engines.

Objectives

The information given in this chapter will help you do the following:

- Explain the function of manufacturing.
- Summarize primary and secondary processing.
- Summarize the six types of secondary processes.
- Explain the four major types of manufacturing systems.
- Summarize the seven steps in developing a manufacturing system.
- Recall the four basic managerial functions.
- Identify the three major resources needed to produce products.
- Recall the purpose of marketing.

Key Words

These words are used in this chapter. Do you know what they mean?

advertising
assembling
casting
conditioning
continuous manufacturing
custom manufacturing
die
finishing
firing
flame cutting
flexible manufacturing
forging
forming
inspection
intermittent manufacturing
machining
manufacturing
molding
pilot run
primary processing
quality
secondary processing
separation
shearing
thermoforming
tooling

Preparing to Read

Look carefully for the main ideas as you read this chapter. Look for the details that support each of the main ideas. Use the Reading Organizer at the end of the chapter to organize the main and supporting points.

A factor that makes people different from other beings is that people design and make tools. This ability has led to objects that help us do things. At first, people made only things they could use themselves. Some people have called this act *useu-facturing* (making things for personal use). Later, we set up systems to make products for other people to use. This is called ***manufacturing***. Manufacturing changes raw materials into useful products. See Figure 23-1. Everyone uses manufactured products daily. We ride in manufactured vehicles and buy CDs. These CDs have been manufactured. Everyone puts on manufactured clothes and enters buildings through manufactured doors. We read newspapers produced on manufactured printing presses and write on manufactured paper with manufactured pencils. Manufactured games, sporting goods, and toys entertain us. This world would be very different without manufactured products. Each of us needs the output of manufacturing systems.

Figure 23-1. Manufacturing makes useful products from various materials. (©iStockphoto.com/Fertnig)

Manufacturing Processes

Manufacturing produces goods inside a factory. See Figure 23-2. Manufactured activities produce goods that can be classified as durable or nondurable. These two classifications are based on the life expectancy of a product or system. Durable goods include cars, kitchen appliances, and power tools. Nondurable goods include toothbrushes, disposable diapers, and automobile tires.

Manufactured goods have life cycles that start with preliminary planning and design and continue to the final disposal of the goods. Issues in this life cycle to think about include how well the products meet a need, what by-products are produced, and how the products will be disposed of at the end of their life cycle. After manufacturing is done, the goods are shipped to stores.

Figure 23-2. Manufacturing produces products, such as this airplane, inside a factory. (©iStockphoto.com/Bim)

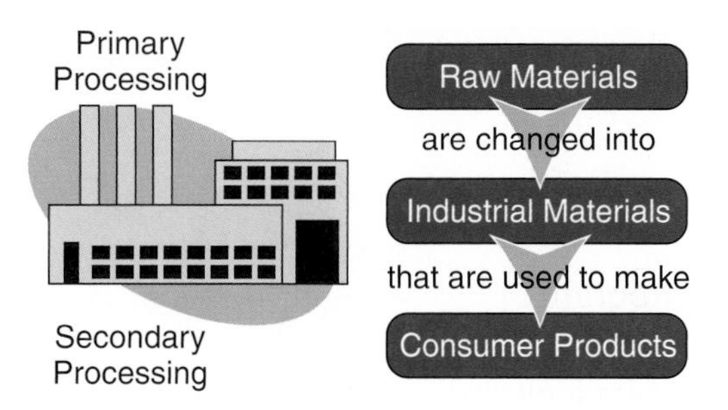

Figure 23-3. Manufacturing includes primary and secondary processes.

There, customers select and buy goods that best meet their needs and wants.

Manufacturing generally includes two major steps. See Figure 23-3. In the first step, raw materials are converted into industrial materials. Trees are made into lumber, plywood, and paper. Ores are converted into sheets of metal. Natural gas is converted into plastics. Wheat is converted into flour. Glass is made from silica sand. These processes are called *primary processes*. See Figure 23-4.

The second step of manufacturing is called ***secondary processing***. The processes involved in this type of manufacturing change the forms of industrial materials into usable products by separating, forming, combining, and conditioning them. Through secondary processing, plywood becomes furniture. Flour becomes bread. Sheets of metal become household appliances. Glass becomes bottles and jars. Plastics become dishes.

Primary Processes

Primary processing includes three major groups of processes. These were discussed in Chapter 5. The following briefly reviews what you learned there:

- **Mechanical processing.** This includes cutting, grinding, or crushing the material to produce a new form. These processes include cutting lumber and veneer, making cement, crushing rock into gravel, and grinding wheat to make flour.
- **Thermal processing.** This is using heat to change the form or composition of materials. These processes include smelting metallic ores (for example, making copper or steel) and fusing silica sand into glass.

Felling a Tree

Moving Logs

Cutting Logs

Removing Bark

Cutting Slabs

Figure 23-4. Converting trees into lumber involves a number of steps. Mature trees are felled. Logs are moved to the mill. The bark is removed from the logs with high-pressure water. The logs are cut into slabs. The slabs are cut into boards of standard widths and lengths. (Weyerhaeuser Co.)

Career Highlight

Computer-Controlled-Machine Operators

Manufacturing

The Job: Computer-controlled-machine operators program and operate computer numerical control (CNC) machines to cut precision products. The machines these operators work with include CNC lathes, milling machines, and electrical discharge machines (EDMs). The operators use knowledge of machining metals and CNC-programming skills to design and make machined parts. They review drawings for the part, determine the machining operations needed, develop a set of instructions for these operations, convert these instructions into computer codes, test the program, and perform the machining operations.

Working Conditions: Most shops that have CNC machines are clean and ventilated. Often, the machines are partially or totally enclosed. This reduces the worker's exposure to noise, chips, and lubricants. The operator often stands for most of the workday. Most CNC operators work 40-hour weeks. They might, however, have to work evening and weekend shifts.

Education and Training: Computer-controlled-machine operators might receive their training on the job, through apprenticeship programs, in vocational programs, or in postsecondary schools. These operators should have a basic understanding of metalworking machines, be able to read blueprints, and have knowledge of computers and electronics. Also, previous experience with machine tools is important.

Career Cluster: Manufacturing

Career Pathway: Production

- **Chemical (and electrochemical) processing.** This is using chemical actions to change resources into new materials. These processes include refining aluminum from bauxite and producing most plastics from fossil fuels.

The output of primary processing is called *standard stock*. The materials are available in standard sizes. Most plywood is produced in 4′ × 8′ sheets. Sheet metal is often sold in 24″ × 96″ sheets. Plastics are produced in standard pellets. Sugar is produced in various granular sizes (such as standard, extra fine, and confectionery). Lumber is sold in a number of standard sizes. Many of us have heard people talk about 2 × 4s. This size is a standard lumber size for the construction industry.

Standard stock must be further processed before it is useful. A sheet of plywood is of little value to a person. This sheet becomes useful when it is made into a desk or doghouse. See **Figure 23-5.**

Figure 23-5. Lumber takes on additional value when it is converted into furniture. (©iStockphoto.com/epixx)

Likewise, few of us have use for plastic pellets. These pellets can be converted, however, into automobile trim, bowls, and fabric. After this happens, most people place a higher value on the plastic.

Secondary Processes

Secondary-processing activities can be grouped under six headings. These are casting and molding, forming, separating, conditioning, assembling, and finishing. See Figure 23-6.

Casting and Molding

The first three secondary-processing activities (casting and molding, forming, and separating) give materials specific sizes and shapes. In one type of process, the material is first made a liquid. The liquid is poured or forced into a mold. The mold has a cavity in the shape of the part or product. Inside the mold, the material hardens before it is removed from the mold. This process is called ***casting***, or ***molding***. See Figure 23-7.

All casting and molding activities follow some common steps:

1. A mold is produced.
2. The material is made liquid.
3. The material is put into the mold.
4. The material hardens.
5. The finished item is removed from the mold.

Figure 23-6. Secondary manufacturing processes.

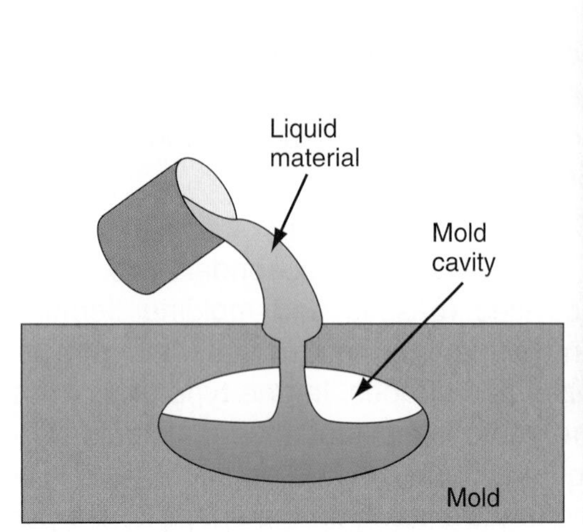

Figure 23-7. Casting and molding processes. Liquid material is poured into a mold, where it hardens. (©iStockphoto.com/wsfurlan)

If you have ever made an ice cube, you have produced a casting. The first step is to make or buy a mold. This mold has a cavity in it. The shape of the cavity gives the finished product its shape.

Some casting processes use a new mold for each product. These molds are usually made of sand or plaster. Such processes are called *one-shot mold casting*, or *expendable mold casting*.

Other casting processes use metal or plaster of paris (ceramic cement made from gypsum) molds. These molds can be used a number of times. This type of casting is called *permanent mold casting*.

Most materials cast are solids. They must be made liquid before they can be cast. Often, they are melted. Other materials, such as clays, are suspended in (mixed with) water.

The liquid material is then put into the mold. In some cases, the material is poured. Gravity draws the material into the mold cavity. In other cases, machines force the material into the cavity.

Once in the cavity, the material must solidify. Hot materials must be cooled to become hard. With other materials, the mold absorbs water to create a solid part. Some materials harden with a chemical action.

Finally, the hardened material must be removed from the mold. One-shot molds are broken away from the cast part. Permanent molds are opened. The finished casting is ejected from the mold.

Casting processes can form many metallic, plastic, and ceramic materials. Typical examples are automobile engine blocks, parts for plastic models, ceramic bathtubs and lavatories, and plaster wall decorations. Candy and other food products can also be cast into shapes.

Forming

Sometimes industrial materials are squeezed or stretched into the desired shape. Processes that use this approach are called ***forming***. See Figure 23-8. Forming includes bending, shaping, stamping, and crushing. All forming processes require two things. They must have a shaping device and an applied force.

One forming process uses a shaping device called a ***die***. This die is usually a set of metal blocks that have cavities

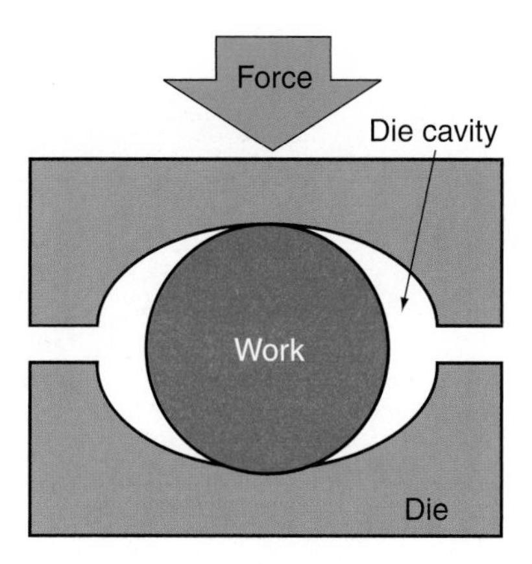

Figure 23-8. A forming process. Here is a simplified drawing of a die. Blacksmithing is a forming process. (©iStockphoto.com/clu)

machined in them. Hot material is placed between the die halves. A hammer or press applies force by moving one die half toward the other half. As they close, the die halves apply force on the material. This causes the material to flow into the die cavities. See Figure 23-9. This process is called ***forging***. Forging is used to make hand tools, automotive parts, and other products requiring a high level of strength.

Another forming process uses a single shaped die or mold. The material can be forced into the die cavity or drawn around a mold. An example of this type of process is ***thermoforming***. A plastic sheet is placed above a mold. The sheet is heated. The plastic sheet is then lowered onto the mold. A vacuum is pulled in the cavity or around the mold. This causes the hot plastic to draw tightly to the sides of the mold. Thermoforming produces plastic parts of all shapes.

Figure 23-9. These workers are using forming dies to make automobile parts. (Honda)

Rolls are also used to form materials. The material is fed between rotating rolls. This action stretches and squeezes the material into a new shape. This process, called *roll forming*, is used to make corrugated roofing and large tank parts. These are just three forming processes. There are many more that use force and shaping devices to form materials.

Separating

Some manufacturing processes shape material by removing excess stock. See Figure 23-10. The extra material is cut, sheared, burned, or torn away. These processes are called ***separation***. They separate or remove the unwanted portion of the workpiece, leaving a properly shaped part.

One type of separation is called ***machining***. See Figure 23-11. This process uses a tool to cut chips of material from the workpiece. All machining requires motion between the tool and workpiece.

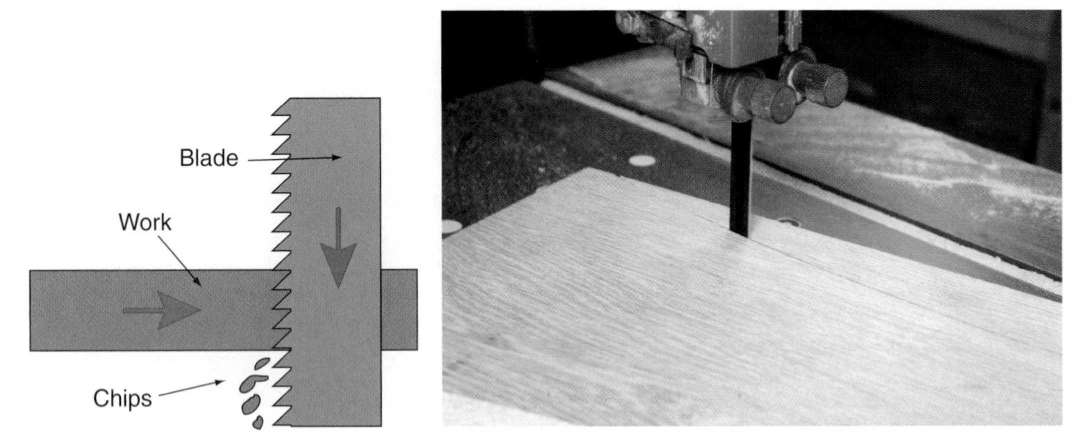

Figure 23-10. Separating processes. Saws, knives, and other tools cut away unwanted material. (©iStockphoto.com/Difydave)

Figure 23-11. An example of machining. Notice the chips being produced as the saw cuts the wood. (©iStockphoto.com/KristianSeptimiusKrogh)

The tool might spin to make the chip. Drilling and many sawing operations use a rotating tool. In other cases, the work is rotated against a solid tool. Lathes use this action. In still other cases, the tool is drawn across the stationary work. The band saw and scroll saw (jigsaw) move the tool across the work to make the cut.

A second separation action is called ***shearing***. This process uses blades. These blades move against each other. The material is placed between the blades.

The moving blades then apply force. This force fractures, or breaks, the material into two parts. Scissors, tin snips, and shears are shearing tools.

The third type of separation process is called ***flame cutting***. Burning gases are used to melt away unwanted materials. Oxyacetylene cutting is an example of flame cutting.

Newer separating processes use beams of light, sound waves, electric sparks, and even jets of water to cut away unwanted materials. Laser machining, ultrasonic (high-sound) machining, and electrical discharge machining are examples of separating. See Figure 23-12.

Conditioning

A fourth type of secondary processing is ***conditioning***. These processes alter and improve the internal structure of materials. This action changes the properties of the material. Conditioning can be done by heating or cooling or with mechanical forces or chemical action. See Figure 23-13.

The most common conditioning activity is thermal conditioning (heat-treating). The material being conditioned is heated to make it harder, softer, or easier to use. Three major types of heat treatments are used on metals. These are hardening, annealing (softening), and tempering (removing internal stress).

Heat can also condition ceramic materials. This process is called ***firing***. Firing involves slowly heating material to a very high temperature. The item is then allowed to cool slowly. During the process, a glass-like ingredient in the ceramic material

Figure 23-13. This pottery will be made hard by heating it to a high temperature and allowing it to cool. This process is called *firing*. (©iStockphoto.com/TerryHealy)

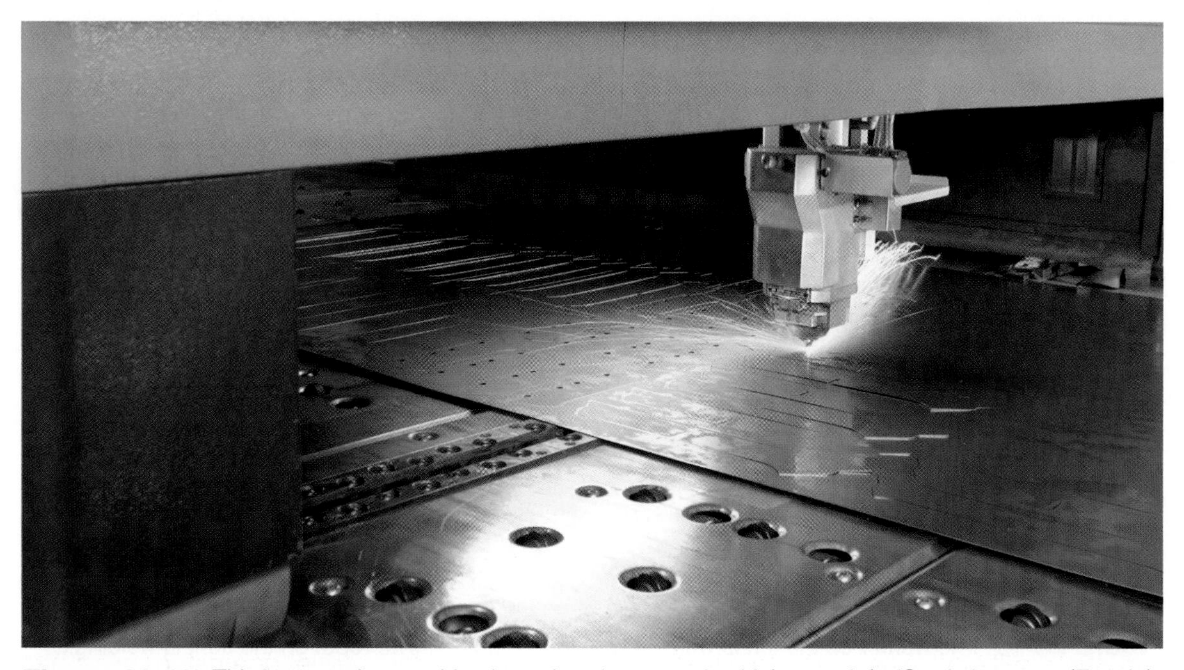

Figure 23-12. This laser cutting machine is cutting sheet metal at high speed. (©iStockphoto.com/Fertnig)

melts and coats the clay particles in the material. As the material cools, it becomes solid. The solid, glasslike materials bond the clay particles into a rigid structure. The result is a very hard, brittle product.

Assembling

Nails, screws, baseball bats, and combs are all one-part products. Most products, however, are made up of several parts. The act of putting parts together is called ***assembling***, or *combining*. Most parts are held together (assembled) using either mechanical fastening or bonding. See **Figure 23-14.**

Mechanical fasteners grip parts and hold them in place. Typical mechanical fasteners are nails, screws, rivets, nuts, bolts, staples, and stitches. Bonding permanently assembles parts together. This can be done either by fusion or by adhesive bonding.

Fusion uses cohesion and the same forces that hold the molecules of the material together. The parts are melted at the joint area. This causes the materials to flow together. When the material cools, a bond is formed. The two parts become one. This assembling process is usually called *welding*. See **Figure 23-15.**

Figure 23-14. The assembling process fastens parts together.

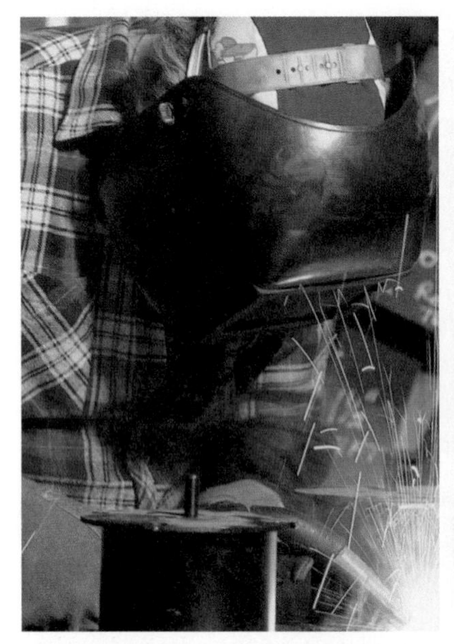

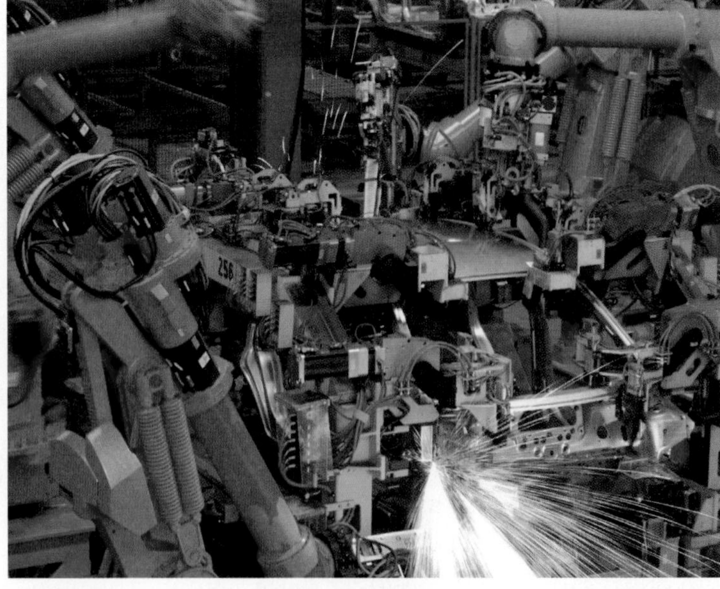

Figure 23-15. The worker on the left is welding two parts together. The robot on the right is welding body parts of a car together. (Coachman Industries and Ford Motor Company)

CAREER HIGHLIGHT

Operations Researchers

The Job: Operations researchers apply analytical techniques from mathematics, science, and engineering to help make decisions and solve problems. They solve problems in several ways and present alternative solutions for management to use in reaching the company's objectives. Operations researchers might encounter a number of different problems, such as forecasting results, scheduling activities, allocating resources, measuring performance, pricing products, and evaluating distribution channels.

Working Conditions: These researchers generally work in a standard office environment. Typically, they work a standard 40-hour week. Operations researchers are under pressure to meet deadlines for their research findings.

Education and Training: Most operations researchers hold a bachelor's degree in computer science or a field such as economics, mathematics, or statistics. Also, they generally have at least a master's degree in operations research, business, computer science, information systems, engineering, or management. In addition, operations researchers must think logically, have good oral and written communication skills, and be able to work with people.

Career Cluster: Business, Management & Administration

Career Pathway: Business Analysis

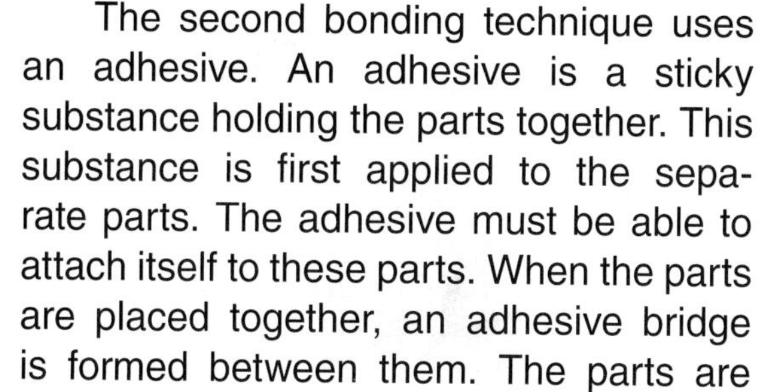

The second bonding technique uses an adhesive. An adhesive is a sticky substance holding the parts together. This substance is first applied to the separate parts. The adhesive must be able to attach itself to these parts. When the parts are placed together, an adhesive bridge is formed between them. The parts are fastened together.

Different materials require different adhesives. A wide range of adhesives is available to adhere almost any two materials together. Plastic trim parts are adhered to automobile bodies. The surface covering of aircraft wings is adhered to the wing structure. Wallpaper is glued to walls of homes and apartments.

Finishing

The final group of secondary manufacturing processes is ***finishing***. This area includes all activities protecting and beautifying the surface of a material. See Figure 23-16.

A finish is usually a surface coat applied to the material. This coating can be an organic material, a metallic material, or a ceramic material. Organic finishes are plastic materials suspended in a solvent. They are applied by brushing, spraying, rolling, or dip coating.

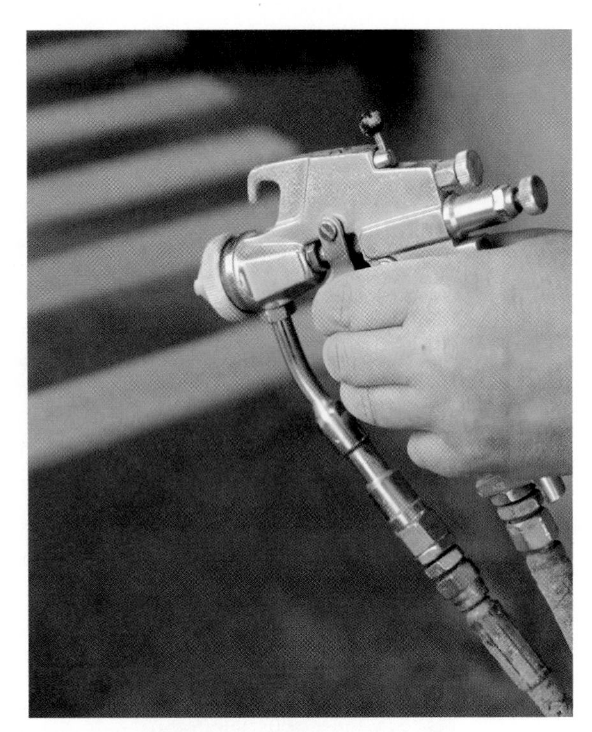

Figure 23-16. A finishing process. Paint can be sprayed onto a product. This coating protects the product and improves the product's appearance. (©iStockphoto.com/OwenPrice)

The finish dries when the solvent (thinner) evaporates. As this happens, the plastic material changes into more complex molecules. This new form produces a hard, uniform coat. The coat keeps water, oil, and other environmental elements from the base material. See Figure 23-17.

Organic finishes are called *paints*, *enamels*, *varnishes*, and *lacquers*. They provide an attractive, protective coating for metals and woods. Metals can be applied as a finish. In this process, a base material is coated with the metal. This can be done in several ways. The following are the most common:

- **Electroplating.** This is using electricity to deposit the metal on the part.
- **Dipping.** This is suspending the part in a vat of molten metal.

Chromium is a common metal coating material. The metal protects and provides a shiny surface. Chromium is generally applied by electroplating the part.

Figure 23-17. The white paint on these kitchen cabinets was applied to protect the wood and add beauty to the manufactured product. (©iStockphoto.com/kourafas5)

Several other metals are applied by dip coating. Zinc coatings protect steel barn siding and roofing, garbage cans, and other steel products. The steel is dipped into molten zinc and then allowed to cool. This process is called *galvanizing*.

A similar process applies tin to steel. The result is a tin-coated steel sheet. This material is widely used to make tin cans.

Ceramic materials also make good coatings. Porcelain and glaze (a glass-like material) are often used. The coating material is applied to the part while the part is cold. The part is then heated to melt and fuse the finish to the product.

Many ceramic products are finished with glaze. This material provides a colorful and water-resistant coating for dishes, planters, and other products. Porcelain is often used to coat metals and ceramic products. Some kitchen appliances are coated with porcelain enamel. Many bathroom fixtures also have a porcelain coating.

Manufacturing Systems

Secondary manufacturing processes are used to make products for everyday use. They must be organized, however, to be effective. These processes must be used as part of a manufacturing system. See Figure 23-18.

Types of Manufacturing Systems

There are four major types of manufacturing systems:

- Custom manufacturing.
- Intermittent manufacturing.
- Continuous manufacturing.
- Flexible manufacturing.

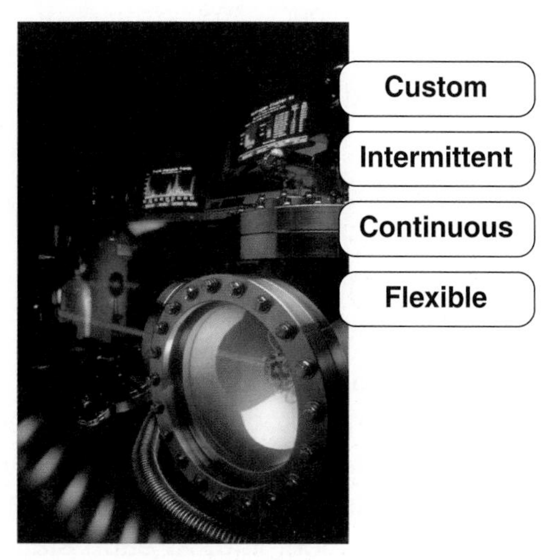

Figure 23-18. The four major types of manufacturing.

Each of these systems is used today to make products. These systems all have advantages and disadvantages.

Custom Manufacturing

Custom manufacturing is the oldest system. In early history, one person made an entire product. This person had all the skills needed to process materials into products.

Most products of Colonial times were custom-made. Silversmiths made silver bowls and candleholders. Cobblers made shoes. Weavers made cloth. Tailors made clothing.

Later, custom-manufacturing systems were used to make very special products. These products were designed for the customer. Only a few products were built to fill a specific need. See Figure 23-19.

Today, spacecraft, ships, some cabinets and furniture, and clothing for special needs are custom manufactured. One person might make simple products. Many people work on complex custom-made products.

Intermittent Manufacturing

As the nation grew, custom manufacturing could not meet customer demand. There were a growing population and increasing wealth. People wanted more and better products. The skilled craft workers could not produce products fast enough.

Small factories were started. Products started to be made in small batches. A dozen or more candlesticks were made at a time. Several pairs of shoes were made in the same size. This system is called ***intermittent manufacturing***.

Intermittent manufacturing is widely used today. In this system, the parts for a product travel in a lot, or batch. For example, suppose 100 birdhouses are needed. One part is the front. First, workers select lumber to make the fronts. Next,

Figure 23-19. This craftsperson is sanding the end of a custom-made table. (©iStockphoto.com/Difydave)

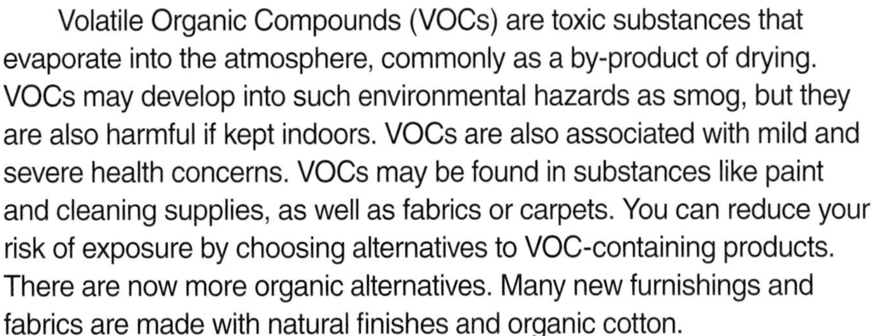

Think Green
Volatile Organic Compounds

Volatile Organic Compounds (VOCs) are toxic substances that evaporate into the atmosphere, commonly as a by-product of drying. VOCs may develop into such environmental hazards as smog, but they are also harmful if kept indoors. VOCs are also associated with mild and severe health concerns. VOCs may be found in substances like paint and cleaning supplies, as well as fabrics or carpets. You can reduce your risk of exposure by choosing alternatives to VOC-containing products. There are now more organic alternatives. Many new furnishings and fabrics are made with natural finishes and organic cotton.

they move the boards to a saw and cut out 100 birdhouse fronts. These parts are put in a tray. This tray moves to a drill press. Workers drill the entry hole for the bird in all 100 parts. The tray of parts travels to another drill press, where the perch hole is drilled in each piece. The parts finally move to another saw, where the roof peak is cut on all 100 parts. In the example, the parts moved from operation to operation in a batch. See Figure 23-20.

Continuous Manufacturing

When many products are needed, ***continuous manufacturing*** is generally used. The parts move down a manufacturing, or production, line. At each station on the line, a worker or automatic machine completes a specific operation. The product

Figure 23-20. To make these birdhouses, the parts moved from one operation to the next in a batch.

Figure 23-22. This flexible-manufacturing system uses robots that perform several functions. (Kalb)

takes shape as it moves from station to station along the line. See Figure 23-21. Completed parts flow to an assembly line, where the parts are put together to form the finished product.

Flexible Manufacturing

A new system of manufacturing is called ***flexible manufacturing***. See Figure 23-22. This system uses complex machines and computers for control. Flexible manufacturing can produce small lots, similar to intermittent manufacturing. This manufacturing, however, uses continuous-manufacturing actions. Thus, flexible manufacturing is the way many modern products are built. This manufacturing produces low-cost products, as the products are needed.

Figure 23-21. One station in a continuous production line that duplicates DVDs. (©iStockphoto.com/Gizmo)

TECHNOLOGY EXPLAINED

propeller: a rotating, multibladed device that drives a vehicle by moving air or water.

FLY·EASTERN AIR LINES
DOUGLAS
DC 3

Figure A. Propellers are used on airplanes, both old and new. (Piaggio Aviation)

All vehicles have five systems: structure, suspension, propulsion, guidance, and control. The propulsion system causes the vehicle to move. This system converts energy into motion.

The propulsion system often uses an engine to create a rotating motion. This motion is transmitted to a device that moves the vehicle. Most land vehicles use wheels to create movement. A different system is required, however, for vehicles traveling on water and in the air. A common propulsion device is a propeller. See Figure A. Propellers are primarily used on airplanes and ships.

A propeller works using two principles of physics. The first principle states the following law of physics: For every action, there is an equal and opposite reaction. The propeller forces air or water backward as it spins. This, in turn, causes the propeller to be pushed forward. Actually, the propeller moves similarly to a screw. The tips of the blades follow a path similar to the threads of a screw. In theory, one revolution of the propeller causes the vehicle to move ahead, as one turn of a nut causes the nut to move on a bolt. A boat or plane does not, however, move that far. Both water and air are yielding fluids. Ships generally move about 60%–70% as much as they should in theory.

The second principle states another law: As a fluid travels faster, its pressure is reduced. See Figure B. When a propeller spins, the water or air travels faster across the front of the blade than it does on the back. This causes the pressure to decrease on the front. The decrease in pressure draws the propeller forward, adding to the propeller's efficiency.

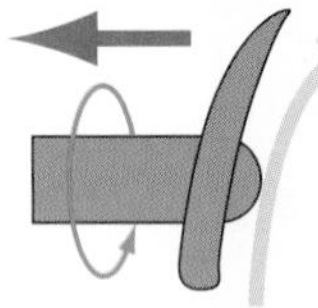
Reaction
Propeller (and vehicle) is pushed forward.
Action
Water or air is pushed backward by the blade.
For every action, there is an equal and opposite reaction.

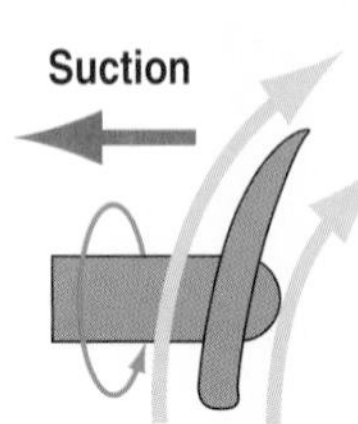
Suction
Air or water travels farther. Pressure is reduced.
Air or water travels shorter distance. Pressure is higher.
When the speed of a fluid increases, the fluid's atmospheric pressure decreases.

Figure B. How a propeller works.

Developing Manufacturing Systems

Manufacturing systems are developed for one purpose: to produce products to meet people's needs and wants. These systems have significantly increased the number of products available, while improving ***quality*** and cutting costs. The manufacturing process includes the designing, developing, making, and servicing of products and systems. This process includes the use of materials, hand tools, human-operated machines, and automated machines. The development of a manufacturing system involves several actions:

1. Selecting operations needed to make the product.
2. Putting the operations in a logical order.
3. Selecting equipment to make the product.
4. Arranging the equipment for efficient use.
5. Designing special devices to help build the product.
6. Developing ways to control product quality.
7. Testing the manufacturing system.

Each of these elements contributes to the efficient production of products. These elements help us use technology wisely.

Selecting Operations

Most decisions about manufacturing-system design are based on product drawings. These documents describe the product that will be built. One of the first system-design steps is to decide which operations are needed. This might sound easy. A hole is needed. What could be simpler? There are many options, however. Should the hole be drilled, punched, cut with a laser, or produced with an electrical discharge machine?

Each feature of the product is studied. Tasks to be performed are listed. The method for doing each task is selected. The result of this activity can be a set of operation sheets. Each sheet lists all the operations needed to make a part.

Sequencing Operations

After the operations are selected, a planner must put them in the proper order. The product must be built efficiently. Also, moving the product from workstation to workstation must be considered. ***Inspections*** must be scheduled. Plans must be made for storing parts and products until the parts are needed or the products are sold.

Remember our example of the birdhouse fronts? The material for the front is first cut to length. Parts are then moved to a drill press. A hole is drilled. Again, the part moves to another drill press. There, another hole is drilled. At this point, the part is inspected. The quality and location of the holes need to be checked. The parts then move to another saw. There, the gable (pointed end) is cut. The finished part is inspected. Finally, the parts move to a storage area, where they can wait for other parts, so assembly can start. This simple example includes the following:

- Four operations (changing the shape or size of the material).
- Four transportations (moving parts from station to station).
- Two inspections (checking the quality of the part).
- One storage (placing the part in a safe place until needed).

All these words describe the order of operations, transportations, inspections, and storage acts. Sometimes, however, words are hard to follow. Charts communicate better. Flow process charts contain this information. See Figure 23-23. These charts are often used to design new manufacturing activities. They are also used to study old procedures. Studying them can

Product Name: Bird House - End	Flow begins: Standard stock	Flow ends: Finished part	Date: 10-17
Prepared by: R.T. Wright Section: R&D		Approved by: Deb	
Process symbols and no. used	○ Operations 4 □ Inspections 2 D Delays 0 ⇨ Transportations 5 ▽ Storages 1		

Task No.	Process Symbols	Description of Task	Machine Required	Tooling Required
	○ ➡ □ D ▽	Move material to saw	stock cart	
	● ⇨ □ D ▽	Cut to length	radial saw	Stop #301
	○ ➡ □ D ▽	Move to drill press	conveyor	
	● ⇨ □ D ▽	Drill large hole	drill press #1	drilling jig 208
	○ ➡ □ D ▽	Move to drill press	conveyor	
	● ⇨ □ D ▽	Drill small hole	drill press #2	drilling jig 308
	○ ⇨ ■ D ▽	Inspect		
	○ ➡ □ D ▽	Move to circular saw	conveyor	
	● ⇨ □ D ▽	Cut gable	circular saw	sawing fixture
	○ ⇨ ■ D ▽	Inspect		
	○ ➡ □ D ▽	Move to storage	stock cart	storage tray
	○ ⇨ □ D ▼	Store		
	○ ⇨ □ D ▽			
	○ ⇨ □ D ▽			
	○ ⇨ □ D ▽			

Figure 23-23. The operation, or flow process, chart has been filled in for the sample birdhouse front.

result in finding a better way to make the part.

Selecting Equipment

Each operation requires equipment. See Figure 23-24. Saws cut lumber. Drill presses drill holes. Conveyors move materials between workstations. Racks are used to store materials and parts.

For each manufacturing system, equipment must be provided. The flow process chart helps engineers determine the equipment they need for the process. Some equipment might already be owned. Other items will be purchased.

Arranging Equipment

The equipment must be arranged for production. Sometimes the machines are used for a number of different products. This is true with intermittent-manufacturing activities. In this type of manufacturing, like equipment is grouped together. A roughing department might be formed to contain all saws, jointers, and surfacers. A finishing department might be equipped to paint several different products.

In other cases, the equipment is set up to make only one product. A continuous-manufacturing system is designed. In this system, the equipment is arranged to make the selected product. In our example, the line to make our birdhouse fronts contains the following equipment: a saw, a drill press, another drill press, and another saw. See Figure 23-25.

Figure 23-24. Look at the equipment on this production line. You can see material-handling equipment (conveyors). There is processing equipment (the shaping machine) on the left, with power controls.

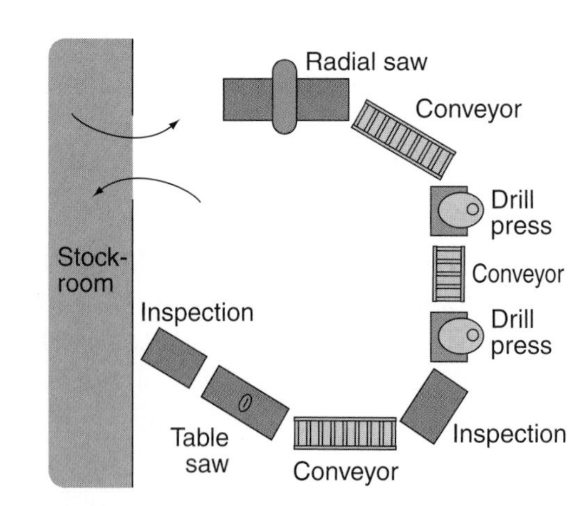

Figure 23-25. A plant layout drawing for the birdhouse-front production line. See how handy both the beginning and the end of the line are to the stockroom.

Career Highlight

Printing Press Operators

The Job: Printing press operators operate and maintain printing presses in newspaper and commercial printing plants. They prepare presses for a printing run, monitor the operation of the presses during the printing process, and keep the paper feeders stocked. Press operators also make adjustments to maintain quality and perform preventative maintenance on the presses.

Working Conditions: These operators are key employees at a printing plant. Their work is physically and mentally demanding. This work can also be tiring because the operators are standing most of the time. Pressrooms are noisy. Workers are exposed to safety hazards as they adjust processes.

Education and Training: Most printing press operators are trained in on-the-job programs. In these programs, they work as helpers for experienced press operators. Operators can be trained through apprenticeship. Apprenticeship combines on-the-job training with classroom instruction. There are also training opportunities at colleges and technical institutes that offer printing programs.

Career Cluster: Arts, Audio-Video Technology & Communications

Career Pathway: Printing Technology

Designing Tooling

Many times, special devices make manufacturing more efficient. These devices might hold a part so the part can be machined. See Figure 23-26. Some devices might hold several parts for welding. In other cases, they might be special dies for forming the material. All these items are called ***tooling***.

These items are designed to make operation more efficient. They should make the operation faster, easier to complete, and safer. For our birdhouse front, several pieces of tooling can be used. They might include devices to do the following:

- Hold the part so the entry hole is always drilled in the same place.
- Hold the part so the perch hole is correctly placed.
- Hold the part so the gable ends are accurately cut.

Controlling Quality

Codes and standards address issues, including safety. In many cases, codes and standards establish specifications to ensure that manufacturers produce items consistently and at an acceptable level of quality. In manufacturing, the International Organization for Standardization (ISO) supports and manages international standards for production. ANSI and ASTM create additional standards.

Product quality is a major concern in manufacturing. Everyone wants products

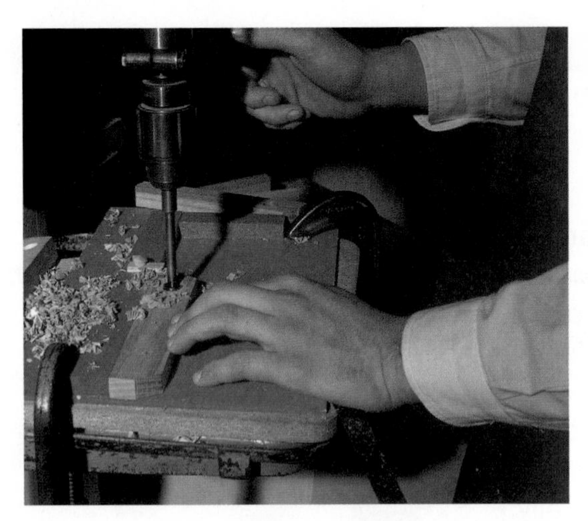

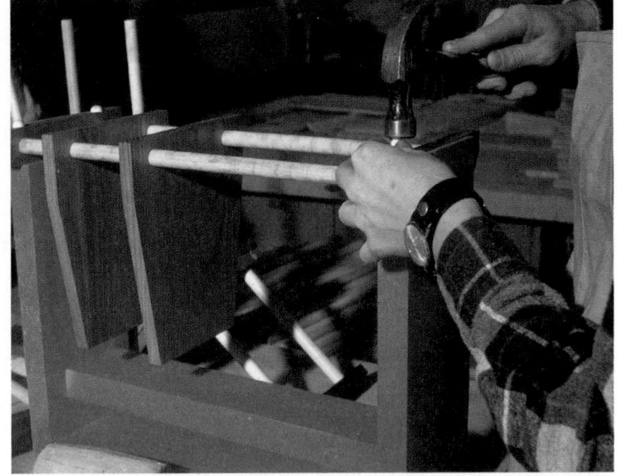

Figure 23-26. Note the simple drilling tooling on the left and the assembly fixture on the right.

that work well and look good. Therefore, parts and products must be checked for quality. This action is called *inspection*. See Figure 23-27. The parts are compared to the drawings. Each part and product must meet the standards designers and engineers set.

Parts failing the inspection might be scrapped. Other parts might be reworked. The defect might be removed from the part. Scrapping and reworking parts add cost to the product. Therefore, every effort is made to make the product right the first time.

Quality control efforts also encourage workers to build good products. These efforts reduce scrap and waste. They also help ensure that customers receive quality products.

Testing the System

The last step in manufacturing-system design is testing the system. The parts of the system must be put in place. The machines must be positioned. Conveyors and other material-handling devices must

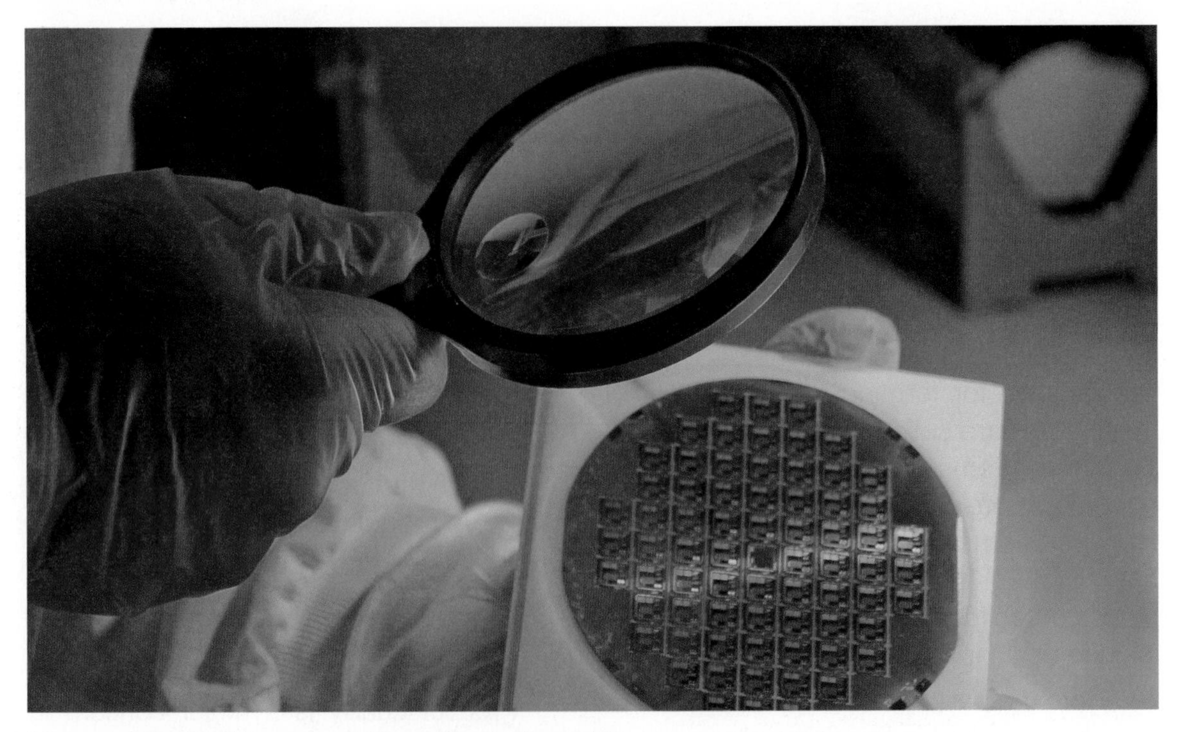

Figure 23-27. This worker is inspecting a silicon chip.

be installed. Tooling must be attached to the machines. The system can then run.

People produce test products using the manufacturing system. This is called a ***pilot run***. Engineers check to see that the operations are working correctly. They observe the flow of material. Also, product quality is carefully checked.

A pilot run is important. This run shows where changes are needed. A pilot run might identify tooling needing to be improved. This run might indicate that the equipment should be reorganized. The pilot run might suggest that the workstations need to be relocated.

After the changes are made, engineers make another pilot run. Changes continue until the system is operating properly. Only then does full-scale production start.

Managing Manufacturing Systems

Most manufacturing is managed. A group of people sees that the system runs properly. These people are called *managers*. They make up management.

Managers do not make the products. They organize the systems so the products are made efficiently. In doing this, they complete the four basic functions:

- **Plan.** The managers set goals.
- **Organize.** The managers divide tasks into jobs.
- **Actuate.** The managers assign jobs and supervise workers.
- **Control.** The managers compare the results to the plan.

Good managers get work done through other people. They provide direction and support. If managers do their jobs well, workers can more easily make products.

Producing Products

The manufacturing of products requires all the major types of resources. People use information and machines. These machines are powered by energy, to change materials into products. See Figure 23-28. In general, machines, many

Figure 23-28. Can you identify the resources being used in this manufacturing line?

Career Highlight

Production and Planning Clerks

The Job: Production and planning clerks maintain records and prepare reports on production activities. They prepare work orders and release them to production departments. These clerks record the materials and parts used to produce products, the number of products produced during a specific period of time, and machine usage. They schedule and monitor the progress of production activities.

Working Conditions: These clerks work closely with supervisors. The supervisors must approve the production and work schedules. Production and planning clerks typically work 40-hour weeks in offices in a manufacturing facility.

Education and Training: Clerks need at least a high school diploma. They learn the job by doing routine tasks under close supervision. Production and planning clerks must learn about the company's operations and priorities before they can begin to write production and work schedules.

Career Cluster: Manufacturing

Career Pathway: Manufacturing Production Process Development

of which are computer controlled, are capable of making higher-quality goods than a skilled craftsperson can make alone.

Marketing Products

Today, there are thousands of products available to each of us. These products must be marketed. Potential customers must be told about them. Generally, ***advertising*** and marketing do this task. See Figure 23-29.

When the idea for a product or service is being developed, the inventors create a marketing plan. The marketing plan specifies the intended customers, expected sales, and possible advertising and sales methods. Marketing a product means telling the public about it, as well as assisting in selling and distributing it. This process involves gauging what the public wants and then advertising and selling products to consumers.

Ads are delivered in several ways. Some are placed in magazines and newspapers. Others are aired on radio and television or delivered over the Internet. Still others are delivered through the mail or displayed on billboards and signs.

Figure 23-29. Catalogs can be used to advertise products.

Advertisements make people act. They bring customers to stores, where the sales effort takes place. Salespeople encourage customers to buy the product.

These people sell expensive products, such as automobiles and computers. They present the value of the products to the customer. Less-expensive products (such as toothpaste and colas) are simply displayed. The customers' actions are often based on the advertising effort, the appearance of the package, or previous experience.

After a product or system is sold or leased, it sometimes needs to be serviced. Servicing, or providing support after a sale has been made, is an important part of the manufacturing process. These services can include installing, troubleshooting, servicing, and repairing.

Summary

Manufacturing provides all the products we use. The food we eat, the clothes we wear, and the vehicles we travel in have all been manufactured. Manufacturing generally includes primary and secondary processing. Primary processes convert raw materials into industrial materials. Secondary processes change industrial materials into usable products.

To meet the large demand for products, complex manufacturing systems have been developed. They let people efficiently make things we need and want. The major types of manufacturing systems are custom manufacturing, intermittent and batch manufacturing, continuous manufacturing, and flexible manufacturing. Custom manufacturing is used to make specialized products designed for a specific customer. In intermittent manufacturing, the parts of a product travel from operation to operation in a batch, so more products can be produced in a short amount of time. Continuous manufacturing uses an assembly line to produce many products quickly. Flexible manufacturing uses continuous-manufacturing actions. This manufacturing can be used to produce small batches of a product, however, very similar to intermittent manufacturing.

There are several steps involved in developing a manufacturing system. To produce products efficiently, operations must be selected and sequenced, equipment must be selected and arranged, tooling must be designed, quality must be controlled, and the system must be tested. Managers assist in efficient production by providing direction and support. After the products are produced, they must be marketed to potential customers.

STEM Connections

Mathematics

Select a simple product. Determine the geometric shapes it contains. Measure and record these shapes in standard and metric units.

Science

Select a manufacturing process, such as sawing or injection molding. Investigate the scientific principles used in the process. Prepare a poster display explaining these principles.

Curricular Connections

Social Studies

Investigate the products manufactured in your city or state. Try to determine why these, and not other products, are made there.

Social Studies

Select a manufacturing job and determine the education or training needed for it.

Activities

1. Select a simple product. List the types of processes used to make it (for example, casting and molding, forming, separating, conditioning, finishing, or assembling).
2. Select a simple product, such as a bookend. List the following:
   A. The operations you would use to make it.
   B. The equipment or tools you think you would need.
   C. The points at which you would check for quality.
   D. Any special tooling you think you would need.
3. Suppose you were to drill a hole in the center of a 4″ × 4″ piece of wood. Sketch the piece of tooling you would use to make the hole accurately in 100 parts.
4. Working in a small group, prepare a marketing plan for an idea, a product, or a service of your choice. Using desktop publishing software, develop advertising flyers. Develop a Web page for your product.

Test Your Knowledge

Do not write in this book. Place your answers to this test on a separate sheet of paper.

1. Manufacturing changes _____ into useful _____.
2. There are two types of manufacturing:
   A. Changing raw materials into industrial materials is called _____.
   B. Changing industrial materials into usable products is called _____.
3. Match the terms and descriptions. Give an example of each.

_____ Cuts materials.	A. *Chemical processing.*
_____ Heats materials to change them.	B. *Mechanical processing.*
_____ Uses chemicals to make new materials.	C. *Thermal processing.*

4. Name the six secondary manufacturing processes.
5. Identify the processes described below:
   A. Uses hollow forms to shape liquid materials. _____
   B. Forces or squeezes materials into new shapes. _____
   C. Cuts away excess stock by burning, shearing, or cutting. _____
   D. Alters interior structures of materials. _____
   E. Fastens parts together by any means. _____
   F. Protects or beautifies the outsides of products. _____
6. Describe each type of manufacturing system:
   A. Custom.
   B. Intermittent.
   C. Continuous.
   D. Flexible.
7. Flow process charts are sometimes used to show the order of operations. True or false?
8. Tooling is (select all correct answers):
   A. A device that holds a part while the part is being manufactured.
   B. Designed to make manufacturing more efficient.
   C. Sometimes a special shape for forming the material.
9. Label the following basic managerial functions:
   _____ Setting goals.
   _____ Dividing tasks into jobs.
   _____ Assigning jobs and supervising workers.
   _____ Comparing the results to the plan.
10. What are the three major inputs (resources) used in producing products?
11. Finding customers and telling them about a new product is known as _____.

Reading Organizer

Draw a bubble diagram for each main idea in the chapter. Make each of the main ideas the central bubble, while using details in smaller bubbles to surround the main points. An example from this chapter is shown.

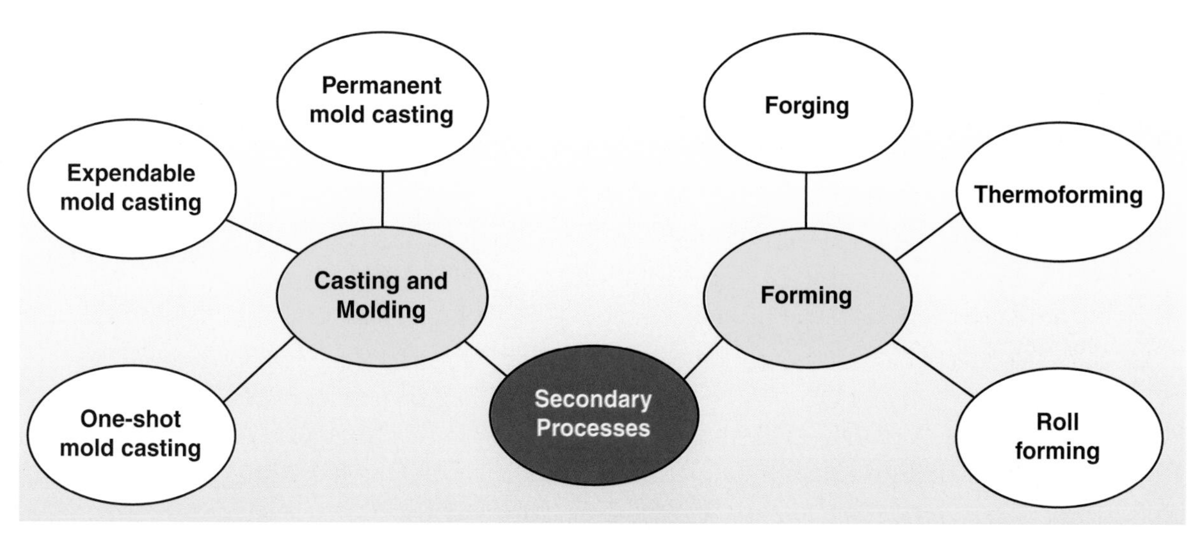

CHAPTER 23 ACTIVITY A

MANUFACTURING TECHNOLOGY

INTRODUCTION

Manufacturing is a major context in which technology is used. This context changes the form of materials to make them worth more. This can be done using custom techniques. People can make a product all by themselves. They can make it for their own use. More often, however, a manufacturing system called *continuous manufacture* is used. The product is made on a production line. In this activity, you will use such a line. Working on the line, you and your classmates will change the form of materials. You will change strips of wood into a CD holder. See Figure 23A-1.

During the activity, notice the use of the resources. See if you can see tools, materials, energy, information, time, and people being used. Materials are specified in a bill of materials. See Figure 23A-2. Most people will think the CD holder is worth more than the strips of wood and nails used to make it.

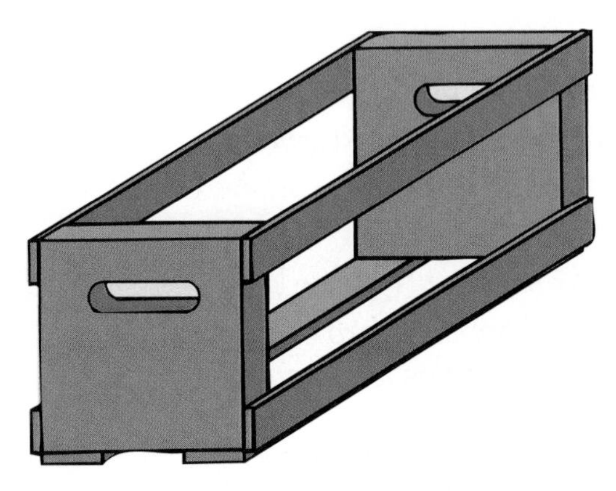

Figure 23A-1. An isometric drawing of the CD holder.

Quantity	Description	Size	Material
2	Ends	3/4″ x 5″ x 5″	Pine
4	Side strips	1/4″ x 3/4″ x 10″	Pine
2	Bottom strips	3/4″ x 1 3/8″ x 10″	Pine
24	Wire nails	3/4 x 18	

Figure 23A-2. The bill of materials for the CD holder.

Equipment and Supplies

- Wood strips:
  - 1/4″ × 3/4″ × random length (side-strip stock).
  - 1/4″ × 1 3/8″ × random length (bottom-strip stock).
  - 3/4″ × 5″ × random length (end stock).
- 5/8″ × 18 wire brads.
- A backsaw or band saw.
- A scroll saw.
- A hammer.
- A brace or drill press.
- A 3/4″ auger or speed bit.
- A disc or an oscillating sander.
- Abrasive paper.
- A sanding block.
- Special tooling:
  - An end cutoff jig.
  - A side-strip cutoff jig.
  - A bottom-strip cutoff jig.
  - An end-hole drilling jig.

Procedure

1. Study the operation process chart. See Figure 23A-3. Determine the steps needed to make the CD holder.
2. Read the explanation of the major steps to determine how to complete each step:

   E-1 Set up a band saw or backsaw (miter box and backsaw) to cut a 5″ piece from the end stock. A special cutoff jig can be built to guide the backsaw during the cutoff operation.

   E-2 Drill two 3/4″ holes in the end. The centers of the holes are 1″ from the top edge of the end. Locate the holes' centers 1 1/4″ from one another.

   E-3 Draw a line connecting the tops of the two holes and another connecting the bottoms of the holes.

   E-4 Insert a scroll saw blade into one of the 3/4″ holes. Tighten the blade in the saw. Cut out the marked section between the two holes.

   E-5 Smooth and straighten the saw cuts with a file.

Caution

Do not damage the curved portion at the ends of the slots.

I-1 Inspect the part for size and quality of the finger slot.

E-6 Sand the edges of the end parts on a disc or an oscillating sander or with abrasive paper and a sanding block.

End

- E-1 Cut the strip to length
- E-2 Drill finger holes
- E-3 Draw lines between holes
- E-4 Cut out finger slot
- E-5 File the cut straight
- I-1 Inspect
- E-6 Sand edges
- E-7 Sand ends
- E-8 Sand faces
- I-2 Inspect
- A-1 Assemble bottom strips to ends
- A-2 Assemble side strips to ends
- I-5 Inspect
- A-3 Apply finish material
- I-6 Inspect

End Strip

- S-1 Cut the strip to length
- S-2 Sand faces
- S-3 Sand edges
- S-4 Sand ends
- I-3 Inspect

Bottom Strip

- B-1 Cut the strip to length
- B-2 Sand faces
- B-3 Sand edges
- B-4 Sand ends
- I-4 Inspect

Figure 23A-3. The operation process chart. How many steps are there to produce the CD holder?

CAUTION

Do not round the edges because the side strips must mount flat on these surfaces.

E-7 Sand the ends of the end parts on a disc or an oscillating sander or with abrasive paper and a sanding block.

CAUTION

Do not round the ends because the bottom strips must mount flat on these surfaces.

E-8 Sand the faces with an oscillating sander to remove mill marks and smooth the surfaces.

I-2 Inspect the results of the sanding steps. Look for square, flat edges and ends.

S-1; B-1 Set up a band or miter saw to cut a 10″ piece from the stock for the end strip **(S-1)** or the bottom strip **(B-1)**. A special cutoff jig can be built to guide a backsaw during the cutoff operation.

S-2; B-2 Sand the faces with an oscillating sander to remove mill marks and smooth the surfaces.

CAUTION

Be sure not to round these surfaces because one face must fit flat against the ends of the holder.

S-3; B-3 Sand the edges of the end parts on a disc or an oscillating sander or with abrasive paper and a sanding block.

S-4; B-4 Sand the ends of the end parts on a disc or an oscillating sander or with abrasive paper and a sanding block.

CAUTION

Do not round the ends because they must be flush with the ends after assembly.

I-3; I-4 Inspect the results of the sanding steps. Look for square, flat edges and sides.

A-1 Locate the bottom strips. Nail both ends of each strip with two 3/4″ × 18 wire nails.

NOTE

A locating fixture can be used to speed this operation.

A-2 Locate the side strips. Nail both ends of each strip with two 3/4″ × 18 wire nails.

NOTE

A locating fixture can be used to speed this operation.

I-5 Inspect the assembled product. Route any defective products to a rework station or scrap.

A-3 Apply appropriate finishing material by brushing, wiping, or dipping.

I-6 Inspect the final product.

CHALLENGING YOUR LEARNING

How could you have made the product easier to make? What changes could you make to improve the quality of the product?

SAFETY RULES

- Keep your hands away from the path of the blade.
- Keep your fingers at least 2″ away from the blade at all times. Use a fixture to hold small workpieces.
- Upper-guide assembly should always be 1/4″–1/2″ above the stock. This reduces the amount of blade exposed.
- Push stock forward, rather than to the side.
- Work only within the band saw's capacity. Thick stock should be fed more slowly than thin stock.
- Students in the lab should observe the safety zone around the band saw. A broken blade occasionally "climbs" out to the right of the operator.
- Should the blade break, step aside. Disconnect power to the machine.

TSA Modular Activity

This activity develops the skills used in TSA's Go Green Manufacturing event.

Go Green Manufacturing

Activity Overview

In this activity, you will obtain discarded materials from local business or industry and design products that can be developed with those materials. Design a manufacturing process to produce the product. Perform market research for the product. Prepare a comprehensive report.

Materials

- Discarded materials from local business or industry (will vary).
- Paper.
- A pencil.
- A three-ring binder.
- A computer with CAD and word processing software.

Background Information

- **Obtaining materials.** Contact local businesses to obtain scrap material. You can write letters, make telephone calls, or visit businesses in person. Your local telephone book will provide many leads.
- **Design.** After obtaining materials, use brainstorming techniques to develop a list of potential products. Create rough design sketches for some of the products. Consider the available tools and materials when selecting the best idea.
- **Drawings.** Create working drawings, using a CAD system. Use multiview drawings with as many views as needed to fully describe the product. If your product is small enough, create the drawings at full scale.
- **Prototype.** Produce a prototype of the product. As you develop the prototype, consider the most effective manufacturing sequence. Are there design changes that could improve the manufacturing process? If design changes are made as you develop the prototype, be sure to update the drawings to reflect those design changes.

- **Production-plan flowchart.** Develop a production-plan flowchart illustrating the manufacturing sequence for the product.
- **Manufacturing.** Manufacture several units of your product, using the production process. Make a trial run first. Adjust the flowchart as needed to improve efficiency. Take photographs to document each step of production.
- **Marketing.** Develop a marketing plan, including an advertisement. Who might want to purchase the product? What is a reasonable price for the product?

GUIDELINES

In the course of this project, you must develop a report containing the following items:

- A cover page.
- Contents.
- A written description of the product.
- A print advertisement for the product.
- Design sketches.
- Working drawings.
- A materials list.
- A list of tools and machines.
- A production flowchart.
- Photographs of the manufacturing process with written explanations for each image.
- Letters of donation from the businesses supplying the materials.

EVALUATION CRITERIA

Your project will be evaluated using the following criteria:

- The report.
- Quality of the product.
- Creativity in design and use of materials.

CHAPTER 24

Medical Technology

DID YOU KNOW?

- One of the most important diagnostic tools in medicine is the stethoscope. This tool is an instrument physicians use to detect sounds in the body, such as a heartbeat. The stethoscope was invented in 1819.
- The first completely artificial heart was installed in a person in 1982. The device was the Jarvik-7® heart.
- The first inoculation for smallpox was developed in China in about 600 BC.
- A British physician, Edward Jenner, made the first major advancement in modern times for preventing smallpox. In 1796, he discovered that people could be immunized by inoculating them with material from cowpox sores.
- Louis Pasteur created the first successful vaccine against rabies in 1885.
- In 1954, Jonas Salk introduced an injectable vaccine for poliomyelitis (polio). Later, Albert Sabin developed an oral vaccine for the disease.

Objectives

The information given in this chapter will help you do the following:

- Recall some major points in the history of medicine and health care.
- Summarize the three major goals of health and medicine.
- Explain the two major thrusts of illness and injury prevention.
- Explain the role of technology in prevention programs.
- Summarize the function of vaccines in illness prevention.
- Recall the two major things health-care professionals can do to help ill or injured people.
- Give examples of ways technological devices are used in diagnosing illnesses and physical conditions.
- Summarize how technological devices are used to treat illnesses and physical conditions.

Key Words

These words are used in this chapter. Do you know what they mean?

clinical test
computerized tomography (CT) scanner
diagnose
disease
drug
electrocardiograph (EKG) machine
endoscope
immunization
inoculation
intensive care
magnetic resonance imaging (MRI)
pathologist
prevent
radiology
surgery
treat
ultrasonics
vaccination
vaccine
wellness
X-ray machine

Preparing to Read

As you read this chapter, outline the details of medical technology. Use the Reading Organizer at the end of the chapter to organize your thoughts.

Diseases have been the concern of people over the ages. Early humans blamed diseases on demons. Treatments for these ailments were based on magic and folk remedies. As societies progressed, treatments became mixes of magic and rational approaches. Treatments for some diseases of the skin and eyes were developed first because these problems were visible. Internal disorders continued to be treated with magic-based approaches.

By about 2600 BC, physicians were a recognized part of society. These people practiced an early form of medical science. They used simple technological devices and instruments to do their work. Over the years, the field of medicine slowly developed. Hospitals started to be built and used during the Middle Ages in Europe. They treated people with many diseases. Hospitals were especially important during the large epidemics of bubonic plague, leprosy, and smallpox that swept the continent.

An important milestone in medical history happened in the seventeenth century. At that time, scientists discovered that blood circulates in the human body. An English physician named William Harvey established that the heart pumps the blood in continuous circulation. Marcello Malpighi's discovery of tiny blood vessels, called *capillaries*, followed Harvey's discovery. Other important work investigated the brain and nervous system. Also, new knowledge of the liver, muscles, and heart was discovered. These and many other discoveries led to a new age in medicine.

Great advances in the diagnosis and treatment of diseases were made during this age. Germs were discovered. Their role in diseases was established. This led to the discovery of the causes for major diseases, such as anthrax, diphtheria, tuberculosis, and plague. Techniques to stop these diseases through vaccines were developed.

New surgical methods were also developed. Aseptic (sterile) surgery became common. Physicians started using sterilized instruments and techniques to avoid infecting patients. By the mid-1800s, anesthesia was being used in surgery.

By the end of the 1800s, new tools, such as X rays and ultraviolet lamps, were developed to diagnose and treat illnesses. Huge advances in medicine have occurred since 1900. They have helped to greatly increase the average person's life expectancy. Longer life has given medicine new challenges. People who live longer suffer from higher rates of heart disease, cancer, and stroke. These conditions have replaced infectious diseases as the leading causes of death. This has given rise to new treatments and the need for new technologies. Progress and improvements in medical technologies are used to improve health care. The use of new medicine and technology helps people live healthier and less painful lives.

Goals of Health and Medicine

There are three major goals of health and medical programs. See Figure 24-1. First, people use technology in health and medicine to help ***prevent*** illness and injury. Second, they use technology to ***diagnose*** diseases and injuries they think they have. Finally, they use technology to ***treat*** illnesses and injuries they could not prevent.

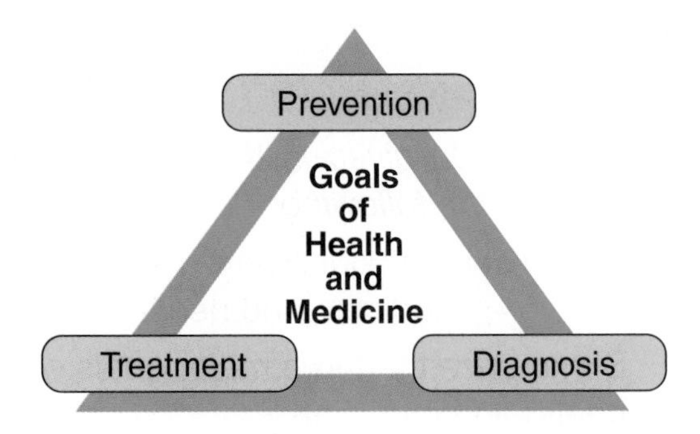

Figure 24-1. The goals of medical science and technology.

Prevention Programs

There is an old saying that "an ounce of prevention is worth a pound of cure." This means it is much better to prevent an injury or disease than to try to cure one after it occurs. There are two major ways people use technological products to help prevent illnesses and injuries. The first is ***wellness***. The second is ***vaccinations***, or inoculations.

Wellness Programs

People are considered being well when their bodies are in good health. Therefore, wellness can be described as a state of personal well-being. Wellness programs have people do things that help keep their bodies healthy. These actions can be considered preventative medicine. They involve at least four major factors. See Figure 24-2. These factors are nutrition and diet, environment, stress management, and physical fitness.

Wellness programs stress that people must be concerned with what they eat. These programs emphasize proper nutrition as an important factor in personal

Figure 24-2. The four areas of wellness programs.

health. Wellness programs focus on the quantity and value of the food people eat. They are concerned with the intake of vitamins, fats, proteins, and other life-sustaining food components. Technology is used to preserve and improve the nutritional value of foods and process farm products in the food we eat. See Figure 24-3.

These programs stress that people must be aware of the environment in which they live. They emphasize the need to control and improve the quality of the air we breathe. Wellness programs indicate the hazards associated with excessive exposure to direct sunlight. Technology can be used to deal with health hazards in the environment, purify water, clean the air, and develop sunscreen lotions and sunlight-filtering clothing.

A wellness approach suggests that people must be aware of the health damage that emotional stress can cause. People should keep their bodies fit through activity and exercise. Technology has also been applied to stress management and physical fitness through two major areas: exercise and sports.

Technology, wellness, and exercise

Exercise requires that people exert their bodies to improve their health. People do two major types of exercises. First, there is anaerobic exercise. Anaerobic exercise involves heavy work using only a few muscles. Examples of anaerobic exercise are weight lifting and sprinting. This type of exercise is maintained for short intervals of time. Anaerobic exercise is done to increase strength and muscle mass. These exercises have limited benefits, however, to cardiovascular health. Cardiovascular health is the major goal of wellness exercises.

Figure 24-3. Technology is used to process the food we eat. This food helps us stay healthy.

The second type of exercise is aerobic exercise. Typical aerobic exercises are walking, jogging, and swimming. These exercises use a large number of different muscles. They use a lot of oxygen to keep the muscles moving continuously. Aerobic exercises place demands on the cardiovascular and respiratory systems to supply oxygen to the working muscles. This exercise reduces the risk of heart disease and increases endurance.

Aerobic exercise can be done without special equipment. People can walk or

Career Highlight

Physical Therapists

The Job: Physical therapists provide treatments and recommend exercises that restore function, increase mobility, and relieve pain for patients who have had injuries or illnesses.

Working Conditions: Most physical therapists work in hospitals, clinics, and private offices that have special equipment designed to exercise and strengthen muscles.

Education and Training: Physical therapists must complete a college or university program with basic science courses and advance into special courses teaching biomechanics, microanatomy, human growth and development, examination procedures, and therapeutic techniques. In every state, after graduating from an accredited physical therapist educational program, physical therapists are required to pass a licensure exam before they can practice.

Career Cluster: Health Science

Career Pathway: Therapeutic Services

jog to improve their health. They can swim laps in a pool or lake. Even these activities, however, often require technology. Special shoes and clothing have been designed and produced to help in these activities. Swimming pools and walking paths in parks are built using construction technology.

Also, many people use exercise equipment to improve their health and well-being. This equipment is a result of technological design and production activities. Typical exercise equipment includes the following:

- **Treadmills.** These are moving belts allowing people to walk or jog in place to provide an aerobic fitness workout. See Figure 24-4.
- **Stationary bikes.** These are nonmoving bicycles allowing people to obtain the benefits of bicycling without leaving home. Some models of the bikes have handlebars that move. This added feature creates resistance to provide an upper-body workout, as well as the lower-body workout. See Figure 24-5.

Figure 24-4. A typical treadmill used in exercise programs. (SportsArt)

Figure 24-5. This stationary bike can work many muscle groups. (StarTrac)

Figure 24-6. This climber is an exercise machine simulating the work required to climb stairs. (StarTrac)

- **Stair climbers.** These machines are devices allowing people to obtain the benefits of climbing without using stairs. They provide lower-body workouts. See Figure 24-6.
- **Rowing machines.** These machines simulate actual rowboats with oars. They allow people to simulate rowing a boat with oars, using their arms and legs.
- **Home gyms.** These are multistation exercise machines allowing people to work on many different muscle groups.

All these machines have sensors and other technological devices to help people stay fit. These devices measure and display information, such as speed and heart rate. They might graph the amount of energy used and the effects on the body. See Figure 24-7.

Technology, wellness, and sports

Sports are another way to promote wellness. They are games or contests involving skill, physical strength, and endurance. Sports can be played as an economic activity, in which players are paid and fans pay admission fees. This discussion, however, deals with sports played for the sake of fitness, personal health, and enjoyment.

All sports require technological products. These products might be constructed playing fields, or venues. Venues are part of our built environment. See Figure 24-8.

Figure 24-7. Many exercise machines have technological displays providing information about the effects of exercising. (StarTrac)

The playing fields are the results of construction technology. Construction technology is presented in Chapter 20. For example, if you use softball for a fitness activity, it requires a specific playing venue. Softball needs a field with bases and probably a backstop. In contrast, if golf is the fitness sport, a course with holes and traps is required.

Often, technology is used to improve the natural playing surfaces for sports. For example, special grasses have been developed for playgrounds and playing fields. Fertilizers have been developed and manufactured to encourage grass to grow. Special lawn-grooming and moving equipment has been developed to maintain the fields. Likewise, snow-grooming equipment prepares and renews the surfaces of ski runs. Snowmaking equipment has been developed to supplement natural snow for skiing. These are just a few examples showing how technology has helped develop and maintain sport venues.

Most sports require manufactured equipment and personal protection. Each game or contest, however, uses specific technological products. The game equipment for baseball (such as baseballs, bats, and gloves) is different from the game equipment for golf (for example, clubs and golf balls). Similarly, the personal clothing and protection equipment for baseball is

Figure 24-8. A basketball court is an example of a typical venue for sports.

different from the clothing and protection equipment for golf.

Players in many sports wear special clothing and protective gear. Special shoes might be developed to provide foot support. They might absorb the shock that running on hard surfaces causes. The clothing might be made from fabrics shedding rain and snow. These garments might maintain body heat or wick away perspiration. Participants in some sports wear protective gear to reduce the chance of injury. For example, baseball players wear batting helmets to protect their heads from wild pitches. The game equipment and protective equipment are technological products. They are designed and manufactured for specific purposes.

Prevention through Immunization

Healthy bodies can go only so far in preventing illnesses. Another approach is called ***immunization***, or *vaccination*. Immunization is the process of systematically vaccinating people through a series of shots to prevent ***disease***. See Figure 24-9. This process is the same as ***inoculation*** and is the common method used to promote natural resistance to specific diseases. Over history, people have developed two different approaches to immunization. The first is passive immunity. This approach injects blood from an actively immunized person or animal into a patient. The other system is active immunization. In this system, a disease-causing microorganism or a product of the microorganism is injected into the body. This approach uses modified or killed bacteria or viruses to create protection.

Vaccines are special ***drugs*** created to prevent diseases from affecting people. They do not cause disease. Instead, they cause the body's immune system to build a defense against a disease. This defense allows the body's immune system to immediately respond to a particular disease.

More than 50 ***vaccines*** for preventable diseases have been developed. They are designed to begin with birth and are available throughout a person's lifetime. How they are developed is discussed later in the chapter.

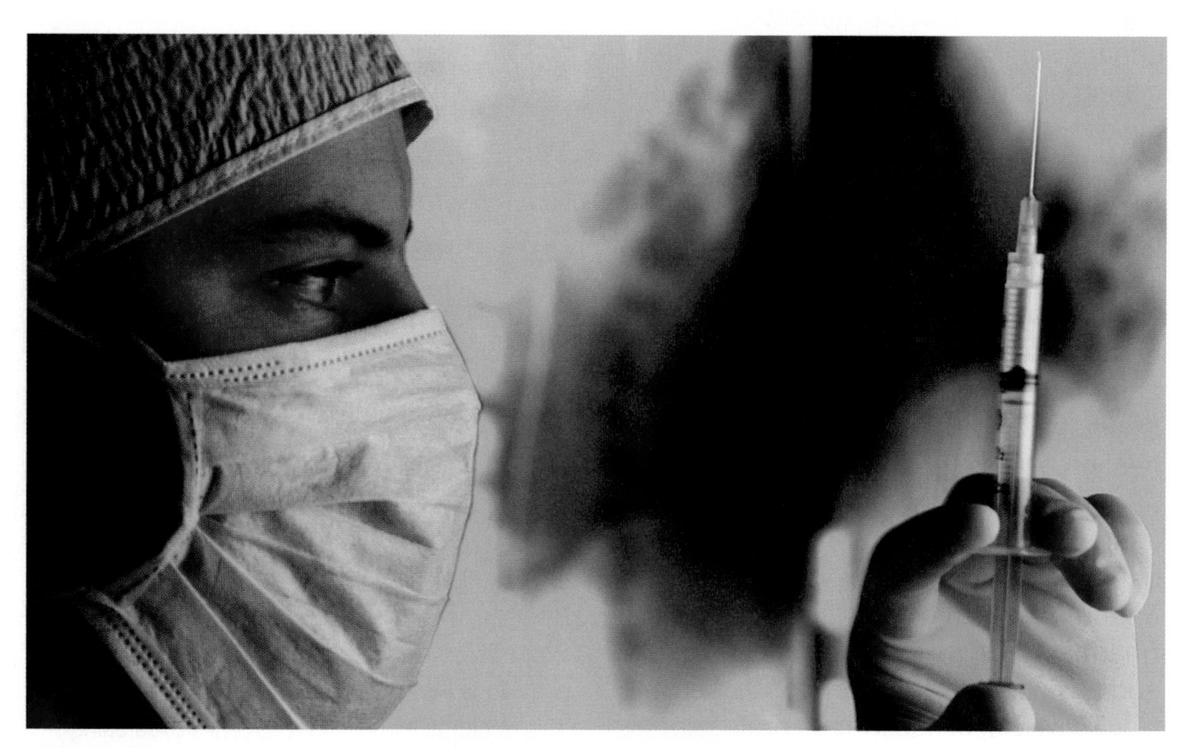

Figure 24-9. Inoculations are given using technological devices.

The vaccines developed for use in immunization require specific equipment to sustain settings in which ample quantities of the vaccines can be created. The technological system designed to produce the right environment in which a vaccine can be cultured is vital to the success of the large amount of the vaccine required for immunization. Increasing the production of a vaccine requires understanding how an organism is modified to produce a vaccine and how a vaccine works. Augmenting the production also requires addressing the quantity needed for all concerned and providing sufficient resources for proper production of the vaccine.

Dealing with Illness and Injury

The second focus of health and medicine deals with people having diseases or injuries. This area of medicine involves diagnosing and treating these diseases and injuries. The goal of this area is to reduce human suffering and physical disability.

A disease can be described as any change interfering with the normal functioning of the body. Treating diseases and injuries requires a number of different health-care professionals. These professionals include physicians. Physicians diagnose and treat diseases and injuries. These professionals also include nurses, who help physicians in their work, and medical technologists, who gather and analyze specimens to assist physicians in diagnosis and treatment. Dentists are health-care professionals who diagnose, treat, and help prevent diseases of the teeth and gums. Dental hygienists assist dentists in surgery and clean teeth. Also on the health-care team are pharmacists. Pharmacists dispense prescription drugs and advise people on the drugs' uses.

This team of health-care professionals uses technology to make its work more effective. Many people seek medical care because they are ill or injured. To help these people, health-care professionals do two major things:

- **Diagnose.** The professionals determine medical problems. They try to establish why the people are ill or what injuries they have. Diagnosis is done using interviews, physical examinations, and medical tests. See Figure 24-10.
- **Treat.** Health-care professionals use medical procedures to cure diseases, heal injuries, and ease symptoms. Treatment can involve the use of surgery, drugs, or other procedures.

Using Technology in Medicine

Diagnosing and treating illnesses and injuries involve tools and equipment. Physicians and dentists use technology to

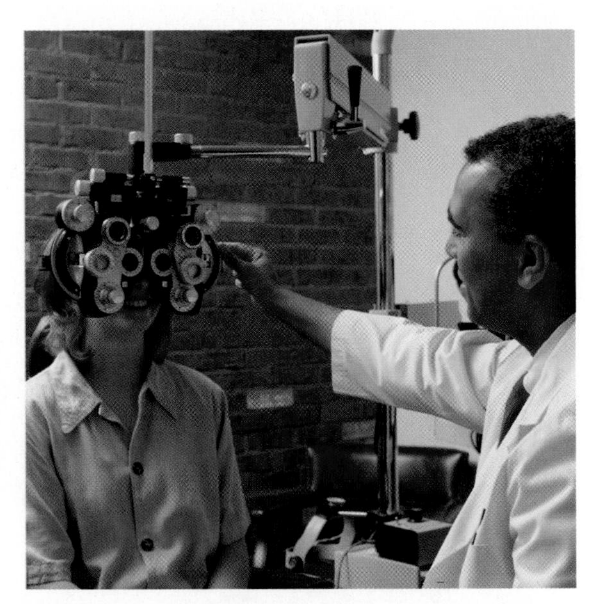

Figure 24-10. Physicians diagnose medical conditions through interviews and examinations. This doctor is examining a patient's eyes using special equipment.

extend the potential to deal with medical problems. Many different types of technology have been developed over the years.

Technology and Diagnosis

In times past, physicians had to depend on people to describe their symptoms. Often, these descriptions were not accurate. Many were hard to understand. The descriptions were used, however, to plan treatments. At times, this led to an inability to cure the diseases or treat the injuries. To deal with these problems, people saw a need for diagnostic equipment. For this discussion, routine, noninvasive, and invasive diagnostic equipment is discussed.

Routine Diagnostic Technologies

Routine diagnostic equipment is used to gather basic information about the patient's condition and general health. See Figure 24-11. This equipment provides a baseline of general information about the patient. Routine diagnostic equipment often includes the following:

- Thermometers to determine body temperature.
- Scales to measure body weight.
- Devices to measure blood pressure.
- Stethoscopes to listen to heartbeats and lung conditions.

Noninvasive Diagnostic Technologies

Noninvasive diagnostic equipment is used to gather information about the patient without entering the body. A typical example of this type of diagnostic is called ***radiology***. Radiology uses electromagnetic waves and high frequency sounds, or ***ultrasonics***, to diagnose diseases and injuries. Diagnostic radiology uses special equipment called *body scanners*. Body scanners produce images of the body without entering it. Examples of body-scanning equipment include the following:

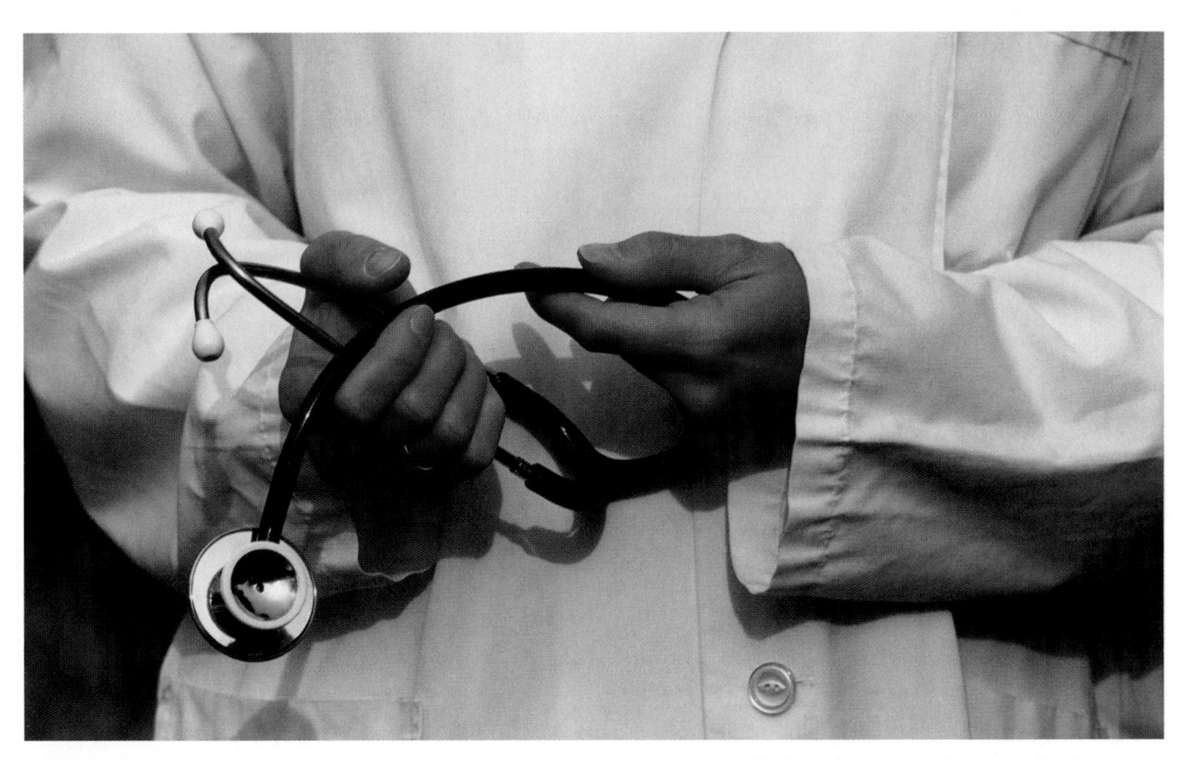

Figure 24-11. Doctors use general diagnostic equipment to start finding the cause for a medical condition. This photo shows a stethoscope.

- ***X-ray machines.*** An X-ray machine is essentially a camera. See Figure 24-12. This machine uses X rays, instead of visible light, to expose the film. These X rays are short electromagnetic waves that can pass through solid materials, such as human tissue. Denser materials, however, such as human bones, absorb some of or all the waves. Assume someone puts a piece of film under your hand and then passes X rays through your hand. Your skin and tissue let most of the X rays pass through them. The film behind that part of your hand is almost completely exposed. The bones in your hand, however, absorb most of the X rays. The film behind them is not exposed completely. When the film is developed, an image of the bones in the hand appears. Any fractures or joint deformities are shown.
- ***Computerized tomography (CT) scanner,* or computerized axial tomography (CAT) scanner.** A major disadvantage of X rays is that the image is flat. The image shows the body in two dimensions. To deal with this shortcoming, CAT scanners have been developed. The scanner sends a thin X-ray beam as it rotates around the patient's body. See Figure 24-13. Crystals opposite the beam pick up and record the absorption rates of the bone and tissue. A computer processes the data and creates a cross-sectional image of the part of the body being scanned.

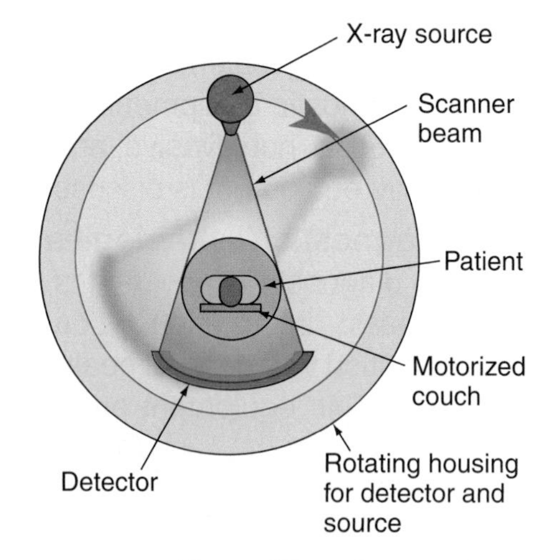

Figure 24-13. How a CT machine works.

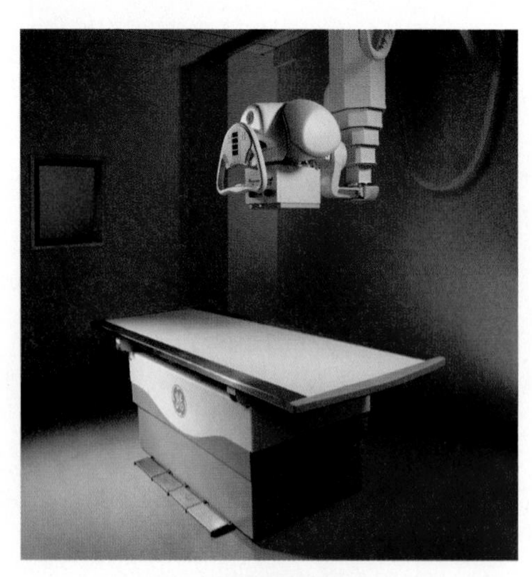

An X-Ray Machine

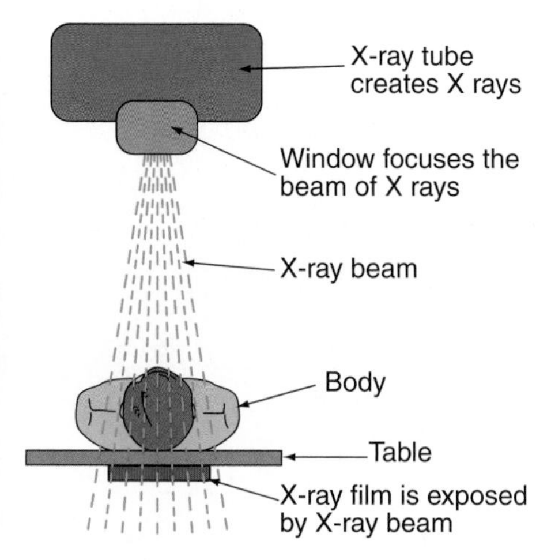

The X-Ray Machine's Operation

Figure 24-12. The X-ray machine uses electromagnetic waves to create an image of a body part. The drawing on the right shows the machine's operation. (GE Medical Systems)

Technology Explained

canal lock: a device allowing ships to change elevation as they travel through a canal.

Canals are part of some inland-waterway transportation systems. They must have three features. First, there must be an adequate source of water to keep them full. The canals can be fed from rivers or reservoirs. Second, there must be a way to cross ravines and streams. A “water bridge” called an *aqueduct* is used. Third, there must be a method for raising or lowering a ship to different canal levels, as the elevation of the land changes. A device called a *lock* fills this need. See **Figure A.**

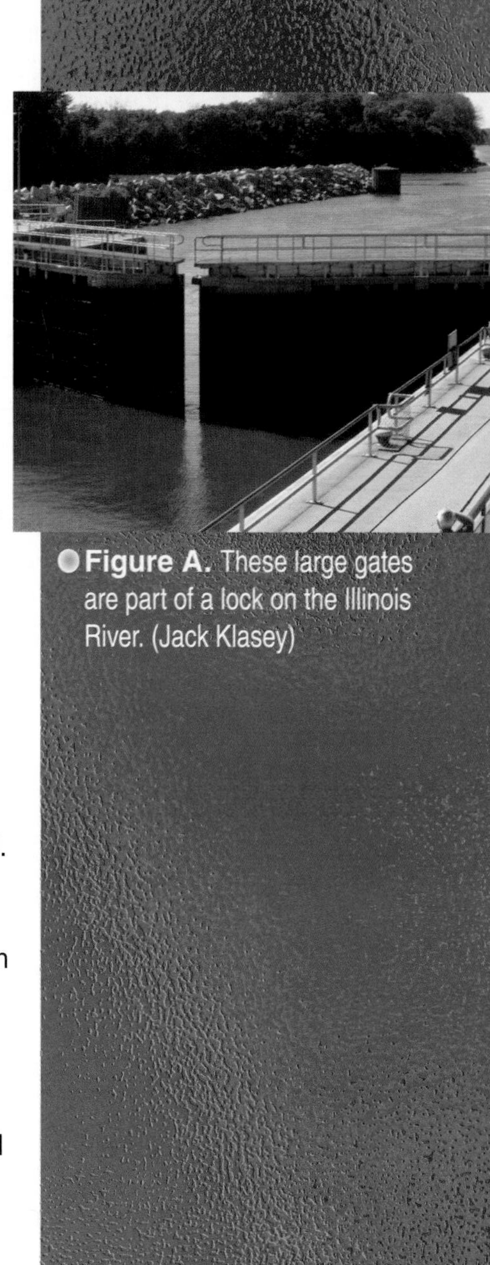

Figure A. These large gates are part of a lock on the Illinois River. (Jack Klasey)

A canal lock is essentially a trough with a gate at each end. The operation of the lock is simple. **Figure B** shows how a boat moves from a higher elevation to a lower elevation in a lock.

First, the lock is filled to the upper water level. The upper gate is opened. The boat enters the lock. Second, the upper gate is closed to seal the lock. Third, sluice gates are opened in or around the lower gates. The sluice gates allow water to flow out of the lock until the level of water inside the lock is equal to the lower level. Fourth, the lower gates are opened, letting the boat sail out of the lock.

To move a boat up to a higher elevation, the procedure is reversed. The boat enters the lock when the water level equals the lower level. The lock is then closed. The sluice gates are shut. The lock fills, lifting the boat to the upper level. The upper gate is opened. The boat sails out of the lock.

The Saint Lawrence Seaway is an inland waterway that has several canal sections. The seaway rises more than 180′ (54 m) from Montreal to Lake Ontario, through two lakes, three separate canals, and seven locks. Once the seaway reaches the Great Lakes, ships pass through Lake Ontario to the Welland Ship Canal. This canal has eight locks to raise ships to the Lake Erie level. The elevation difference between the two lakes is about 330′ (100 m). The channel from Lake Erie to Lake Huron and on to Lake Michigan is the same elevation. No locks are needed. Five more locks are required, however, to raise ships 22′ (7 m), up to the level of Lake Superior.

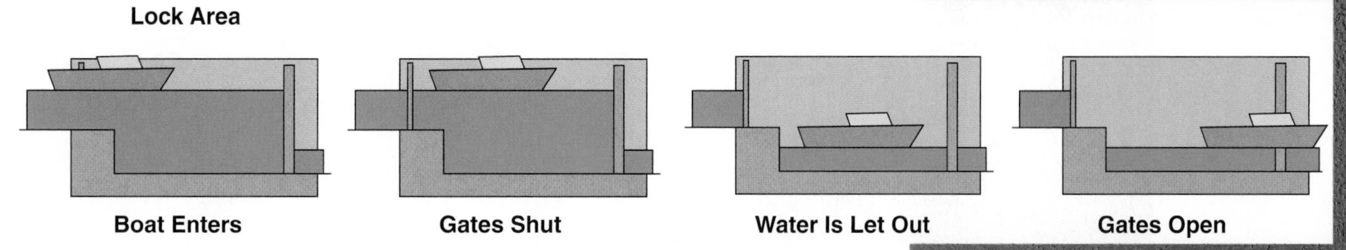

Figure B. How a canal lock works. The boat enters the lock. The upper gates are shut. Water is let out of the lock until the boat is even with the lower water level. The lower gates open. The boat leaves the lock.

- ***Magnetic resonance imaging (MRI).*** X rays can cause damage to body parts. To deal with this hazard, a new imaging technique was developed. This is called *MRI.* See Figure 24-14. This technique uses magnetic waves, rather than X rays, to create the image. MRI can produce a computer-developed cross-sectional image of any part of the body very quickly.
- **Ultrasound.** This technique uses high frequency sound waves and their echoes to develop an image of the body. The ultrasound machine subjects the body to high frequency sound pulses, using a probe. The sound waves travel into your body. There, they hit a boundary between tissue and bone or tissue and fluid. Some of the sound waves are reflected back to the probe. Others travel further, until they reach another boundary and get reflected. A computer processes the reflected sound waves to produce still or moving images.

Not all diagnostic activities involve imaging equipment. There are other technological devices used in diagnosis. One important nonimaging diagnostic device is the ***electrocardiograph (EKG) machine***. See Figure 24-15. This device is used to produce a visual record of the heart's electrical activity. As the heart works, it sends off very small electrical signals. These signals can be detected on the skin. Electrodes are attached to selected locations on the body. These electrodes capture the signals. The EKG machine amplifies the signals. This machine produces a graph of their values. A physician can read this graph to determine how the heart is functioning.

Another important diagnostic device is the ***endoscope***. This device allows a physician to actually look inside the body. An endoscope is a narrow, flexible tube containing a number of fiber-optic fibers smaller in diameter than a human hair. The

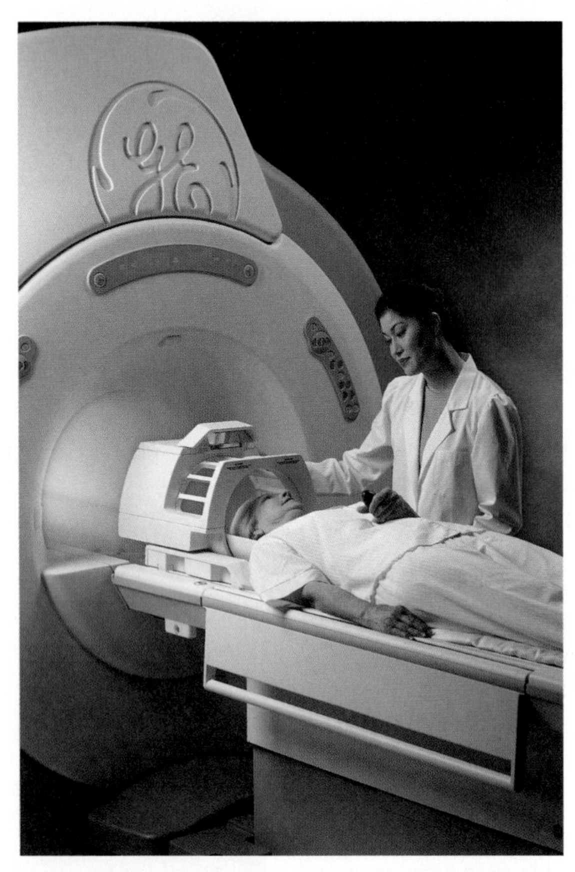

Figure 24-14. An MRI machine in use. (GE Medical Systems)

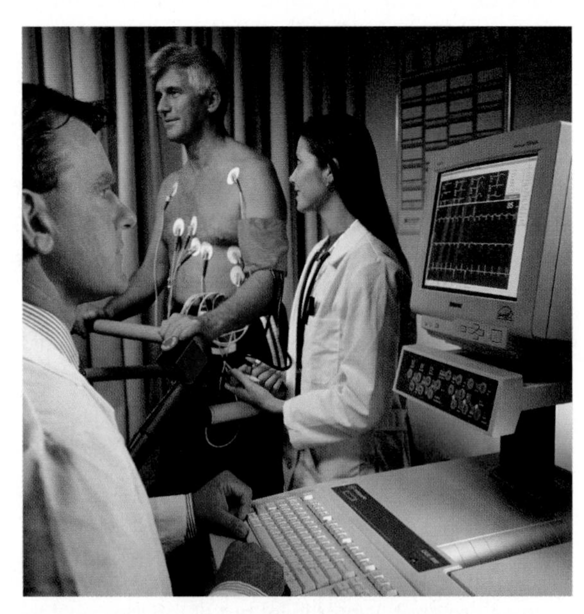

Figure 24-15. An EKG machine senses and records heart actions. (GE Medical Systems)

tube can be threaded through a natural opening, such as the throat, or through a small incision. Light is sent through the fibers. This light shines on an interior part of the body and is reflected back through the fibers. The reflected light forms a series of dots. Each fiber in the tube produces one dot. These dots form a picture of an internal organ or another part of the body. These examples are just a few of the many devices designed and built to help diagnose illnesses and physical conditions. They are examples of the dramatic uses of technology to help reduce human suffering.

Invasive Diagnostic Technologies

Invasive diagnostic technologies involve removing tissue or fluids from the body for analysis. These approaches might include drawing and testing blood samples. They might also include taking tissue samples (biopsies) for laboratory examination.

A blood test is a chemical analysis of a sample of blood. The sample is tested in a laboratory, using a number of different technological procedures. These procedures identify the composition of the blood. The tests can determine the presence of specific chemicals associated with a disease. They can also detect imbalances in the chemical composition of the blood. This data provides health-care professionals with the information needed to treat illnesses or physical conditions.

Many medical conditions, including cancer, are diagnosed by removing a sample of tissue. A ***pathologist*** examines this tissue to determine if it is normal or diseased. See Figure 24-16.

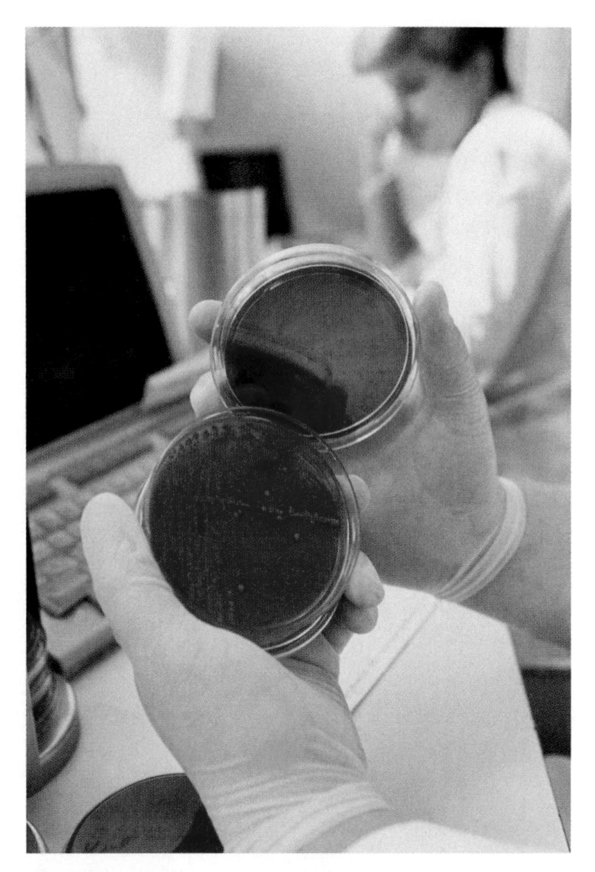

Figure 24-16. This technician is examining a culture that can identify an illness.

Technology and Treatment

Healing of illnesses and physical conditions can require drugs and specialized equipment. Both of these approaches are the results of technology. They are the products of design and production actions.

Treatment can be performed in a physician's office, at an immediate-care (emergency) center, or in a hospital. A hospital is a facility where trained professionals use knowledge and equipment to treat illnesses and injuries. See Figure 24-17. Most hospitals contain a number of special areas:

- **Outpatient treatment center.** This is the area where people are scheduled to receive treatment and return home. Minor surgeries, medical tests and examinations, and other routine treatments are performed here.
- **Emergency room.** This is the area where seriously ill or injured people enter the hospital. Ambulances or family members usually bring these people here.

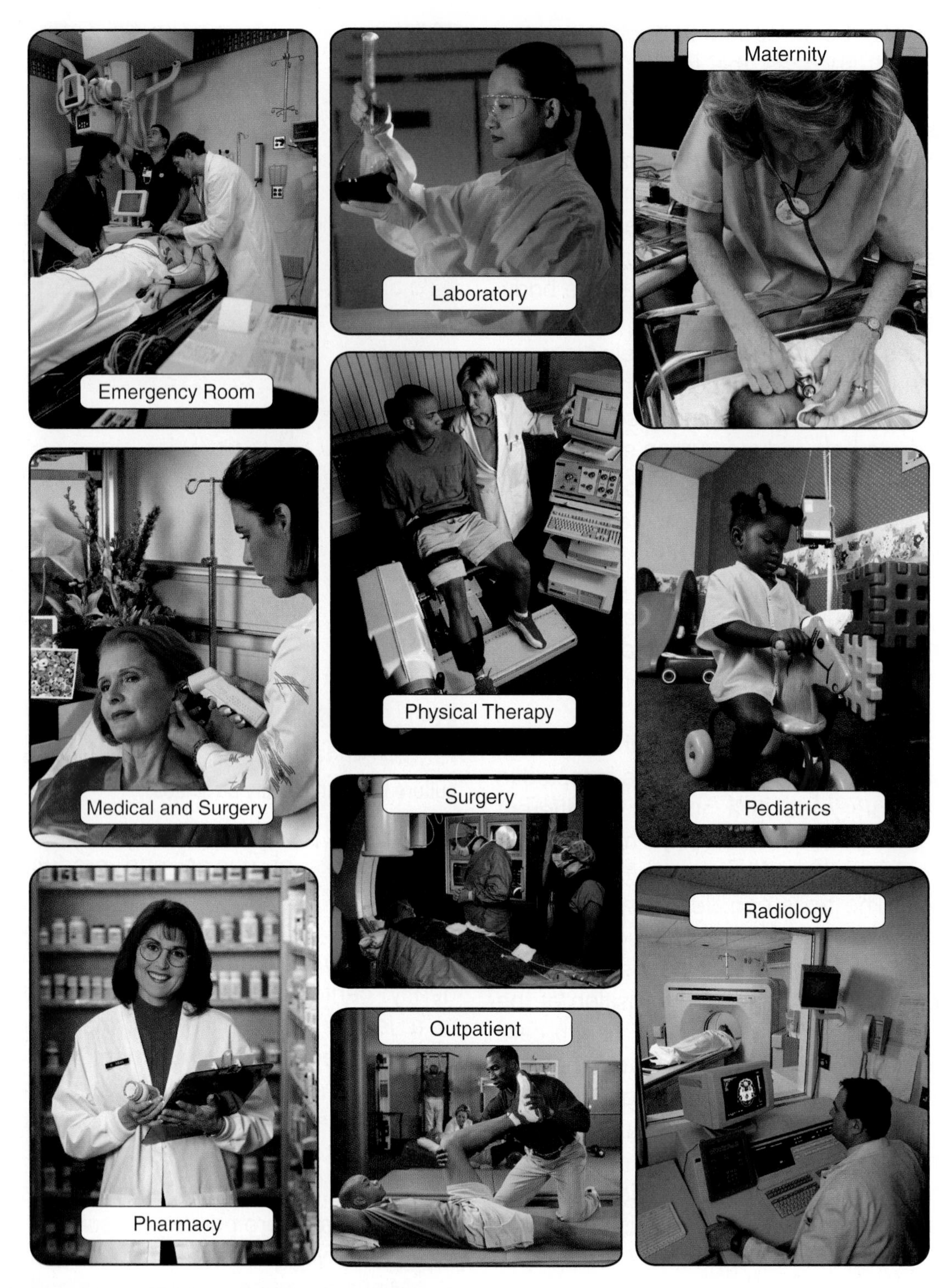

Figure 24-17. These and other areas of a hospital use technology to help patients. Look for all the examples of technology you can see.

- **Operating room.** This is the area where ***surgeries*** are performed. The operating room includes preoperation (pre-op) areas, operating theaters, and postoperation (post-op) recovery areas.
- **Medical and surgical floors.** These are areas of general care for people who are ill or have had surgery.
- **Intensive care unit.** This is the area where seriously ill people receive constant care and monitoring. Often, heart attack and stroke victims and people in serious accidents are placed in ***intensive care***.
- **Pediatric floor.** This is the area where sick and injured children receive general care.
- **Maternity ward.** This is the area where mothers give birth. The maternity ward has sections for caring for both the mothers and newborns.
- **Physical therapy room.** This is the area where people receive treatment to strengthen muscles. Treatment for loss of mobility and pain is received here.
- **Pharmacy.** This is the area for preparing and storing all drugs used in the hospital.
- **Radiology unit.** This is the area where various imaging equipment is used. In this area, the images are also read.
- **Pathology unit.** This is the area where blood and tissue samples are taken and analyzed.

Each of these areas in a typical hospital uses modern equipment and techniques. These areas all depend on technology to treat patients.

Treatment with Medical Equipment

The equipment used to treat patients comes in many shapes and forms. See Figure 24-18. This equipment includes life-support equipment, such as cardiac pacemakers, defibrillators, and artificial kidneys; computer systems to monitor patients during surgery and in intensive care; instruments and devices, such as laser systems for eye surgery and catheters to open blocked blood vessels; and radiology-treatment systems for cancers and other growths.

A description of all the technological devices used to treat diseases would fill books. For this chapter, two major devices are discussed. These provide a quick view of technology as it is applied to treating injuries and illnesses.

Radiology treatment is used for types of cancer. This treatment is called *therapeutic radiology.* Therapeutic radiology uses high-energy radiation to destroy the cancer cells' ability to reproduce. This

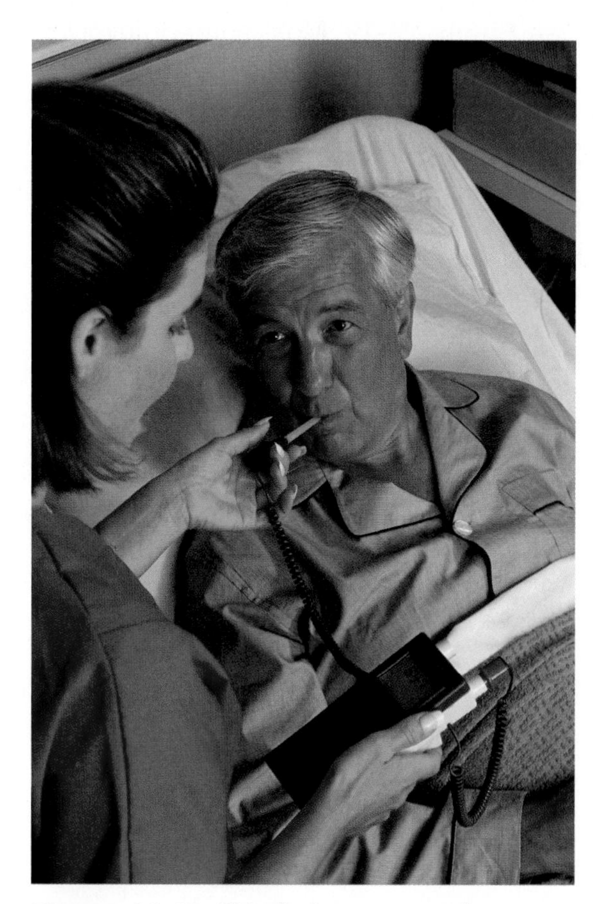

Figure 24-18. This device measures the temperature of the patient.

radiology works because normal cells can recover from the effects of radiation better than cancer cells can.

Sometimes, radiation therapy is only part of a patient's treatment. Patients can be treated with radiation therapy and chemotherapy (chemicals or drugs). Surgery might follow these treatments.

Radiology is also used as a nonsurgical treatment for a number of ailments. This use is called *intervention radiology.* The images that the radiology equipment produces allow the physician to guide catheters (hollow, flexible tubes), balloons, and other tiny instruments through blood vessels and organs. An example of this approach is balloon angioplasty. Angioplasty uses a balloon to open blocked arteries.

Surgery is a very common way to treat diseases and injuries. Operations can be used to remove diseased organs, repair broken bones, and stop bleeding. See Figure 24-19. Most surgery involves manually removing diseased tissue and organs. New technologies, however, are being used for many types of surgery.

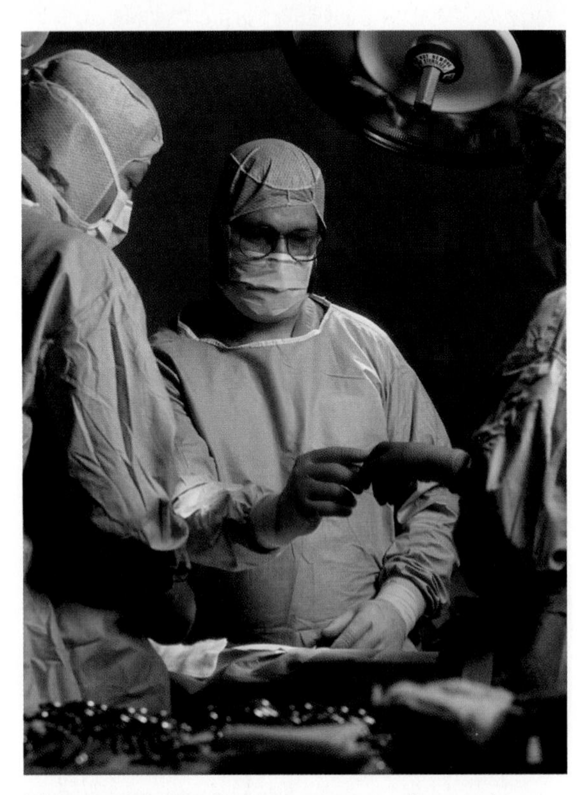

Figure 24-19. Surgery removes diseased tissues or repairs body damage.

High frequency sound waves can be used to break up kidney stones. Lasers use a beam of light to vaporize or destroy tissue. Transplant surgery allows organs removed from one person to be implanted into another person. Also, devices, such as pacemakers, can be implanted.

To prevent diseases from spreading, it is important for all medical equipment to be kept sanitary. Sanitation processes used in the disposal of medical products help to guard people from dangerous organisms and illnesses. These processes shape the principles of medical safety. Appropriate use and management of harmful materials help to shield people from avoidable harm and also help to ensure safe environments.

Treatment with Drugs and Vaccines

Humans have always experimented with substances to treat pain and illness and restore health. These substances are called *drugs.* They are any substances used to prevent, diagnose, or treat diseases. Drugs can also be used to prolong the lives of patients with incurable conditions. A special type of drug is called a *vaccine.* See Figure 24-20. Vaccines are substances administered to stimulate the immune system to produce antibodies against a disease. These substances have helped eliminate diseases, such as measles, whooping cough, and mumps.

Most modern drugs are the products of chemical laboratories. See Figure 24-21. These drugs are called *synthetic drugs.* Using gene-splicing or recombinant DNA has developed a number of new drugs. Genetic engineering entails altering the structure of DNA to create new genetic makeups. This approach joins the DNA of a selected human cell to the DNA of a second organism, such as a harmless bacterium. The new organism can produce the disease-fighting substance

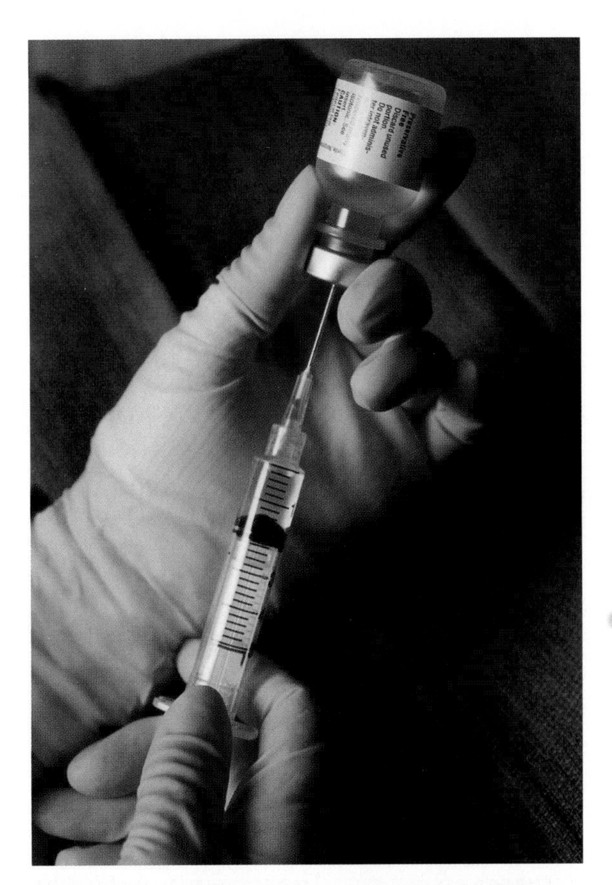

Figure 24-20. A vaccine helps prevent illness.

Figure 24-21. Drugs are developed through chemistry and technology.

that the original human cell could produce. This new substance is extracted from the bacterium and processed into a drug. Genetic engineering is done in a laboratory, using reagents and other tools allowing researchers to make controlled variations in genetic information and structure.

The drug-development process generally starts with a need to treat a disease or physical condition. Researchers start looking for a chemical substance that might have some medical value. These researchers might work with thousands of different substances before they find one that can serve as a drug.

Once a substance that might have medical value is discovered, an extensive testing program starts. See Figure 24-22. The first tests are performed on small animals, such as rats and mice. If the tests are successful, additional tests are conducted on larger animals, such as dogs and monkeys.

The tests are carefully evaluated to see if the drug treats the disease or physical condition. The drug's toxicity (capability to poison a person) is also evaluated. Obviously, a drug must be effective and have low toxicity before it can be used on people. At this point, a request is made to the FDA to conduct ***clinical tests***. These tests are generally conducted in three phases. These phases take many months to complete.

During the first phase, the drug is given to a small number of healthy individuals. These tests are designed to determine the

Figure 24-22. The steps in testing a new drug.

drug's effects on people. If the drug passes this test, it moves into the next phase.

In the second phase, the drug is given to a small number of people who have the disease or physical condition the drug is intended to treat. The test subjects are divided into two groups. The first group is given the drug. The other group is given an inert substance, such as sugar. This inactive compound is called a *placebo*. The test determines if the group receiving the drug fares better than the group not receiving the drug (the placebo group).

During the final phase of testing, the drug is given to a much larger group of people. This test is used to determine dose levels, side effects, and interactions with other drugs. The results of all the tests are submitted to the FDA for approval. The agency weighs the drug's benefits against any risks that might be present. The FDA must decide if the drug is effective and safe. If the FDA determines that the drug meets its criteria, it approves the drug for use.

Summary

Medicine and medical professionals treat ill and injured people. These professionals provide illness-prevention, diagnosis, and treatment programs. They help people prevent illness by promoting good diet and proper exercise. Medical professionals diagnose illnesses and injury using imaging and other equipment. They treat illnesses using equipment and drugs developed through technological actions. By using technology, health-care professionals help people live better lives with reduced levels of pain, injury, and illness.

STEM Connections

Mathematics

Do a series of physical exercises while you are wearing a heart-rate monitor. Graph the results of the exercise over time.

Science

Select a technological device used to treat illnesses. Describe the scientific principles (such as optics or radiation) the device uses.

Curricular Connections

Social Studies

Investigate a major disease and describe the drugs and technological devices that have been developed to treat it. Try to determine by whom and where each development was made.

Activities

1. List five exercises you can do or sports you can play. List the equipment (technological products) required for each.
2. Visit a hospital, nursing home, or retirement center. List all the technological devices you see used to treat illnesses or physical conditions.
3. Prepare a display or poster explaining how a major medical device works. Include the inventor, the operating principles, and the device's uses in the presentation.

Test Your Knowledge

Do not write in this book. Place your answers to this test on a separate sheet of paper.

1. What are capillaries?
2. Construct a simple timeline of milestones in the history of medical technology.
3. The three goals of medicine are to _____, _____, and _____ diseases and injuries.
4. Paraphrase the saying "An ounce of prevention is worth a pound of cure."
5. Give two examples of ways to prevent illness and injury.
6. List five types of exercise machines or devices.
7. Technology is used in sports to construct game _____; _____ equipment and clothing; and playing fields, or _____.
8. Describe the procedure of immunization.
9. Drugs promoting natural resistance to diseases are called _____.
10. Determining the cause of a medical problem is called _____.
11. Procedures curing medical conditions are called _____.
12. _____ use short electromagnetic waves to expose a film.
13. An MRI machine uses _____ waves to create an image of a body part.
14. Removing diseased tissue is called _____.
15. _____ are substances used to treat pain and illness.

Reading Organizer

On a separate sheet of paper, create a detailed outline based on what you've read about medical technology.

Example:

I. Prevention programs
   A. Wellness programs
      1. Exercise
      2. Sports

CHAPTER 24 ACTIVITY A

MEDICAL TECHNOLOGY

THE CHALLENGE

Develop a device that allows people to pick objects up off the floor or ground without bending over.

INTRODUCTION

Some older people and many individuals who suffer from arthritis, back problems, and other conditions find it very difficult to bend over to pick objects up off the ground. See Figure 24A-1. Bioengineering and technology can address this problem. Devices can be developed to allow people to grasp objects at their feet without bending over.

Figure 24A-1. Some people find it extremely hard to bend over to pick items up off the floor.

DESIGN BRIEF

Develop a device or system people can use to grasp and lift objects at floor level without bending over, using the following materials:

- Wood strips.
- A small-diameter metal rod or wire.
- Masking or duct tape.

- A garden hose.
- Juice cans.
- String.
- Wood glue.
- Wire nails or brads.
- Bolts
- Nuts.
- Poster board.

PROCEDURE

Address the design challenge by completing the following steps:
1. List the design limitations.
2. Develop several possible solutions to the problem.
3. Select a promising solution.
4. Refine the solution.
5. Build a prototype of the solution.
6. Test and evaluate the solution.
7. Indicate ways to improve the solution.

CHALLENGING YOUR LEARNING

What are the strengths and weaknesses of your product? How could you change the design to make it better?

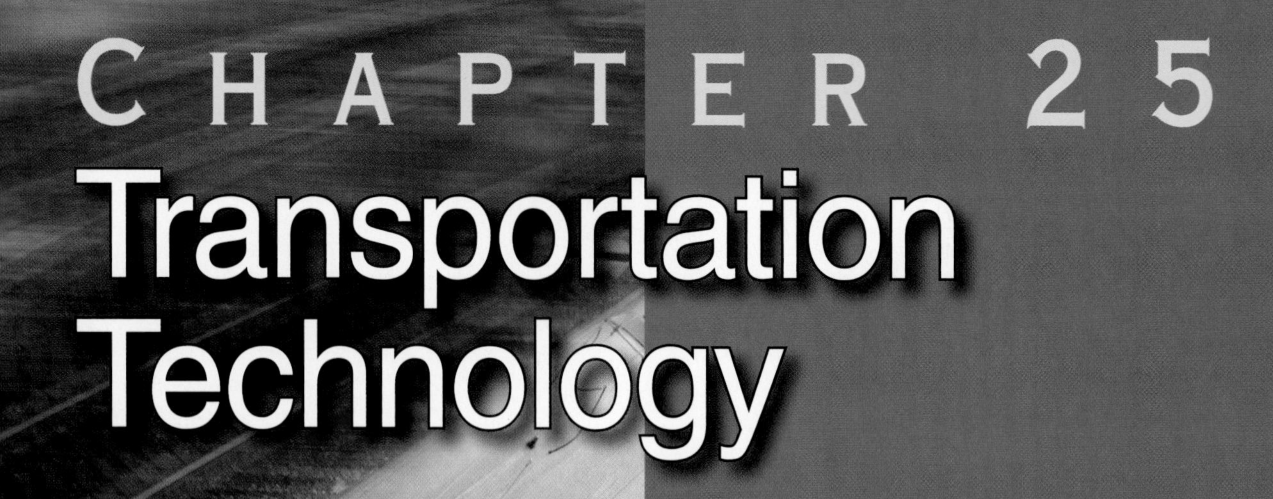

CHAPTER 25
Transportation Technology

DID YOU KNOW?

- The first human-powered aircraft was the *Gossamer Condor* airplane. This aircraft flew a figure eight course in 1977.
- The tunnel connecting England with the rest of Europe is designed for trains only. Special trains are designed to carry cars and trucks. Passenger lounges carry people.
- In Europe, trains that have hydraulic tilting mechanisms help them make curves at high speed. In Italy, the tilting train, called the *Pendolino*, can travel at 156 mph (250 kilometers per hour).
- Synthetic automobile fuel made from coal is produced in South Africa.

Objectives

The information given in this chapter will help you do the following:

- Recall some important events in the history of transportation.
- Explain what a transportation system is.
- Summarize the three types of transportation systems.
- Explain the six main processes of transportation.
- Give examples of the four modes of travel.
- Recall the meaning of intermodal transportation.
- Summarize the two main parts of a transportation system.
- Summarize the five vehicular systems.
- Explain the functions of support facilities in transportation systems.
- Explain how quality control is used to ensure that passengers are transported safely and efficiently.

Key Words

These words are used in this chapter. Do you know what they mean?

air transportation
cargo
control
conveyor
ferry
freighter
guidance
intermodal transportation
land transportation
loading
ocean liner
pipeline
propulsion
roadway
routing
scheduling
shipping lane
space transportation
storage
structure
suspension
tanker
terminal
transportation
tugboat
unit train
unloading
water transportation

Preparing to Read

Think of the four different modes of transportation available. As you read the chapter, write examples of each mode. Use the Reading Organizer at the end of the chapter to organize your thoughts.

Many people think nothing of jumping into a car and going to the supermarket or mall. Other people catch planes to faraway places. Transportation was not always so easy. In the earliest times, people had to walk everywhere they went. To improve transportation, people started to develop new ways to move loads. They developed sleds, rafts, and other crude devices. See Figure 25-1. These people started to use pack animals to carry loads. Later, they developed canoes and other boats. By 4000 BC, the sailboat was in use. About 5000 years ago, the wheel was invented. See Figure 25-2. These devices laid the foundation for new land-transportation systems. People developed ships to carry themselves on the water and wagons to move loads on land. By the 1400s, sailing ships were used to explore the world. A steam-power carriage was developed in 1769. The Montgolfier brothers flew the

Figure 25-1. This sled, or stoneboat, is based on the same principle as the earliest sleds. The stoneboat was used to move rocks from farmers' fields.

first hot air balloon in 1783. The first railroad locomotive was developed in the early 1800s. Rail lines were used for streetcars during the same period. In 1885, Karl Benz built the first practical automobile. The Wright brothers flew their airplane in 1903. In 1926, Robert Goddard flew the first liquid-fuel rocket. The first commercial jet, the *Comet*, flew in 1949. The Russians launched *Sputnik*, the first human-made satellite, in 1957. Neil Armstrong became the first man to walk on the Moon in 1969. The space shuttle started flying in 1981. As you can see, transportation has advanced greatly over time. Much of this change has come in the last 250 years.

What Is Transportation?

We all use transportation devices and systems. Transportation systems move people or cargo from an origin (a starting point) to a destination (an ending point). Most transportation systems include a way to carry a load and a dedicated pathway. See Figure 25-3. The load is carried in a vehicle (such as a car or plane) or fixed container (such as a pipeline). The path might be a permanent structure (for

Figure 25-3. Notice the load carriers (the train cars) and the pathway (the rails).

A Wheel

A Wagon

Figure 25-2. This wheel was made from a cross section of a log. An iron band was placed around it. The wheel was used on an early logging wagon in Oregon.

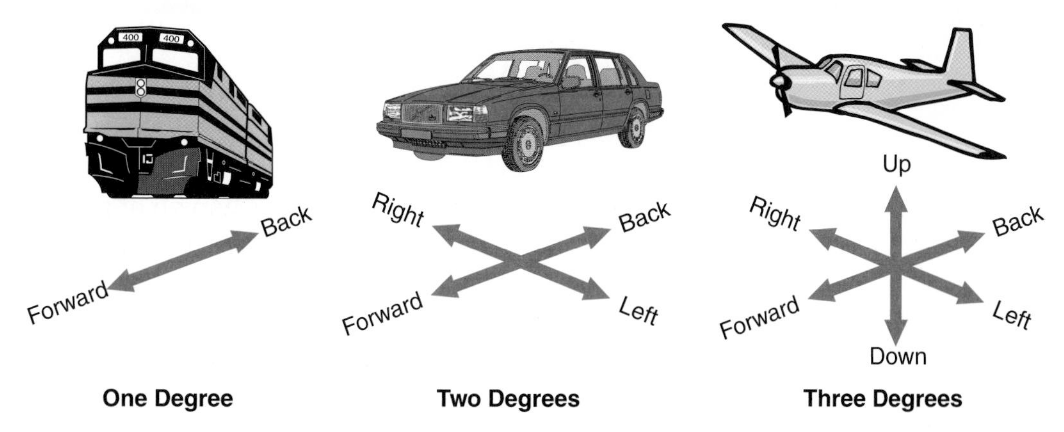

Figure 25-4. The degrees of freedom to operate for various vehicles.

example, a roadway or railroad) or an invisible path (for example, an air route). Transporting people and goods requires a combination of individuals and vehicles.

While carrying a load, the carrier (vehicle) has degrees of freedom to move people and cargo. See Figure 25-4. For example, a pipeline or railroad track has one degree of freedom. These carriers travel in a fixed path from the origin to the destination. Once the material is entered into the system, this kind of vehicle can only go one way. By contrast, cars, boats, and bicycles have two degrees of freedom. They can go forward or backward. These vehicles can also turn to the left or right. The operators can decide to travel a number of different routes on the land or water. These vehicles cannot, however, go up or down, as an airplane can. Aircraft and spacecraft have three degrees of freedom: forward or back, left or right, and up or down. They can travel in all directions and many altitudes as they move between points.

Types of Transportation

Over the years, three types of ***transportation*** have evolved: personal, public, and commercial transportation systems. See Figure 25-5. Personal

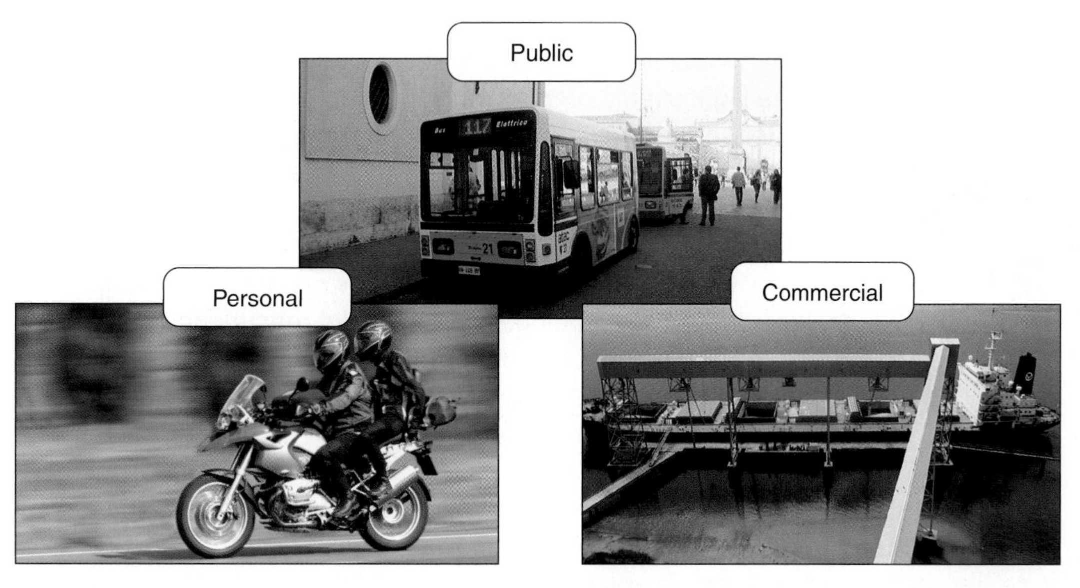

Figure 25-5. Types of transportation systems. (©iStockphoto.com/M_D_A)

Technology Explained

flight simulator: a computer-controlled device allowing pilots to practice flying an airplane without actually using a plane.

The first time you rode a bicycle, you did not hop on and ride away. You had to learn how to balance the bicycle and pedal. Before you mastered these skills, you might have fallen over a few times. You made mistakes as you practiced how to ride. Making mistakes with a bicycle was not a big deal. You got up, dusted yourself off, picked up your bike, and tried again.

Pilots who are learning to fly also need to practice. Making a mistake with an airplane, however, can be dangerous and costly. In order to practice without risk, pilots use simulators. Simulators range from simple training devices for small aircraft to sophisticated units for large airliners. See Figure A. We will discuss large simulators here.

A simulator mimics the sounds, motions, and views from an airplane in flight. Pilots can improve their skills and learn to fly new models of aircraft. A trainee pilot can develop the ability to take directions from flight controllers and practice takeoffs, cruising, and landings. Emergency procedures can be practiced over and over. The safety of the pilot, the public, and a very expensive aircraft is not put at risk.

A simulator has several major systems. The first is the control system. The instructor uses the control system's computer to set up a desired flying condition.

The second is a capsule-motion system. See Figure B. This is a series of hydraulic cylinders moving the capsule. The movements of the capsule imitate the motions of an aircraft in a particular situation. The next part is an audiovisual system. The audiovisual system uses projectors, screens, and speakers. This system conveys to the pilots what they would see and hear if they were flying real airplanes. Finally, the simulator has a set of controls and instruments very similar to those in a real aircraft. The pilot uses them to "fly" the simulator. They are connected to the control-system computer.

The flight simulator uses a closed loop control system. The computer sets up a situation. This control-system computer directs the audiovisual system to show the pilots what they would see and hear in a real situation. The computer also directs the motion system to provide the physical feel of the situation. The control-system computer causes the instruments to show readings that would be expected in the situation. When the pilot reacts, the computer makes the simulator act similarly to the way a real airplane would. The computer changes the view, sounds, instrument readings, and motions to simulate the actions of an aircraft. If the pilot makes a mistake, the computer can simulate an emergency. The pilot can practice how to get the aircraft back to a safe condition. No one is injured. The pilot can find out how to avoid a crisis.

Figure A. This simulator is used to train pilots to fly the Boeing 767 aircraft.

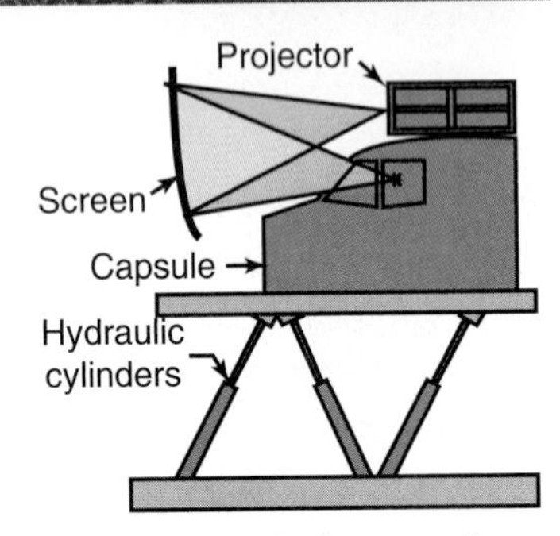

The Capsule's Suspension and Movement

The Capsule's Instruments and Controls

Figure B. The capsule is suspended on hydraulic cylinders and moves to mimic the motions of an aircraft. This compartment contains a complete set of instruments and controls.

transportation involves people or families traveling in their own vehicles. Driving to work and riding a bicycle to the store are examples of personal transportation. Public transportation involves systems that governmental agencies operate. This includes riding on buses and commuter trains built with public (tax) money. Profit-making companies operate commercial transportation systems. These systems provide transportation services in the hopes of making money. Airlines, trucking companies, and taxi services are examples of commercial transportation systems.

Transportation Processes

Transportation involves a number of different processes, or actions. The major processes are routing, scheduling, loading, moving, unloading, and storing. See Figure 25-6. These processes are vital for the whole transportation system to operate efficiently. They can be used alone or in a variety of combinations to transfer cargo and people.

Routing determines the path the load travels. For example, people can decide the path they take on a vacation from Chicago to Los Angeles. An airline develops a route map for its operations. This system plans the origin and destination for each flight in its daily schedule.

Scheduling is assigning a time for the travel. Again, the person on vacation decides how many miles to travel each day. Meal and lodging stops are determined. In the airline example, each flight is assigned a departure time and an arrival time.

Loading is the physical placement of cargo and people on a vehicle or allowing people to enter the vehicle. This process can take place in a terminal or at some other site. In the vacation example, the family climbs into the family car. The airline loads its passengers at the airport terminal.

Moving involves transporting the people or cargo to the destination. See Figure 25-7. In the vacation example, the driver guides the family car along streets and highways until it reaches the destination. The pilot of the airplane guides the airliner through takeoff, flight, and landing.

Unloading is the opposite action to loading. This action allows passengers to disembark at the destination. See Figure 25-8. ***Cargo*** is removed from the vehicle. The family on vacation gets out of the car at the vacation spot. The airline passengers leave the airplane at the destination terminal.

Storage is required in some transportation acts. A container can be loaded and stored until it is scheduled to leave a port. Bulk materials can be stored until a ship arrives to receive the cargo. Products can be placed in a warehouse so they are secure.

Routing
Scheduling
Loading
Moving
Unloading
Storing

Figure 25-6. The processes of transportation.

Figure 25-7. This gondolier propels the craft with cargo along a route.

Figure 25-8. These students are getting on a school bus (a public transportation vehicle). (©iStockphoto.com/LUGO)

Modes of Transportation

There are several ways of moving people and goods from one point to another. For example, you can move a load of computers on a truck. This load can be sent on a ship. The computers can also be flown to the destination. The way the computers are moved is called the *mode of travel.* The modes of travel include the following:

- **Land.** This is using transportation means on the surface of or below the earth to move people and cargo.
- **Water.** This is using transportation means on or below the water to move people and cargo.
- **Air.** This is using transportation means traveling though the atmosphere to move people and cargo.
- **Space.** This is using transportation means traveling beyond the atmosphere to move people and cargo.

Land Transportation

Transportation systems operating on or beneath the earth's surface are known as ***land transportation***. They move over or through constructed pathways. These systems include the following types:

- **Fixed-path systems.** These systems have one degree of freedom. They move from one origin to one destination. Fixed-path systems include railroads, pipelines, and conveyors. They also include on-site systems, such as elevators, moving sidewalks, and escalators.

- **Variable-path systems.** These systems use vehicles that can be guided through two degrees of freedom. These vehicles include automobiles, bicycles, buses, trucks, forklifts, and motorcycles.

Fixed-Path Transportation

Fixed-path systems provide a single path from the origin point to the destination. These systems include railroads, pipelines, and conveyors. Railroads are efficient systems for moving people and goods. They can carry many passengers and huge amounts of cargo. The vehicles and their tracks are built to carry heavy loads. The trains move cargo quickly, without traffic jams, because schedules control the number of trains on a track.

Cargo is carried on two major types of trains. Freight trains are made up of individual cars carrying many different products and materials. A freight train traveling along its route drops off cars and picks up others along the way. This is usually done in central locations called *freight yards*. A ***unit train*** is used when large quantities of a single material are being hauled. This train moves the cargo from the same origin to the same destination, time after time. For example, coal is moved from Wyoming coal mines to midwestern power plants on unit trains. These trains also move grain from America's heartland to ports along the coasts.

People are moved on two types of trains. Commuter and light-rail trains take people to and from their jobs. Long-distance travelers might ride on standard railroad passenger trains, such as Amtrak® railway cars. See Figure 25-9.

Some materials are moved on systems that do not use vehicles. These systems include pipelines and conveyors. In pipelines, pumps suck or push material through the system. Petroleum and petroleum products, natural gas, and coal are often moved this way. Belt and bucket conveyors are used to move grain, gravel, and wood chips.

Pipelines are generally buried. This conserves valuable land. The cargo they carry encounters little danger from the weather and thieves. The materials being transported will not likely be damaged or contaminated.

Figure 25-9. This modern passenger train can move people between major cities quickly and comfortably.

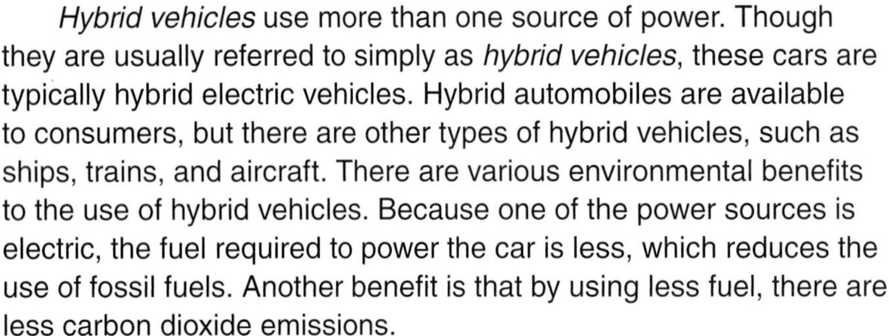

THINK GREEN

Hybrid Vehicles

Hybrid vehicles use more than one source of power. Though they are usually referred to simply as *hybrid vehicles*, these cars are typically hybrid electric vehicles. Hybrid automobiles are available to consumers, but there are other types of hybrid vehicles, such as ships, trains, and aircraft. There are various environmental benefits to the use of hybrid vehicles. Because one of the power sources is electric, the fuel required to power the car is less, which reduces the use of fossil fuels. Another benefit is that by using less fuel, there are less carbon dioxide emissions.

Conveyors are stationary, built-in structures that move materials and products. They are often used in manufacturing. These structures move materials along manufacturing lines. They are used to transmit materials from mine shafts, or pits, to processing operations. Conveyors lift grain in elevators. Special conveyors are also people movers. See **Figure 25-10.** They transport people from one part of a large building to another on moving sidewalks.

Variable-Path Transportation

Cargo and people can be moved in variable-path (steerable) vehicles. These vehicles travel on highways, on streets, and even inside buildings. See **Figure 25-11.**

Figure 25-10. A moving sidewalk, or people mover, can be considered a "people conveyor." (Elke Wetzig)

On a Street

In the Countryside

Inside a Building

Figure 25-11. Variable-path vehicles move cargo and people on streets, in the open countryside, and inside buildings. (Caterpillar, Inc.)

People using these vehicles have flexible schedules and routes. Drivers can move people and cargo day or night. They can take a direct route or an alternate route to avoid traffic. Variable-path vehicles include automobiles, trucks, buses, and taxis. In addition, forklifts and tractors are variable-path vehicles used in factories, on construction sites, and at airports.

Water Transportation

Waterways provide another type of transportation. These waterways can be natural oceans, lakes, or rivers. Some waterways are human-made channels, such as the Suez and Panama Canals. See Figure 25-12.

A Natural Waterway

A Human-Made Waterway

Figure 25-12. Waterways can be natural features, such as oceans, lakes, or rivers, or human made, such as the historic B & O Canal.

Water transportation is generally cheaper than land transportation. This transportation can be used, however, only where rivers, lakes, and other bodies of water are navigable. The bodies of water must be wide enough and deep enough for heavily loaded watercraft to travel on them.

There are two major types of waterways used for transportation. These are oceans and inland waterways. Oceans and seas are the masses of water separating continents and major landmasses. Inland waterways are the rivers and lakes within a landmass.

Ocean transportation carries freight and people on ships. Ships carrying people are called ***ocean liners***. ***Freighters*** carry products and solid materials. Some freighters carry cargo in containers that are easily loaded and unloaded. See Figure 25-13. ***Tankers*** carry liquids such as petroleum and chemicals.

There are no constructed paths or highways on the open seas. Ships do, however, keep to regular routes over the water. These routes are known as *sea-lanes*, or ***shipping lanes***.

A number of vehicles are used for inland shipping. Barges are large floating cargo boxes without engines. See Figure 25-14.

Figure 25-13. This freighter is loaded with containers. (U.S. Department of Agriculture)

Figure 25-14. Barge transportation is a familiar sight on canals and rivers. (U.S. Park Service, Natchez Trace Parkway)

Tugboats, or towboats, are small vessels that move the barges over the water. They can be seen as the "locomotives" of the water. ***Ferries*** are used to move people and vehicles across bodies of water.

Air Transportation

A mode of transportation developed in the last century is ***air transportation***. Air transportation uses aircraft to move people and cargo to their destinations. Aircraft include all vehicles traveling within Earth's atmosphere.

Airplanes are the most commonly used aircraft. These airplanes might be commercial airliners, company planes, or private aircraft. Helicopters can be used for air transportation. They are usually used for short, commuter-type hops.

Airliners are used to move large numbers of people along set routes. Airlines (air-transportation companies) operate them. Businesses operate company planes to move their employees and cargo. Private aircraft are individually

Figure 25-15. Smaller aircraft are used for private and company travel.

owned airplanes used for personal use. They are called *general aviation craft*. See Figure 25-15.

Similar to the oceans, the sky has no highways. To be safe, aircraft follow established routes. In addition, air lanes are at various heights. For example, planes flying west travel at a different height than planes flying east do.

Space Transportation

Space transportation is a very new transportation mode. This transportation includes unmanned and manned flights. Unmanned flights have rockets traveling far into outer space. They are used to explore the universe. Cameras aboard the space vehicles send back photographs of unexplored space.

Manned flights have taken astronauts to the Moon. See Figure 25-16. The space shuttle ferries humans between space stations and Earth.

Figure 25-16. Space transportation has taken people to the Moon. (NASA)

Intermodal Transportation

Often, moving people and goods from one point to another requires more than one transportation mode. For example, travelers might use land transportation to travel from home to the airport in a taxi. These people might fly to a port city by using air transportation. There, they might board a cruise ship for a holiday vacation. In all, they used three modes of transportation. Likewise, a shipment of apples might be loaded in a refrigerated container. The container might be transported to a port on a train. There, it is loaded on a ship bound for Japan. This use of more than one mode of transportation is called ***intermodal transportation***. See Figure 25-17.

Transportation Systems

A transportation system includes two main elements. There is a vehicle, or mechanism to carry people or cargo. Also, there are support facilities. These support components include a constructed pathway or an assigned pathway; one or more terminals; and subsystems to provide life, legal, operational, maintenance, and economic support.

Vehicle Systems

Most transportation systems include vehicles. These vehicles are designed to move people and cargo safely. Passengers must be protected and comfortable. The vehicles must be able to operate in one or more of the modes described earlier. Transportation vehicles are made up of subsystems that must operate together for a system to work efficiently. See Figure 25-18. Every vehicle has five subsystems:

- ***Structure.*** This provides a rigid framework to protect the vehicle's contents and support other systems.
- ***Propulsion.*** This provides the means to move the vehicle.

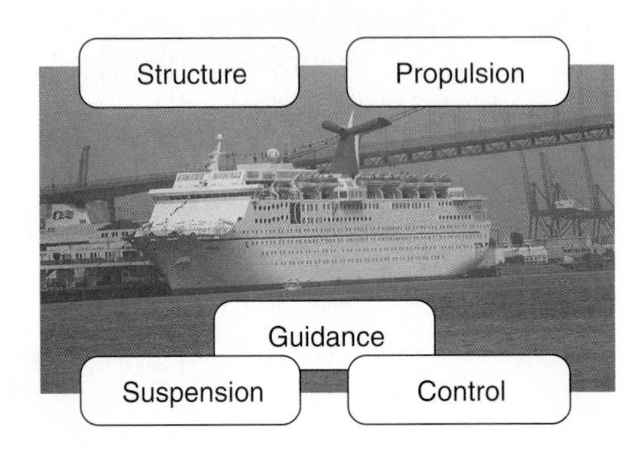

Figure 25-18. The subsystems in a vehicle.

Figure 25-17. Intermodal transportation means use of more than one type of transportation. Three are shown here: ship and tug (water transportation), train (fixed-path land transportation), and truck (variable-path land transportation).

Career Highlight

Air-Traffic Controllers

The Job: Air-traffic controllers coordinate the movement of airplanes along flight paths to ensure that the planes remain a safe distance from one another. They must maintain safety while they efficiently direct flights to and from airports. Airport-tower controllers use radar and visual observation to coordinate the flow of aircraft into and out of the airport. En route controllers coordinate the aircraft in one of 21 air-route traffic-control centers located around the country.

Working Conditions: Controllers work a 40-hour week and might work some overtime. They rotate day, night, and weekend shifts, so the control centers can be operated round-the-clock. There is considerable mental stress associated with the job.

Education and Training: Air-traffic controllers must complete an education program that the FAA has approved and pass a test that measures the ability to learn a controller's duties. Also, applicants must have three years of full-time work experience, four years of college, or a combination of both. Once hired, controllers must complete a 12-week training program at the FAA Academy. After graduation from the Academy, the controller must complete several years of work experience, classroom instruction, and independent study to become a fully qualified controller.

Career Cluster: Transportation, Distribution & Logistics
Career Pathway: Transportation Operations

- ***Suspension.*** This maintains the vehicle on the pathway.
- ***Guidance.*** This receives information needed to operate the vehicle.
- ***Control.*** This enables the vehicle to change speed and direction.

Structure

You ride in many vehicles each year. Probably, you have ridden in your family car. You might ride in a bus to school. To get to a faraway place, you might ride on a train or in a plane.

Each of these vehicles has a structure. Structural systems are the framework and body of a transportation vehicle or system giving it shape. This structure provides a number of important areas. There is a power area containing the engines, transmission, and other systems. In addition, there is an area for the driver to operate the vehicle. This area contains a place to sit or stand and a series of operating controls. Finally, there is a place for the load. This area might contain and protect cargo. This place keeps the cargo away from the weather. This area keeps unwanted people from stealing or damaging the cargo. This place might keep the cargo from freezing or keep the cargo cool so it does not spoil. Railroad boxcars, semi-truck trailers, and shipping containers are examples of load-carrying structures.

If passengers are transported, the structural system contains seats. See Figure 25-19. There are lighting and climate controls (heating and air-conditioning). The structure has windows and doors. Railroad cars and the passenger compartments of airplanes are examples of people-carrying structures. The operator and passenger areas are located together in automobiles and buses.

Propulsion

Transportation vehicles must move from place to place. To do this, they require a propulsion (power) system. This system includes a power source, an energy converter, a power-transmission system, and a drive mechanism. For example, a car has an engine (the power source), a transmission or transaxle (the power transmission), and wheels and tires (the drive mechanism).

Figure 25-19. Vehicles transporting people have attractive and comfortable areas in which passengers can ride.

Heat engines are the most common way to power vehicles. These engines, as discussed in Chapter 21, are gasoline, diesel, gas turbine, and rocket engines. Electric motors are also used to power some vehicles.

Gasoline engines use a piston and cylinder to generate the power. Fuel and air are drawn or injected into the cylinder. The piston compresses the fuel and air mixture. A spark ignites the fuel. The rapidly burning fuel is turned into a gas. The expanding gas drives the cylinder down, creating power. Gasoline engines are used in smaller vehicles, such as cars, delivery trucks, small airplanes, and pleasure boats.

Diesel engines operate similarly to gasoline engines. They have a piston and cylinder moving through fueling, compression, power, and exhaust strokes. The engines differ in the way the fuel is ignited. In gasoline engines, a spark plug causes the fuel to burn. A diesel engine uses the heat that compressing the fuel generates. Diesel engines operate at higher compression ratios (up to 20 to 1) and produce more power. They are used in large vehicles, such as heavy-duty trucks, railroad locomotives, and ships. See Figure 25-20. Also, because of the better fuel economy, these engines are used in some automobiles and delivery trucks.

The gas turbine and jet engines pass air through a compressor. The compressed air is sent into a combustion area, where fuel is added and ignites. The burning fuel becomes hot gas. This gas passes through turbine blades. The gases producing the power spin these blades. Turbine and jet engines are used primarily in aircraft.

Rocket engines use solid or liquid fuels to create power. The burning fuels become gases. These gases are then ejected through a nozzle. The force of

Figure 25-20. Diesel engines power this car-and-passenger ferry.

the gas leaving the nozzle creates the power needed to move the vehicle. Rocket engines are primarily used to power spacecraft.

Finally, electric motors can be used to power vehicles. The motors are used to turn the wheels on commuter trains, streetcars, and some automobiles. See Figure 25-21.

A combination of more than one power type is used in some vehicles. For example, a railroad locomotive is called a *diesel-electric locomotive*. This locomotive uses a diesel engine to drive an electric generator. The electricity produced powers electric motors driving the wheels of the locomotive. Nuclear-powered ships use nuclear reactors to produce steam. The steam powers turbines driving the ship.

Suspension

Vehicles must move along a path from an origin to a destination. To keep them on the path, they use suspension systems. These systems connect, or associate, vehicles with their surroundings. Suspension systems use direct contact, lift, or buoyancy to suspend the vehicles. Wheels on land vehicles are in direct contact with ***roadways*** and rail lines. They suspend the vehicle as it follows the path.

Ships and boats use buoyancy to suspend themselves in water. This principle of science tells us that a force equal to the weight of the water an object displaces sustains the object in water. See Figure 25-22. If a boat weighs 1000 lbs.,

Figure 25-21. The mechanism shown in this picture picks up electricity to power a commuter train.

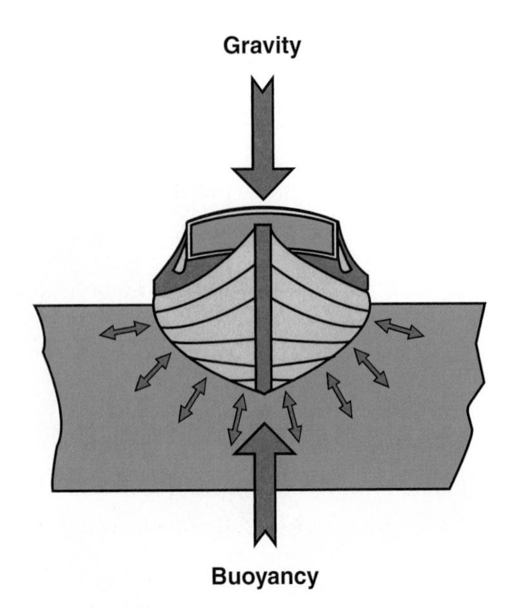

Figure 25-22. A boat floats if the force of the displaced water equals the force of gravity.

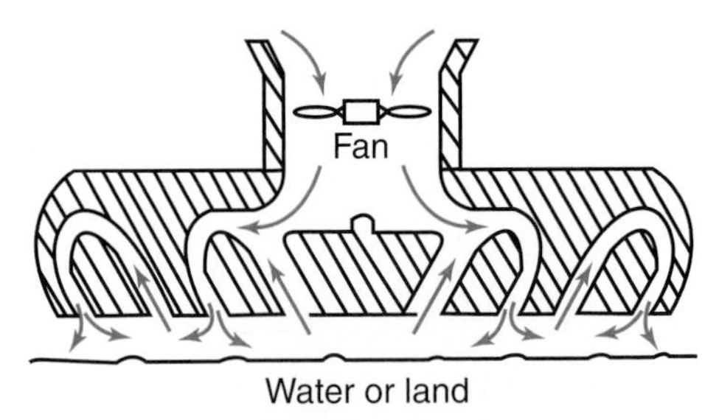

How a Hovercraft Floats

A U.S. Navy Hovercraft

Figure 25-24. A hovercraft floats on a high-pressure air bubble. A fan at the top compresses air and sends it to the bottom of the craft. The U.S. Navy developed this hovercraft. The hovercraft was tested in the Arctic. Someday, this hovercraft might also be used for oil exploration there. (Standard Oil of Ohio)

the boat sinks into the water until it displaces 1000 lbs. of water. This boat sinks if its shape does not displace 1000 lbs.

Aircraft and hydrofoils use lift to create suspension. See **Figure 25-23.** The airfoil has a relatively flat bottom and a curved top. When air flows over the shape, the air on top has farther to travel. This causes it to increase speed. The faster-moving air has less pressure. The difference in pressure gives the wing lift.

There are less common suspension systems. One of these is known as a *ground-effect machine*, or *hovercraft*. A fan in the vehicle creates a thin cushion of high-pressure air. The machine moves about on this cushion of air. See **Figure 25-24.**

Another system uses magnetic forces to suspend the vehicle. This system is called *magnetic levitation (maglev)*. The

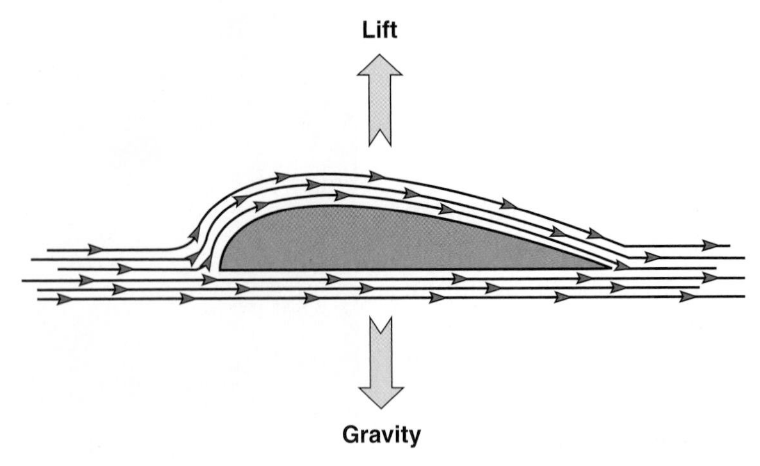

Figure 25-23. The faster-flowing, lower-pressure air across the top of a wing creates lift.

vehicle has a set of lifting magnets, causing it to rise above the pathway. Another set of propelling magnets causes the vehicle to move forward and backward.

In all vehicles, there are forces trying to stop forward movement. For example, wind resistance hampers forward motion. The flatter the surface is, the more wind resistance it has. This resistance is called *drag*. Semi trucks and buses have more wind resistance than automobiles have. Their flat front surfaces provide large areas of wind resistance. Many vehicles are designed to overcome this wind resistance. They are said to be aerodynamically designed. See Figure 25-25.

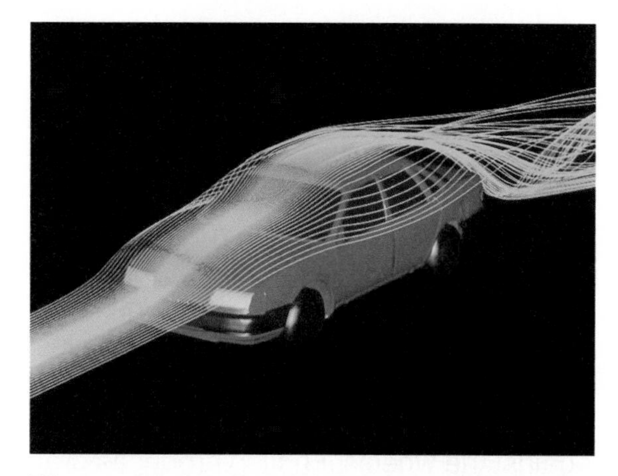

Figure 25-25. The aerodynamics of this vehicle design are being tested. (Ford Motor Company)

Also, gravity pulling down on the vehicle reduces forward motion. Heavier vehicles require more power to move than lighter ones do. Therefore, the heavier the vehicle is, the larger power source it requires. Trucks require larger engines than minivans do.

Friction is another factor reducing forward motion. There is friction within the vehicle. The vehicle's moving parts resist movement. Also, land vehicles experience rolling resistance. There is friction between the tires or wheels and the roadway or rail line. These forces use power.

Guidance

Guidance systems provide information to the vehicle's operator. They help the person operate the vehicle safely and correctly. See Figure 25-26. Guidance can be onboard or outside the vehicle. Many vehicles use electronic guidance equipment. Ships, for example, have navigational instruments to stay on course while crossing large bodies of water. The instrument panel of an automobile has many gauges providing guidance. Many newer automobiles carry electronic maps that operate off GPSs.

Outside guidance systems include the control towers giving pilots instructions about takeoff and landing. Railroads have signals along their tracks. Highways have traffic signals and travel information on road signs.

Control

Control systems allow operators to direct and regulate a vehicle's travel. They obtain information from the

Lights

Gauges

Figure 25-26. Traffic lights and the gauges on an automobile dash provide guidance information. (Buick)

guidance system to determine the variation in speed, direction, or altitude of a vehicle. The operator must be able to start, stop, speed up, slow down, and turn the vehicle. These acts are managed through systems of control. Control systems vary somewhat from vehicle to vehicle.

Stopping a vehicle is done with braking systems. Trains, highway vehicles, and airplanes have wheel brakes. Air or hydraulic pressure pushes pads or blocks against the wheels. Friction produces drag on the wheels to slow or stop the vehicle. Airplanes also have wing flaps. These flaps produce air drag and help slow the plane. Aircraft engines are also used to slow the plane after landing. Reversing the propellers slows or stops ships.

Acceleration and deceleration controls are used to vary a vehicle's speed. Controlling the amount of fuel delivered to the engine varies the speed. These controls are called *throttles*. The gas pedal in a car is a throttle.

Vehicles also must have directional control. Variable-path land vehicles have wheels that turn left and right. Ships and airplanes have rudders for left and right movement. Airplanes also control up and down movement with an elevator. The elevator is part of the tail assembly.

Most vehicles have other controls. These include switches to turn on lights, windshield wipers, windshield washers, defrosters, heaters, and radios. The controls are usually simple electrical switches. They control electrical current to the devices.

Support Systems

A transportation system is more than vehicles moving people and cargo. This system also includes a number of support facilities. Support systems include vehicle pathways and terminals. They also supply life, legal, operational, maintenance, and economic support for safe and efficient operation.

Figure 25-27. Roadways, bridges, and other structures support transportation systems.

Transportation systems require pathways. There are roadways, bridges, and overpasses on highways. See Figure 25-27. Tunnels allow roads and rail lines to pass through mountains. Towers support pipelines as the pipelines cross ravines. These facilities must be maintained so the transportation system is efficient. See Figure 25-28.

Also, most systems have stations, or ***terminals***. See Figure 25-29. Most terminals provide offices for employees of the transportation system. There might be areas to receive cargo or sell tickets to passengers. Usually, there is a place to secure the cargo or allow passengers to wait. Other services might be provided, including food areas, rest rooms, and shops.

Stations and terminals are needed to make intermodal travel convenient. Buses and automobiles bring passengers to and from airports. There must be facilities for unloading. Dockside facilities provide

Figure 25-28. This crew is using equipment to repair a railroad track.

Figure 25-29. Passenger terminals provide services for the traveler.

needed services for cargo handling. Vehicles and cranes move seagoing containers from ships. Containers are transferred to and from land transportation. Escalators, elevators, moving sidewalks, stairs, and conveyors are used at terminals. They help travelers reach airplanes, buses, and other modes of transportation.

Many terminals provide space for vehicle service. See Figure 25-30. Trucks, airplanes, ships, and other vehicles need routine attention. This service keeps them in good operating condition. Lubricants must be changed. Vehicles need to be refueled. Tires might need replacing. The interiors are cleaned. The outsides are washed.

Areas for vehicle storage are provided at many terminals. Park-and-ride lots let people park their cars and take public transportation (buses and commuter trains). Airports and railroad stations must have space for long-term parking facilities. Most terminals also have space to store cargo. Airports have baggage sorting and storage areas. Warehouses are built next to docks and airports.

Besides providing pathways and terminals for transportation systems, the support systems also must provide support for safe and efficient operation. Providing legal support is one way they can do this. Governmental regulations often affect the design and operation of transportation systems. State organizations control the use of highway systems, establish speed limits, and monitor other operating conditions. The Federal Aviation Administration

Figure 25-30. A plane is serviced and refueled at the boarding gate.

(FAA) regulates airspace and air safety and issues licenses to pilots. Other agencies regulating the transportation industry set highway-construction standards, control the safety of interstate trucking, and endorse shipping on the oceans.

Quality Control in Transportation Systems

Quality control in transportation involves inspection and maintenance of three separate components: transportation vehicles, transportation structures, and transportation processes. Transportation vehicles, such as automobiles, locomotive engines, and boats, are manufactured using a quality control process that produces an acceptable product. These vehicles also require periodic maintenance to ensure proper operation.

The construction of transportation structures includes quality control methods typical of the construction industry. These include material testing and on-site inspections. Following construction, structures such as roads, bridges, tunnels, and railways are continually inspected and repaired as needed.

The quality of a transportation process is based on the effectiveness of the system in transporting goods or people in a timely and safe manner to the intended destination. The process is controlled by monitoring the system and making adjustments as needed. For example, overcrowded roads can produce unacceptable delays and unsafe conditions. To remedy this situation, the city planner might recommend adding more lanes to the road, improving traffic-control systems, developing additional routes, or providing mass-transit options.

SUMMARY

Transportation is the moving of people and goods from one place to another. The three main types of transportation are personal, public, and commercial transportation. Transportation involves many processes, including routing, scheduling, loading, moving, unloading, and storing. Modern transportation includes land, water, air, and space travel. When more than one system is employed, it is called *intermodal transportation*.

A transportation system is made up of structures and support systems. The structures are buildings and vehicles. Vehicles are made up of several subsystems. The subsystems include structure, propulsion, suspension, guidance, and control systems. Support systems include pathways, terminals, and other forms of support for safe and efficient operation.

STEM CONNECTIONS

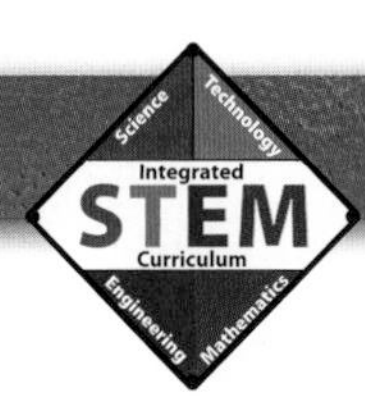

Science

Develop a display explaining the scientific principles of lift, buoyancy, and drag.

Mathematics

Select a city in your state you would like to visit. Calculate the distance to this town. Determine how long it would take to travel there, using set speed limits. Calculate the cost of gasoline for the trip, using a predetermined fuel-economy figure.

CURRICULAR CONNECTIONS

Social Studies

Research the types of transportation systems used at a specific time in history. Explain the reasons they were used.

ACTIVITIES

1. Select a vehicle. Describe its subsystems (structure, propulsion, suspension, guidance, and control).
2. Obtain a map of your state. Select a city you would like to visit. From the map, determine the most direct route and an alternate, more scenic route.
3. Interview a parent or an older friend about travel experiences as a youth. Write a report on the interview.

Test Your Knowledge

Do not write in this book. Place your answers to this test on a separate sheet of paper.

1. When was the wheel invented? Why is it important to the history of transportation?
2. Select the one best answer. Transportation is:
   A. Airplanes, automobiles, and ships.
   B. Carrying people from one place to another.
   C. Moving people and goods from one place to another.
   D. Vehicles.
3. Match each statement with the right type of transportation. You will use some answers more than once.

   _____ Flying on American Airlines to Chicago.
   _____ Riding a bicycle to school.
   _____ Riding in the family car on a vacation.
   _____ Riding in a taxi to the train station.
   _____ Riding the city bus to work.

   A. Personal transportation.
   B. Public transportation.
   C. Commercial transportation.

4. The process of assigning a time to travel is called _____.
5. The process of placing cargo in a secure place before loading it on a vehicle is called _____.
6. The four modes of transportation are _____, _____, _____, and _____.
7. The two types of land-transportation vehicles are _____ and _____.
8. Give an example of intermodal transportation.
9. Describe the two main components of a transportation system.
10. The five vehicular systems are _____, _____, _____, _____, and _____.
11. Roads and terminals are part of the _____ system in transportation.
12. Give an example of a quality control breakdown in a transportation system.

Reading Organizer

On a separate sheet of paper, list the different modes of transportation. Give examples of each type.

Mode of Transportation	Examples of Transportation Mode
Example: Land transportation	Automobiles

CHAPTER 25 ACTIVITY A

TRANSPORTATION TECHNOLOGY

INTRODUCTION

Every day, millions of people and countless tons of cargo are moved. They are transported from one place to another. Technological systems are used to move these goods and people. Vehicles use energy to haul cargo. People apply power to move these vehicles over land, across water, and through the air.

These transportation systems include vehicles and pathways. The pathways allow the vehicles to move from one place to another. They can be highways, railroad tracks, rivers, or oceans. Air and space provide pathways for airplanes and spacecraft.

People design each of these systems. The systems allow us to extend our potential for movement. This activity allows you to design and test a vehicle for a transportation system. You will use your creative ability and manual skill to design and test a boat hull.

EQUIPMENT AND SUPPLIES

- A waterway or long narrow channel that holds water.
- A 2″ × 4″ × 9″ foam block (hull).
- 2 1/2″ × 3 3/4″ × 28-gauge sheet metal (keel).
- A small eye screw.
- Sail material:
  - A 6″ × 6″ square of fabric.
  - Two 3/16″ × 6″ dowels.
  - A 3/8″ × 7 3/4″ dowel.
- A rule.
- A square.
- A scratch awl.
- Tin snips.
- A mill file.
- Needle-nose pliers.
- Tailor's chalk or a fine-tipped marker.
- Fabric glue or a needle and thread.
- A hot-wire foam-material cutter or band saw.
- A coping saw.

- A rasp or Surform® tool.
- Coarse abrasive paper.
- Compressed air—15 pounds per square inch (psi)—or a fan and cardboard tube.
- A stopwatch or other timepiece that reads in seconds.

PROCEDURE

Making Hulls

1. Your teacher will divide you into groups. Each group should have three students.
2. Each group will build three basic hulls. See **Figure 25A-1.**
   a. Student 1 will build a round-bottom hull.
   b. Student 2 will build a V-bottom hull.
   c. Student 3 will build a flat-bottom hull.

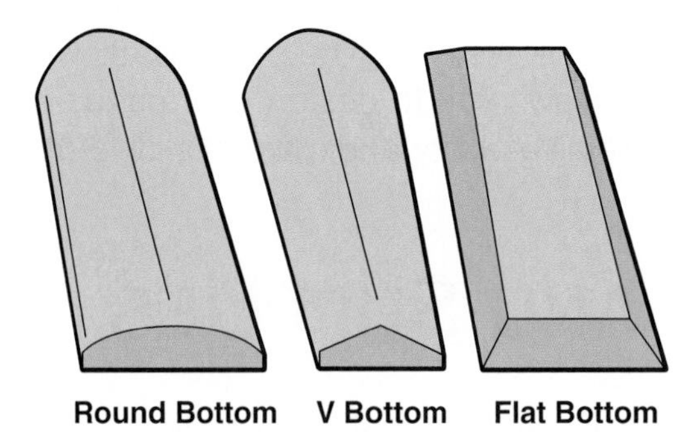

Figure 25A-1. The boat-hull designs. Each group will make one of each design.

3. Carefully watch your teacher demonstrate safe techniques for cutting and shaping foam material.
4. Each group should get three foam blocks.
5. Each member of the group will complete the following steps:
   a. Design a hull meeting the assigned shape.
   b. Cut the foam block into the basic shape, using a hot-wire cutter or band saw.
   c. Smooth the hull shape using a rasp, a Surform® tool, or abrasive paper. Be sure both sides of the centerline are alike.

Making Sails

1. Select the materials listed under "Equipment and Supplies" for each sail.
2. Lay out and mark the fabric, using tailor's chalk or a fine-tipped marker. See **Figure 25A-2.**
3. Fold the fabric at each end to form a sleeve.
4. Sew or glue the fabric in place.
5. Cut two pieces of 3/16″ dowel, 6″ long, using a coping saw.

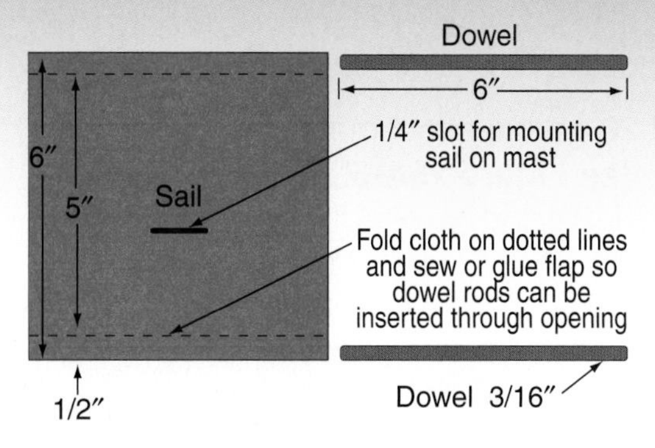

Figure 25A-2. Make the sail from cloth and dowels.

6. Insert the dowels in the top and bottom sleeves of the sail.
7. Cut a piece of 3/8″ dowel, 7 3/4″ long, using a coping saw.
8. Shape the dowel to make the mast. See Figure 25A-3.

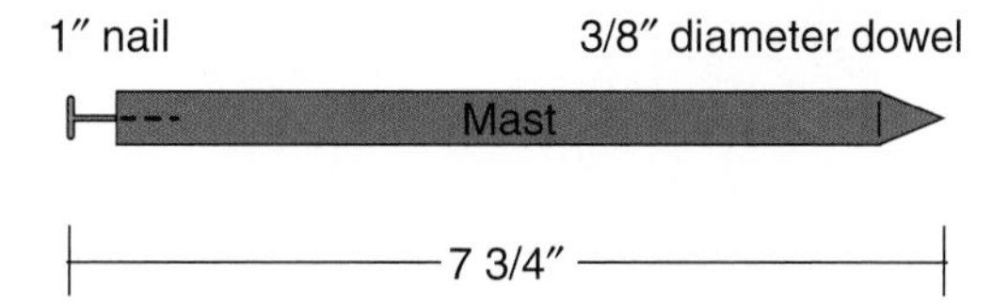

Figure 25A-3. The mast should be tapered at one end for insertion into the foam hull.

9. Insert the mast into the sail. See Figure 25A-4.

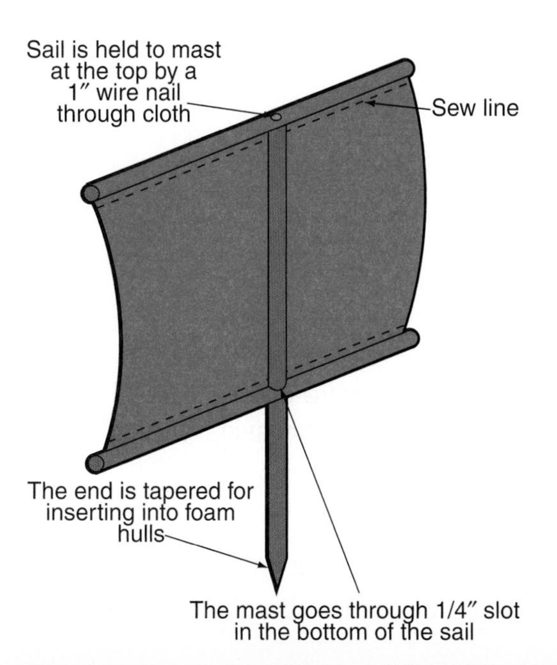

Figure 25A-4. Assemble the sail as shown.

Making a Keel

1. Select a piece of 28-gauge sheet metal.
2. Lay out the keel. See Figure 25A-5.

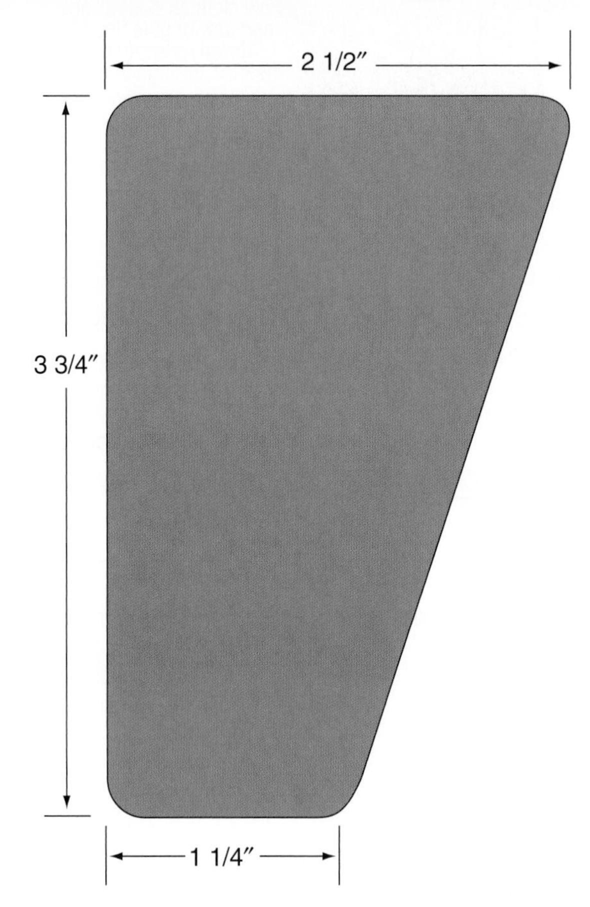

Figure 25A-5. Shape the keel out of sheet metal.

3. Cut out the keel, using a pair of tin snips.
4. File all edges with a mill file to remove burrs.

Assembling the Boat

1. Insert the keel into the center of the bottom of the boat.
2. Attach the sail to the top of the boat.
3. Slightly open the eye of the eye screw with a pair of pliers.
4. Install an eye screw into one side of the boat. This screw will attach to a string in the waterway. The string will keep the boat traveling in a straight line in the waterway.

Testing the Boats

1. Your teacher will provide the class with a waterway in which to test the boats. The waterway will have a string or fishing line stretched along its length.
2. Place the boat at one end of the waterway.
3. Hook the screw eye over the guideline.
4. Give the sail a continuous blast of 15-psi compressed air. Alternatively, you can use a fan to blow air through a cardboard tube.
5. Clock the time it takes the boat to reach the end of the waterway.
6. Record the hull type of the boat.
7. Log the time it took the boat to reach the other end.
8. Determine the average travel time for each of the following groups:
   a. The flat-bottom boats.
   b. The round-bottom boats.
   c. The V-bottom boats.
9. Compare the results. Determine which shape is most efficient.

Challenging Your Learning

Why was the winning hull shape the best? For what application is this hull shape best suited? Why would you not use this hull shape in all applications?

Safety Rules

The tools and equipment needed for this activity can cause injuries, if not handled properly. Your instructor will provide safety rules and demonstrate safe procedures. Do not use any tool or piece of equipment unless you know how to do so safely.

Section 5 Technology and Society

Technology is very much a part of our lives. The more we use technology to control and change our environment, the more technology affects how we live. Let's consider how technology affects our values. Suppose you had to live in a tent because your community had to move often. Would you value the same possessions you value now? Consider the things you have in your bedroom. How many of them would you want to keep if you were moving every month? Would you prefer a bed or a sleeping bag? Do you think you would want a chest of drawers or a duffel bag?

You might live all your life in one community. If you move, moving vans, trains, and airplanes are available so you can take all your belongings. You have read in the preceding chapters about how technology has become interwoven with our day-to-day living.

Not all the effects of technology have been good. Technology has had some negative impacts as well. We have to look ahead and decide what is best for the future. Constantly, we are learning how to use technology to our advantage and prevent additional damage to our environment. In the following two chapters, you have a chance to discuss the impacts of technology. You will learn how society is looking at the effects of technology on the environment.

Technology Headline:

Nanowire Battery-Powered Electric Cars

With gas prices continually on the rise, car manufacturers are under increasing pressure to build reliable, fuel-efficient vehicles. Many hybrids—cars powered by a combination of electricity and gas—are already on the market. Ideally, car manufacturers will be able to create an entirely electric car capable of traveling long distances on a single charge. Some electric cars exist, but are only able to travel short distances. One of the main challenges automotive engineers face is designing a safe battery powerful enough to make extended trips.

Developers hope a new model of the lithium-ion battery will be the solution to limited battery life in electric cars. Lithium-ion batteries are currently used in many common electronic devices such as laptops, digital music players, and cell phones. They are composed of a lithium-ion core with graphite anodes. An *anode* is the end of the battery that accepts current, which then flows toward the cathode end of the battery. The *cathode* end of the battery then releases current. As lithium-ion batteries are charged, the lithium expands—it can grow up to four times its original size. If a battery becomes too large, graphite anodes would not be able to contain the swelling lithium. This is not usually a problem with normal lithium-ion batteries, but when creating one strong enough to power a vehicle for several hundred miles, the battery would have to be larger and anode material redesigned. The required room for the lithium to swell would take up a significant amount of space; lacking additional area, battery safety becomes a concern. Switching from a graphite anode to a nanowire anode would substitute a rigid material with one that would better accommodate the lithium expansion. As compared to graphite, silicon nanowire anodes can expand enough to accommodate nearly ten times more lithium.

Electric cars typically have large batteries, which limit the size of the vehicle's interior. Reducing the size of the battery and making it more efficient with increased longevity would improve the overall driving experience. With the advent of nanowire batteries, electric cars would have smaller, more powerful batteries. These cars would have a higher energy output than traditional lithium-ion batteries and take up far less space.

A nanowire battery would be able to power an electric motor for an extended period of time. These cars would not only be more fuel-efficient, they would also be cheaper than traditional gasoline engine vehicles and hybrids because electricity is an inexpensive alternative to gasoline. The development and safe use of this battery will happen sometime in the future, and when it does it will change the face of the automotive industry.

CHAPTER 26

Technological Impacts

Did You Know?

- Charles Kettering said, "We should all be concerned about the future because we will spend the rest of our lives there."
- Nearly 40% of the people in developing countries live in cities. This places large demands for jobs, housing, and public services.

Objectives

The information given in this chapter will help you do the following:

- Give examples of some reasons for technological change.
- Explain how technology has made life better for humankind.
- Give examples of some undesirable effects of technology relating to resources, the environment, and people.

Key Words

These words are used in this chapter. Do you know what they mean?

acid rain
air pollution
engineered material
global village
inexhaustible resource
noise pollution
productive
scarcity
soil erosion
technological unemployment
water pollution

Preparing to Read

As you read this chapter, outline the details of the impacts technology has had on society. Use the Reading Organizer at the end of the chapter to organize your thoughts.

Technology affects everyone and impacts the way we live, think, and act. For example, many people feel they must be in constant contact with other people. This feeling is much different than the feelings people have had at other times in history. This new attitude is a result of television, cellular phones, and other rapid communication devices. People also think of time and distance differently than they once did. A trip across the country in the 1800s took weeks or even months. Travelers on the Oregon Trail took up to six months to travel from Missouri to Oregon. Now, we can drive the distance in a few days. You can fly from St. Louis, Missouri to Portland, Oregon in a matter of a few hours. What was once a trip for only the adventurous and brave is now routine for many people. Changes in transportation technology have changed this attitude about travel and distance. See **Figure 26-1.**

Let's consider a simple technology and its impact on people and the surroundings. Think about traveling to school or a friend's house. Using a technological device (the bicycle) can make the trip easier and quicker. Consider, however, the bicycle's effects on the surroundings. The bicycle operates best on a smooth path with a hard surface. The bike path uses land. This land now cannot be used for other purposes. The bicycle needs a storage place at home and a space at school or the friend's house. The storage might be a building or an outside bike rack. Both of these have some effect on the environment. They take up space and are built from natural resources.

Consider another option to make the trip to school or a friend's house. An

Figure 26-1. Trips that once took weeks to complete now take a matter of hours.

automobile gets you there faster than the bicycle does. Also, it also protects you from the weather. The automobile can be used in hot or cold weather or when it is raining or snowing. The impacts of the automobile on the environment, however, are much greater than those of the bicycle. The path for the automobile must be wider than a bike path, for the automobile to travel safely. The narrow bike path must be replaced with a street or highway. See Figure 26-2. The parking space for the vehicle takes up more space. To protect it, a garage is often built onto a home. In other cases, a parking garage is provided with an apartment building. These structures take up more space and use more materials than bicycle storage units do. The exhaust from the automobile's engine can cause damage to the environment. Also, the car makes noise as it moves from place to place.

These are just two examples of the impacts of technological devices on people and the environment. They show how using technology changes people and their surroundings. There are many more examples because each application of technology has its own impacts.

The World of Change

All technological advancements cause change. They change the people who use them. People, through their buying habits and political actions, also cause technology to change. New technologies create new lifestyles and opinions about what is needed and good. For example, in earlier days, most men worked in stores and factories. The women stayed home and cared for the family. Housewives had time to prepare meals using fresh meats, fruits, and vegetables. As times changed, many women started working outside the home. This caused many technological changes in the workplace. Also, having less time to prepare meals created a need for processed

Figure 26-2. Building roads uses land that now can no longer be used for other uses.

foods and preprepared meals. Packaged and frozen dinners reduce the time needed to cook meals. Technology was used to reduce meal-preparation time and increase time available for other activities, such as work, recreation, and family events.

For another example, consider the physical condition of people. In earlier history, people did a lot of manual labor. The types of jobs and work around the home kept these people in good physical condition. Now, many people spend most of their time in a chair at a desk or in front of a computer. Children spend less time playing games requiring running and other physical exercise. They entertain themselves with video games and television. Many modern people are out of shape and overweight. This has given rise to the development and use of exercise equipment. This technology was almost unheard of a century ago. See Figure 26-3.

Some of the changes technology causes are good. Others are not. Some of the effects have been intended. They were planned. Other effects have been unhappy surprises. They were unforeseen and unplanned. Some impacts are recognized immediately. Others happen at a later date. See Figure 26-4.

Developing and using technology produces change. Why do we change the

Figure 26-3. The frozen dinner and the treadmill are two examples of technological innovations developed to meet changing needs and lifestyles. (©iStockphoto.com/kcline; ©iStockphoto.com/JonasSanLuis)

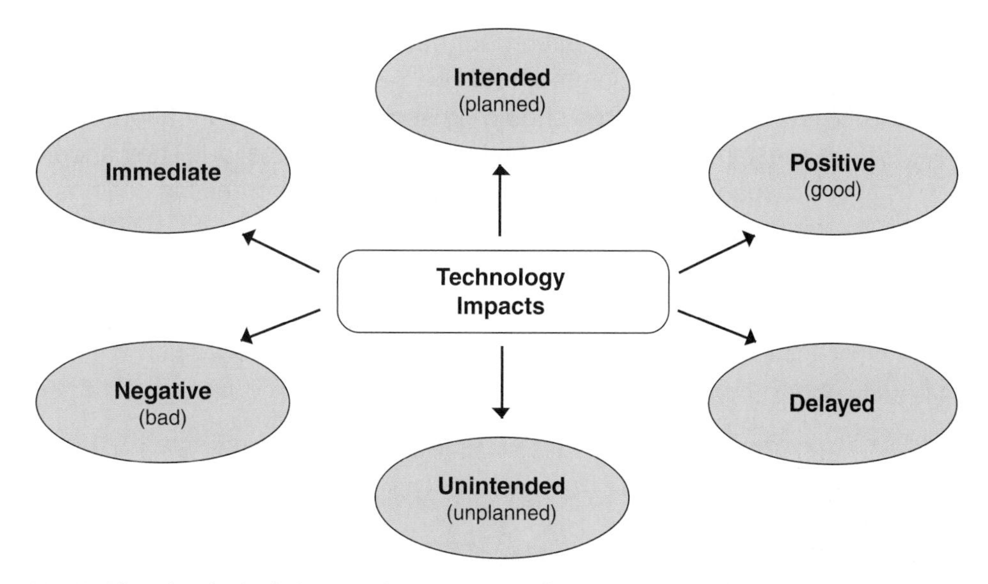

Figure 26-4. All technological changes have impacts. Some are good. Others are bad.

Technology Explained

earthquake protection: actions people take to make buildings safer during natural disasters.

The structures people build are not immune to natural forces. For example, earthquakes can strike suddenly, violently, and without warning. See Figure A. In early days, builders relied on the weight of a structure to keep the structure in place. Often, the structures were not bolted to the foundations. The walls were nailed only to the floor systems. History, however, has shown that this protection is inadequate.

Figure 26-A. Two views of the damage an earthquake can cause. (Federal Emergency Management Agency)

To reduce the risk of damage and injury, people can do a number of things in their homes. These actions can reduce the dangers of serious injury or loss of life. Some of these actions include fastening shelves and bookcases securely to walls. See Figure B. Wood strips or metal bars should be installed to keep objects from vibrating off shelves. Large or heavy objects should be placed on lower shelves. Items such as bottled foods, glass, and china should be stored in closed cabinets that have doors with latches.

Secure the bookcase to the wall with angle brackets.

Add wood strips or steel rods to keep books from falling forward.

Figure 26-B. How to make a bookcase earthquake resistant.

Heavy items, such as pictures and mirrors, should be hung away from beds and areas where people sit. Pictures and hanging plants should be hung off securely installed screw eyes. The eyes should be closed to keep the objects from falling.

In addition, water heaters should be secured by strapping them to the wall. See Figure C. They should also be bolted to the floor. All overhead lights should be braced so they do not swing in large arcs. Gas lines to heating units and other appliances should have flexible connections.

Homes can be altered for earthquake protection in a number of ways. The structure should be securely bolted to the foundation. The superstructures should be secured to the foundation with straps. Freestanding chimneys have metal liners installed. They should be strapped to the side of the house. Any deep cracks in ceilings or foundations should be filled.

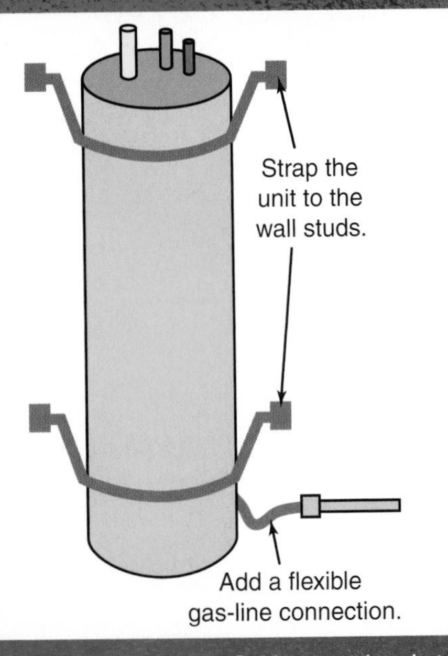
Strap the unit to the wall studs.
Add a flexible gas-line connection.

Figure 26-C. How to make a hot water heater earthquake resistant.

way we do things? What is the reason we develop and use technological products and services? Usually, there are many reasons for change:

- We want to improve the world around us and make our countries, cities, and homes better places to live.
- We want to do work more easily and apply energy more efficiently. Most of us want to be able to process information and material more quickly and with less effort.
- We want to travel more easily and receive information more quickly. People want to be connected to other people and places around the nation and world.
- We want more time to enjoy ourselves. This extra time is called *leisure.* Most of us want to have a variety of ways to spend our free time.
- We want to help others. People want to help other people live better lives with less poverty.

Technology has given us many good things. We have better food, better houses, better ways to get information, and better ways to travel because of technology. See **Figure 26-5.** We have better ways to communicate with family and friends because of technology. Our health is better because we have better nutrition (diets). Better medicines help cure illnesses and treat physical conditions.

TECHNOLOGY AND CHANGE

All technology is interrelated. New materials allow us to build better products and buildings. Knowledge to work with these new materials requires new ways to communicate with workers. Also, new products require new ways to repair and service them. This requires new training programs. People have to be taught how to use these new products. Therefore, owners' manuals are developed. Users' videos are prepared. With each new advancement, new production and communication techniques are required.

Technological change is everywhere. This change impacts us in every facet of our lives. Let's look at some of these changes.

Materials and Change

At one time, people used only the mat-erial they found around them. See **Figure 26-6.** These people used trees,

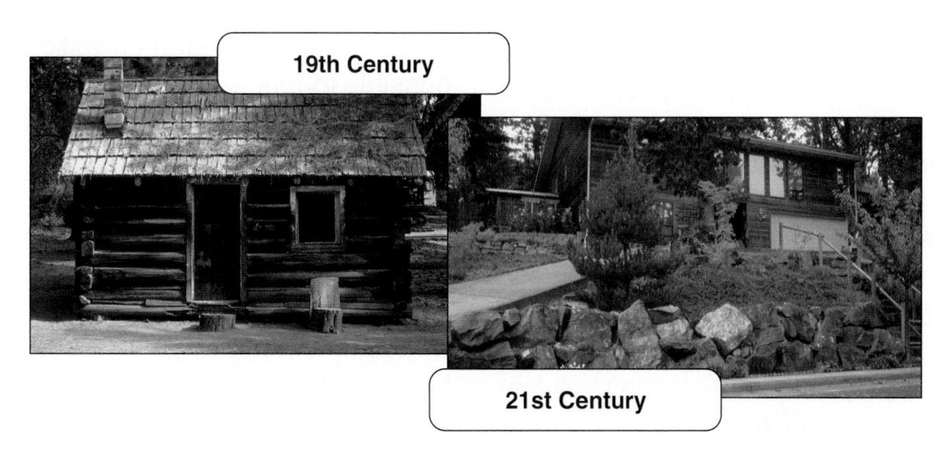

Figure 26-5. Think of the way people lived in these homes. How has technology changed our quality of life over time?

Figure 26-6. Look at this log-cabin interior. What materials do you see being used? How is this different from a modern home?

stones, and clay. Plant fibers and animal parts were vital materials. Water provided a major source of power. Wood, coal, and animal oils were the fuels people used to produce heat, light, and power. Humans were tied to the natural world in which they lived.

Today, we have materials our ancestors never knew about. Technology has allowed us to use natural resources in new ways. We can smelt (refine) metals, laminate wood particles together into sheets, and produce complex ceramic materials. Also, we have learned to make new materials. We have created a new class of materials called ***engineered materials***. These materials are commonly called *plastics*. They are developed to meet a specific need. This is in contrast to designing a product around an existing material, such as wood or steel. Today, the need can be identified. A material (plastic or composite) can then be developed to meet the special need.

Plastic was once considered a cheap, inferior material. Through technological advancements, however, today, many things are made of plastic. Wrinkle-free, easy-care clothing is made of fibers developed from plastic. Automobile and aircraft parts are made of composites (plastic-based materials). Sheet plastics are the shatterproof "glass" in storm doors. Human body parts can be replaced with tough, long-lasting, plastic parts.

Tools and Change

Have you ever visited a museum showing old tools and machines? If so, you have seen how different they are from the tools we use today. See Figure 26-7. Over time, people have developed new and improved tools and machines. These technological advancements have allowed people to do several things:

- Reduce the amount of labor needed to complete a job. See Figure 26-8.
- Produce higher-quality, more consistent products.
- Reduce the scrap and waste created in the process.

These new machines let people build better products more cheaply. They have made work on farms, in factories, and on construction sites easier. People can do more work with less effort using modern machines.

Communication and Change

In recent years, communication technology has changed drastically. It is possible to touch a series of numbers on a telephone and speak to people anywhere in the world. News from around the world enters our homes on radio or television waves or on cable. Signals bouncing off satellites orbiting in space beam sound and pictures into our homes.

Likewise, the way printed communication is produced has changed dramatically. At one time, type was set, pages were laid out, printing plates were produced, and pages were printed. This took large

Figure 26-7. Look at these old tools. How many have you seen recently? Why do you think they are not used much today? (©iStockphoto.com/GlobalP)

Figure 26-8. This modern green bean–picking machine has replaced labor-intensive handpicking.

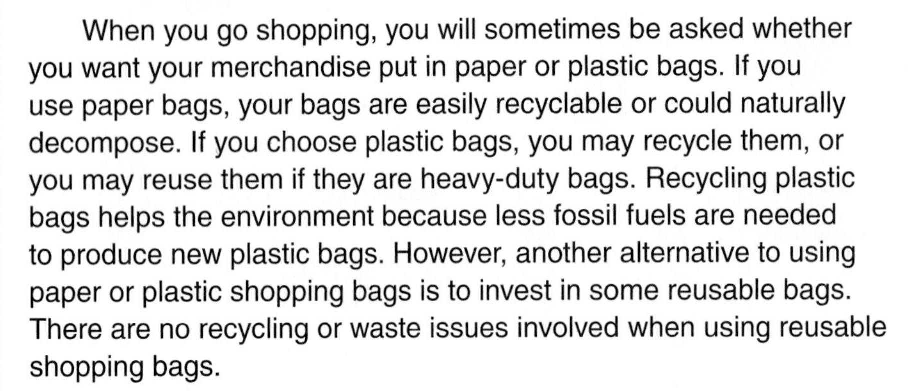

Think Green

Shopping Bags

When you go shopping, you will sometimes be asked whether you want your merchandise put in paper or plastic bags. If you use paper bags, your bags are easily recyclable or could naturally decompose. If you choose plastic bags, you may recycle them, or you may reuse them if they are heavy-duty bags. Recycling plastic bags helps the environment because less fossil fuels are needed to produce new plastic bags. However, another alternative to using paper or plastic shopping bags is to invest in some reusable bags. There are no recycling or waste issues involved when using reusable shopping bags.

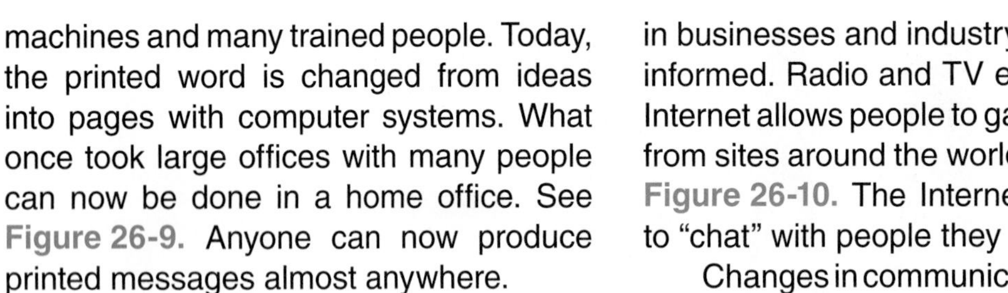

machines and many trained people. Today, the printed word is changed from ideas into pages with computer systems. What once took large offices with many people can now be done in a home office. See Figure 26-9. Anyone can now produce printed messages almost anywhere.

Communication media has become a major part of our lives. This media is used in businesses and industry to keep people informed. Radio and TV entertain us. The Internet allows people to gather information from sites around the world with ease. See Figure 26-10. The Internet allows people to "chat" with people they have never met.

Changes in communication technology have created a feeling of a ***global village***. We are closely connected to everyone

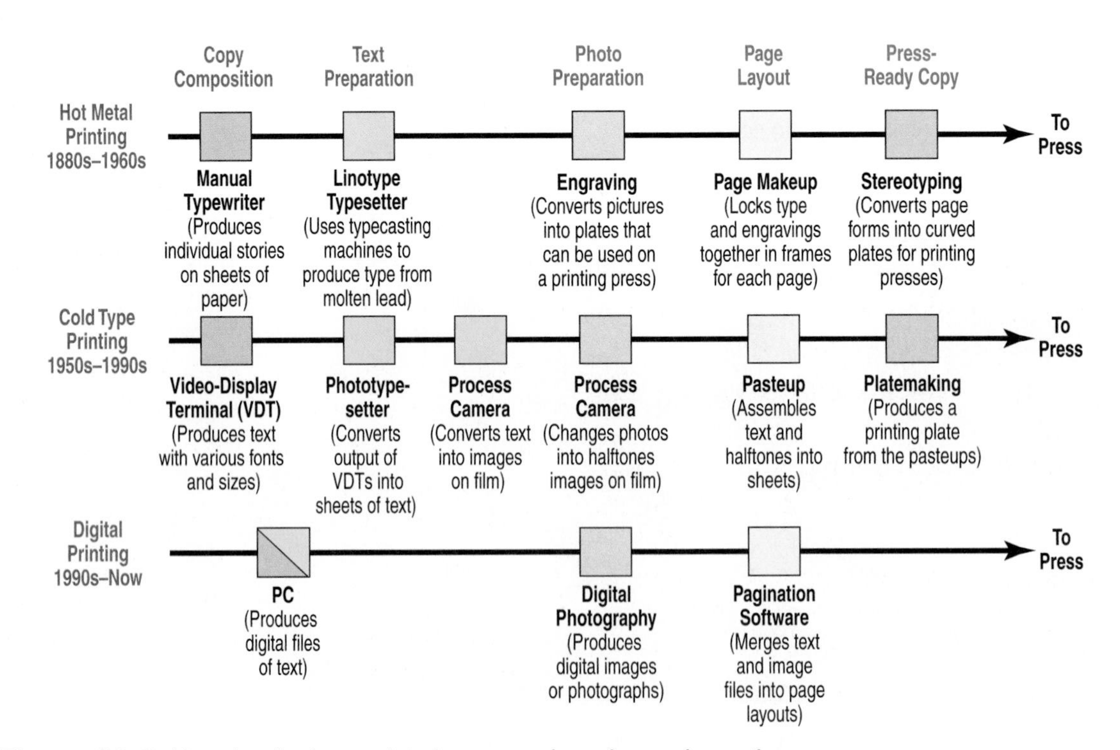

Figure 26-9. How developing a printed message has changed over time.

Figure 26-10. The Internet allows us to access information from any location—at home, at work, or on the road. (©iStockphoto.com/Nikada)

on the globe. This is similar to how our early ancestors were closely connected to everyone in their villages. This closeness makes floods, storms, and wars more personal and real.

Transportation and Change

In early times, people seldom traveled more than a few miles from their birthplaces. Rivers and dirt trails were the avenues of travel. People seldom traveled for fun.

Before 1900, "getting away" probably meant hitching a horse to a buggy and taking a drive in the country. Now, it might mean taking a quick trip to a faraway place. Getting away can be going on a getaway at some distant resort or park. The trip might start with a bus or an automobile ride to an airport. A plane can then whisk you across the country in a matter of hours. People travel more often and with greater ease than ever before.

Transportation technology has shrunk the world. Technological advancements moved people from the steam-powered train to the jet passenger plane in little more than 100 years. See Figure 26-11. The speed of travel changed from less than 50 mph to more than 500 mph. What was once a long trip is now a short hop. Today, in the time it once took to travel 1000 miles, rockets have carried astronauts to the Moon and back to Earth.

Fast trains make travel between cities quick and comfortable. For example, the Eurostar Italia moves people in Italy from Florence to Rome at speeds up to 185 mph. The 200-mile trip takes about 90 minutes. See Figure 26-12.

Steam Engine

Jet Engine

Figure 26-11. In slightly more than 100 years, people advanced from steam-powered travel to jet-powered travel. (©iStockphoto.com/dwetzel; ©iStockphoto.com/narvikk)

Figure 26-12. Fast European trains travel at speeds reaching nearly 200 mph. (©iStockphoto.com/hfng)

Construction and Change

Technology has changed the types of structures we build. This application has also changed the way we build them. At one time, construction projects were limited to the materials at hand. If people had lots of stones, they built stone structures. Places that had forests allowed people to build wood structures. Sod houses were built on plains without trees.

Also, construction techniques used lots of labor, with most work being done by hand. People cut, shaped, lifted, and assembled materials into buildings and other structures. Technology, however, has changed all this. We are building many more structures in all parts of the world. New materials allow us to build all kinds of structures in almost any location. See **Figure 26-13.**

Figure 26-13. Notice the use of copper, wood, and ceramic materials on this roof.

Also, new building techniques allow people to build very unique structures. For example, lightweight materials and new techniques were used to build the *International Space Station*. See Figure 26-14.

Production and Change

At one time, people worked most of their waking hours. They hunted, gathered plants, cooked meals, and did other productive work. People had little time to enjoy themselves. Over time, technology changed this. People could produce more goods or services in less time. This is being more ***productive***. Being more productive allows people to have better lives and leisure time to do whatever they want.

Productivity also contributes in other ways. In our society, people are paid for the work they complete. With this income, we can purchase services and products that others produce. The work of these people supports the efforts of other workers.

Production also allows people to feel they are of value. Their work is needed. They are doing something that helps themselves, their family, and other people.

Negative Impacts

In the previous discussion, the positive aspects of technology were discussed. Some of the ways technology has improved our lives were presented. There are thousands of other ways technology has made life better. Not all impacts of technology, however, are positive. The use of technology can also have negative impacts on resources, the environment, and people. The progress and use of technology present moral concerns. Individuals often question whether the use of some technologies is ethically acceptable.

This is not because technology is bad. Rather, it is because we have not always understood technology. At times, we have not used technology wisely. Technologically educated people should use facts to evaluate and understand trends. They should recognize the positive or negative effects of a technology. These people should be able to fulfill their personal and public responsibilities to assess technology.

Figure 26-14. The *International Space Station.* (NASA)

Impacts on Resources

One negative impact of technology is that it can create ***scarcity***. Materials are made from natural resources, many of which have a limited supply. When these resources are used up, there will be no more. These are called *nonrenewable resources*. They are sometimes also called *exhaustible resources*. See **Figure 26-15.** For example, iron ore, coal, and petroleum are examples of nonrenewable resources. Other materials can be replaced. Farm crops, forests, and fish are examples of resources that can be replaced. They are called *renewable resources*. See **Figure 26-16.** Still other resources have an inexhaustible supply. Wind, solar energy, and ocean currents are examples of ***inexhaustible resources***. To deal with the possible scarcity of resources, people should do the following:

- Reduce the use of nonrenewable resources in products and energy-conversion activities.
- Develop and use renewable resources.
- Promote the use of inexhaustible resources, particularly in energy-conversion technologies. See **Figure 26-17.**
- Recycle used materials to produce new materials.

For example, consider our use of petroleum. The United States imports about half of its petroleum from foreign countries. About one-third of all oil is used in transportation activities. Much of this fuel is burned in automobiles.

People need to be aware of the impacts of their use of transportation technologies. They can address their consumption of gasoline by demanding automobiles that

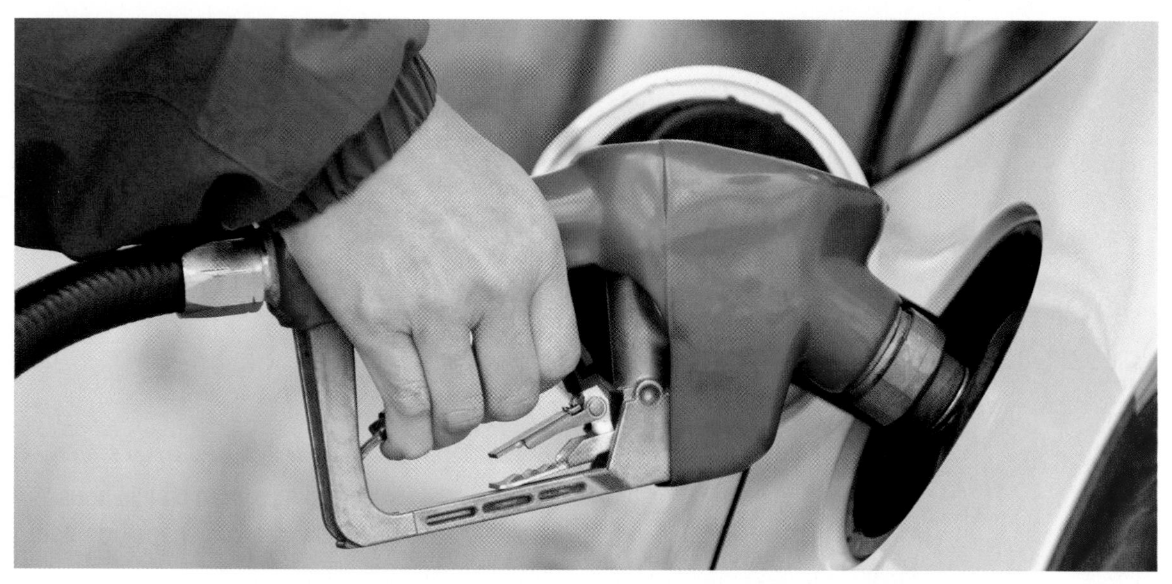

Figure 26-15. When people fill up their cars with gasoline, they are using an exhaustible resource. (©iStockphoto.com/WendellFranks)

Figure 26-16. These wood beams are made from a renewable resource—trees.

have more efficient engines and lighter bodies. People can use mass transportation to reduce their use of personal automobiles. Electric and hybrid (combination fossil fuel–and–electric power) vehicles lessen the impacts of travel on the environment. See Figure 26-18.

Likewise, people can reduce the use of natural gas and petroleum in their homes. Taking advantage of the heating effect of winter sunlight with windows facing south

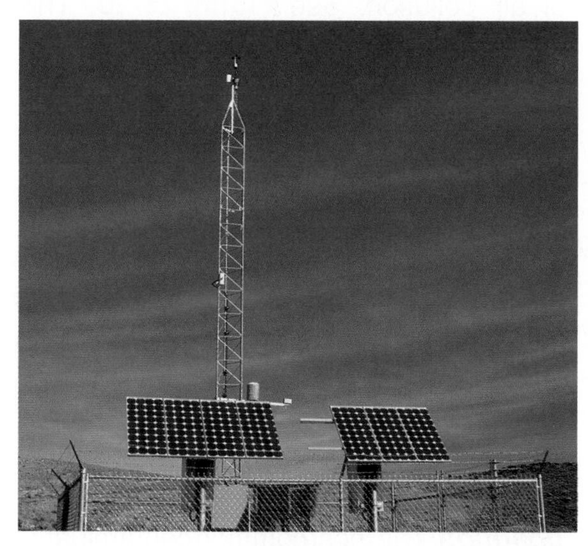

Solar Cells

A Wind Turbine

Figure 26-17. The solar cell and the wind turbine are examples of using inexhaustible energy sources.

Figure 26-18. Using this battery-powered bus can help reduce the air pollution in a city.

Figure 26-19. The plastic wrap on this house seals air leaks that allow heat to escape from the building during the winter.

can reduce fuel use. Likewise, putting more insulation in walls and ceilings can reduce heat loss. Also, installing windows with two and three panes of glazing (glass) in them reduces energy use. Sealing up cracks in walls and around doors and windows reduces heating and cooling requirements. See Figure 26-19.

Impacts on the Environment

Decisions to develop and use technologies frequently put environmental and financial concerns in direct opposition with one another. Some technological activities that have improved how we live have also damaged our surroundings. This includes the following types of damage to the environment:

- Air pollution.
- Water pollution.
- Noise pollution.
- Soil erosion.

Air Pollution

Some technological activities put dust, fumes, smoke, gases, and other materials into the air. This damage is called ***air pollution***. Air pollution can cause forests to lose their leaves, metals to corrode, and people to suffer illnesses.

Nature causes some pollution. Winds stir up plant pollen and dust. Plants and animals release gases into the air. The pollution humans cause, however, can be more serious.

Burning fossil fuels in power plants and vehicles are some of the greatest sources of air pollution. See Figure 26-20. This burning action releases many chemicals into the atmosphere.

Chemicals can be harmful when they are introduced into the air. Industrial accidents and spraying crops with insecticides can cause air pollution. Some pollutants in the air are captured by moisture and return to Earth in what is known as ***acid rain***. This rain can cause major damage to forests and farm crops.

Water Pollution

Disposal of wastes from various technological activities has caused ***water pollution***. The most common contaminants are wastes from factories, individuals, and farms. See Figure 26-21. Factories use

Figure 26-20. Smokestacks at a coal-burning power plant.

Figure 26-21. This waste lagoon is used to treat industrial waste. (American Petroleum Institute)

water to cool and clean materials. People use sewer systems. Runoff from farmland contains fertilizers, weed killers, and insecticides. A common household pollutant is laundry detergent. The detergent enters water supplies through sewers. People applying too much fertilizer to lawns and landscape plants also can cause water pollution. Polluted waters, besides being unfit to drink, are harmful to both humans and wildlife. See Figure 26-22. Pollution particularly affects fish and waterfowl.

Figure 26-22. Thoughtless actions by people cause pollution. An example is this personal waste (coffee cups) that washed up on the shore of a lake. (©iStockphoto.com/Creativel)

Noise Pollution

Another unwanted output of technological activities is ***noise pollution***. This is actually "silence pollution." Unwanted noise pollutes the silence. This noise can be from factories, transportation vehicles, musical concerts, and other loud technologies.

Soil Erosion

Many human activities can cause ***soil erosion***. Inefficient farming operations can allow rain and irrigation water to wash away valuable soil. Poorly planned earthmoving activities on construction sites and in mining can cause erosion. Human carelessness can cause forest and range fires. These fires can leave land exposed to erosion. Tracks made by recreational and off-road vehicles have damaged desert lands and croplands.

Proper planning in many outdoor technological activities can reduce soil erosion. No-till farming, terraces, and grass buffer zones can help preserve farm soil. Grading mine sites can reduce water runoff damaging adjacent land. See Figure 26-23. Allowing fairly frequent natural fires to maintain forestland and rangeland health can reduce the chance for erosion damage. These natural fires clear away dead plant materials and thin the area so healthy plants remain. The resulting area is less likely to support a major fire that kills nearly all plant life, exposing the soil to serious erosion force. See Figure 26-24.

Impacts on People

Technology affects everyone. By itself, however, technology is neither good nor bad. Technology's impacts are felt as people make decisions about its development, production, and use. These human decisions can result in both desirable and undesirable consequences. Technology can make life better. This capability can, however, also have negative impacts. These impacts can affect a person's outlook on life, job opportunities, and a host of other issues.

Figure 26-23. Carefully planned mining activities can reduce soil erosion and water pollution.

Figure 26-24. This land that a forest fire has swept is susceptible to erosion.

Impacts on Jobs

Computers and robots are replacing many machine operators. This shift has led to a demand for new types of workers: technologists and technicians. These workers need to be better educated and better trained. The low-skill, low-education jobs are rapidly disappearing because of technology. This has led to ***technological unemployment***. Technological unemployment means people with little education have lost their jobs to computer-driven machines and robots. Also, there is another impact called *technological unemployability*.

This means many people cannot become employed because they lack the proper education. No matter how many jobs are available, these people cannot find work. These people do not have the skills needed for the available jobs. They are unemployable. These people are the high school dropouts who have few skills and lack the knowledge needed for modern jobs.

Transportation and communication technologies have also added to the negative impacts on people's jobs. With modern transportation systems, low-skill work can be moved to less developed countries. See **Figure 26-25.** These countries have many people who will work for low wages. Communication technology allows for designs to be developed in one country. The products can be made in another country. Transportation technology allows the output to be shipped anywhere around the globe.

Impacts on People's Lives

In some cases, technology can impact people's lives without the people realizing it. For example, communication technology has advanced so much that instant communication is a normal part of life. People have become accustomed to communicating with others instantly through e-mail and by cellular telephones. For many people, instant communication is something they value even more than their own leisure time. It is common to see people talking on cell phones, receiving text messages, or sending e-mails while on vacation, **Figure 26-26.** At the same time, while instant communication is possible, some feel that personal relationships suffer because of it. A great deal of communication is based largely on e-mail, text messages, and short phone conversations. Some people believe that we do not get to know others as well as in the past and, therefore, do not have as many one-on-one communication skills.

Impacts on Economics and Politics

Technology has large impacts on the economy and economic growth of countries around the world. These impacts are both positive and negative. Technology enables some areas of the world to be

Figure 26-25. Modern transportation systems allow jobs and products to move freely around the world.

Figure 26-26. This photo showing two young women text messaging at a movie theater shows that many people value instant contact with friends and family. (©iStockphoto.com/LeggNet)

productive and contribute goods and services from their economy to the global economy. For example, irrigation systems have enabled some African countries to grow crops to support their own citizens and to sell to other countries. Mining technology has created opportunities for some nations to build strong economies by mining their resources. Technology can create new political systems and power struggles, however, between those nations and civilizations that have technology and those that do not. Access to technology can also divide societies into upper and lower classes. In some societies, these classes have become the ruling class and the working class.

In the United States and many other countries around the globe, technology has advanced the potential for citizens' participation in government. Citizens can be better informed about government actions through communication technologies such as newspapers, government Web sites, Internet Web logs (blogs), and television channels covering government actions. These sources of media become extremely powerful in times of elections, as candidates use the technology to persuade and convince voters to support their positions. The technologies involved in polling also allow citizens a voice in government. The polls that different agencies conduct influence government actions.

Impacts on Culture

Technology affects the way people of different cultures live, the kinds of work they do, and the decisions they have to make. Advances in technology often raise questions about cultural beliefs. Medical technology is an area that often raises cultural dilemmas. Many cultures have specific beliefs about the use of medicine and other medical technologies and rely on either natural sources or religious sources of healing. When people with these cultural

Career Highlight

Airline Pilots

The Job: Pilots fly airplanes to carry people and cargo to their destinations. They plan the flights, check the aircraft to ensure it is functioning properly, and make sure passengers and cargo are loaded correctly. Pilots also review weather conditions en route and at the destination, choose a route, and fly the airplane to the destination.

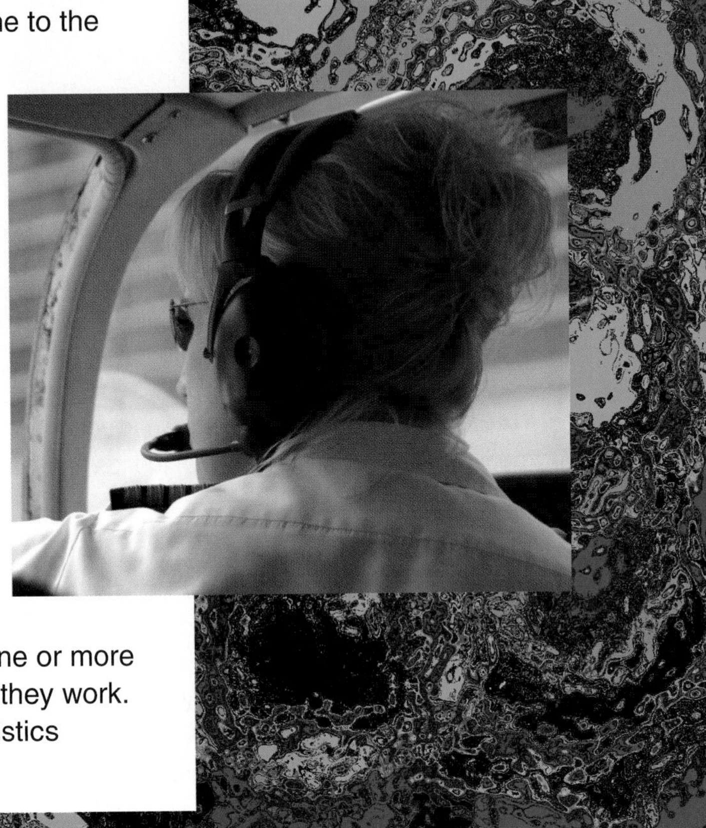

Working Conditions: Airline pilots cannot fly more than 100 hours a month or 1000 hours a year. Most pilots have variable work schedules, in which they work several days in a row and then have a number of days off. The majority of flights involve overnight layovers, causing pilots to be away from home a number of days each month. Pilots encounter mental stress from being responsible for the safety of the passengers, cargo, and aircraft.

Education and Training: These pilots hold an airline-transport pilot's license. This license requires the person to be at least 23 years old and have a minimum of 1500 hours of flying experience. The flight experience must include both night and instrument flying. Additionally, airline pilots must pass written and flight examinations given by the FAA. Most pilots have one or more advanced ratings, depending on the company for which they work.

Career Cluster: Transportation, Distribution, & Logistics

Career Pathway: Transportation Operations

beliefs become seriously ill, the opportunity to use a new technology must be weighed against their cultural beliefs.

A decision must be made before the use of any technology. Often, the decision is a simple and relatively easy choice. Other times, however, the decision of whether or not to use a technology involves ethical issues. Cloning, nuclear power, space research, and genetically engineered foods raise ethical and cultural questions. See Figure 26-27. Even the creation of software that allows for the downloading of copyrighted materials requires ethical choices to be made. Understanding the impacts technology has on culture and ethics is an important part of being technologically literate.

Figure 26-27. Genetically modifying plants is one of many technological issues that raises cultural and ethical questions. (©iStockphoto.com/chemicalbilly)

Summary

Technology has contributed to our way of life. This capability has given us many useful products and services. We live better because of technology. People have more comfortable homes and better health care. We can travel and communicate with more ease.

Technological advancement is not, however, without a price. There are negative impacts. Poor use of technology can cause air and water pollution. Poor farming, forestry, and mining practices can damage soil. Jobs change or disappear as a result of technological advancements.

We can never go back to a life with a low level of technology. Most people would not want to do that. To eliminate or reduce our technology's undesirable effects, however, we need to use it wisely. We need to keep ourselves prepared to produce and use new technologies.

STEM Connections

Science

Collect samples of water from areas in the community. Test them for purity. Chart your results.

Mathematics

Research the length of the workday for 10-year periods (decades), starting in 1800. Graph the results. Complete the same activity for average income (wages) from factory workers, starting in 1900.

Curricular Connections

Social Studies

Collect news articles about the impacts of technology on people and society. Arrange them in terms of positive and negative impacts. Develop a poster communicating your findings.

Activities

1. On your way to school or home, list examples of pollution you see or smell. Describe what caused them and how they can be reduced.
2. Prepare a scrapbook or poster showing ways technology is benefiting humankind. Use the Internet, newspapers, and magazines as sources of information.
3. Prepare a scrapbook or poster showing types and sources of pollution. Use the Internet, newspapers, and magazines as sources of information.
4. Visit a farm or forest. Discuss soil-erosion problems with the owner.
5. Choose a culture different from your own. Research ways technology has influenced its development. Write a short report on your findings.

Test Your Knowledge

Do not write in this book. Place your answers to this test on a separate sheet of paper.

1. What are five reasons to develop and use new technologies?
2. Describe the importance of engineered materials.
3. Give two examples of benefits people get from new and improved tools.
4. Paraphrase the meaning of a global village.
5. Summarize the main advancements in transportation over the past century.
6. How has technology changed the types of buildings we can build?
7. Producing more in less time means you are more _____.
8. Identify three things people can do to avoid scarcity of resources.
9. List four types of negative impacts of technology.
10. Losing your job because of technology is called _____.

Reading Organizer

On a separate sheet of paper, create a detailed outline based on what you've read about technology's impacts on society.

Example:

I. Technological change
   A. Materials
      1. Refined materials
      2. Engineered materials

CHAPTER 26 ACTIVITY A

TECHNOLOGICAL IMPACTS

INTRODUCTION

The technology we use impacts our daily lives. The clothes we wear, the types of vehicles we ride in, and the food we eat all require technological decisions. These technological decisions must be made with great care because they affect more than just our own lives. The technology we use affects society. When we use and dispose of technological products, they affect other people, our resources, and the environment. We must consider the impacts of the use and disposal of each product of technology we find in our daily lives. This activity gives you an opportunity to consider the impacts of several products you use on a daily basis.

MATERIALS AND SUPPLIES

- A pencil.
- Paper.
- Assorted classroom materials.

PROCEDURE

Work by yourself or with one partner. During the activity, complete the following steps:

1. Prepare a form similar to the one below. See **Figure 26A-1.** Your form should fill a piece of paper.

Technological-Impact Survey						
	Use			Disposal		
Item	People	Resources	Environment	People	Resources	Environment
1						
2						
3						
4						
5						

Figure 26A-1. A technological-impact survey form.

2. Select five objects from around the classroom, the school, or your home. Try to pick objects that are different from each other (for example: a school bus, a stapler, and a peanut butter–and–jelly sandwich).
3. Complete the form by doing the following:
   A. List each object in the first column.
   B. Record the positive and negative impacts on people, resources, and the environment that exist while the object is being used. Place a *P* after each positive impact and an *N* after each negative one.
   C. List the positive and negative impacts on people, resources, and the environment that exist when the object is disposed of at the end of its life cycle. Place a *P* after each positive impact and an *N* after each negative one.
4. Select one of the objects. Think of ways to decrease the negative impacts.
5. Create a report including the following items:
   A. A description of the object.
   B. A drawing of the improved product.
   C. An explanation of the changes made.

TSA Modular Activity

This activity develops the skills used in TSA's Environmental Focus event.

ENVIRONMENTAL FOCUS

Activity Overview

In this activity, you will identify an environmental problem in your community. You will then develop a solution to the problem and implement the solution. Finally, you will analyze the effectiveness of your solution and create a multimedia presentation and report outlining the project.

Materials

- Presentation software.
- A three-ring binder.
- A computer with word processing software.

Background Information

- **Identifying a problem.** Begin by identifying at least three environmental problems. Use research techniques to identify these problems. Some methods you can use to learn about environmental problems include the following:
  - Read articles dealing with environmental issues in your local newspaper.
  - Contact local environmental organizations.
  - Interview people at your school.

  You might want to consider problems related to the following topics:
  - Air-pollution issues, such as automobile exhaust, smokestack emissions, and cigarette smoke.
  - Water-pollution issues, such as a polluted creek or stream.
  - Land issues, such as littering in public areas and construction development in natural areas.
  - Animal-protection issues, such as a hazardous situation for native animals.
  - Waste issues, such as lack of recycling in homes, businesses, or schools.

For each problem you identify, use brainstorming techniques to list possible solutions. Select a problem-solution combination you will be able to research and implement effectively.

- **Research.** Using books, periodicals, and the Internet, research the problem. Clearly define the problem. What is causing the problem? Is the problem creating hazards? If so, what are they? What types of solutions have others used for similar problems?
- **Implementing your solution.** If your solution depends on involving others, you need to find volunteers or create a method of informing people of your solution. You might need to contact individuals or post printed flyers. Before implementing your solution, be sure to consider how you will analyze it.
- **Analyzing your solution.** Analyze your solution by comparing the situation before the solution was implemented to the situation after the solution was implemented. If possible, select several evaluation criteria. These might include a survey of opinions or measurable values. Photographs might also be helpful in analyzing your solution.

Guidelines

- The multimedia presentation must include your method of identifying the problem, your research, and your data collection and analysis.
- Your report must be created using a word processor and presented in a three-ring binder. This report must include a title page, a table of contents, problem identification, research (including photographs, if applicable), data collection and analysis, strategies, conclusions, recommendations, and a copy of the multimedia-presentation slides. Your report must not exceed 12 pages (not counting the presentation slides).

Evaluation Criteria

Your project will be evaluated using the following criteria:

- Problem identification.
- Solution functionality, practicality, and impact.
- The multimedia presentation.
- The report.

CHAPTER 27

Technology and the Future

Did You Know?

Science fiction writers often create imaginary future worlds in their stories. Some writers have actually been successful at predicting the future. Gene Roddenberry is one such writer. His *Star Trek* television series and movies have shown devices that, years later, have been invented in some form. The communicators used in *Star Trek* are very similar to the cell phones used today. The computer used in the ship has functions similar to the capabilities of our current computers. Other science fiction books and movies feature technologies such as homing beacons and sensors. These technologies are now realities. Today, some filmmakers even hire futurists so their films are as accurate as possible.

Objectives

The information given in this chapter will help you do the following:

- Explain the role of a futurist.
- Explain the concept of a forecast.
- Summarize the four methods of forecasting.
- Identify several potential future technologies that might exist in the seven applications of technology.

Key Words

These words are used in this chapter. Do you know what they mean?

biomass fuel
Delphi Method
expert survey
forecast
futurist
hypersonic flight
manufactured panel
scenario
trend

Preparing to Read

Look carefully for the main ideas as you read this chapter. Look for the details that support each of the main ideas. Use the Reading Organizer at the end of the chapter to organize the main and supporting points.

Technology is constantly changing the way we live our lives. See **Figure 27-1.** The new high-tech devices will soon be commonplace or even obsolete. Look around and imagine how your life would be without the technology you have. You would be affected in ways you probably cannot imagine.

It is often helpful, when looking into the future, to look back on the past to understand how far we have come. Imagine that, instead of being born when you were, you were born about 200 years earlier, in 1800. Life would be very different in many obvious ways. There would be no Internet or television. Automobiles and microwave ovens did not exist. There are many things you take for granted that would not be around. It would be another 6 years until your parents could drink coffee from a coffeepot that strained the grounds. You would be 44 years old before you were able to communicate with people in other towns through the telegraph. More than likely, you would not even know anyone from another town. Even if the town was only 10 miles away, it would take four hours to get there by wagon. See **Figure 27-2.** You would not need to go shopping for designer clothing because the sewing machine was not invented until 1833. Your clothes would not have zippers until you were 93 years old. You would not have lived to see airplanes, rockets, frozen food, or nylon fabric.

Figure 27-1. Technology, such as robotic devices, is becoming more advanced. As seen in the image, Advanced Step in Innovative Mobility (ASIMO) can descend stairs. (Honda)

If that was life 200 years ago, imagine life 200 years from now. Would you even know how to use the technology of the time, if you were transported 200 years into the future? What types of technology will people have? See **Figure 27-3.** Will people still be using computers? Do you think cars will still have wheels and use gasoline? Will there even be cars? We know life in the future will be different.

Figure 27-2. The horse and surrey is a mode of travel from many years ago. You can still find it used today in Amish communities.

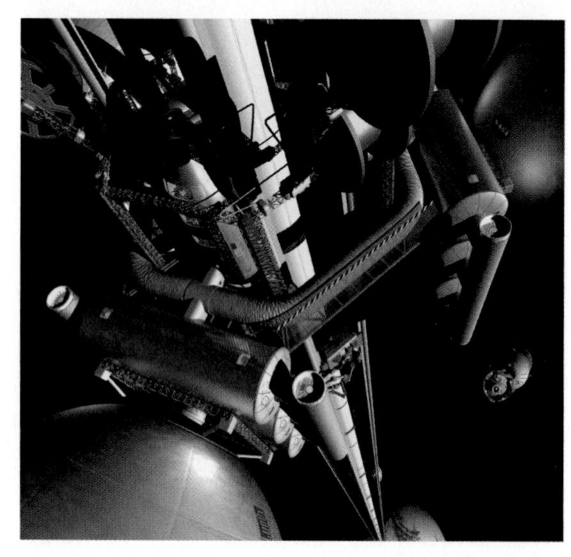

Figure 27-3. The space elevator is one possible future technology. (NASA's Marshall Space Flight Center and Science@NASA)

The world is always changing. The only thing constant is change. There is no way to know the exact answers to all our questions about the future. There are, however, methods of predicting what will come in the future.

Futurists

Have you ever tried to predict the future? Maybe you predicted your score on a test. You might have speculated which teams would make the play-offs. If you have done anything similar to this, you might be a beginning futurist. ***Futurists*** are people who spend their time studying the future. They are not fortune-tellers and do not have psychic powers. These people are much more scientific. They use data and information about what is happening now to create descriptions of what might be coming in the future.

Futurists are found in many careers and fields. Economists are often futurists. They use the current events in the financial markets to predict what will happen to the economy. Science fiction writers are futurists. They write about the future and technologies people will be using in many years. A futurist does not, however, have to look years into the future. Some futurists look into the future only several hours or days. One of the most common kinds of futurist is a meteorologist. These futurists predict the weather hours into the future. A long-range prediction for a meteorologist is one week.

The job of a futurist is very important. There are very practical uses for the reasonable guesses of futurists. The predictions of futurists help change some courses of action. If a futurist sees that the economy will go into a recession, lawmakers and consumers can make changes to help the economy. Another use of a futurist is to determine the jobs that will be needed in the future. A futurist might help a college or university change the classes it offers, so the school can prepare students for the jobs that will be needed when the students graduate.

Forecasting

The job of the futurist is to make predictions. The predictions are known as *forecasts*. ***Forecasts*** are estimates of the possible future. A weather forecast is one example. This forecast is not 100% accurate.

Career Highlight

Conservators

Science, Technology, Engineering & Mathematics

The Job: Conservators care for and preserve art, artifacts, and specimens displayed in museums and used in research. They use a variety of equipment and techniques, including X rays, microscopes, special lights, and chemical testing, to examine objects and determine the condition, the need for restoration, and a method of preservation. Conservators usually specialize in a particular material or group of objects, such as printed materials, paintings, textiles, decorative arts, furniture, or metals.

Working Conditions: Generally, conservators work under contract to treat particular items in a museum or another institution. They might be self-employed or work for a company specializing in conservation of artifacts.

Education and Training: Most conservators have a master's degree in conservation or a closely related field and considerable work experience. A few individuals enter conservation through apprenticeships with museums or private conservators. Apprenticeships are often supplemented with courses in history, chemistry, and art.

Career Cluster: Science, Technology, Engineering & Mathematics

Career Pathway: Science and Math

The weather forecast is the forecaster's most scientific guess. This forecast can change and be incorrect at times.

Weather forecasts, similar to all forecasts, are more accurate for the near future. A forecast for the use of electric automobiles in the next two weeks is fairly easy to develop. An estimate for the use of these automobiles in 50 years is much more difficult. There are basically three types of forecasts: short-range, midrange, and long-range. In most types of forecasting, short-range forecasts are projections for the next 2 years. See Figure 27-4. Midrange forecasts can look into the future about 10 years. Long-range forecasts can extend through the next 50 years. It is hard to even imagine what life will be like in 50 years, let alone to try to predict the technology that will be used.

Futurists use different methods of forecasting because it is hard to simply guess what the future holds. These methods help the futurists more accurately predict what the futures holds. The most popular methods include the following techniques:

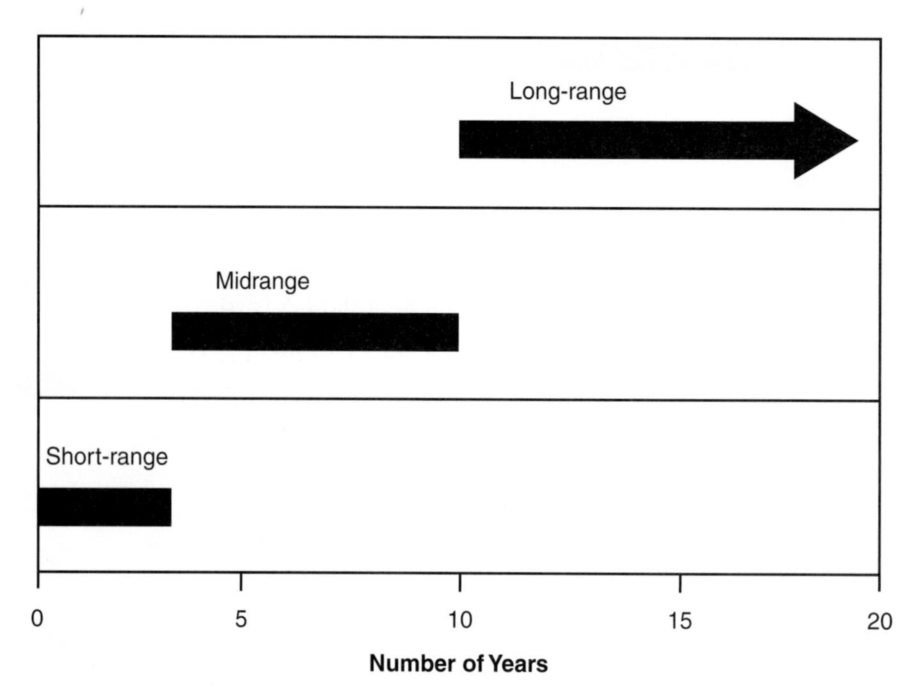

Figure 27-4. The ranges of forecasts.

- Trending.
- Surveying.
- Creating scenarios.
- Modeling.

Trending

Trending is a very popular method of forecasting. This method is used in many types of forecasts and is based on trends. A ***trend*** is an activity that occurred in the past, is occurring at the present, and has an outcome expected to occur in the future. Trends are patterns of technological activities showing a tendency or taking a general direction. They are used to offer guidance in determining if a product or system should be used. See **Figure 27-5.** Futurists examine trends and determine in which direction they are moving. They also estimate how long the trends will last and monitor possible consequences of technological progress. For example, in the past, cellular phones were much larger in size than they are today. This trend suggests cellular phones will be even smaller in the future. There is a limit, however, to how small these phones can be and still work. A futurist will estimate the year the cellular phones will reach this size limit.

Surveying

There are several ways futurists conduct surveys. The first is called an ***expert survey***. In this method, several experts on certain topics are asked for their opinions. They can be asked a series of questions or asked to write an essay

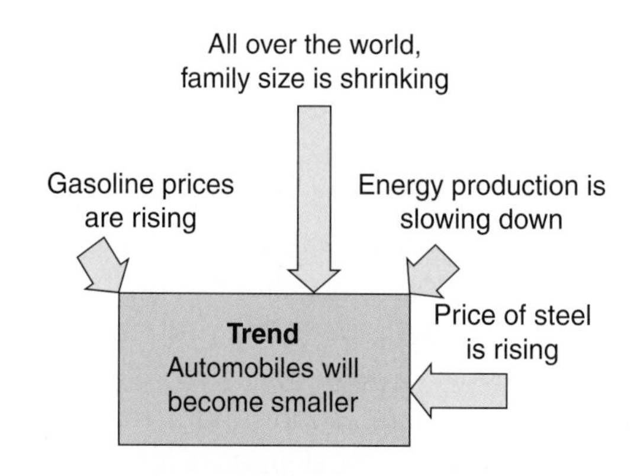

Figure 27-5. Futurists use past and present events to forecast the trends of the future.

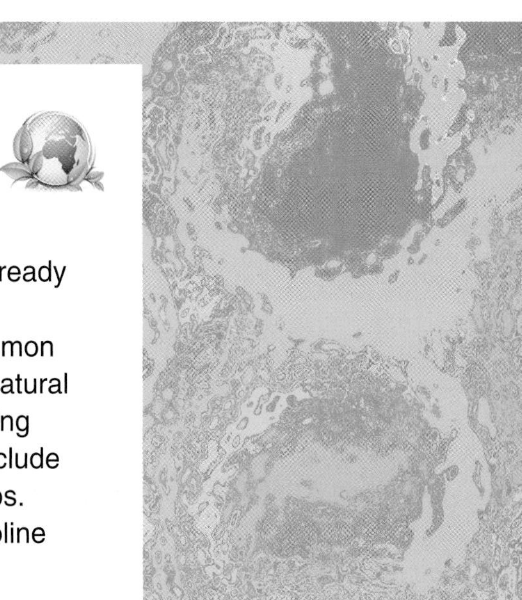

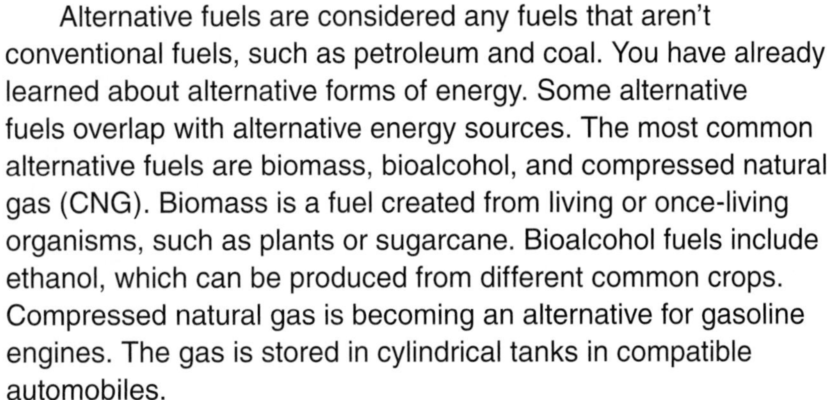

THINK GREEN

Alternative Fuels

Alternative fuels are considered any fuels that aren't conventional fuels, such as petroleum and coal. You have already learned about alternative forms of energy. Some alternative fuels overlap with alternative energy sources. The most common alternative fuels are biomass, bioalcohol, and compressed natural gas (CNG). Biomass is a fuel created from living or once-living organisms, such as plants or sugarcane. Bioalcohol fuels include ethanol, which can be produced from different common crops. Compressed natural gas is becoming an alternative for gasoline engines. The gas is stored in cylindrical tanks in compatible automobiles.

on their thoughts about the future. This method results in a number of different outlooks on the future.

Some surveys, however, are used to develop a single view on the future. One type of survey used for this is the ***Delphi Method***. The RAND Corporation developed the Delphi Method to forecast the future. The Delphi Method begins by sending a number of experts a survey about the future. These experts respond to the questions and send the surveys back. All the responses are gathered and sent back to the experts, along with a new survey. The experts again complete and return the surveys. This process continues until they all reach agreement on the final conclusion.

Creating Scenarios

A ***scenario*** is an outline of a series of events. See Figure 27-6. The events can be either real or imagined. Futurists create scenarios for planned and proposed events. A scenario is divided into three sections. The first is the explanation of the event. A possible future event might be that there is not enough farmland to grow the crops that the growing population needs. The next section of the scenario is a possible solution to the problem. A solution to the problem of

By the year 2040, city planners found that our major cities needed to solve serious problems. Moving about in the central city was a problem. Streets were clogged with traffic. Expressways had not helped the problem to any degree. Even worse, air pollution from the traffic was threatening the health of citizens.

Planners decided on a bold plan to reduce traffic. First, all personal-vehicle traffic was banned from the cities. Mass-transit systems replaced the automobile. Only trucks delivering goods, foodstuffs, and services were allowed within the city limits. Bicycle traffic was allowed anywhere.

Neighborhood shopping areas were located where they were within easy walking distance for shoppers. Streets were no longer congested with traffic. Air pollution was reduced. Many commuters were attracted back to the central city by the quality of life they found there.

Figure 27-6. A scenario can be an imagined event. This outline looks at related happenings and attempts to predict what the solution might be. Usually, several scenarios are written for the same set of events.

less farmland might be to develop aquaculture. Aquaculture is the growing of plants in the ocean. The last section of a scenario is the expected outcome of the solution. In this example, the outcome might be that the use of aquaculture has replaced the need for crops that previously required farmland.

This scenario contains only one possible solution and outcome. Most futurists create several scenarios for each event. By creating many scenarios, they are able to choose the best outcome. They can then suggest the best solution to the problem.

Modeling

Models are helpful in controlling events in the future. They are used to change the outcome of a system or process. Computer, mathematical, and graphic models are all used in forecasting the future. Weather forecasting often uses computer models. These models are used to show the movements of warm and cold fronts. Computers are also used to calculate large mathematical models. See Figure 27-7. Economists use mathematical models to show how the economy will change over time. The data from mathematical models is often put into graphs.

Graphs are helpful because they are usually easy to understand. It is also easy to compare two different graphs. Futurists often create two different graphs by changing a single variable. By making different graphs, the researcher can ask, "What if?" For example, futurists might be studying the likelihood of developing cars that steer automatically. They can develop one graph showing the likelihood of cars that ride on rails being invented. Another graph can be created to show the likelihood of cars that are steered with a GPS being invented.

Figure 27-7. National Hurricane Center (NHC) Director Bill Read and Hurricane Specialist Eric Blake study Hurricane Dolly at landfall. (NOAA)

Graphic models can also help organize the future. Charts are used to plan future events. Building contractors use charts to plan the construction of a building. See Figure 27-8. These contractors list all the different processes that must be completed in the form of a timeline. They use the chart to organize and control the process.

The Future of Technology

When we think of the future, it is common to think of the technology that will be used. You might wonder what types of houses you will live in or what foods you will eat. Maybe you think about the vehicles you will drive or the medications that will be available. Every application of technology will see changes in the future.

Agriculture and the Future

Agriculture is the application of technology dealing mostly with growing foods we eat. Throughout history, agriculture has become much more advanced. Agriculture will continue to become more advanced in the future. There are several areas of advancement we can forecast for the future of agriculture:

- The farming and harvesting of crops will be done in new places. Currently, crops are mainly grown in soil, on land. In the future, they will be grown without soil, using hydroponics. They will also be grown in the ocean, a method known as *aquaculture*. Crops will probably be grown in space, if people begin to live there for long durations of time.
- Crops will be genetically altered. One reason for this is so they can grow without the use of pesticides. They will be altered to withstand bugs and other pests. Crops will also be altered to provide additional vitamins and minerals.
- The process of farming will become more automated and computer based. See Figure 27-9. Farmers will be able to farm more land in less time. Computers will analyze and process data to help select the correct crops, apply fertilizers, and harvest outputs.
- The technology will exist to genetically engineer animals. The purpose of the engineering will most likely be to create animals producing high yields, with less feed.

June	July	Aug	Sep	Oct	Nov	Dec	Jan
Documents					Drywall		
	Survey/Grade					Flooring	
		Site work/Foundation				Interior trim	
			Rough carpentry			Final inspection	
				Plumbing			
				HVAC			
				Electrical			
				Roofing			

Figure 27-8. An example model used to organize the construction of a home.

Figure 27-9. Today, many types of farm equipment are equipped with computers and GPSs. In the future, the computers will be more advanced and control more functions. (NASA's Marshall Space Flight Center and Science@NASA)

Construction and the Future

In the past, constructed works were built mainly from wood, stone, and steel. In the future, this might be much different. Future buildings might be built with new lightweight materials, such as plastics. While building materials are a large change, other changes will include the following:

- Buildings will be built to take advantage of computer technology. Homes will be built to include a number of sensors that interact with a central computer. They will also be built so computers can control the rooms. A central computer will control the lights, heating, cooling, music, and many other features.
- Homes will be built with ***manufactured panels***. These panels include the interior and exterior surfaces of the walls. See Figure 27-10. The electrical and plumbing chases are prebuilt inside the panel. Robots will be able to assemble the pieces to complete the construction of the home.

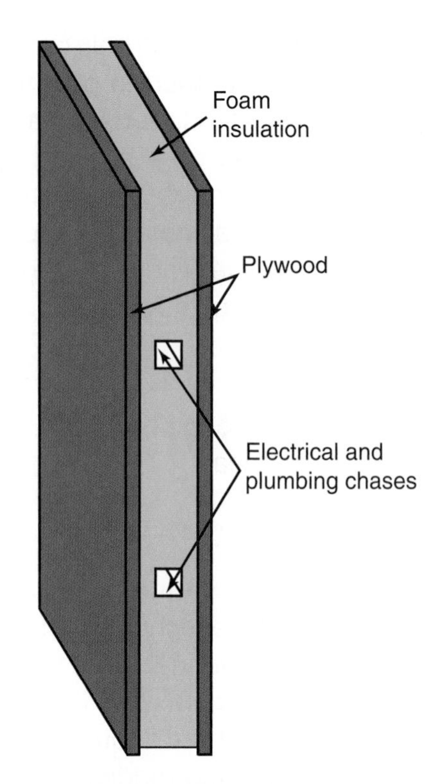

Figure 27-10. Building panels can be used to replace typical frame construction.

- Buildings will be constructed in the sea and underground because cities are getting larger and there is less land.

Energy and the Future

The energy used in the future will be much different from what is used today. The current trend shows that the world will run out of fossil fuels. In order to keep this from happening, several events will take place in the future:

- Energy conservation will be a major design factor. Products will be designed to use as little energy as possible. Homes will be better insulated and will use energy more efficiently.
- New types of fuels will be used as alternatives to fossil fuels. ***Biomass fuels*** are an example of a new type of fuel. They are mainly plants that absorb energy from sunlight when

they grow. The plants (for example, trees, corn, and sugarcane) release the energy when they are burned. Biomass fuels can also include garbage. They are cleaner and cheaper than fossil fuels.

- Other existing sources of energy will be used more in the future. Solar, wind, and nuclear energy will be used more widely. See Figure 27-11. Tidal energy will be collected from the waves of the ocean.

Figure 27-11. Wind energy is one form of energy that will be used more in the future. (AMEC Wind)

Communication and the Future

The years since the mid-1980s have seen great advancements in the areas of information and communication, with the evolution of the PC. In the years to come, those advancements will be made obsolete. Computers will become more powerful and smaller in the future. The spaces for the storage of information will also become smaller. Entire libraries of information will be held on devices the size of an eraser. The future world of communication will include technologies such as the following:

- The use of holography as a form of media storage and projection.
- Videophones, interactive video screens, and virtual reality. These technologies will be common in homes. See Figure 27-12.

Figure 27-12. Advances in electronic paper have made the invention of the electronic book reader possible.

- Printers and fax machines able to print in three dimensions.
- Voice recognition. This technology will replace the keyboards on computers and will allow communication with a number of devices and appliances.

Medicine and the Future

Medical technology is an area of great concern. Due to advances in medical technology and medicine, people are living longer than in the past. The longer people live, the more important medical technologies are. As people age, they need devices such as eyeglasses and hearing aids. Some people even need new organs because of age or illness. In the future, doctors will be able to create new organs for those in need. They will also be able to do the following:

- Take three-dimensional, color X rays of patients. See Figure 27-13.
- Replace limbs with prosthetics that use electricity to function similarly to human limbs.
- Manipulate and use genes to treat diseases.

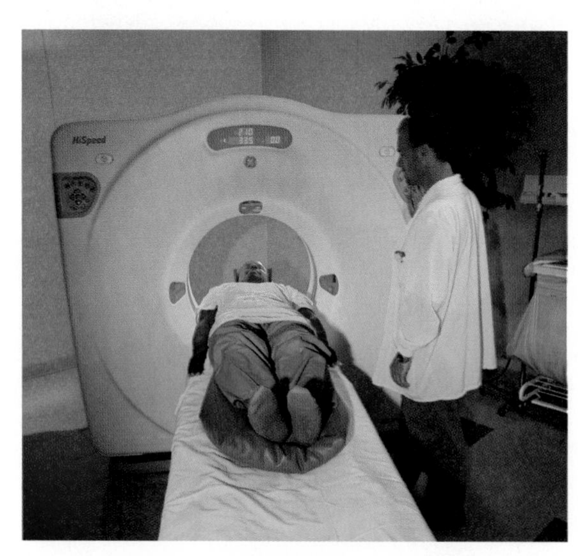

Figure 27-13. Digital imaging will advance. Doctors will be able to review more advanced images of the body. (NASA)

- Implant chips containing medical records into patients. The patients will be scanned to find all their information. This will be very helpful in emergency situations. In these situations, doctors do not have time to locate the patients' records.

Manufacturing and the Future

The manufacturing of goods will be different in the future. Manufacturing will be much more automated than it is now. Robots and machines will replace low-skilled workers. They will be used to complete repetitive or dangerous tasks. The use of these robots and machines will also create jobs for people who can design and repair robots. Other advances in manufacturing will include the following changes:

- New materials will change the way products are manufactured. They will be lighter and stronger than current materials. The materials will be designed and created at the molecular level using nanotechnologies.
- Products will become smaller and smarter. More products will contain computers and computer chips. This will require products to be manufactured on a smaller scale.

Transportation and the Future

There are several problems facing transportation in the future. One is the growing congestion on roadways. Designers will design solutions to this problem. High-speed commuter trains and vehicles better designed for city traffic will be developed. Future transportation vehicles will also include the following:

TECHNOLOGY EXPLAINED

nanotechnology: the design of products at the molecular level.

Nanoparticles are often only several nanometers in length. This is extremely small. A piece of paper is around 100,000 nanometers thick. The use of nanotechnology can change the behavior and characteristics of a material. For example, a tennis racket can be built stronger and lighter with the use of carbon nanotubes. Computer chips can be built smaller and can run cooler with the use of nanotechnology.

The design and construction of nanotechnology structures can occur in two ways. The first is a top-down approach. Using this method, the materials are removed from a larger source by stamping, milling, or printing at an extremely small scale. A bottom-up approach uses nanomaterials to build new structures. This can create revolutionary products. A bottom-up approach is often impractical, however, because of the time it takes to build a structure made from such small particles. Scientists are working on ways for the particles to assemble themselves.

Nanotechnology is presently being used, and will be used in the future, to create useful and efficient products. Applications of nanotechnology include the delivery of medications directly to cancer cells and the creation of better vitamins. Self-cleaning glass and stain-resistant clothing can be created with nanotechnology. See **Figure A.** Nanotechnology can be used to create water-filtration systems and ink with unique properties. This design will certainly be used in the future to create stronger, smaller, and more efficient products.

Figure A. Nanotechnology was used to enhance this fabric for spill and stain resistance. (Nano-Tex)

- Vehicles designed to routinely travel between Earth and outer space. See **Figure 27-14.** Outer space will become a vacation spot for many travelers.
- Airplanes that can fly at hypersonic speeds. ***Hypersonic flight*** occurs at five times the speed of sound.
- Cars guided by rails. These cars will decrease the amount of traffic accidents. They will also increase the speed at which cars can travel.

Figure 27-14. Reusable space vehicles will make space travel more economical. (NASA)

Summary

The present is much different from the past. We know technology will continue to change. The future will be very different from today. It is interesting to try to predict the future. Predictions of the future are known as *forecasts*. Making these predictions is the main job of futurists. Futurists use several different methods to forecast the future. They base their forecasts on the events of the present and how the events will impact the future.

Forecasting helps us plan for the future. We are able to control and shape the future by making changes today. If a forecast shows we will have a problem in the future, we can take steps to make sure the problem never arises. Changes will occur in all areas of technology. We must prepare for them. The changes cannot be stopped. We can, however, control them.

Curricular Connections

Language Arts

Write a creative essay about the technology that will be used in the future.

Social Studies

Write a short paper on the way society will change in the future. Explain where and how people will live and work. Also, discuss how people will communicate with each other.

Activities

1. Imagine your life 50 years from now. Choose an event that you are forecasting will happen in your life. Create a scenario of this event.
2. Select a career in forecasting. Write an essay on the requirements, duties, and education required for the job.
3. Choose a future technology from one of the contexts of technology. Create a display showing how the technology will function and be used.
4. Prepare a report on an emerging and innovative technology. You can choose one of the emerging technologies described in this chapter or research elsewhere to find a topic. Use a variety of sources in preparing your report. Web sites, magazines, and television programs are good sources of information on emerging technologies.

Test Your Knowledge

Do not write in this book. Place your answers to this test on a separate sheet of paper.

1. Identify five ways your life would be different if you were born 200 years ago.
2. A(n) _____ is someone who uses information about the present to forecast the future.
3. Give three examples of careers in which futurists can be found.
4. A(n) _____ forecast is a projection about 2 to 10 years into the future.
5. Name four ways to forecast the future.
6. A trend is an activity that _____.
   A. occurred in the past.
   B. is occurring in the present.
   C. has an outcome expected to occur in the future.
   D. All of the above.
7. The Delphi Method is one example of a model. True or false?
8. Summarize the process of creating scenarios.
9. List three examples of potential future technologies in the area of agriculture.
10. Buildings will be built exactly the same in the future as they are today. True or false?

Reading Organizer

Draw a bubble diagram for each main idea in the chapter. Make each of the main ideas the central bubble, while using details in smaller bubbles to surround the main points. An example from this chapter is shown.

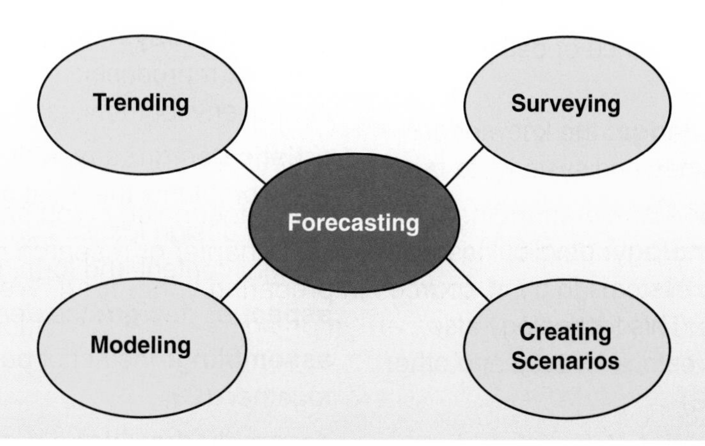

Technical Terms

A

acetate: a thin plastic used in product and architectural models. (15)

acid rain: pollutants in the air captured by moisture and returned to Earth. (26)

acoustical property: a quality of a material governing how the material reacts to sound waves. (5)

acrylic: a plastic material that comes in sheets and can be purchased at either a hardware store or a hobby store. (15)

active solar system: a system using moving parts to capture and use solar energy. This system can be used to provide hot water and heating for homes. (21)

adaptation: using an invention for something other than the intended purpose. (9)

advertising: the act of making a product or business known to the public. (23)

aesthetic: the appearance or design of a product. (12)

agricultural knowledge: the knowledge about using machines and systems to raise and process foods. (1)

agricultural technology: developing and using devices and systems to plant, grow, and harvest crops. This technology also includes raising livestock for food and other useful products. (3)

agriculture: using materials, information, and machines to produce the food and natural fibers needed to maintain life. (19)

air pollution: dust, fumes, smoke, gases, and other materials that damage the air. (26)

air transportation: a mode of transportation using aircraft to move people and cargo to their destinations. (25)

alloy: a mixture of two or more elements. (5)

alphabet of lines: technical lines used in multiview drawings. (17)

animal husbandry: the breeding, feeding, and training of animals. (19)

animal science: the science of crossbreeding livestock. (19)

anode: the negative area of a battery. (21)

apartment building: a building that has self-contained housing units, each occupying only part of the building. (20)

appearance: a characteristic people see and have feelings toward. (14)

aquaculture: the growing and harvesting of fish, shellfish, and aquatic plants in controlled conditions. This type of agriculture uses ponds, instead of soil, to grow its crop. (19)

architectural drawing: a drawing showing buildings and structures to be constructed. (17)

artificial ecosystem: a human-made complex reproducing some facets of the natural environment. (19)

artistic design: a type of design used to solve problems the artist chooses. This type of design focuses on two aspects of design: content and form. (10)

aspect of design: the design criteria. (14)

assembling: the act of putting parts together. (23)

assembly drawing: a drawing made to show how to assemble a product. (17)

assess: to decide or determine. (1)

assessment: the final stage in the testing of a solution. (16)

audience assessment: gathering information about a group of people to be reached by a message. (22)

audio recording: the recording of sound, such as talking, singing, or music, on tapes, records, or CDs. (22)

B

balance: the weight of the elements on each side of a design. (14)

baler: a machine that picks up a windrow and conveys it into a baling chamber, where the hay is compressed into a cube. (19)

bar code reader: a system used to read numerical codes on packages and tags. (22)

bill of materials: a list of all the parts needed to make one product. This list gives the part names, sizes, and quantities of materials to be used. (17)

bioenergy: energy from organic matter. (6)

biogas: a gas produced by processing animal and plant waste. (6)

biomass: the sum of all organic matter in an area. (6)

biomass fuel: a new type of fuel that will be used as an alternative to fossil fuels. (27)

biotechnology: a part of technology dealing with using biological agents in an industrial process to produce goods or services. (19)

boundary: a property line that a survey establishes. (20)

box: a shape that can be either a cube or a rectangle. (13)

brainstorming: a group process for solving a problem. (13)

brand name: the title of a line of products. (18)

bristol board: similar to common poster board, this product can easily be curved. This board is not, however, very rigid. Usually found in sheets 1/16″ thick, it is smooth and shiny on both sides. (15)

broadcast message: communication that radio and television stations carry. (22)

Bronze Age: the stage in human history that took place after the Stone Age. (3)

building code: a standard or regulation regarding the safe construction of a structure. (18)

bullet point: a single piece of information, such as one item from a list. (12)

C

canal lock: a device allowing ships to change elevation as they travel through a canal. (25)

career ladder: the process people go through to work their way up to better jobs through experience, additional training or education, and perseverance. (8)

cargo: materials or products carried on transportation vehicles. (25)

carrier: a channel for carrying electronically generated messages. A carrier can be radio waves of the electromagnetic spectrum, a wire conductor, or optical fibers. (22)

casting: a process in which an industrial material is first made into a liquid. (23)

cathode: the positive area of a battery. (21)

central processing unit (cpu): the part of the computer processing and interpreting all the information. (22)

ceramic: inorganic matter made up of crystals. (5)

challenge: an obstacle or a goal needing to be met. (9)

chart: a graphic aid used to model how a set of information is arranged or highlight a series of events. Charts can also show sets of data. (15)

chemical converter: a technology that converts energy in the molecular structure of substances to another energy form. (21)

chemical energy: a reaction between two substances when mixed. (6)

chemical processing: using chemicals to change the form of materials. (5)

chemical property: a quality controlling how a material reacts to chemicals. (5)

chipboard: a product often used for backers in calendars, notepads, and packages of paper. Chipboard is gray in color and is more rigid than bristol board. (15)

civil structure: a structure designed for public use, such as a road, a bridge, a monument, or another public project. (20)

clay: a natural material used to create models. (15)

client: a customer who buys designers' work. (14)

clinical test: a test conducted in three phases to decide if a drug is safe and effective. (24)

closed-loop system: a control system that uses feedback and is a built-in part of the device. (2)

coal: a solid form of fossil fuel made up mainly of carbon. (6)

color: a property of light determined by the light reflected off an object. (14)

combine: a machine used to harvest grains. (19)

commercial building: a building used for professional offices, a shopping center, a supermarket, a lodging establishment, or a repair facility. (20)

communication: the act of exchanging ideas, information, and opinions. (22)

communication and information technology: developing and using devices and systems to gather, process, and share information and ideas. (3)

communication satellite: a broadcasting station in outer space that can receive messages from Earth and then relay those messages back to Earth. (22)

communication skills: skills that include the ability to use speech and writing to explain and present ideas clearly. (8)

communication technology: the use of equipment and systems to send and receive information. (22)

compact disc read-only memory (CD-ROM): a CD that stores information and can be accessed, but not altered, by the user. (22)

company: an economic organization that changes resources into products and services. (8)

compass: a drawing tool used to create circles and arcs. (17)

competition: two or more companies selling the same types of products and striving to gain customers. (18)

competitor: a company selling the same types of products as another company. (18)

composite: a solid material combining two or more materials, yet each material retains its own properties. (5)

comprehensive: a complete sketch describing the arrangement of the elements, including type, illustration, and white space, when preparing to produce a published message. (22)

computer: a programmable electronic device that can store, retrieve, and process data. (22)

computer-aided design (CAD): a type of design using computers to create and store technical drawings for products. (17)

computerized tomography (CT) scanner: a scanner that rotates around the patient's body, as a computer processes the data and creates a cross-sectional image of the body part being scanned. (24)

computer model: a three-dimensional view generated by a computer screen. (15)

computer testing: a form of testing that can generate very accurate data. This testing requires a computer model and testing or simulation software. (16)

conclusion: the findings of research after all data has been gathered and analyzed. (12)

conditioning: an action altering and improving the internal structure of materials. This action changes the properties of the material. (23)

condominium: a living unit joined together with another, but owned by a separate family or individual. (20)

conductor: a material or an object permitting an electric current to flow easily. (21)

cone: a shape that is round at one end and comes to a point at the other. (13)

conservation: making better use of the available supplies of any material. (21)

constraint: a limitation or boundary for a design. A constraint tells designers how far they can go with a design. (11)

construction: the process of using manufactured goods and industrial materials to build structures on a site. (20)

construction knowledge: the knowledge about using machines and systems to erect buildings and other structures. (1)

construction technology: using systems and processes to erect structures on the sites where the structures will be used. (3)

consumer: a user of a product. (18)

consumer safety: a form of government regulation that requires businesses to disclose information about product use and safety. (16)

contact paper: a vinyl material that is sticky on one side. (15)

container shipping: using sealed containers to group and contain items for bulk shipment. (2)

content: the topic, information, or emotion an artist is trying to communicate. The content is often the problem being solved. (10)

continuous manufacturing: a system in which the parts move down a line. (23)

contractor: one who hires workers and directs building processes. (20)

contrast: a juxtaposition adding variety to a design. (14)

control: a subsystem enabling a vehicle to change speed and direction. (25)

conveyor: a stationary structure that moves material. The structure supports rollers or belts, over which the material is transported from one place to another. (25)

cornflakes: a breakfast cereal processed from corn. (19)

corrugated cardboard: a very sturdy material ranging in sizes from 1/8″ to 1/2″ thick. This cardboard has outer sheets of heavy paper, and the inner core is made of heavy paper bent into ridges. (15)

cost: the money spent to design a product and the money that will be spent manufacturing the solution. (16)

cost-benefit analysis: a method in which designers attempt to determine if the benefits of the design are worth the costs. (16)

creative thinking: the human ability to think in creative ways. (9)

criterion: an element of the problem needing to be solved. Criteria are some of the requirements, or parameters, placed on the development of the product or system. (11)

critical thinking skills: skills that include the ability to evaluate different sides of an argument and draw conclusions from this evaluation. (8)

crop: a grain, vegetable, or fruit that has been agriculturally cultivated. (19)

cultivator: a machine used on farmland to remove weeds and open the soil for water. (19)

custom manufacturing: a process in which products are designed and built to individual specifications. (23)

cylinder: a round shaft with one circle on each end. (13)

D

data: a collection of facts, numbers, and ideas. (7)

decode: to put a meaning to a message. To decode is to understand the message, so proper action can be taken. (22)

Delphi Method: a survey the RAND Corporation developed to forecast the future. (27)

descriptive knowledge: the knowledge of relationships, values, shapes, and forms used to describe objects and events. (1)

descriptive research: information gathered by measuring and describing products and events. This research describes something as it is. (7)

design: a common technique used to create products or systems. A design is a plan for a device or product. (1)

design brief: an outline the designer creates to guide the entire design process. This outline includes the requirements, or the criteria and constraints, of the design. (10)

design process: a series of steps to design a new product or system. (10)

design report: a portfolio containing documents related to a design project. This report includes the design brief, sketches, testing results, drawings, and a final recommendation. (17)

design sheet: created by designers, this sheet acts as a summary of possible solutions to a design problem. (14)

design style: a period of design in which many of the same appearance features are evident. (14)

desired output: something produced that had been hoped for. (2)

detailed drawing: generally, an orthographic drawing giving the size and shape of an individual part. (17)

detailed sketch: a sketch containing detailed information about the size and shape of the product. A detailed sketch shows length, width, squareness, and roundness. (10)

diagnose: to identify a medical problem. (24)

die: a set of metal blocks used to cut out, form, and stamp material. (23)

digital videodisc (DVD): a CD used for storing information, such as high-resolution video material. (22)

digitize: to put waves into the form of a series of numbers, so a computer can process them.

dimension: a measurement showing three different types of information: size, location, and shape. (14)

dimensioning: a process using two types of lines in a drawing: extension and dimension lines. (17)

disc: a tilling machine consisting of a series of cupped discs mounted on one axle. (19)

discovery: the act of first noticing something that occurs naturally. (9)

discrimination: an expression of a biased attitude, act, or behavior toward another person. (8)

disease: any change interfering with the normal functioning of the body. (24)

distance multiplier: a simple machine, such as a lever, that can change the amount of movement applied to it. (4)

double-sided tape: tape that is sticky on both sides and used to stick one material on top of another. (15)

downlink: the signal relayed back to an Earth station. (22)

drawing board: a smooth surface on which to draw. This board includes one straight edge for the T square to move along. (17)

drilling machine: a machine tool with a rotating drill feeding into the workpiece and producing a hole. (4)

drip irrigation: a type of irrigation that uses main lines to bring water near the plants. (19)

drug: any substance used to prevent, diagnose, or treat a disease. (24)

ductility: the ability of a material to be pulled, stretched, or hammered without breaking. (5)

durability: the amount of time a solution will work. (16)

E

earthquake protection: actions people take to make buildings safer during natural disasters. (26)

electrical energy: the energy of moving electrons. (6)

electrical property: the quality controlling a material's reaction to electrical current. (5)

electricity: the movement of electrons through materials called *conductors*. (21)

electrocardiograph (EKG) machine: a device used to produce a visual record of the heart's electrical activity. (24)

electromagnetic induction: a physical principle causing a flow of electrons when a wire moves through a magnetic field. (21)

electromagnetic radiation: energy moving through space in waves. (22)

elegant solution: a product meeting a human need in the simplest, most direct way. (13)

elements of design: the basic components that are used as part of any composition and include lines; forms, or shapes; texture; and color. (13)

encode: to convert information into a code that can be used in a communication system. (22)

endoscope: a narrow, flexible tube, containing a number of fiber-optic fibers, allowing a physician to look inside the body. (24)

energy: the ability to do work. Energy comes from inexhaustible, renewable, and exhaustible sources, and it allows technological systems to operate. (2)

energy and power technology: developing and using systems and processes to convert, transmit, and use energy. (3)

energy converter: a device that changes one type of energy into a different energy form. (21)

energy knowledge: the knowledge about using machines and systems to convert, transmit, and apply energy. (1)

engineer: a person who designs processes, products, and structures. (8)

engineered material: a material, commonly called *plastic*, developed to meet a specific need. (26)

engineering design: a plan to create solutions for problems. (10)

engineering drawing: a sketch used to illustrate products to be manufactured. (17)

engineering material: solid matter that has a set, rigid structure. (5)

equality: a state in which all people are alike in status. (8)

ergonomics: human factors engineering. Ergonomics is used to make sure people can use the design solutions. (14)

evaluation grid: a chart used to record data. (14)

exhaustible resource: a material that, once used, can never be replaced. (5)

experimental research: **1.** the kind of research scientists conduct, which structures activities so changes or improvements can be measured. **2.** the use of tests to learn new information. (7)

expert survey: a survey in which several experts on certain topics are asked for their opinions. (27)

external combustion engine: an engine with the energy source outside its motion-producing mechanism. (21)

external selection: a selection someone outside of the design process makes. (14)

F

feedback: the process of giving back data on how well a system is operating. This process provides information so adjustments can be made to improve the system's outputs. In the design process, feedback is the use of information from a later step to improve an earlier step. (2)

feedback system: a control system. (2)

ferry: a boat used to move people and vehicles across bodies of water. (25)

field test: a test conducted in the environment in which the solution will be used. (16)

film message: a photograph, transparency, movie, slide, or filmstrip used for entertainment or for presenting information. (22)

finances: money needed to pay for other inputs of a system. (14)

financial invention: an invention created to make money. These inventions often make things quicker or easier to do. (9)

finishing: protecting and beautifying the surface of a material. (23)

firing: using heat to condition ceramic materials. (23)

First World country: a region that has resources and a desire for technology. (3)

fission: a nuclear reaction that causes the atom's nucleus to split apart. (6)

flame cutting: using burning gases to melt away unwanted materials. (23)

flexible manufacturing: a system that uses complex machines, which computers control. This manufacturing can produce small lots, similar to intermittent manufacturing. Flexible manufacturing, however, uses continuous-manufacturing actions. (23)

flexography: a type of printing done from a rubberlike plate with raised letters on it. (22)

flight controls: a series of moveable control surfaces that can be adjusted to cause an airplane to climb, dive, or turn. (8)

flight simulator: a computer-controlled device allowing pilots to practice flying an airplane without actually using a plane. (25)

flooring: wood, carpeting, linoleum, or ceramic tile used to cover the subfloor. (20)

fluid converter: a device that converts moving fluids, such as air and water, to another form of energy. (21)

foam: a type of sculpting material that is sturdy, but easy to shape. (15)

foam-core board: a sturdy material that has an inner layer of foam, covered on each side with paper. This board is used for final mock-ups and can be used for some prototypes. (15)

force multiplier: the quality of a simple machine that makes it capable of exerting more force to a load than is applied to the simple machine. (4)

forecast: a prediction. (27)

forestry: the growing of trees for commercial use, such as lumber and timber products, paper and pulp, and chips and fibers. (19)

forging: forming hot material by heating it and shaping it with a hammer. (23)

form: the space an object takes up. (10)

format: the size and shape of a message carrier. (22)

forming: squeezing or stretching materials into the desired shape. Forming also includes bending, shaping, stamping, and crushing. (23)

formula: the most common type of mathematical model. Formulas help us to understand complex math, science, and technology concepts. (15)

45° triangle: a triangle that has one 90° corner and two 45° corners. (17)

fossil fuel: the remains of once-living matter. (6)

foundation: the part of a structure tying the structure to the ground. The foundation also supports the weight of the structure. (20)

four-cycle engine: a heat engine having a power stroke every fourth revolution of the crankshaft. (21)

freighter: an oceangoing vessel designed to carry products and materials. (25)

fuel cell: an energy converter that converts chemicals directly into electrical energy. (21)

function: a characteristic describing how a solution works. (10)

fusion: a nuclear reaction causing parts of hydrogen atoms to join. (6)

futurist: a person who uses different methods of forecasting to predict the future. (27)

G

gas turbine: a type of engine that creates power from high-velocity gases leaving the engine. (21)

general job skills: skills that can be applied to many different jobs. (8)

gene-splicing: the process of using enzymes to cut the DNA chain at any point and splice the desirable parts back together. (19)

genetic engineering: selective breeding and pollinating, which allow people to develop animals and plants with desirable traits. (19)

geothermal energy: energy derived from the natural heat present in the earth. (5)

global village: the world viewed as a community in which distance has been drastically reduced by electronic media. (26)

goal: a reason or purpose for a system. (2)

grain drill: a machine developed in the early 1700s to plant seeds. This machine is also called a *seed drill*.

graph: an illustration showing how two or more items compare to each other. A graph is a visual representation of data. (12)

graphic communication: a message, such as a picture, graph, photograph, or word, placed on a flat surface. (22)

graphic model: a visual representation on paper. These models are used to organize and communicate information. (15)

graphic organizer: a diagram that helps to organize thoughts and develop ideas. (13)

grinding and sanding machines: machine tools that use abrasives to cut materials for the workpiece. (4)

guidance: a subsystem of a transportation vehicle that receives information needed to operate the vehicle. (25)

H

harassment: an offensive and unwelcome action against another person. (8)

hardboard: a sheet made by compressing and rolling wood fibers together. Hardboard is used in a number of ways in models. These sheets are dark brown in color, very hard, and stiff. (15)

hardware: the equipment and components making up an entire computer system. (22)

harrow: a frame with teeth, dragged over the ground to give the soil tilth. (19)

harvest: the process of gathering in crops. (19)

hazardous material: any substance that exposes people to a health risk. (5)

hazardous waste: an unwanted by-product of technology that exposes people to a health risk. (5)

heat energy: a form of energy. This energy is present in the increased activity of molecules in a heated substance. (6)

heat engine: an energy converter that converts energy, such as gasoline, into heat and then converts the energy from the heat into mechanical energy. (21)

high-rise building: a multistory residential or commercial building that has a skeleton frame. (9)

historical research: information gathered from already-existing information. (7)

hot-melt glue: a type of glue used in modeling. This glue dries fast and clear. Hot-melt glue can be used on most materials, except certain foams. (15)

human engineering: the management of humans that examines the way people interact with a solution. (16)

human factors engineering: ergonomics. This engineering is used to make sure people can use a solution. (14)

humanities knowledge: knowledge about the society around us. This knowledge includes religious beliefs; governmental organizations; and the history of families, communities, and the world. (1)

hydroelectric: making electricity using waterpower. (6)

hydroponics: growing plants in nutrient solutions without soil. (19)

hypersonic flight: the ability to travel in the air at five times the speed of sound. (27)

hypertext-markup language (HTML): a special type of computer-database system used to create documents on the WWW. The system links objects (text, pictures, music, and programs) to each other. (22)

I

ideation: creating a number of new ideas to solve a problem. (13)

illustration board: a type of paperboard material designers use that comes in a variety of colors. (15)

immunization: the process of systematically vaccinating people through a series of shots to prevent disease. (24)

inclined plane: a surface placed at an angle to a horizontal surface. This plane makes the moving of a heavy object from one level to another possible with less force than lifting straight up. (4)

independent learning skills: skills that include the ability and desire to develop new knowledge and skills outside formal educational settings. (8)

Industrial Age: a period in which most Western countries changed from rural to urban, as new machines and sources of power were developed to support the industry of the time. (3)

industrial building: a structure that houses a company making products or providing an important service. (20)

Industrial Revolution: the time period from 1750 to 1850. (9)

industry: a number of companies producing competing products. (8)

inexhaustible resource: a resource that has an endless supply. (26)

information: data that has been sorted and arranged. Information is facts and opinions people receive during daily life from a variety of sources. (7)

Information Age: the period of time in which technology changed rapidly. (3)

information and communication knowledge: the knowledge about using machines and systems to collect, process, and exchange information and ideas. (1)

information skills: skills that include the ability to locate, select, and use information using appropriate technology. (8)

infrastructure: the basic facilities needed for a building. (20)

in-house design team: a group of people working for a company who specialize in designing products for only that company. (14)

innovation: the process of altering an existing product or system to improve it. (9)

inoculation: the process of vaccinating people through a series of shots in order to promote natural resistance to specific diseases. (24)

input: a resource a system uses to meet the identified goal. (2)

input device: any piece of hardware, such as a scanner, that can be used to enter information into a computer. (22)

inspection: the act of checking parts and products for quality. (23)

instrument landing system (ILS): a navigational aid used to help pilots guide aircraft onto runways for safe landings. (22)

insulation: a material that resists heat passage. (20)

intended output: the reason for which a system was designed. These outputs are produced in response to human needs and wants. (2)

intensive care: the area of a medical facility where seriously ill people receive constant care and monitoring. (24)

interference: any outside force that makes a message less clear to the person receiving it. (22)

intermittent manufacturing: a system that produces parts of a finished product in batches. (23)

intermodal transportation: a system of travel using more than one transportation system. (25)

internal combustion engine: a heat engine that burns fuel inside its cylinders. (21)

internal selection: a selection of the best solution by the design team, after a review and an evaluation of the designs using a set of criteria. (14)

internal source: a person or team of people working within a design company who provides improvement information for a design. (18)

Internet cookie: a small text file that a Web site saves on a computer. (12)

invention: a new and unique product an inventor creates. (9)

invention process: the use of imagination and knowledge to turn ideas into devices, products, and systems. (9)

Iron Age: the period from 1000 to 500 BC. (3)

irrigation: artificial watering to maintain plant growth in areas too dry for successful farming. (19)

isometric drawing: a type of drawing that represents three-dimensional objects in two dimensions. (17)

isometric sketch: a sketch showing the front, top, and sides of an object, just as the eye sees them. (13)

J

jet engine: an engine that obtains oxygen for thrust. (21)

joist: a parallel beam extending from one wall to the other to support the weight of the floor and furniture that will be on it, as well as the people who will be on it. (20)

K

kinetic energy: the energy an object has because of its motion. (6)

knowledge: specific information known about various subjects. (2)

L

laboratory: an area where temperature, moisture, and amount of light can be controlled. (16)

landscape plant: a plant grown in a controlled environment. These plants provide flowers and vegetables for home gardeners and landscape contractors. (19)

land transportation: a transportation system operating on or beneath the earth's surface. (25)

language: the signs and symbols people use to communicate with one another. (1)

laser: a device that emits a beam of coherent, monochromatic light. (4)

layout: the process of arranging printed material, graphic illustrations, and photographs on a page to prepare the material for print. (22)

leadership skills: skills that include the ability to influence people to work toward a common goal. (8)

leisure invention: an invention created for the pleasure of inventing. This type of invention is often patented and very creative. (9)

lever: a mechanism, or simple machine, that multiplies the force applied to it. A lever consists of a long arm, to which the force can be applied, and a fulcrum, on which the arm rotates. (4)

library research: research used to get background information on a problem. (12)

line: the shortest distance between two points. In design, a line is described as a stretched dot. (14)

listening skills: skills that include the ability to listen and understand what a person is saying. (8)

loading: physically placing cargo and people on a vehicle. (25)

M

machine: a device, or mechanism, that changes the amount, speed, or direction of a force. A machine is also a framework to which mechanisms or tools can be attached to make work more efficient. (2)

machineable wax: a very dense material cut and shaped in milling machines and lathes. This wax is used for small models or parts of prototypes. (15)

machine tool: a machine that makes other machines. Machine tools change raw materials into parts. (4)

machining: using a tool to cut chips of material from a workpiece. (23)

magnetic property: the quality describing a material's reaction to magnetic forces. (5)

magnetic resonance imaging (MRI): a technique that uses magnetic waves, rather than X rays, to create an image. (24)

manager: a person who organizes resources to produce products or services efficiently. (8)

manufactured panel: a panel including the interior and exterior surfaces of a wall. (27)

manufacturing: changing raw materials into useful products for public use. (23)

manufacturing knowledge: the knowledge about using machines and systems to convert natural materials into products. (1)

manufacturing technology: developing and using systems and processes to convert materials into products in a factory. (3)

market: a group of people buying and selling a product. (12)

marketing mix: the focus on the four *P*s of marketing: product, price, place, and promotion. (18)

market research: research performed in person, over the phone, or through the mail that helps designers determine what types of people like their products. (12)

matboard: museum board. Matboard comes in many colors and a variety of textures. (15)

material: a substance from which useful products or items are made. (2)

material property: a characteristic affecting how a material reacts to outside conditions. (16)

mathematical model: the use of symbols, along with numbers and letters, to help us understand complex math, science, and technology concepts. (15)

mechanical converter: a device that converts kinetic energy to another form of energy. (21)

mechanical drawing: a very accurate and precise drawing a designer uses. (10)

mechanical energy: the energy present in moving bodies. This energy is sometimes called *kinetic energy*. (6)

mechanical processing: changing material by cutting, crushing, pounding, or grinding it into a new form. (5)

mechanical property: the quality of a material affecting how the material reacts to mechanical force and loads. This property affects how the material reacts to twisting, pulling, and squeezing forces. (5)

mechanical system: a system that provides convenience and comfort inside structures that provide shelter. (20)

mechanism: a basic device that controls or adds power to a tool. Mechanisms are designed to multiply the force applied or the distance traveled. (4)

medical knowledge: the knowledge about using machines and systems to treat diseases and maintain the health of living beings. (1)

medical technology: developing and using devices and systems promoting health and curing illnesses. (3)

medium: any material that carries a message from a sender to a receiver. (22)

memory: the place where the information is stored on the computer. (22)

metal: an inorganic material, which is usually in a solid form, that has the qualities of opacity, ductility, and conductivity. (5)

milling and sawing machines: machines that use a motor to make a straight or circular saw blade move. (4)

mock-up: an appearance model of a product. This model shows how the object will look in real life. A mock-up, however, does not operate. An appearance model is made of easily worked materials, such as clay, wood, or cardboard. (10)

model: a three-dimensional copy of a new product. (10)

modular design: design that incorporates premade components. (14)

molding: shaping a part or product with the use of a mold. (23)

monitor: the primary device for displaying information from a computer. (22)

mounting tape: a thin piece of foam that is sticky on both sides. (15)

multiview drawing: a drawing that uses several views to describe an object. This drawing is also called an *orthographic drawing*. (17)

N

nanotechnology: the design of products at the molecular level. (27)

natural gas: a gas mostly made up of a gas called *methane*. This gas is a simple chemical compound made up of carbon and hydrogen atoms. Pumped from underground, it is generally used for heating buildings and producing electric power. (6)

natural resource: a material that appears in nature. (5)

natural system: everything that appears in nature without human interference. (2)

need: a requirement to live, such as shelter, clothing, and food. (11)

noise pollution: unwanted noise from factories, transportation vehicles, musical concerts, and other loud technologies. (26)

nonrenewable resource: a material made from natural resources, many with a limited supply, that once used up, cannot be replaced. (6)

nuclear energy: heat released when atoms are split. (6)

O

oblique drawing: a sketch used to show objects for which one view is the most important. (17)

oblique sketch: a drawing similar to an isometric sketch, except only the side view is at an angle. These sketches show the front view in its true shape. (13)

ocean liner: a large ship designed to carry passengers on sea voyages. (25)

open-loop system: a feedback system that has external controls and requires human intervention. (2)

operating system: a group of devices that manage a computer system by controlling the hardware and software. An operating system manages such things as the processor, memory, and disc space. This system provides a consistent way for application software to deal with the hardware. (22)

opportunity: the use of a new product or system or an existing product or system in a new way. (11)

optical property: the quality governing a material's reaction to light. (5)

orthographic drawing: a drawing that has several views used to describe the object. This drawing is also called a *multiview drawing*. (13)

output: the result of any system. (2)

output device: a piece of hardware allowing the user to get information from a computer. (22)

outsourcing: sending designs outside of the company to be built. (14)

P

paraffin wax: a mineral wax that can be carved for small models. This wax can also be heated and cast into a mold. (15)

parametric sketch: a computer drawing defined by certain relationships. (17)

passive solar system: a system that does not use moving parts to capture and use solar energy. (21)

patch: a small piece of software that fixes a known problem. (18)

patent: an exclusive right given to an inventor of a new product, design, or plant so others cannot make or sell the product without the permission of the inventor. (9)

pathologist: a medical professional who examines tissue to determine if it is normal or diseased. (24)

peanut butter: nuts processed into a spread. (3)

peat: decayed matter compressed into solid material. This matter is used to heat homes in some parts of the world. (6)

performance: how well a solution works. Performance also can be the quality of a picture or sound. (16)

personal problem: a problem affecting an individual. (11)

perspective drawing: a drawing most similar to what the eye sees. These drawings use vanishing points and vertical lines to convey the image. (17)

pest control: a spray used to control insects and weeds that might damage farm crops. (19)

petroleum: the liquid form of fossil fuel made up mainly of carbon and hydrogen. (6)

photographic system: a method of storing visual material by capturing the material on film and paper that have been made sensitive to light. (22)

physical model: an actual three-dimensional replica of a design. These models are built so the designers and clients are able to see how the design will look. (15)

physical property: a basic feature of material, such as density, moisture content, and smoothness. (5)

pictorial drawing: a drawing of an object that appears as the eye sees the object. (13)

pilot run: the use of a manufacturing system to produce test products. (23)

pipeline: a large-diameter pipe usually laid underground. This pipe is designed to move liquids and loose, solid material from one place to another. (25)

pivot sprinkler: a sprinkler that uses one long line attached at one end to a water source. (19)

planing and shaping machines: machines that cut metal using a single-point tool. These machines produce a flat cut on the surface of the work. (4)

plant: to place seeds in the ground to grow. (19)

plant science: the science of cross-pollinating crops. (19)

plastic: a synthetic organic material designed and produced to meet specific needs. (5)

plastic cement: a very strong adhesive used on plastic pieces. This cement dries clear and fast. (15)

plotter: a printer able to print on large paper. (17)

plow: a blade-shaped plowshare that cuts, lifts, and turns over soil. (19)

plywood: a material made by gluing thin sheets of wood together. Plywood is often used as a base for models and normally used in places that will not be seen. (15)

pollution: an output harming the environment. (2)

polymer: organic matter usually made from natural gas or petroleum. (5)

polystyrene: a foam used in construction as sheet insulation. Polystyrene can be cut and shaped with everything from saws to sandpaper, but hot glue and other types of model glue melt it. (15)

polyurethane: a foam commonly known as *urethane foam*. Polyurethane is often used for large or thick models because it breaks easily if it is thinly shaped. (15)

potential energy: energy at rest. Potential energy is, however, able to do work. (6)

power: **1.** the amount of work done in a set period of time, or the rate at which work is done. **2.** the rate at which energy is changed from one form to another or moved from one place to another. (6)

prevent: to keep something from happening. (24)

primary processing: converting raw materials into industrial materials. (23)

printing system: the methods and machines used to produce words and pictures on untreated paper and other materials. (22)

problem: anything that can be made better through change or improvement. (9)

problem solving: the process of beginning with a problem and ending with a solution. (9)

problem-solving skills: skills that include the ability to identify a problem, find possible solutions, and choose the best solution. (8)

problem statement: a clearly defined, open-ended statement stating a problem, not a solution. (11)

process: the actions taken to put resources to use. A process includes the steps taken to produce products and services. (2)

produce: to make or grow. (1)

product design: the creation of products meeting human needs and wants. (10)

production: the building of a solution. (14)

productive: accomplishing more work in less time. (26)

profit: the amount of money left over after a company's bills are paid. (18)

propeller: a rotating, multibladed device that drives a vehicle by moving air or water. (25)

property: a characteristic of a material. (5)

proportion: the size and shape of an object. (14)

propulsion: a vehicle subsystem that provides the means to move a vehicle. (25)

prototype: a working model of a product. This model is made to test the design. Usually, a prototype is made from the material that will be used in the manufactured product. (10)

public building: a building designed to meet public needs. The building is usually paid for with tax money. (20)

published message: a message produced by printing. Published messages are produced in newspapers, magazines, books, and greeting cards. (22)

pulley: a small wheel with a grooved rim. A pulley can also be a wheel (attached to a shaft) that transfers motion by way of a belt. (4)

pyramid: a shape that comes to a point at one end and a square at the other end. (13)

Q

quality: in manufacturing, the degree, or grade, of excellence in a product. (23)

questioning: a process in which the designer asks why things are done the way they are in an existing product. (13)

R

radiant energy: the movement of atoms present in sunlight, fire, and any matter. (6)

radiology: the use of electromagnetic waves and high frequency sounds to diagnose diseases and injuries. (24)

rafter: a sloping member of a roof frame running between the ridge board and an outer wall of a building. (20)

rapid prototyping: computer-controlled technologies used to create models and parts. (15)

recall: a request that those who purchased a product bring it back because a safety problem was found. (18)

receive: to convert incoming radio waves into perceptible signals. (22)

receiver: an electrical or electronic device that gathers and processes messages that a transmitter generates. (22)

refined sketch: a freehand drawing that adds detail and develops design ideas shown in rough sketches. (13)

regulation: a rule, or law, that a government organization makes. (18)

reliability: the extent to which a product, service, or system yields the same results every time it is used. Reliability is dependability. (16)

religious building: a building designed and built to house public worship. (20)

rendering: a colored or shaded sketch. This sketch shows the final appearance of a product. (10)

renewable resource: a material that can be replaced. (5)

research: to scientifically seek and discover facts. (12)

research process: a set of steps used to find all the facts about a problem. This process is also called the *scientific method*. (12)

residential building: a structure used for living. (20)

rhythm: the effect of motion, created with repetition of elements. (14)

roadway: a permanent structure used for vehicle transportation. (25)

rocket engine: a reaction engine based on a principle of physics. (21)

roof: the exterior surface on the top of a building that protects the building and its contents from the effects of weather. (20)

rough: an initial sketch showing basic ideas. (22)

rough sketch: a pictorial, freehand drawing showing only basic ideas of the size and shape of a product. These sketches are done quickly and without detail to capture ideas that come to the designer. (10)

routing: determining a specified path for a vehicle to travel from a starting point to a destination. (25)

S

safety: the act of making a product safe for consumers to use, for workers to produce, and for the environment. (16)

salary: money earned from work and paid on a weekly or monthly basis. (8)

scale: a specialized type of ruler used to make measurements. (17)

scarcity: the state of being in short supply. Scarcity is a condition in which there is not enough of an item to meet the demand. (26)

scenario: a method of predicting the future in which a person sets up a series of related events that might happen in the future. (27)

scheduling: setting a time for travel. (25)

schematic: the use of pictures and symbols, rather than words, to show a process. (15)

schematic drawing: a drawing used to show the relation of parts to one another and how the product flows. (17)

scientific discovery: the act of first noticing something that occurs naturally, usually done by a scientist or researcher. (9)

scientific knowledge: knowledge about the world. This knowledge explains the laws and principles governing the universe. (1)

scientific method: a set of steps used to find all the facts about a problem. (12)

screw: a simple machine in which an inclined plane is wrapped around a shaft. A screw is a force multiplier used to fasten parts. (4)

secondary processing: changing industrial materials into usable products. (23)

self-management skills: skills that include the ability to accept responsibility for contributing to a goal and manage time effectively to reach the goal. (8)

separation: a type of process that removes unwanted portions of a workpiece. (23)

shading: the use of markings made within outlines to suggest degrees of light and dark in a drawing. Shading relies on a light source and helps to show how an object will look in either sunlight or interior lighting. This use of markings also allows the designer to see the shape and form of the object. Shading adds depth and dimension to a drawing to make the drawing appear more realistic. (13)

shadowing: the use of shadows placed on the opposite side of the light source and following the rough shape of the object being drawn to add depth and dimension to a flat sketch. (13)

shape: the space made by enclosing a line. Shapes can be flat or three-dimensional and have a variety of lengths and widths. (14)

shearing: a separation action that removes excess material by breaking the material into two parts. (23)

shearing machine: a device that slices materials into parts by using opposed edges to cut the workpiece. (4)

shipping lane: a sea lane. This lane is the regular route ships take over the water. (25)

simple machine: a basic device controlling or adding power to a tool. (4)

sketch: a tool that helps designers communicate their ideas on paper. (13)

social invention: an invention created to help the inventor or other people. These inventions make our lives better and easier and impact society. (9)

social problem: a problem dealing with society. (11)

social skills: skills that include the ability to understand and respect what people say, think, and do. (8)

soil erosion: the wearing away of the earth's surface, primarily due to water and wind. (26)

solar cell: a device that converts the energy from sunlight into electric energy. (21)

solar converter: a device that converts energy from the Sun to another energy form. (21)

solar energy: energy of the Sun given off as heat and light. (6)

solid model: the most complete three-dimensional computer model. These models are used to represent the insides and outsides of designs. (15)

space transportation: a new mode of transportation that uses unmanned and manned flights to explore the universe. (25)

specification: a written direction for the builder who controls the quality of work and materials on a construction project. (20)

specification sheet: a list describing items in detail that cannot be shown in drawings. (17)

specific job skills: the skills needed to do a particular job. (8)

sphere: a perfectly round object. (13)

spin-off: an invention that has been adapted to solve another problem. (9)

spray adhesive: a type of glue applied by spraying it out of an aerosol can. (15)

sprinkler irrigation: a type of irrigation that provides moisture for crops by shooting water through the air, in an effort to simulate rainfall. (19)

standardization: a process used to make standard parts that can be used for more than one product. (18)

step-down transformer: a transformer reducing the voltage of electric current. (21)

step-up transformer: a transformer increasing the voltage of electric current. (21)

Stone Age: the period of time up to 3500 BC, when stones were the main material for making tools. (3)

storage: a space in which items are placed for safekeeping. (25)

storyboard: the layout for a television commercial. A storyboard consists of sketches of each shot, accompanied by the script for the commercial. (22)

structure: a vehicle subsystem that provides a rigid framework to protect the vehicle's contents and support other systems. (25)

stud: a vertical member of a wall frame. Studs are usually made of wood 2″ × 4″ or 2″ × 6″ thick and spaced 16″ or 24″ apart. (20)

superstructure: the part of a structure placed on and above the foundation. (20)

surface model: a three-dimensional computer model showing how a certain design will look. This model can be shown in different lights and angles. A surface model cannot be used to show interior details, however, because the inside is hollow. (15)

surgery: a medical procedure used to remove diseased organs, repair broken bones, and stop bleeding. (24)

survey: to measure to determine boundaries of lots and acreage. (20)

survey research: information gathered to find people's reactions to designs and events. (12)

suspension: the subsystem that maintains a vehicle on a pathway. (25)

sustainable design: the design and creation of products in ways that are less harmful to the environment and the people who use the products than the processes used in designing and creating standard products. (10)

swather: a machine that cuts and windrows hay in one pass over a field. (19)

system: a set of related parts. (2)

system design: the organization of different parts to solve a problem. (10)

T

table: a graphic model made of rows and columns, often used to compare sets of data. (15)

tanker: a large oceangoing vessel carrying liquid materials, such as petroleum and chemicals. (25)

teamwork skills: skills that include the ability to work with other people to achieve a goal. (8)

technical drawing: a drawing engineers and architects use to show how products are to be made or buildings are to be built. These drawings can also show how communities are to be developed or parks are to be landscaped. (1)

technician: a specialist who works with an engineer or a scientist. In engineering occupations, technicians work between an engineer and a skilled artisan. (8)

technological knowledge: knowledge of the human-built world used to design, produce, and use tools and materials. (1)

technological problem: a problem that can affect individuals and groups of people. These problems can be solved with devices or systems. (11)

technological system: an organized grouping of inputs, processes, outputs, feedback, and goals. (2)

technological unemployment: a situation in which people with little education lose their jobs to computer-driven machines and robots. (26)

technologist: a specialist in a manufacturing enterprise or some other enterprise. A technologist works under an engineer or a scientist. (8)

technology: the use of knowledge, tools, and systems to make life easier and better. (1)

telecommunication system: a system of exchanging information over a distance. (22)

template: a piece of plastic with shapes and symbols cut into it that designers use to trace around the edge. (17)

terminal: a support system of the transportation system used to make travel more convenient. (25)

test: an experiment or examination used to generate results. (16)

texture: the surface of an object. The texture can be both the appearance and the feel of a product. (14)

thermal conductivity: the measurement of heat moving through a material. (5)

thermal converter: a device that converts heat energy into another form. (21)

thermal energy: energy from the Sun's heat. (6)

thermal expansion: a state in which materials are heated and become longer and wider. (5)

thermal processing: using heat to change material into a more useful form. (5)

thermal property: a material's reaction to heat. (5)

thermoforming: a forming process that produces plastic parts of all shapes. (23)

Third World country: a region where technological advancement is slow, due to poor resources and an absence of human will. (3)

30°-60° triangle: a triangle that has one 90° corner, one 30° corner, and one 60° corner. (17)

thumbnail sketch: a small, rough sketch used to generate ideas. (13)

tidal energy: energy that uses the mechanical energy of moving water. (6)

tillage: the operation of breaking and pulverizing soil. (19)

time: a key resource in developing and operating technological systems, measured in machine time, work hours, product life, or service intervals. (2)

tool: a manufactured device designed to perform specific tasks. Generally, this device is an instrument worked by hand. (4)

tooling: a special device that holds a part while the part is being manufactured. Tooling is designed to make manufacturing more efficient. (23)

tower crane: a self-raising crane used to lift materials on high-rise building projects. (20)

tractor: a machine that provides power to pull all types of farm equipment. (19)

trade-off: a preference or an exchange for one feature or idea in favor of another. (16)

transformer: a device that can increase or decrease the voltage of electric current. (21)

transmission: the act of moving energy to where it performs work. (21)

transmit: to send a message from one person or place to another. (22)

transmitter: an electrical or electronic device that develops and sends messages over air waves. (22)

transportation: the movement of people and materials from one place to another. (25)

transportation knowledge: the knowledge about using machines and systems to move people and cargo. (1)

transportation technology: developing and using devices and systems to move people and cargo from an origin point to a destination. (3)

treat: to cure diseases, heal injuries, and ease symptoms. (24)

trend: the general course of events. In technology, a trend is the direction taken by inventions and the producing of new products and machines. (27)

Trombe wall: an 8″- to 16″-thick masonry wall with a layer of glass mounted about 1″ in front of the wall. (21)

T square: a tool made up of two pieces: the head and blade. This tool is used to draw all horizontal lines in a drawing. (17)

tugboat: a strongly built, powerful boat used for towing and pushing loads in the water. (25)

turbine: a device that transforms energy from a flowing stream of fluid into rotating mechanical energy. (21)

turning machine: a machine tool that uses a stationary tool to cut a material rotating around an axis. (4)

U

ultrasonics: high frequency sounds. (24)

undesirable output: scrap, waste, and pollution. These outputs are produced by most technological systems and can be harmful to the environment. (2)

unit train: a train that carries only one product to a single destination. (25)

unity: a condition of artistic harmony. Unity is the opposite of contrast. This condition helps a design blend together. (14)

universal design: the design of products so all people can use them without adaptations. (11)

unloading: removing cargo from a vehicle or passengers disembarking at their destination. (25)

uplink: the upward signal sent from an Earth station. (22)

use: to select an appropriate product and determine which technological service to put into action. (1)

utility: a service, such as electricity, water, or fuel, brought into a building. (20)

V

vaccination: a series of shots to prevent disease. (24)

vaccine: a special drug created to prevent diseases. (24)

value: the degree of light and dark in a design. (14)

variable: any condition that can be changed. (16)

veneer: a thin sheet of wood that comes in many different widths. Veneer is used to give the appearance of a high-quality wood. (15)

video recording: capturing motion pictures to be replayed through a television set. (22)

visualization skills: skills that include the ability to see three-dimensional objects and systems shown by drawings and schematics. (8)

vocabulary: words developed to describe new technological devices, tools, and actions. (1)

W

wage: money earned for each hour of work completed. (8)

want: something people desire. Wants are not needed to survive. They, however, help make life better and easier. A want can be small or large and can make life more enjoyable. (11)

water pollution: waste from factories, individuals, and farms that contaminates the water. (26)

water transportation: transportation through, and supported by, water. This transportation includes ships, sailboats, rafts, barges, tugboats, and submarines. (25)

water turbine: a device that uses the energy of moving water to make electricity. (21)

wave communication: electronic communication that uses radio waves of the electromagnetic spectrum to carry signals. (22)

wave energy: energy that uses the forces present in the coming and going of waves near the shore. (6)

wax: a material that can come from several different sources (animals, vegetables, or minerals). Wax is used as a sculpting and molding material. (15)

wedge: a simple machine that combines two inclined planes. This machine is the basis for a number of hand tools, including the knife, chisel, and ax. (4)

wellness: a state of personal well-being. (24)

wheel and axle: one of the six simple machines. This machine is a type of lever. (4)

white glue: a safe, water-based, very easy-to-use adhesive. This glue can be used on paper products, foam, and even some wood. (15)

wind energy: an indirect form of solar energy. (6)

wireframe model: a three-dimensional model that uses lines to represent the edges of objects. (15)

wood glue: an adhesive used to glue pieces of wood. This glue dries quickly (in 20–30 minutes) and is easy to use. (15)

work: the application of force applied to move an object. (6)

worker: a person who produces products and services that have been designed. (8)

working drawing: the most complete drawing a designer produces and engineers and architects use to display all the information needed to build a solution. These drawings also show how the products are assembled. (17)

World Wide Web (WWW): a computer system of Internet servers that uses specially formatted documents. (22)

X

X-ray machine: a type of machine that examines, treats, or photographs with short, electromagnetic waves that can pass through solid materials. (24)

Index

C

D

E

F

G

H

I

J

K

L

M

N

O

P

Q

R

S

T

U

V

W

X